Plant Genomics for Sustainable Agriculture

Ram Lakhan Singh • Sukanta Mondal •
Akarsh Parihar • Pradeep Kumar Singh
Editors

Plant Genomics for Sustainable Agriculture

Springer

Editors
Ram Lakhan Singh
Nilamber-Pitamber University
Medininagar, Jharkhand, India

Sukanta Mondal
Animal Physiology Division
ICAR-National Institute of Animal Nutrition & Physiology
Bengaluru, Karnataka, India

Akarsh Parihar
Department of Genetics & Plant Breeding
Anand Agricultural University
Anand, Gujarat, India

Pradeep Kumar Singh
Department of Biochemistry
Dr. Ram Manohar Lohia Avadh University
Ayodhya, India

ISBN 978-981-16-6973-6 ISBN 978-981-16-6974-3 (eBook)
https://doi.org/10.1007/978-981-16-6974-3

This Springer imprint is published by the registered company Springer Nature Singapore Pte Ltd.
The registered company address is: 152 Beach Road, #21-01/04 Gateway East, Singapore 189721, Singapore

Dedicated to our students who refined our knowledge of biology by their intelligent questions, queries, and discussions over the years

Preface

Global food demand is ever increasing and expected to be double by the year 2050 due to exponential increase in population while the agricultural productivity is declining, largely due to shrinking of resources particularly cultivable land. The United Nations (UN) predicted that the world population will exceed nine billion by 2030, improving the quality and quantity of food production is an inevitable necessity. According to the UN, this doubled food requirement must come from virtually the same land area as today. With tremendous increase in world population, plants that are the major food source must be produced at higher rate to fulfill the food requirement. The agricultural lands are limited; therefore, there is need to explore some other alternative methods to fulfill the food requirement. Over the past few decades, advancement in modern breeding programs and the genome-related technologies exploit the use of genes from all the viable sources to develop and produce genetically modified crops. These crops offer improved crop yield, nutrition potential/quality and minimize yield losses due to biotic and abiotic stresses. Similarly, to cope up with huge population, it will be quite useful to use the recent modern tools of genomics like molecular breeding and marker-assisted selection for crop improvement. Marker-assisted breeding involving trait introgression for biotic and abiotic resistance, breaking the genetic plateau and quality improvement will require a prior attention. The marker-assisted selection/breeding, DNA markers, and mapping populations have massive potential to perk up the effectiveness and accuracy of conventional plant breeding. The approaches to study genotype–phenotype associations like QTL mapping, GWAS/Association mapping, transcriptomics, and other techniques will be frequently used to enhance the crop yield. The advent of novel genomics methods including NGS (Next-Generation Sequencing) has massively changed traditional breeding into next-generation breeding. Genome editing is a promising technique to alter specific genes to improve trait expression. Several genomic approaches such as transcriptomics, metabolomics, and proteomics help in identification of novel metabolites in a particular cell. Development in these omics-based approaches and their utilization can enhance the crop yield by regulating various mechanisms involved in stress biology. The plant epigenetic modification plays a vital role in acclimatization, stress tolerance, adaptation, and evolution processes. The important traits of crop life such as flowering time, fruit development, risk avoidance from environmental factors, and crop immunity are attained by the plant epigenetic modifications.

This book has been developed with the objective of providing an updated source of information on the plant genomics to enhance the food requirement for sustainable development. The text of this book includes various genomic tools/methods that are used to improve quality and yield of crop plants. This book is a priceless resource for graduate and postgraduate students, biotechnologists, plant breeders, and others who are concerned with genomic studies. Every chapter has been planned and prepared by the authors in such a way that it presents the subject thoroughly following a reader-friendly approach. The contributors to the book are internationally recognized experts in their field, and they represent reputed institutions across the globe. Overall, this book is a timely addition since the interest in plant genomics has been growing considerably during the last few years.

Key Features

- Identifies and explores biotechnological approaches for sustainable agricultural production
- Encompasses modern genomic tools like molecular breeding and marker-assisted selection, RNA interference technology, whole genome sequencing, next-generation sequencing for crop improvement
- Focuses on strategies for enhancing agricultural productivity and sustainability

Organization of the Book

This book consists of 15 chapters that focus on current approaches and strategies involved in plant genomics to enhance the global plant-derived food production to achieve food security for growing population.

Chapter 1 covers the brief introduction, scope, and applications of biotechnology and genomics for sustainable agricultural production.

Chapter 2 deals with elucidation of structure and organization of plant nuclear genome that leads to availability of genome sequences of different plant species and has opened newer avenues for genetic engineering including genome editing for crop improvement.

Chapter 3 focuses on the use of direct and indirect techniques/methods for gene transfer in plants for the production of transgenic plant that will allow us to fulfill the food requirement of growing population.

Chapter 4 describes about genetically modified crops and their applications.

Chapter 5 provides information about the processes and tools involved in plant transcriptome analysis including study design, RNA isolation, library preparation, sequencing platforms and bioinformatics analysis for annotation, pathway mapping, and differential gene expression.

Chapter 6 focuses on recent modern tools of genomics like molecular breeding and marker-assisted selection for crop improvement. Marker-assisted breeding involves trait introgression for biotic and abiotic resistance. The chapter outlines the conventional breeding techniques, molecular breeding involving marker-assisted selection/breeding, DNA markers, and mapping populations that have massive potential to perk up the effectiveness and accuracy of conventional plant breeding through marker-assisted selection (MAS), advantages of marker-assisted selection and its commonly used applications in plant breeding.

Chapter 7 describes the novel genomics methods including NGS (Next-Generation Sequencing) and breeding tools that have massively changed traditional breeding into next-generation breeding. It also describes about genome editing which is a promising technique to alter specific genes to improve trait expression for the improvement of crop quality and yield.

Chapter 8 covers the whole genome sequencing of plants during the past, present, and future, and this will speed up and lower the cost of future sequencing technologies. It also highlights the genome databases, tools that are mostly used over the globe.

Chapter 9 discusses about the role of model plants in genomics.

Chapter 10 highlights the RNA interference technology in plants for nutritional improvements, increased defense against biotic and abiotic stresses, alteration in morphology, reduced content of food allergens, crafting male sterility, enhanced secondary metabolite synthesis, and production of seedless plant varieties.

Chapter 11 deals with the use of genomics to improve stress tolerance in plants.

Chapter 12 summarizes the current studies and findings in the study of the organellar (chloroplasts and mitochondria) genome concerning their structure, organization, distribution, regulatory mechanism, and gene transfer technologies.

Chapter 13 deals with the DNA barcoding in plants. DNA barcoding can be an effective and efficient tool for exploring and protecting biodiversity, expedite bioprospecting, and defending against bio-piracy.

Chapter 14 covers the advances in epigenetics for crop improvement and sustainable agriculture. It also emphasizes on the epigenetic changes and its effect on crop growth, yield, and herbicide resistance.

Chapter 15 highlights the ethical aspects and public perception on plant genomics.

Medininagar, India — Ram Lakhan Singh
Bengaluru, India — Sukanta Mondal
Anand, India — Akarsh Parihar
Ayodhya, India — Pradeep Kumar Singh

Acknowledgements

We are indebted to all contributing authors for their tremendous efforts to finish their work on time and for their gracious tolerance of our repeated reminders for revisions. We wish to express special appreciation to the editorial and production staffs of Springer Nature for their excellent work. Special thanks are due to Ms. Aakanksha Tyagi, Associate Editor-Life Sciences, Springer India, with whom we started the project at proposal level and got constant critical advice throughout the project. Finally, we wish to extend appreciation to our family and friends for providing strong moral support and encouragement.

We are sure that this book will prove of equally high value to advanced undergraduate and graduate students, research scholars, teachers, crop scientists, capacity builders, and policy makers. We would like to receive your valuable feedback to improve the content of this book in the next edition.

Ram Lakhan Singh
Sukanta Mondal
Akarsh Parihar
Pradeep Kumar Singh

Contents

1 **Introduction, Scope, and Applications of Biotechnology and Genomics for Sustainable Agricultural Production** 1
Akarsh Parihar, Shiwani, Sukanta Mondal, Pradeep Kumar Singh, and Ram Lakhan Singh

2 **Structure and Organization of Plant Nuclear Genome** 15
Monika Singh, Era Vaidya Malhotra, Vartika Srivastava, and Jyoti Ahlawat

3 **Transgenesis in Plants: Principle and Methods** 41
Pradeep Kumar Singh, Pankaj Singh, Rajat Pratap Singh, and Ram Lakhan Singh

4 **Genetically Modified Crops and Their Applications** 71
Minal Garg

5 **Transcriptomics in Plant** 99
Pratik Satya, Sougata Bhattacharjee, Debabrata Sarkar, Suman Roy, Laxmi Sharma, and Nur Alam Mandal

6 **Molecular Breeding and Marker-Assisted Selection for Crop Improvement** 129
Akarsh Parihar and Shiwani

7 **Bioinformatics in Plant Genomics for Next-Generation Plant Breeding** 165
Pratibha Parihar

8 **Whole-Genome Sequencing of Plants: Past, Present, and Future** . . . 191
Talambedu Usha, Prachurjya Panda, Arvind Kumar Goyal, Anusha A. Kukanur, Arunagiri Kamala, M. K. Prasannakumar, K. Sidhalinghamurthy, and Sushil Kumar Middha

9 **Model Plants in Genomics** 241
Soham Ray, Pratik Satya, Laxmi Sharma, Suman Roy, Amit Bera, Sayantan Santra, and Srinjoy Ghosh

10 **RNA Interference Technology in Plants: Mechanisms and Applications in Crop Improvement** 265
Priyanka Gautam, Rishabh Kumar, Zainab Feroz, Sivakumar Vijayaraghavalu, and Munish Kumar

11 **Use of Genomics to Improve Stress Tolerance** 291
Ritu Mahajan, Nisha Kapoor, and Bijender K. Bajaj

12 **Genetics of Plant Organelles: Plastid and Mitochondrial Genomes** 313
Samar Singh, Jogindra Naik, and Ashutosh Pandey

13 **DNA Barcoding in Plants: Past, Present, and Future** 331
Pradosh Mahadani, Majusha Dasgupta, Joshitha Vijayan, Chandan Sourav Kar, and Soham Ray

14 **Advances in Epigenetics for Crop Improvement and Sustainable Agriculture** 351
Venkateshwaran Krishnaswami, Munish Kumar, and Sivakumar Vijayaraghavalu

15 **Ethical Aspects and Public Perception on Plant Genomics** 371
S. C. Srushti, Shivaleela Biradar, S. J. Shreedevi, M. Naveen Kumar, S. Mondal, and R. L. Babu

Contributors

Jyoti Ahlawat School of Agricultural Sciences, G. D. Goenka University, Gurugram, Haryana, India

R. L. Babu Department of Bioinformatics and Biotechnology, Karnataka State Akkamahadevi Women's University, Vijayapura, Karnataka, India

Bijender K. Bajaj School of Biotechnology, University of Jammu, Jammu, Jammu & Kashmir, India

Amit Bera ICAR-Central Research Institute for Jute & Allied Fibres, Kolkata, West Bengal, India

Sougata Bhattacharjee ICAR-Vivekananda Parvatiya Krishi Anusandhan Sansthan, Almora, Uttarakhand, India

Shivaleela Biradar Department of Bioinformatics and Biotechnology, Karnataka State Akkamahadevi Women's University, Vijayapura, Karnataka, India

Majusha Dasgupta Department of Biotechnology, Gauhati University, Jalukbari, Guwahati, Assam, India

Zainab Feroz Department of Biochemistry, University of Allahabad, Prayagraj, Uttar Pradesh, India

Minal Garg Department of Biochemistry, University of Lucknow, Lucknow, Uttar Pradesh, India

Priyanka Gautam Department of Biochemistry, University of Allahabad, Prayagraj, Uttar Pradesh, India

Srinjoy Ghosh ICAR-Central Research Institute for Jute & Allied Fibres, Kolkata, West Bengal, India

Arvind Kumar Goyal Department of Biotechnology, Bodoland University, Kokrajhar, Assam, India

Arunagiri Kamala Department of Biochemistry, Maharani Lakshmi Ammanni College for Women, Bangalore, Karnataka, India

Nisha Kapoor School of Biotechnology, University of Jammu, Jammu, Jammu & Kashmir, India

Chandan Sourav Kar ICAR-Central Research Institute for Jute & Allied Fibres, Kolkata, West Bengal, India

Venkateshwaran Krishnaswami Department of Allied Health Sciences, Noorul Islam Center for Higher Education, Kumaracoil, Kanyakumari, Tamil Nadu, India

Anusha A. Kukanur Department of Biochemistry, Maharani Lakshmi Ammanni College for Women, Bangalore, Karnataka, India

Munish Kumar Department of Biochemistry, University of Allahabad, Prayagraj, Uttar Pradesh, India

Rishabh Kumar Department of Biochemistry, University of Allahabad, Prayagraj, Uttar Pradesh, India

M. Naveen Kumar Department of Biotechnology and Genetics, M.S. Ramaiah College of Arts, Science and Commerce, Bengaluru, India

Pradosh Mahadani Bioinformatics Centre, National Tea Research Institute, Kolkata, India

Ritu Mahajan School of Biotechnology, University of Jammu, Jammu, Jammu & Kashmir, India

Era Vaidya Malhotra Tissue Culture and Cryopreservation Unit, ICAR-National Bureau of Plant Genetic Resources, Pusa Campus, New Delhi, India

Nur Alam Mandal ICAR-Central Research Institute for Jute and Allied Fibres, Barrackpore, Kolkata, West Bengal, India

Sushil Kumar Middha Department of Biotechnology, Maharani Lakshmi Ammanni College for Women, Bangalore, Karnataka, India

Sukanta Mondal ICAR-National Institute of Animal Nutrition and Physiology, Adugodi, Bengaluru, Karnataka, India

Jogindra Naik National Institute of Plant Genome Research, Aruna Asaf Ali Marg, New Delhi, India

Prachurjya Panda Faculty of Health Science, Ben-Gurion University of the Negev, Be'er Sheva, Israel

Ashutosh Pandey National Institute of Plant Genome Research, Aruna Asaf Ali Marg, New Delhi, India

Akarsh Parihar Department of Genetis & Plant Breeding, Anand Agricultural University, Anand, Gujarat, India

Pratibha Parihar N.V. Patel College of Pure and Applied Sciences, CVM University, Vallabh Vidyanagar, Anand, Gujarat, India

M. K. Prasannakumar Department of Plant Pathology, University of Agricultural Sciences, Bengaluru, Karnataka, India

Soham Ray ICAR-Central Research Institute for Jute & Allied Fibres, Kolkata, West Bengal, India
Institute of Applied Microbiology, Collage of Agriculture, Guangxi University, Nanning, Guangxi, People's Republic of China

Suman Roy ICAR-Central Research Institute for Jute and Allied Fibres, Barrackpore, Kolkata, West Bengal, India

Sayantan Santra Institute of Applied Microbiology, Collage of Agriculture, Guangxi University, Nanning, Guangxi, People's Republic of China

Debabrata Sarkar ICAR-Central Research Institute for Jute and Allied Fibres, Barrackpore, Kolkata, West Bengal, India

Pratik Satya ICAR-Central Research Institute for Jute and Allied Fibres, Barrackpore, Kolkata, West Bengal, India

Laxmi Sharma ICAR-Central Research Institute for Jute and Allied Fibres, Barrackpore, Kolkata, West Bengal, India

Shiwani Centre of Excellence in Biotechnology, Department of Agricultural Biotechnology, Anand Agricultural University, Anand, Gujarat, India

S. J. Shreedevi Department of Bioinformatics and Biotechnology, Karnataka State Akkamahadevi Women's University, Vijayapura, Karnataka, India

K. Sidhalinghamurthy Department of Biochemistry, Bangalore University, Bengaluru, Karnataka, India

Monika Singh Division of Genomic Resources, ICAR-National Bureau of Plant Genetic Resources, Pusa Campus, New Delhi, India

Pankaj Singh Department of Biotechnology, Dr. Rammanohar Lohia Avadh University, Ayodhya, Uttar Pradesh, India

Pradeep Kumar Singh Department of Biochemistry, Dr. Rammanohar Lohia Avadh University, Ayodhya, Uttar Pradesh, India

Rajat Pratap Singh Department of Biotechnology, Guru Ghasidas Vishwavidyalaya, Koni, Bilaspur, Chhattisgarh, India

Ram Lakhan Singh Department of Biochemistry, Dr. Rammanohar Lohia Avadh University, Ayodhya, Uttar Pradesh, India
Nilamber-Pitamber University, Medininagar, Jharkhand, India

Samar Singh National Institute of Plant Genome Research, Aruna Asaf Ali Marg, New Delhi, India

Vartika Srivastava Tissue Culture and Cryopreservation Unit, ICAR-National Bureau of Plant Genetic Resources, Pusa Campus, New Delhi, India

S. C. Srushti Department of Bioinformatics and Biotechnology, Karnataka State Akkamahadevi Women's University, Vijayapura, Karnataka, India

Talambedu Usha Department of Biochemistry, Bangalore University, Bengaluru, Karnataka, India

Joshitha Vijayan ICAR-Central Island Agricultural Research Institute, Port Blair, Andaman & Nicobar Islands, India

Sivakumar Vijayaraghavalu Advanced Research Centre, Narayana Medical College, Nellore, Andhra Pradesh, India
Narayana Translational Research and Incubation Center, Narayana Medical College, Nellore, Andhra Pradesh, India

Introduction, Scope, and Applications of Biotechnology and Genomics for Sustainable Agricultural Production

1

Akarsh Parihar, Shiwani, Sukanta Mondal, Pradeep Kumar Singh, and Ram Lakhan Singh

Abstract

Population in the globe is estimated to exceed 9 billion by 2050 putting a great challenge to all crop scientists to meet this growing demand. There should be 60% more agricultural production required in 2050 as it was in 2007. The huge shifts from investment in inputs related to fertilizer and pesticides driven technologies to technology based on genetic modifications increasing yields with fewer inputs are indicated. Food and nutrition security have become burning issues in the international discussions at all levels of government as plans are being made to cope up with a changing global climate and increasing global population. One of the most important environmental challenges faced by the developing world is how to meet current food needs without undermining the ability of future generations to meet their needs. Crop production should be adequate to feed the population now and, in the future, also. The current status

A. Parihar (✉)
Department of Genetis & Plant Breeding, Anand Agricultural University, Anand, Gujarat, India

Shiwani
Centre of Excellence in Biotechnology, Department of Agricultural Biotechnology, Anand Agricultural University, Anand, Gujarat, India

S. Mondal
ICAR-National Institute of Animal Nutrition and Physiology, Adugodi, Bangalore, Karnataka, India

P. K. Singh
Department of Biochemistry, Dr. Rammanohar Lohia Avadh University, Ayodhya, Uttar Pradesh, India

R. L. Singh
Department of Biochemistry, Dr. Rammanohar Lohia Avadh University, Ayodhya, Uttar Pradesh, India

Nilamber-Pitamber University, Medininagar, Jharkhand, India

R. L. Singh et al. (eds.), *Plant Genomics for Sustainable Agriculture*,
https://doi.org/10.1007/978-981-16-6974-3_1

of agricultural technologies would not be sufficient to meet up the production challenges in future. Innovative technologies have to be exploited in order to enable sufficient food availability in the future. However, even after years of constant efforts by breeders, there are several unanswered issues using traditional methods. In this regard, biotechnology plays an immense role in agriculture by providing better feed and fuel to the growing world.

Keywords

Genomics · Next generation sequencing · Marker assisted selection · Whole genome sequencing · Phenomics

1.1 Introduction

Critical issues facing agriculture globally include delivery of human health care, reduction in hunger, and increasing energy supply, all in a sustainable manner with optimum animal welfare and minimal negative impact on the environment. The United Nations (U.N.) predicted the world population will exceed nine billion by 2050, improving the quality and quantity of food production is an inevitable necessity. According to the U.N., this doubled food requirement must come from virtually the same land area as today. The U.N. Food and Agriculture Organization (FAO) further stated that 70% of this additional food must come from the use of new and existing agricultural technologies. Therefore, agricultural production faces an exceptional challenge to feed the increasing global population. The existing food production system is under huge pressure to double their food productivity to meet the demands of ever-increasing global population. The annual yield gain reported for major crops like rice, wheat, maize, and soybean (1.2% average) is still less than what is required (2.4%) to reach the goal of doubling global production by 2050. As the world is experiencing high demands for crop production, by 2050 global agricultural production may require to be increased by 60–100% to meet these burgeoning demands as well as there is need to provide food security to the approx 870 million now chronically undernourished (FAO 2012). Food production also suffers from dramatic changes and rapid climate changes including drought, floods, and other disasters. About 80% of world's population are poor and lives in rural areas which typically rely on local agriculture for their survival needs (FAO 2019). Global yields of major crops are projected to be reduced on average, according to forecasts. For every degree Celsius rise in global mean temperature, wheat, rice, maize, and soybean yields decrease by 6.0%, 3.2%, 7.4%, and 3.1%, respectively (Zhao et al. 2017). CGIAR system (https://www.cgiar.org/) has initiated a 'Two Degree Initiative for Food and Agriculture' with the aim of assisting 200 million small scale food producers across the globe to adapt at the speed and scale needed for the current pace of climate change. A significant increase in food production has to be achieved with finite or even depleting land resources and water systems while meeting the demand for ecosystem preservation.

The dynamics of pest pathogens are likely to be influenced by extreme weather conditions, undermining the plant defense response (Atlin et al. 2017). For decades, traditional plant breeding systems have produced a variety of widely accepted high-yielding crop cultivars all over the world. Longer time spent on variety growth and breeding cycles, on the other hand, is a roadblock to plant breeders' ability to respond quickly to increasing food production demands (Lenaerts et al. 2019). Improving crop productivity rates by breeding entails making significant improvements to our existing plant breeding activities and decisions (Santantonio et al. 2020). Crop improvement has made a major contribution to food security and breeding climate-smart cultivars is thought to be most sustainable way to boost food production. Though, recent studies have argued that current food production practises are insufficient, and that the food system must be transformed. A fundamental change is required within the conventional breeding structure so as to reply satisfactorily to the growing food demands. Crop improvement for food and nutritional security has become a major global concern, particularly in light of population growth and challenges such as climate change and water scarcity. Plant breeding has been very successful in developing improved varieties using conventional tools and methodologies. The success of plant breeding has relied in the utilization of natural and mutant induced genetic variation and in the efficient selection, by using suitable breeding methods, of the favourable genetic combinations. However, existing crop breeding strategies alone will not deliver a high enough rate of crop improvement to satisfy demand in the short or long term. The combination of conventional breeding techniques with genomic tools and approaches is leading to a new genomics-based plant breeding. Genomics-assisted breeding is considered to have the greatest potential for overcoming these challenges and ensuring a sustainable increase of food production by adapting available crops to biotic and abiotic stresses and breeding novel crop varieties. Plant genomics provides breeders with a new set of tools and techniques that allow the study of the whole genome, and which represents a paradigm shift, by facilitating the direct study of the genotype and its relationship with the phenotype. Recombinant DNA technology can help to design almost any desirable characteristic in plants by controlled targeted gene expression. In this new plant breeding context, genomics will be essential to develop more efficient plant cultivars, which are necessary, according to FAO, for the new 'greener revolution' needed to feed the world's growing population while preserving natural resources. Plant genomic data is being utilized in genetic engineering to ensure that better, and fitter varieties of crops are available to ensure food security to the population. Recent progress in genomics technologies has imparted greater strength to the breeders' toolbox. Using latest techniques such as genomics, biotechnological interventions, speed breeding, genomic selection and genome editing, limitations of traditional breeding could be overcome.

1.2 Crop Improvement and Plant Genetic Resources

One of the most sustainable methods for conserving valuable genetic resources over time while also increasing agricultural production and food security is to use plant genetic resources (PGR) in crop improvement, followed by adoption, cultivation, and consumption or marketing of the improved cultivars by farmers. A more productive use of plant genetic diversity and utilization of plant genetic resources may be a prerequisite to meeting the challenges of growth, food security, and poverty alleviation, according to the Food and Agricultural Organization of the United Nations (FAO) (FAO 1996).

Sources which are available for the improvement of a cultivated plant species are commonly referred to as genetic resources. In traditional plant breeding, genetic resources are the materials which do not require selection for adaption to the target environment and are not immediately useful to the breeders. According to the gene pool concept, genetic resources are categorized into primary gene pool, secondary gene pool, tertiary gene pool, and isolated genes. The primary gene pool consisted of those crop species either itself or other species which are easily crossable with it while the secondary gene pool is consisted of related species that are difficult to cross with the target crop, meaning crossing is less effective and crossing progenies are partly sterile. The tertiary gene pool consists of those species which can only be utilized by advanced techniques like embryo rescue or protoplast fusion. The fourth form of genetic resource may come from related or unrelated plant species, animals, or microorganisms.

The value of various types of genetic resources for crop enhancement largely depends on the crop species. One of the main reasons for the inadequate use of genetic resources in traditional plant breeding is the lack of environmental adaptation of plant genetic resources. Other barriers to PGR use in crop improvement include large performance differences between PGR and actual breeding materials for complex inherited traits such as lack of inbreeding tolerance and unknown affiliation to heterotic pools, as well as genetic issues such as pleiotropy, linkage between desired and undesired PGR alleles, and gene co-adaptation within both breeding populations and plant genetic resources. Genome science tools may be able to finally unlock the genetic potential of our wild and cultivated germplasm resources for the benefit of humanity. The utility of molecular markers and genome research can help the utilization of PGR for crop development in a better way. The revolutionizing advances in plant genomics has evolved from the enrichment and advances made in conventional genetics and breeding, molecular biology, molecular genetics, molecular breeding, and molecular biotechnology in the land of high-throughput DNA sequencing technologies powering the plant research to sequence and understand the genetic compositions, structures, architectures, and functions of full plant genomes. Recent technological development and challenges faced in the field of agriculture have led to the emergence of various genomic tools that can be used to explore and exploit the plant genomes for crop improvement. Next Generation Sequencing (NGS) technologies are allowing the mass sequencing of genomes and transcriptomes, which is producing a vast array of genomic information.

1.3 Biotechnological Interventions

Advances in genomics have allowed scientists to decode genomes for any crop species, as well as knowledge on genes responsible for essential agronomic traits, in the modern period. Gene information can be used to speed up breeding programmes and develop better, higher-yielding varieties. In some cases, the only crop improvement options are genetically modified (GM) crops. The success of Bt cotton has already been realized in our country, as India has gone from being an importer to a major exporter in just a few years. Biotech crops will help farmers in increasing productivity according to a study from international service for the acquisition of agri-biotech applications, which also claims that biotech crops have traditionally been the fastest-adopted crop technology, with farmers satisfied with their benefits and high adoption rates.

Biotech crop plantings have increased 113-fold since 1996, covering 2.5 billion hectares, demonstrating that biotechnology is the world's fastest-growing crop technology. Adoption rates of major crops are close to 100% in countries with long histories of high adoption, such as the USA, Brazil, Argentina, Canada, and India, suggesting that farmers prefer this crop technology to traditional varieties. In 2018, more versatile biotech crops with different traits became available on the market to meet the needs of farmers and consumers. Globally, from 1996 to 2018, biotech crops generated economic gains at the farm level worth US$ 167.8 billion. Pesticide use has been decreased by 620 million kilogrammes by utilizing biotech crops. In 2016 alone, fewer insecticides spray reduced carbon dioxide emissions by 26.7 billion kilograms, equivalent to taking 11.9 million cars off the road for a year. Biotech crops have helped 18 million small farmers and their families to overcome the cycle of poverty. According to the Global Status of Commercialized Biotech/GM Crops in 2018, biotech crops were grown and imported in 70 countries in 2018, marking the 23rd year of continuous biotech crop adoption. 26 countries (21 developing and 5 developed) planted 191.7 m ha of biotech crops in 2017, up 1.9 m ha from the previous year's total. Farmers around the world are continuing to embrace biotech crops, indicating that biotech crops are already helping to address global problems such as poverty, malnutrition, and climate change.

Global agricultural productivity gains are showing a change away from heavy investments in increased fertilizer and pesticide inputs and toward technology-driven changes (including genetic modification) that increased yields with fewer units of input. Increasing population, dwindling agricultural land and water bodies, decreasing productivity, and growing environmental and agricultural issues all contribute to the need for long-term technical interventions in the sector to ensure global food and nutrition stability.

Agricultural biotechnology has been successfully used for decades to increase food production and productivity by creating insect, disease, and herbicide tolerant varieties as well as environmentally sustainable biological products. These products reduce the use of inorganic materials and chemicals in agriculture while increasing productivity and crop nutrition through modern nutrient absorption, their enhancement. Extremely promising novel techniques like Next Generation Sequencing

(NGS), Cisgenics, and Genome Editing are being extensively used by the many countries. Some of the 'high-tech' novel technologies considered to be very promising in near future in agricultural biotechnology include genome sequencing technologies for crop breeding, RNAi-based gene silencing technologies, new plant breeding techniques (NPBT) including site specific mutagenesis and deploying genes from cross-compatible species through transgenesis (gene transfer), breeding with transgenic inducer line, RNA-dependent DNA methylation (RdDM), reverse breeding, agro-infiltration, grafting techniques, and speed breeding.

1.4 Genomics for Crop Improvement

Genomics is the field of science which correlates genomes at its various levels of structure, function, and evolution. It is aimed towards mapping of genes, their interaction and editing for the betterment of humans. Its function is accomplished with the use of techniques like sequencing and in silico analysis. The emerging field of genomics came into existence during the end of last century with revolutionary vision towards understanding of living forms and it reflects to be the most promising approach in upcoming decades also (Lander 1996; Lander and Weinberg 2000). The advantage towards the approach was founded with the discovery of nucleic acids sequencing by Sanger and coworkers in UK and by Maxam and Gilbert in USA in 1977. The discerning technique led to unlock the whole plan of the improvement of an organism by deciphering the sequence of bases of DNA. The revolution on genomics research was commenced with the crucial phase of Human Genome Project and now has become vital for human welfare in relation to pharmaceutical industry as well as agriculture. The completion of genome and expressed sequence tag (EST) sequencing and gene discovery projects for several crop species like rice, *Arabidopsis*, sorghum, maize, and soybean based on first-generation Sanger sequencing methods have generated a wealth of genomic and genic sequence information including fully characterized known and candidate genes, transcription factors, and regulatory sequences. With the combination of traditional and high-throughput sequencing platforms, there has been a tremendous increase in genomic resources available, including expressed sequence tags (ESTs), BAC end sequence, genetic sequence polymorphisms, gene expression profiling, whole genome (re)-sequencing, and genome wide association studies.

Plant genomics, the study of whole plant genome, their organization and evolutionary patterns along with the functional analysis has become the heart of crop improvement programme. The invention of DNA chip technology is graceful as it allows studying genome wide gene expression patterns with the ease as it simultaneously checks thousands of genes. The gene expression and their regulation with reference to the growth, development, and defense of plants are now the area to be focused with use of translational genomics. The accumulation of huge genomic data with the progress of sequencing technologies can fulfill the need of manipulation in gene expression in order to develop or stimulate responses towards various biotic and abiotic stresses.

Environmental influences, such as extremes of drought, salinity, and temperature, which impose water deficit stress, place significant restrictions on plant productivity (Boyer 1982). More stress tolerant crops need grow to overcome these constraints and increase production quality in the face of a budding world population. Traditional breeding strategies that intended to use genetic variation resulting from varietal germplasm, interspecific or intergeneric hybridization, induced mutations, and somaclonal variation in cell and tissue cultures have only had limited success; very few new plant introductions have resulted in increased stress tolerance under field conditions (Flowers and Yeo 1995). The complexity of stress tolerance characteristics, low genetic variation of yield components under stress conditions, and the absence of successful selection techniques restrict conventional approaches. In addition, quantitative trait loci (QTLs) associated with tolerance at one stage in development can vary from those associated with tolerance at other stages. Desirable QTLs along with the introgressed tolerance trait may require extensive breeding to restore desirable characteristics. Nonetheless, as the resolution of the genetic and physical chromosome maps of major crops strengthens, marker assisted selection of definite secondary yield-related characteristics (e.g. anthesis-silking interval, osmotic adjustment or alteration, membrane stability, or physiological tolerance indices) will prove to be very useful. This strategy could be used in conjunction with pyramiding strategies or sequential selection for accumulation of physiological yield-component traits.

Previous attempts to strengthen drought, high salinity, and low temperature tolerance by breeding and genetic modification have had limited success due to the genetic complexity of stress responses. Progress is now expected through comparative genomics studies of an evolutionarily diverse range of model species, and by the utilization of techniques including expressed sequence tag high-throughput analysis, large-scale parallel gene expression analysis, targeted or random mutagenesis, and gain of function or complementation of mutants.

Agricultural challenges and recent technological advances led to the introduction of high-throughput instruments to explore and manipulate plant genomes for crop improvement. The goal of these genomics-based approaches is to decode the entire genome, including genetic and intergenic regions, in order to gain insights into plant molecular responses, which in turn will provide specific crop improvement strategy. Genomics approaches for crop improvement against stresses are functional, structural, and comparative genomics. Advances in genomics technologies have provided a more thorough analysis of crop genomes and a deeper understanding of stress tolerance mechanisms dynamics. Apart from stress tolerance mechanism, research has also centred on molecular mechanisms regulating stress mediated signalling in plants and the underlying regulatory network of interacting proteins. In order to generate stress resistant crops, functional elucidation of genes involved in these regulatory pathways is intended. Next generation sequencing (NGS) technologies allow mass sequencing of genomes and transcriptomes, creating a broad range of genomic knowledge. Through the advancement of bioinformatics, the study of NGS data has made it possible to discover new genes and regulatory sequences regulating important traits. Many genomic regions associated with significant traits linked to abiotic stress tolerance have also been established with the generation of countless numbers of markers and their use in genome-wide association studies. The discovery of new genes, the determination of their patterns of expression in response to abiotic stress, and an enhanced understanding of their functions in adaptation to stress (acquired by the use of functional genomics) would provide the foundation for successful engineering strategies that lead to greater tolerance to stress.

1.4.1 Whole Genome Sequencing

With the fact of growing global population, changing climate, and environmental pressure, there is an urgent need to accelerate breeding novel crops with higher production, drought or heat tolerance, and less pesticide usage. Advances in genomics offer the potential to speed up the process of developing crops with promising agronomic traits. The recent advent of high-throughput next-generation whole genome and transcriptome sequencing, array-based genotyping, and modern bioinformatics approaches have enabled to produce huge genomic and transcriptomic resources globally on a genome-wide scale in diverse crop genotypes. Moreover, the integration of structural, functional, and comparative genomics including epigenomics with marker-assisted breeding (MAB)/genomics-assisted breeding has been implicated to be an effective approach for identification of genes/QTLs and expressed QTLs (eQTLs) and their regulatory sequences involved in expression of an individual trait in crop plants. The integration of available traditional and modern -omics resources/approaches comprehensively with genomics assisted breeding will certainly decode the molecular and/or gene regulatory networks for identification of functionally relevant novel gene-associated targets and alleles controlling the complex quantitative yield and stress tolerance traits in crop plants.

Whole genome sequencing (WGS) is a laboratory technique which determines the whole DNA sequence of an organism's genome at once. DNA sequencing methods and computer technique assemble the tremendous biological sequence data that uncovers the order of bases in a whole genome of an organism (Saraswathy and Ramalingam 2011). In 1979, whole genome shotgun sequencing was used for small genomes which range from 4000 to 7000 base pairs (Staden 1979). The first genome of *Haemophilus influenzae* was sequenced in 1995 (Fleischmann et al. 1995). Further in 2000, the sequencing of almost an entire human genome was completed (Lander et al. 2001). These previously used techniques for WGS were slow, labour-intensive, and costly (Kwong et al. 2015). Wandering-spot analysis was a method reported by Gilbert and Maxam that sequence 24 bp and it was time consuming, and labour required (Gilbert and Maxam 1973). But Sanger sequencing changed the whole scenario when it came into the play as this is technical advances automated, dramatically speed up, also termed as the chain-termination or dideoxy method.

The advancement of WGS helped in identification of the disease associated variants such as complex genomic regions, inaccurate variant calling, detection of SNP, and the phase of the locus, etc. In the process of genomic annotation and analysis, important point is the causation between genomic variants and disease association. It helps to obtain explicit understanding on variation effects of individuals (Albert and Kruglyak 2015). Quantitative assay of biological interactions, downstream effectors such as transcription factors are now feasible with the help of next generation sequencing for genome wide sequencing. Due to advancement in computational and experimental science provided a better understanding of transposable elements sequences in genomic assays and a renewed idea for the importance of TE biology (O'Neill et al. 2020). As a result, plant breeders can use NGS data to discover regulatory sequences and their relative positions as well as can establish molecular markers for marker assisted selection (MAS).

The decoding of entire genomes for a number of plant species has become possible thanks to with the advances in DNA sequencing technology. The sequencing of multiple genomes opens up new possibilities for pan-genomic studies aimed at identifying essential and core genes in crop species. Genomic technologies make effective use of germplasm stored in global repositories for their characterization and utilization.

For crop improvement programmes genetic diversity plays a major role and genetic variations from landraces in crop breeding have been successfully exploited in crops like rice for dwarfing genes, wheat, and barley for mlo alleles (Mascher et al. 2019). Domestication and modern breeding lead to the narrow genetic variations of current crop breeding programmes. Genome-scale studies of large germplasm panels have emerged as a valuable resource for understanding genomic variation dynamics during domestication and selective breeding in recent years (Zhao et al. 2015). For example, recent sequencing of multiple accessions in various crop species in conjunction with genome-wide association studies (GWAS) has aided in the identification of key genomic regions linked to crop domestication and improvement (Varshney et al. 2017). The availability of a reference genome sequence has prompted the sequencing of several accessions of a plant species in order to conduct genome-scale research.

1.4.2 Marker Assisted Selection and QTL Mapping

Crop improvement for various traits depends upon the identification of desirable genes and genotypes harbouring such genes. Identification of such genes and genotypes is facilitated by mapping QTL, finding the tightly linked marker with the QTL and eventually utilizing that markers/QTL in marker assisted selection. Marker-assisted selection is an indirect selection method in which a trait of interest is chosen based on a trait-linked marker. A good MAS necessitates the mapping and close association of a gene to a marker otherwise, it is difficult to analyse or evaluate using traditional methods.

Molecular markers have been generally used to facilitate target gene introgression using the backcross scheme (marker-assisted backcrossing, MABC). MABC also facilitates the recovery of recurrent parent genotype and the elimination of donor parent genome flanking the target gene for minimizing linkage drag. MABC is well suited for introgression of oligogenes and large effect QTLs for defect correction of an otherwise superior variety that is used as the recurrent parent. The identification of diverse strains or hybridization with elite cultivars is needed, to expand the genetic base of core breeding material. Numerous studies have been conducted on the evaluation of genetic diversity in breeding material for all crops. DNA markers have proven to be an invaluable tool for describing genetic tools and providing breeders with more accurate knowledge to aid in parent selection. MAS has been employed for the improvement of many traits in different crops. It has been extensively used for improved access and utilization of germplasm resources, QTL mapping, gene pyramiding, and backcross breeding.

1.4.3 High Throughput Phenotyping

Despite recent advances in genomics, a lack of appropriate phenotyping data (phenomics data) has resulted in weak gene/QTL discovery, which has hampered progress in genomics-assisted crop improvement programmes. As a consequence, high-throughput, reliable, and comprehensive trait data are needed to understand the genetic contribution to phenotype variation. Sustaining and rising crop yields with the benefits of modern genetics methods now depend on phenomics rapid advancement. In a single day, modern phenomics tools intend to record data on characters such as plant production, architecture, growth, biomass, photosynthesis, and so on for hundreds to thousands of plants. As a result of automation, remote control, and data analysis pipelines acquiescent to high throughput phenotyping platforms permitted screening of huge plant populations, germplasm collections, breeding content, and mapping populations with improved precision and accuracy in phenotypic trait acquisition while reducing labour input.

The environment plays a critical role in plant phenomics because most of the essential traits in plants are quantitative in nature and heavily influenced by environmental factors. Plant breeders also want to create crop varieties that have good buffering and stability and can perform well in a variety of environments. As a result, any crop phenomics strategy needs accurate documentation of the experimental environmental conditions (e.g., rainfall, temperature, photoperiod, and soil characteristics).

Phenotyping strategies are categorized into forward phenomics and reverse phenomics. Forward phenomics helps in selecting and identifying superior genotypes while reverse phenomics dissects the best genotypes to discover why they are superior. The forward phenomics offers immediate candidate germplasm for breeding, while the reverse phenomics is a long-term strategy for developing improved crop ideotypes. Reverse phenomics entails applying a variety of new methods to a small set of germplasm in order to uncover common techniques that are responsible for stress tolerance or yield capacity. The phenomics data has been used to find genes/QTL via QTL mapping, association mapping, and genome-wide association studies (GWAS) for crop improvement using genomics-assisted breeding (GAB).

1.4.4 Bioinformatics for Next-Generation Plant Breeding in Plant Genomics

Numerous bioinformatics-based analytical methods are well known in many areas of plant genomic science including comparative genomic analysis, phylogenomics and evolutionary analysis and genome-wide association research. Based on NGS technologies, many autonomous and ultra-high-throughput platforms have recently been developed by big companies such as Roche, Illumina, Applied Biosystems, and so on. All of them are well-fitted for the broad sequence requirements of the present and even future. When entire genomes have been sequenced, it is a vital method to identify and explain the gene and non-coding material in these sequences. For this reason, comparative genomic analysis of plants has emerged as a new area of modern biotechnology, because its main purpose is to predict functions for many unknown genes through the study of significant differences and similarities between organisms. However, in the available datasets of orthologs formed from the same ancestor, these genes are expected to appear. Phylogenomics is known as molecular phylogenetic analysis, in which genomic database sets are used to predict gene function and investigate the evolutionary relationships between organisms. GWAS has a powerful plant species application to classify phenotypic variability in loci correlated with characteristics, as well as allelic variation in candidate genes that resolve quantitative and complex characteristics.

Multiple techniques, including microarrays, expressed cDNA sequence tag (EST) sequencing, serial gene expression analysis (SAGE) tag sequencing, massively parallel signature sequencing (MPSS), RNA-Seq, also known as 'Whole Transcriptome Shotgun Sequencing', can determine the expression of many genes by measuring mRNA levels (WTSS). Protein microarrays and high throughput (HT) mass spectrometry (MS) may provide a snapshot of the proteins found in a biological specimen. In making sense of the protein microarray and HT MS data, bioinformatics is very involved. Gene regulation is the dynamic phenomenon of events that ultimately lead to an increase or decrease in the activity of one or more proteins by a signal, possibly an extracellular signal such as a hormone. To explore different steps in this method, bioinformatics techniques have been applied. For instance, gene expression can be regulated in the genome by nearby elements. The analysis of promoters includes the identification and review of sequence motifs in the DNA surrounding a gene's coding region. The degree to which the region is transcribed into mRNA is influenced by these motifs. Enhancer elements which are located far away from the promoter regions can also regulate gene expression via 3D (three dimensional) looping interactions. Such associations can be determined using bioinformatic analysis of chromosome conformation capture experiments.

In short, along with advances in bioinformatics, the recent wealth of plant genomic resources has enabled plant researchers to gain a fundamental and systematic understanding of economically important plant and plant processes that are critical for advancing crop improvement. Despite these exciting achievements, there remains a critical need for effective tools and methodologies to advance plant biotechnology, address issues that are difficult to solve using current approaches, and facilitate the translation of this newly discovered knowledge to improve the productivity of plants in next-generation plant breeding.

1.5 Advances in Genomics

Crop improvement has become easier with the advent of genomics technologies. Plant breeding for sustainable crop improvement has gained new momentum with the availability of next generation sequencing technologies, modern plant breeding approaches like association mapping, genome wide association selection, advanced backcross QTL analysis, modern genotyping technologies like mass spectrometry allowing SNP discrimination and identification of SNPs based on difference in mass to charge ratio of amplified fragments, SNP arrays.

Identification of large number of single nucleotide polymorphisms (SNPs) with the help of high-throughput genotyping is boosting up the execution of genome wide association studies (GWAS), which relate DNA variants to phenotypes of study. GWAS has enabled the mapping of the genomic loci in diverse set of population which are associated with economically important characters including resistance to biotic and abiotic stress, yield, and quality. There are many genotyping methods available, but among them whole genome resequencing (WGR), reduced representation sequencing (RRS), and SNP arrays are the three most commonly used genotyping techniques in GWAS.

WGR and RRS are based on NGS technologies and bioinformatics pipelines to facilitate alignment of reads to a reference genome while SNP arrays are made up of allele specific oligonucleotide (ASO) probes (which include target SNP loci and flanking regions) that are fixed on a solid support and used to cross-examine complementary sequence from DNA samples and deduce genotypes based on the hybridization signal. Choosing the most suitable (cost-effective) genotyping method for crop GWAS generally requires a thorough review of several factors, including the study's intent and scope, crop genomic features, and technical and economic issues associated with each genotyping method.

1.6 Conclusion and Future Perspective

Biotechnology has the ability to play a part in securing food and nutrition. The completion of whole genome sequencing of crop plants resulted in a plethora of molecular markers, some of which have the potential to speed up the plant breeding process while also solving genetic purity and adulteration issues. Biotechnology tools deserve to be embraced in the fight against food and nutrition insecurity. This can be made available to small-scale farmers with little or no risk to human health or the environment if reasonable biosafety legislation and policies are in place. As a result, in a world where inaction leads to the deaths of thousands of children, we must not overlook any aspect of a potential solution, like agricultural biotechnology.

Recent advances in genomics research have provided geneticists, biologists, and breeders with a range of cutting-edge tools and technologies that help breeding programmes be more precise and effective. As reference genome assemblies become more widely accessible, gene discovery and trait modulation methods have changed dramatically. In addition to advance genomic researches, gene editing methods in plants for elucidating candidate genes and genetic interactions can be used.

According to Varshney et al. (2019) breeding strategies such as marker-assisted back crossing (MABC) are better suited to removing defects from mega-varieties; however, increasing genetic gains per unit time necessitates rapid population improvement driven by genome-wide predictions and associations. The growing availability of multi-omics data and high-dimensional phenotypic data is exposing the potential challenges associated with data handling and interpretation. Plant breeders must be properly trained, and this will play a key role in adopting more advanced methods for crop improvement, such as systems biology-driven breeding (Lavarenne et al. 2018). To provide solutions for sustainable agriculture, such concerted initiatives involving several disciplines will be critical.

References

Albert FW, Kruglyak L (2015) The role of regulatory variation in complex traits and disease. Nat Rev Genet 16(4):197–212

Atlin GN, Cairns JE, Das B (2017) Rapid breeding and varietal replacement are critical to adaptation of cropping systems in the developing world to climate change. Glob Food Secur 12:31–37

Boyer JS (1982) Plant productivity and environment. Science 218:443–448

FAO (1996) Global plan of action for the conservation and sustainable utilization of plant genetic resources and the Leipzig declaration, adopted by the international technical conference on plant genetic resources. Food and Agricultural Organization of the United Nations, Leipzig, Germany, Rome

FAO WFP, IFAD (2012) The state of food insecurity in the world 2012. Economic growth is necessary but not sufficient to accelerate reduction of hunger and malnutrition. FAO, Rome

FAO (2019) Agriculture and climate change – challenges and opportunities at the global and local level – collaboration on climate-smart agriculture. Food and Agriculture Organization (FAO) of the United Nations, Rome. 52 pp.

Fleischmann RD, Adams MD, White O, Clayton RA, Kirkness EF, Kerlavage AR, Merrick JM (1995) Whole-genome random sequencing and assembly of *Haemophilus influenzae* Rd. Science 269(5223):496–512

Flowers TJ, Yeo AR (1995) Breeding for salinity resistance in crop plants: where next? Aust J Plant Physiol 22:875–884

Gilbert W, Maxam A (1973) The nucleotide sequence of the lac operator. Proc Natl Acad Sci 70(12):3581–3584

Kwong JC, McCallum N, Sintchenko V, Howden BP (2015) Whole genome sequencing in clinical and public health microbiology. Pathology 47(3):199–210

Lander ES (1996) The new genomics: global views of biology. Science 274(5287):536–539

Lander ES, Weinberg RA (2000) Journey to the center of biology. Science 287(5459):1777–1782

Lander ES, Linton LM, Birren B, Nusbaum C, Zody MC, Baldwin J, Proctor MJ (2001) Initial sequencing and analysis of the human genome

Lavarenne J, Guyomarc'h S, Sallaud C, Gantet P, Lucas M (2018) The spring of systems biology-driven breeding. Trends Plant Sci 23:706–720

Lenaerts B, Collard BCY, Demont M (2019) Improving global food security through accelerated plant breeding. Plant Sci 287:110–207
Mascher M, Schreiber M, Scholz U, Graner A, Reif JC, Stein N (2019) Gene bank genomics bridges the gap between the conservation of crop diversity and plant breeding. Nat Genet 51: 1076–1081
O'Neill K, Brocks D, Hammell MG (2020) Mobile genomics: tools and techniques for tackling transposons. Philos Trans R Soc B 375(1795):20190345
Santantonio N, Atanda SA, Beyene Y, Varshney RK, Olsen M, Jones E et al (2020) Strategies for effective use of genomic information in crop breeding programs serving Africa and South Asia. Front Plant Sci 11:353
Saraswathy N, Ramalingam P (2011) Concepts and techniques in genomics and proteomics. Elsevier
Staden R (1979) A strategy of DNA sequencing employing computer programs. Nucleic Acids Res 6(7):2601–2610
Varshney RK, Saxena RK, Upadhyaya HD, Khan AW, Yu Y, Kim C et al (2017) Whole-genome resequencing of 292 pigeon pea accessions identifies genomic regions associated with domestication and agronomic traits. Nat Genet 49:1082–1088
Varshney RK, Thudi M, Roorkiwal M, He W, Upadhyaya HD, Yang W, Bajaj P et al (2019) Resequencing of 429 chickpea accessions from 45 countries provides insights into genome diversity, domestication and agronomic traits. Nat Genet 51:857–864
Zhao Y, Li Z, Liu G, Jiang Y, Maurer HP, Wurschum T, Mock HP et al (2015) Genome-based establishment of a high-yielding heterotic pattern for hybrid wheat breeding. Proc Natl Acad Sci U S A 112:15624–15629
Zhao C, Liu B, Piao S, Wang X, Lobell DB, Huang Y, Huang M, Yao Y, Bassu S, Ciais P et al (2017) Temperature increase reduces global yields of major crops in four independent estimates. Proc Natl Acad Sci U S A 114:9326–9331

Structure and Organization of Plant Nuclear Genome

2

Monika Singh, Era Vaidya Malhotra, Vartika Srivastava, and Jyoti Ahlawat

Abstract

A genome refers to the complete set of DNA, including all the genes present in an organism. The genetic material in plants comprises of deoxyribonucleic acid (DNA), which is present either as organellar (chloroplast or mitochondrial) genome or nuclear genome. The knowledge of plant nuclear genome can be elaborated as genome size, gene content, extent of repetitive sequences, and polyploidy/duplication events. In this chapter, structure and composition of plant nuclear genome are described. Plant nuclear genome comprises of DNA distributed among the chromosomes containing coding sequence of genes as well as regulatory sequences, repetitive DNAs, and different classes of tandemly repeated sequences. The nuclear DNA is organized as chromatin where DNA is wrapped around the histone proteins to form nucleosomes. The resulting chromatin is further organized into linear chromosomes. Each plant species possesses a characteristic number, size and morphology of chromosomes. A 2350-fold range of diversity in genome sizes in plants has been observed and the haploid number of chromosomes may vary from 2 to 600. Polyploidy, genome/chromosome duplication and amplification of DNA motifs to form repetitive DNA may be attributed to genomic size variations. Some of the structural features of

M. Singh (✉)
Division of Genomic Resources, ICAR-National Bureau of Plant Genetic Resources, Pusa Campus, New Delhi, India
e-mail: monika.singh@icar.gov.in

E. V. Malhotra · V. Srivastava
Tissue Culture and Cryopreservation Unit, ICAR-National Bureau of Plant Genetic Resources, Pusa Campus, New Delhi, India

J. Ahlawat
School of Agricultural Sciences, G. D. Goenka University, Gurugram, Haryana, India

R. L. Singh et al. (eds.), *Plant Genomics for Sustainable Agriculture*,
https://doi.org/10.1007/978-981-16-6974-3_2

chromosomes are well conserved such as centromeres, telomeres and chromatin packaging. The structure and organization of plant genome are a key to replication, transcription and transmission of the genome, which allows duplication and physical reorganization of the whole genome. With the advancement of sequencing technologies, draft or whole genomes of many plant species have been sequenced. The availability of genome sequences of different plant species has opened newer avenues for genetic engineering including genome editing for crop improvement.

Keywords

Chromosomes · DNA · Gene · Nuclear genome · Plants · Regulatory sequences

2.1 Introduction

Genome, the basis of heredity in all living organisms, consists of genetic material in the form of deoxyribonucleic acid (DNA). In plants, the genome comprises nuclear as well as organellar DNA. Organellar genomes, found in chloroplast and mitochondria, are circular molecules of DNA coding for organellar genes only and show cytoplasmic or maternal inheritance, an example of non-Mendelian inheritance. Plant genome consists of coding and non-coding sequences along with regulatory elements and repetitive DNA. The plant nuclear genome consists of DNA and associated proteins, organized into discrete chromosomes. Structurally, DNA is wrapped around the histone proteins to form nucleosomes and the resulting chromatin is organized into linear chromosomes. The amount of DNA present in plant genomes remains almost constant within one species. Both the size of the plant genome and the number of chromosomes vary widely between species (Box 2.1). Polyploidization, duplication of genome or chromosome, and amplification of DNA motifs to give repetitive DNA may contribute to size variation. Plant genomes can be described in terms of genome size or nuclear DNA content, extent of repetitive sequences, and polyploidy/duplication events. This chapter mainly focuses on the structure and organization of plant nuclear genome including evolutionary aspects, coding and regulatory sequences, fine structure of nuclear genome, and genomic size. The revolution in plant genome studies has also highlighted as sequencing in plants.

Box 2.1: Genome Size or Nuclear DNA Content (*C*-Value)

Each plant species has a characteristic number of base pairs in its nuclei, which is referred as its genome size or nuclear DNA content. The *C*-value refers to the amount of DNA in the haploid genome of an organism, which varies from species to species. Nuclear genome size is generally estimated by flow cytometry and is widely measured in picograms (pg) of DNA. However, in

(continued)

Box 2.1 (continued)
context of molecular biology, genomic size is commonly expressed as base pairs for the 1C content. Arumugnathan and Earle (1991) reported genome sizes and *C*-values of more than 100 plant species as estimated by flow cytometry. The Plant DNA *C*-values Database, version 7.1 (https://cvalues.science.kew.org/), provides a comprehensive catalogue of *C*-value data for land plants and algae. Bennett and Leitch (2011) reported variation in the genomic sizes among angiosperms, and a 2350-fold range of diversity was measured among measurements of 6288 species.

2.2 Evolution of Plant Nuclear Genome

In plants, the evolution of chloroplast genome has been widely explored as compared to nuclear genes though nuclear genes are responsible for availability of range of phenotypes well adapted in diversified conditions. The processes involving nucleotide substitution, insertion or deletion of strings of nucleotides, and recombination/conversion between gene copies may govern evolution of plant genes (Clegg et al. 1997).

Clegg et al. (1997) reported evolutionary aspects of three plant nuclear multigene families, encoding the small subunit of ribulose-1,5-bisphosphate carboxylase (*rbcS*), the enzyme chalcone synthase (*Chs*), and alcohol dehydrogenases (*Adh*). The *rbcS* gene family has been proposed to be originated from a prokaryotic ancestor followed by the expansion of the *rbcS* gene family by subsequent processes leading to duplication. Chloroplasts and mitochondria evolved from the prokaryotic endosymbionts that transferred most of their genes to the host plant nuclear genome during evolution. Wang et al. (2011) proposed a model for evolution of plant nuclear genome. According to this model, coordination of activities in mitochondria and chloroplast and other cellular functions may contribute to a strong selection pressure for the differential acquisition of telo-box between mitochondrial and chloroplast ribosomal protein genes (mtRPGs and cpRPGs).

The possible factors responsible for evolution of nuclear genome in plants can be summarized as follows:

(i) **Polyploidy or whole-genome duplication:** In polyploidy, two or more genomes are fused within one nucleus in each cell containing more than two pairs of homologous chromosomes, resulting in immediate genomic and genetic variation. It occurs in several plant species including *Arachis hypogaea* (peanut), *Avena sativa* (oat), *Fragaria ananassa* (strawberry), *Triticum aestivum* (wheat), *Musa* sp. (banana), *Solanum tuberosum* (potato). Polyploidy can be categorized into two types—autoployploidy resulting from whole genome duplication, and allopolyploidy resulting from interspecific or intergeneric hybridizations followed by chromosome doubling (Doyle et al. 2008; Chen 2010).

(ii) **Transposable elements:** Transposable elements are ubiquitous components of plant nuclear genomes comprising of genetic units capable of making copies or moving around the genome. They are considered as main drivers of genome evolution in plants by working at structural level, being essential components of centromeric chromatin (Talbert et al. 2002; Hollister and Gaut 2009; Wicker et al. 2018). The role of transposable elements at the structural level has been demonstrated in bread wheat in order to further reveal their role in high-order chromatin arrangement, chromosome territories, and gene regulation (Wicker et al. 2018).

(iii) **Horizontal gene transfer:** Horizontal gene transfer (HGT) refers to the movement of genetic material from one organism to another, other than its offspring. The classical example of HGT in plants is the transfer of the tumour-inducing genes (T-DNAs) from *Agrobacterium tumefaciens* into the genome of host plant. This ability of *A. tumefaciens* has been exploited in genetic transformation of plants in order to introduce trait of interest in the host genome. The role of HGT in plant evolution has been reviewed by Quispe-Huamanquispe et al. (2017).

(iv) **Alternative splicing:** Alternative splicing is another important factor to enhance the functional diversity of genes, which can provide multiple gene products from a single nucleotide sequence. Specific exons of a gene are retained in the messenger RNA (mRNA) by this mechanism. The conservation and evolutionary aspects of alternative splicing have been studied (Zhang et al. 2016).

2.3 Composition of Plant Nuclear Genome

The plant nuclear genome comprises of DNA and associated proteins organized into discrete chromosomes.

2.3.1 Nuclear DNA

The nuclear DNA of plants consists of coding and regulatory sequences for genes, introns (non-coding sequences), along with various classes of repetitive DNA motifs (Schmidt and Heslop-Harrison 1998). The typical plant gene structure is illustrated in Fig. 2.1.

2.3.1.1 Coding and Non-coding Sequences

Coding sequences also referred to as ***exons*** represent DNA sequences that are transcribed to mature forms of ribonucleic acid (mRNAs) followed by translation into final product, i.e., the encoded proteins (Fig. 2.2). The exons may be interrupted by non-coding sequences called ***introns***, which are removed in mRNA by splicing.

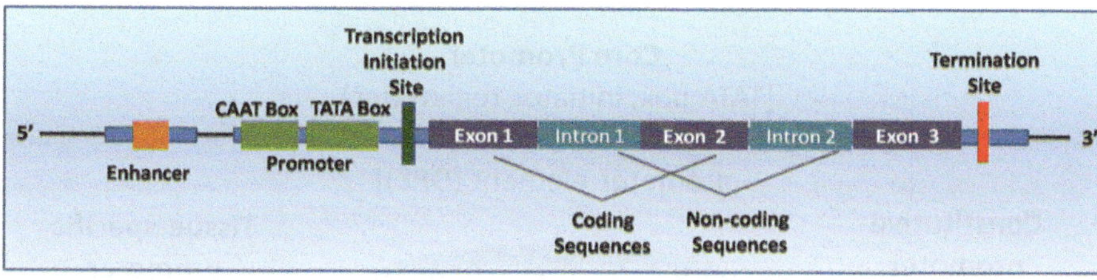

Fig. 2.1 Basic components of gene in plant nuclear genome

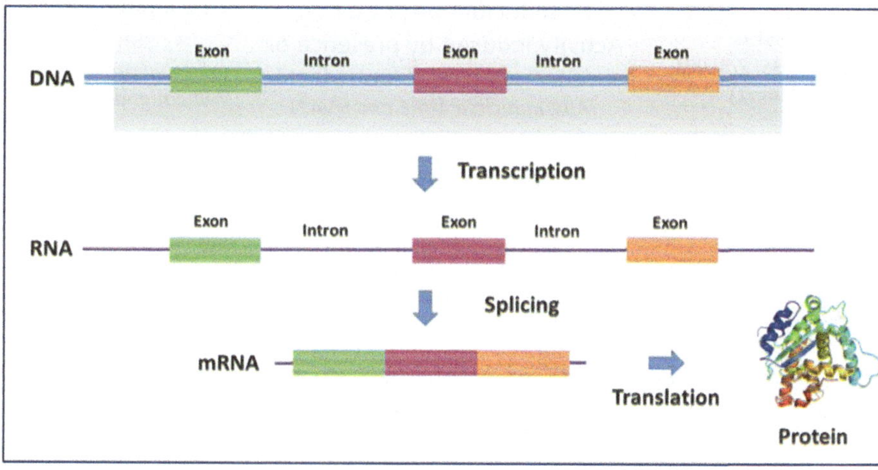

Fig. 2.2 Coding and non-coding sequences

2.3.1.2 Regulatory Elements

Regulatory sequences control the expression of genes, which may be classified as cis sequences and trans-factors. Cis regulatory sequences represent the linear nucleotide portions of non-coding DNA, which may include promoters, enhancers, silencers, and insulators. Trans-factors generally refer to proteins that bind to the cis-acting sequences to control the gene. Transcription factors are proteins that influence the transcription of genes by binding to defined regions of the genome (Latchman 1998).

Promoters: Promoters are the regulatory sequences that drive the expression of a gene. Promoters in plants can be classified as follows (Fig. 2.3):

(i) ***Core promoter:*** The core promoter regulates the initiation of transcription by RNA polymerase II (Juven-Gershon and Kadonaga 2010). TATA box is a defined element of core promoter in plants, which is 25–30 bp upstream of the transcription initiation site (TIS) (Lewin 2001). GC box and CAAT box may also be present, contributing in the initiation of transcription. The conservative promoter element, CAAT box, is located approximately 80–150 bp upstream of TIS. The GC box, GGGCGG, has been located in the upstream promoter of many plant groups.

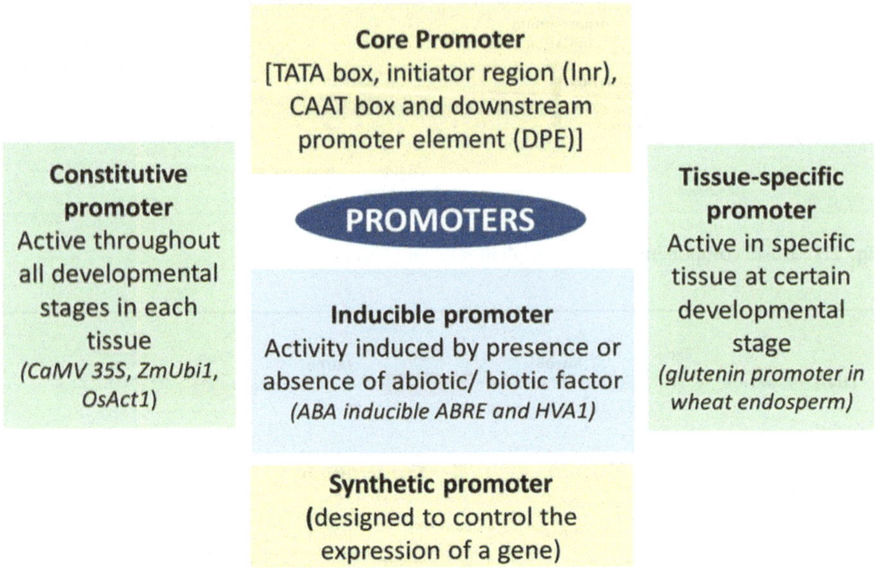

Fig. 2.3 Types of promoters in plants

(ii) ***Constitutive promoters:*** They are active in all the tissues and throughout all the developmental stages. Some of the common constitutive promoters are *Cauliflower Mosaic Virus* 35S (*CaMV*35S), maize ubiquitin (*ZmUbi1),* rice actin *(OsAct1).*

(iii) ***Tissue-specific promoters:*** These promoters are active in particular tissues or developmental stage.

(iv) ***Inducible promoters:*** These promoters are induced in response to stress conditions.

The constitutive, tissue-specific, and inducible promoters are of importance in plant biotechnology (Table 2.1) for expression of transgenes in genetically modified (GM) crops based on the requirement of efficiency, site, or period of action. For instance, in Golden rice, *psy* (phytoene synthase) and *crtI* (phytoene desaturase) genes are driven by an endosperm-specific promoter, so that they are only expressed in the endosperm.

Enhancers: Enhancers are located upstream or downstream of the promoter sequence. They enhance the expression of genes with the help of specific transcription factors.

2.3.1.3 Repetitive DNA

Repetitive DNA is either dispersed throughout the plant nuclear genome or restricted at specific locations in a tandem manner. The repetitive DNA sequences have structural role at the telomeres or chromosomal ends and the centromeres where

Table 2.1 Some promoters being employed in genetic modification in plants[a]

Promoter Constitutive promoters	Expression
(a) *Constitutive and tissue-specific promoters*	
Cauliflower mosaic virus 35S	All tissues and developmental stages
Maize ubiquitin 1 (ubi1)	
Rice actin 1	
Figwort mosaic virus (FMV) 35S	
Tissue-specific promoters	
Phaseolin of kidney bean	Seeds
β-Conglycinin promoter of soybean	Seeds
Zein in maize	Seeds
α-Globulin of cotton	Seeds
Glutelin in wheat/ rice	Seeds
LA22CD07, and LesAffx.6852.1.S1 in tomato	Fruit
Tomato polygalacturonase	Fruit
Histidine decarboxylase A of tomato	Fruit
ISOFLAVONE synthase gene promoters (IFS1 and IFS2) of soybean	Roots
MsPRP2of alfalfa	Roots
PHT1 in *Arabidopsis*	Roots

Promoter Induced by abiotic stress	Inducer	Organism
(b) *Inducible promoters*		
HSP18.2	Thermal shock	*Arabidopsis thaliana*
Rd29	Osmotic stress	*A. thaliana*
Adh	Dehydration and cold stress	*A. thaliana*
rbcS-3A	Light	*Pisum sativum*
Chn48	Ethylene	*Nicotiana tabacum*
PvSR2	Heavy metals	*Phaseolus vulgaris*
Cgmt1	Heavy metals	*Casuarina glauca*
HVADhn45	Drought stress	*Hordeum vulgare*
PtDrl02	Methyl jasmonate	*Populus* sp.
Induced by biotic stress		
CaPrx	Nematode infection	*Coffea arabica*
R2329 and *R2184*	Blast fungus infection	*Oryza sativa*
OsNAC6	Fungus infection	*Oryza sativa*
PPP	Pathogens	*Arabidopsis* sp.

[a]Source: Grunennvaldt et al. (2015)

the chromatids are attached (Fig. 2.4).Some repetitive DNA referred as ribosomal DNA (rDNA) encodes 45S and 5S ribosomal RNA genes.

Repetitive DNA can be broadly classified into two major groups on the basis of their genomic organization and localization on the chromosomes:

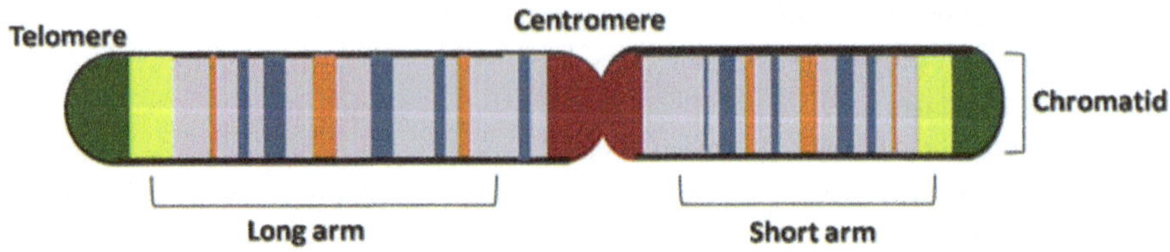

Fig. 2.4 Illustration of distribution of repetitive sequences on plant chromosomes
Tandem repeats – Red: Centromeric, Green: Telomeric, Yellow: Sub-telomeric; Blue: Dispersed sequence; Orange: Gene sequences

(i) ***Repetitive DNA in tandem repeats:*** Individual copies of DNA are arranged adjacent to each other to form a tandem array of monomeric units. These sequences are generally found at specific locations on the chromosomes such as pericentromeric, sub-telomeric, telomeric or intercalary regions.
   Tandem repeated DNA sequences can be classified as:
   (a) *Microsatellites or simple sequence repeats (SSRs)* with motifs of 2–6 base pairs (bp)—SSRs being ubiquitous in nature, are being employed as molecular markers and for DNA fingerprinting in plants.
   (b) *Minisatellites* with monomeric units of 10–40 bp.
   (c) *Satellite DNA (satDNA)* with a variable AT-rich repeat unit (150–400 bp in length).

(ii) ***Dispersed repetitive DNA elements:*** They are scattered throughout the genome, interspersed with other sequences and distributed along the chromosomes (Kubis et al. 1998). Dispersed sequences include transposable elements (jumping genes). Transposable elements discovered by McClintock are mobile genetic elements that can insert into new chromosomal location and are often capable of self-replication.
   Transposable elements are classified as class I (retrotransposons) and class II (DNA transposons) elements based on their structure and mode of transposition (Monden and Tahara 2015). Retrotransposons transpose by the 'copy and paste' method involving reverse transcription of an RNA intermediate and integration of a cDNA fragment. DNA transposons transpose by 'cut and paste' method involving excision and integration into new genomic locations.

2.4 Structure of Plant Nuclear Genome

2.4.1 Overview of Nucleus (Site of Plant Nuclear Genome)

The nucleus serves as the control centre of the cell and hence is considered as the brain of cell. It is the major cell organelle that distinguishes prokaryotic cells from the eukaryotic cells. The nucleus is the main reservoir of the cell's genetic information and various processes for genetic information including DNA replication, transcription, and RNA processing occur within the nucleus and only translation takes place in the cytoplasm.

Robert Brown first observed the nucleus in plant cells and named these structures 'nuclei'. A typical nucleus is composed of the following structures (Fig. 2.5):

(i) Nuclear membrane/nuclear envelope
(ii) Nucleoplasm
(iii) Nucleolus
(iv) Nuclear pores
(v) Chromosomes

2.4.1.1 Nuclear Membrane/Nuclear Envelope

Nuclear envelope acts as a fence between nucleus and cytoplasm to stave off free transmission of molecules. The nuclear envelope has a complex structure consisting of two nuclear membranes (outer and inner), an underlying nuclear lamina, and nuclear pore complexes (Fig. 2.6).

(i) ***Outer nuclear membrane (ONM)*:** The ONM is continuous with endoplasmic reticulum (ER) as shown in Figs. 2.5 and 2.6, therefore the lumen of nuclear membrane is directly connected with lumen of ER. It has ribosomes on the surface that are linked to ER.
(ii) ***Inner nuclear membrane (INM)*:** Proteins specific to the nucleus are present in INM, which include the proteins that bind the nuclear lamina. Integral proteins of INM are synthesized on the rough ER and reach the membrane by lateral diffusion in the connected ER and nuclear envelope membranes.

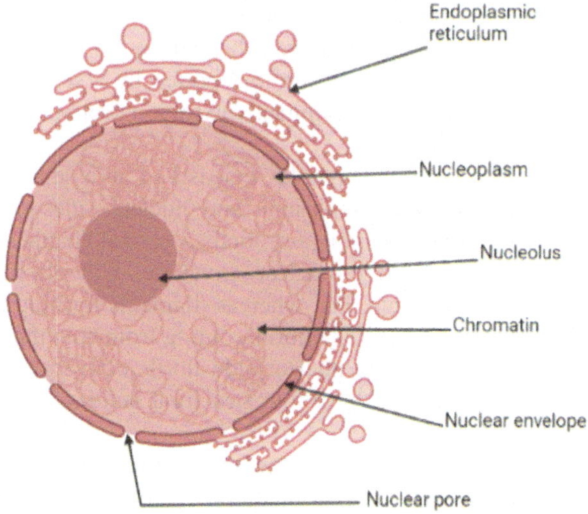

Fig. 2.5 Detailed structure of nucleus

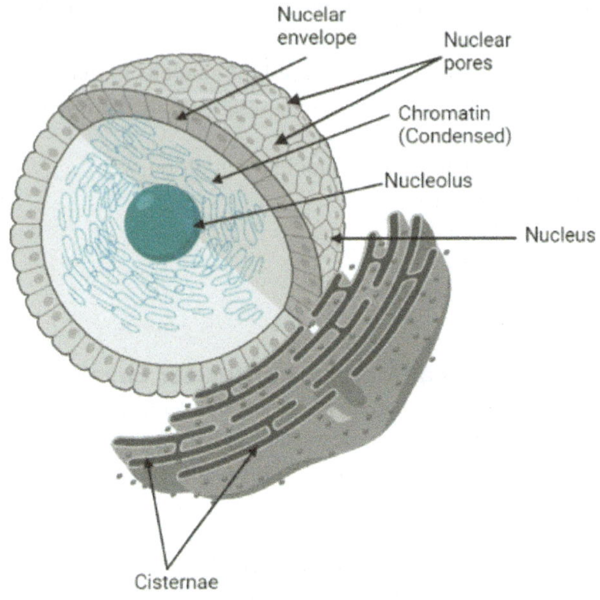

Fig. 2.6 Structure of nuclear envelope

(iii) ***Perinuclear space*:** Space between ONM and INM is called perinuclear space or lumen of envelope. The thickness of each nuclear membrane is 7–8 nm, whereas the perinuclear space is 20–40 nm thick.

(iv) ***Nuclear pores and nuclear pore complex (NPC)***: The phospholipid bilayer of nuclear membrane is permeable only for non-polar micromolecules. The NPC allows transmission of polar micromolecules and macromolecules. NPCs are the points where INM and ONM are continuous. Small molecules are able to pass rapidly through open channels in the NPC by passive diffusion. However, the macromolecules are transported by a selective, energy-dependent mechanism that acts predominantly to import proteins to the nucleus and export RNAs to the cytoplasm.

Structurally, the complex consists of an assembly of eight spokes attached to rings on the cytoplasmic and nuclear sides of the nuclear envelope. The spoke-ring assembly surrounds a central channel, through which proteins and RNAs cross the nuclear envelope. Cytoplasmic filaments protrude from the cytoplasmic ring, whereas filaments forming the nuclear basket arise from the nuclear ring. The fibrils coming from both the cytosolic and nuclear sides of the NPC, on the nuclear side fibrils converge to form basket-like structures (Fig. 2.7). NPC have four structural building blocks:

- *Column subunit* forming the bulk of the pore wall.
- *Annular subunit* that is centrally located.

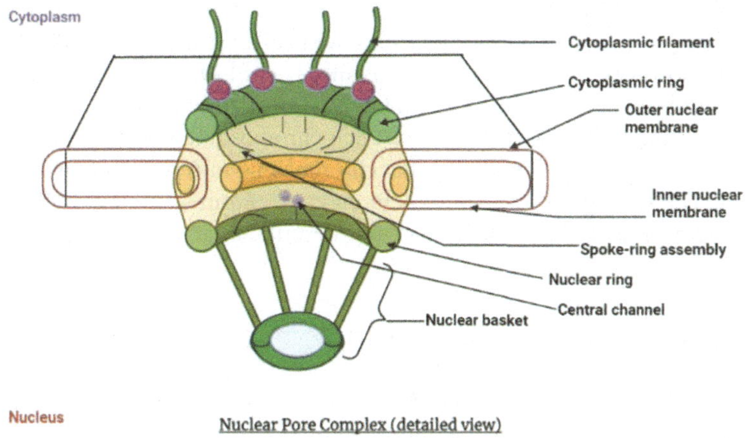

Fig. 2.7 Structure of the Nuclear Pore Complex (NPC)

- *Lumenal subunit* containing transmembrane protein that anchors the complex to the nuclear membrane.
- *Ring subunit* forming the cytosolic and nuclear faces of the complex.

(v) ***Nuclear lamina:*** Nuclear lamina is the attachment of proteins and fibres that are linked to the INM. It contains 60–80 kilo Dalton (kDa) fibrous proteins called lamins. Two lamins interact to form a lamin dimer, which further associate with each other to form nuclear lamina. Some associated proteins are also present. The nuclear lamina provides structural support to the nucleus, helps in DNA repair, controls cell cycle events such as cell division and DNA replication.

2.4.1.2 Nucleoplasm

The term "nucleoplasm" was coined by van Benedenin (1875) while 'karyoplasm' by Flemming (1878). Nucleoplasm is a type of protoplasm, present inside the nucleus to perform specialized operations. The nucleoplasm is composed of water, dissolved ions, and a mixture of other molecules. Nucleotides (required for DNA replication) and enzymes (responsible for the activities that take place in the nucleus) are dissolved in the nucleoplasm. The soluble, liquid portion of the nucleoplasm is called the ***nucleosol*** *or* ***nuclear hyaloplasm***.

2.4.1.3 Nucleolus

Nucleolus the most conspicuous nuclear body, is the site of ribosomal RNA (rRNA) transcription and processing as well as for ribosome assembly. The nucleolus is designed for regulated and efficient production of rRNAs and assembly of the ribosomal subunits, and hence, is referred as a ***ribosome production factory***. The nucleolus may also play a role in RNA modification and several types of RNA move in and out of the nucleolus at specific stages during their processing.

2.4.1.4 Chromosomes

In the nucleus of each cell, DNA molecules are packaged into thread-like structures called chromosomes. Each chromosome is made up of DNA tightly coiled several times around proteins called histones that provide structural support and play a role in controlling the activities of the genes. Each chromosome has a constriction point called the centromere, which divides the chromosome into two sections, or "arms" (Fig. 2.4). The location of the centromere on each chromosome gives the chromosome its characteristic shape, which can also describe the location of specific genes.

2.4.2 Structure of Plant Nuclear DNA

The double-helical structure of DNA was first proposed by Watson and Crick (1953). They used the X-ray diffraction techniques to find out the three-dimensional structure of DNA. The Watson and Crick model of DNA structure is now accepted worldwide. Diagrammatic representation of DNA structure is given in Fig. 2.8.

The various features of this model are discussed below:

(i) ***DNA is made of two polynucleotide chains:***
DNA is a polymer of nucleotides, which are arranged in the form of two chains. The nucleotides present in DNA are deoxyribonucleotides comprising (a) deoxyribose, a pentose sugar, (b) a nitrogenous base, and (c) one phosphate group.

One of the four nitrogenous bases is present in a DNA nucleotide: Adenine (A), Guanine (G), Cytosine (C), and Thymine (T). The nucleotides are linked by phosphodiester bonds to form two polynucleotide chains. Each of these two chains has a 3′ and a 5′ end. The end of the polynucleotide chain having a free hydroxyl group at the third carbon of deoxyribose sugar is called the 3′ end. The other end of the chain containing the free phosphate group attached to the fifth carbon of the deoxyribose sugar is called the 5′ end.

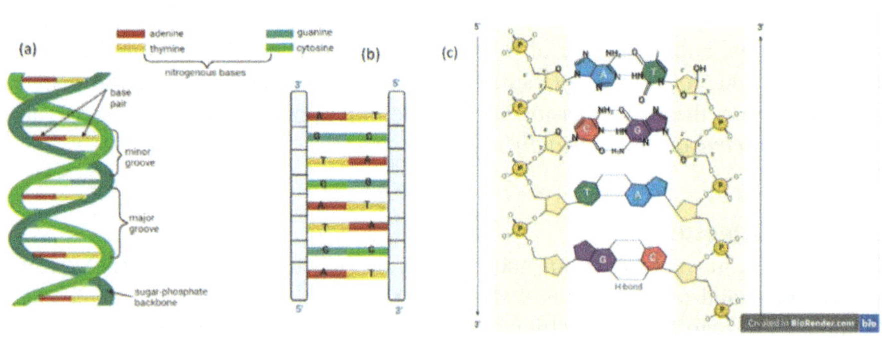

Fig. 2.8 Watson and Crick proposed the double helix model for DNA. (**a**) DNA double helix; (**b**) anti-parallel strands of DNA; (**c**) chemical structure of DNA

(ii) ***Two polynucleotide chains are coiled around each other.***
The two polynucleotide chains of DNA are coiled around one another to form a DNA double helix.

(iii) ***The polynucleotide chains are anti-parallel.***
The two polynucleotide chains of the DNA double helix are arranged parallel and opposite to each other. One chain runs in 3′–5′ direction with its sugar and phosphate molecules directed upwards, whereas the other chain in 5′–3′ direction with the sugar and phosphate molecules faces in downward direction.

(iv) ***DNA double helix has phosphate-sugar backbone.***
The backbone of the double helix is made by the sugar and phosphate molecules. The polynucleotide chains in DNA are arranged in a manner that the sugar and phosphate molecules are present on the outer side of the double helix with the nitrogenous bases facing inwards.

(v) ***The two chains are held together by hydrogen bonds.***
The two polynucleotide chains are held together by hydrogen bonds between their nitrogenous bases. The purines (A or G) of one chain pair with the pyrimidines (C or T) of the other chain via hydrogen bonds. 'A' always pairs with 'T' with two hydrogen bonds, whereas 'G' always pairs with 'C' with three hydrogen bonds.

(vi) ***The ratio of purines and pyrimidines in DNA are equal.***
As A always pairs with T, and G with C, so the amount of A is equal to the amount of T, and the amount of G is equal to that of C. Thus, purines and pyrimidines are always present in equal ratio within a molecule of DNA.

(vii) ***Each turn of helix is made of ten base pairs.***
According to the Watson and Crick model, each turn of DNA double helix has ten base pairs in it. One turn of helix measures around 34 Å units ($1\ \text{Å} = 10^{-10}$) or 3.4 nm. Thus, in DNA double helix each base pair is 3.4 Å units or 0.34 nm apart.

(viii) ***DNA double helix has two grooves.***
The two polynucleotide chains are not symmetrical. When the two strands are wound around each other, spaces are left behind in the form of grooves. Two types of grooves are present in the DNA double helix.
- *Major Groove*: It is the widest groove measuring around 22 Å units.
- *Minor Groove*: The width of this groove is less than that of the major groove. It measures around 12 Å units.

These grooves in DNA double helix provide space for the attachment of enzymes and transcription factors, etc.

2.4.3 Organization of Plant Nuclear Genome

As mentioned earlier, the plant nuclear genome consists of DNA and associated proteins, which are organized into discrete chromosomes. Each non-replicated chromosome and metaphase chromatid comprises a single linear DNA molecule that is unbroken from one end to the other. At the metaphase of mitosis, the DNA is condensed into short rod-like mitotic chromosomes, whereas at interphase, the chromosomes are decondensed within the interphase nucleus.

2.4.3.1 Chromatin Structure and DNA Packaging

Chromatin consists of histone proteins, namely, H1, H2a, H2b, H3, and H4 as the primary protein components along with non-histone proteins and RNA. The histone proteins help to organize DNA into 'bead-like' structures called *nucleosomes* by providing a base on which the DNA can be wrapped around. The core nucleosome particle is composed of 146 bp of DNA wrapped around an octamer of four core histone proteins. These nucleosomes fold into 30 nm chromatin fibres, which are the components of the chromosomes. Dense, compact regions of chromosomes are referred as ***heterochromatin*** consisting of mostly untranscribed and inactive DNA. Regions called ***euchromatin*** are less compact and consist of more highly transcribed genes.

2.4.3.2 Nucleosome

The nucleosome is the fundamental unit of chromatin, which is composed of a core particle and a linker region (as shown in Fig. 2.9).

(i) ***Core particle:*** The core particle is highly conserved between species and is composed of 146 base pairs of DNA wrapped 1.7 turns around a protein octamer of two each of the core histones H3, H4, H2A, and H2B.

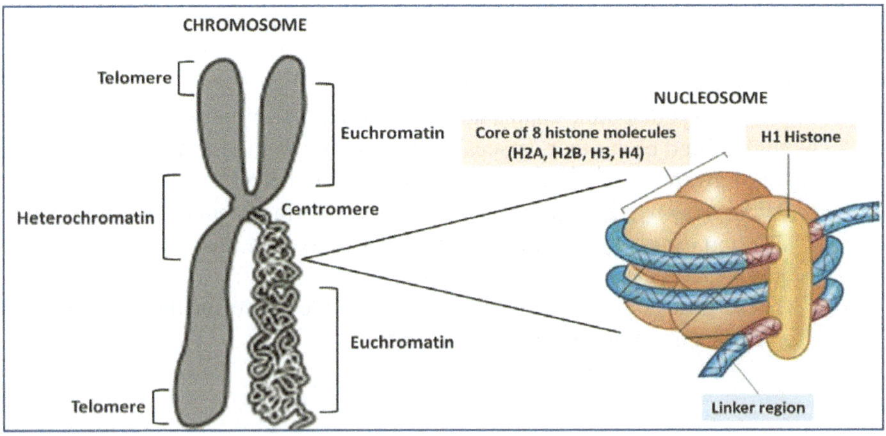

Fig. 2.9 Structure of chromatin

(ii) ***Linker region (or internucleosomal region) that joins adjacent core particles***: The length of the linker region varies between species and cell type. The variable linker histones are incorporated within this region. Therefore, the total length of DNA in the nucleosome may vary among species ranging from 160 to 241 base pairs.

(iii) **Histone proteins:**Two types of histone proteins are present in nucleosomes that play a major role:

(a) ***Core histones:*** The core histones, H3, H4, H2A, and H2B, are small, basic proteins. Central domain is the most conserved region of these histones, which is structurally composed of the "*histone fold domain*" comprising three α-helices separated by two loop regions. In contrast, the N-terminal tails of each core histone are more variable and unstructured. The tails are rich in lysine and arginine residues making them extremely basic. This region is the site of numerous post-translational modifications that are proposed to modify its charge and thereby alter DNA accessibility and protein/protein interactions with the nucleosome. It is significant to note that other proteins that interact with DNA also contain the '*histone fold domain*.

(b) ***Linker histones:*** Linker histones associate with the linker region of DNA between two nucleosome cores and, unlike the core histones, they are not well conserved between species. They are generally composed of three domains: a globular, non-polar central domain essential for interactions with DNA and two non-structured N- and C- terminal tails that are highly basic and proposed to be the site of post-translational modifications. The linker histones assist in spacing nucleosomes and can control higher order compaction by providing an interaction region between adjacent nucleosomes.

An overview of DNA packaging has been summarized in Fig. 2.10.

2.5 Genome Sequencing in Plants

Genome sequences, i.e., the order of nucleotides in a polynucleotide chain, can be considered as the genetic blueprint of an organism. These sequences are the instructions for generating structural and regulatory RNA molecules as well as encoding proteins. Sequencing of the genetic information is an important step in understanding the function and evolutionary relationships of different organisms. The first plant genome sequenced was that of *Arabidopsis thaliana* (Arabidopsis Genome Initiative 2000). Till date, more than 600 complete plant genome assemblies are now available in public repositories. Sequences of some of the important plant species are summarized in Table 2.2.

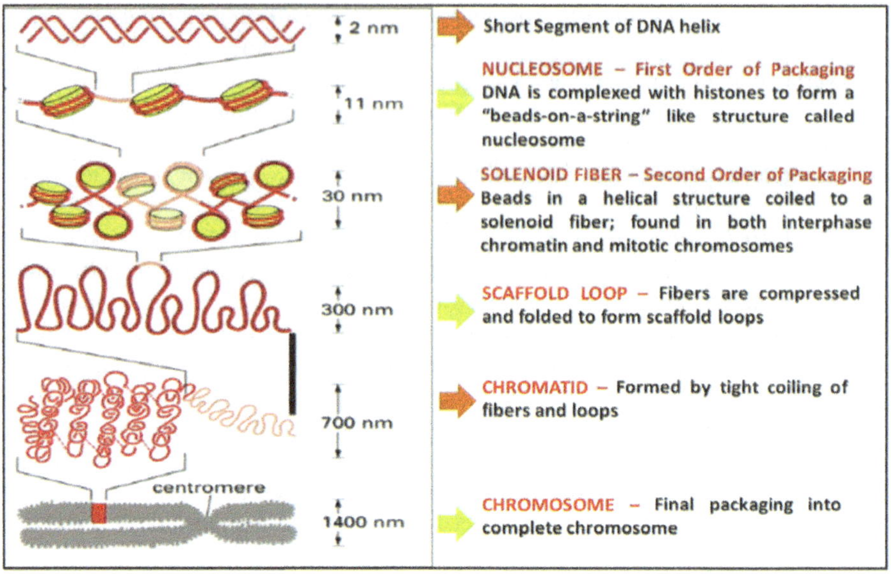

Fig. 2.10 An overview of DNA packaging

2.5.1 First Generation DNA Sequencing

Rapid progress in genome sequencing has been made with the emerging technologies time to time and in late 1970s, two separate groups described new techniques, Coulson and Sanger's plus and minus system and Maxam and Gilbert's chemical cleavage technique, to sequence any nucleotide sequence (Sanger and Coulson 1975; Maxam and Gilbert 1977).

2.5.1.1 The Maxam and Gilbert Technique (Also Known as the Chemical Cleavage Method)

This is based on the chemical modification of DNA followed by cleavage at specific bases. The DNA fragment to be sequenced is purified and radiolabelled at one end. The DNA is then treated with chemicals that break the DNA chain at specific bases, generating a series of labelled fragments from the radiolabelled end to the cut site of each molecule. The labelled fragments are then run on a polyacrylamide gel and the length of the cleaved fragments is determined and thus the sequence is inferred.

2.5.1.2 Sanger's Plus and Minus Technique

Radiolabelled nucleotides are incorporated using a DNA polymerase, followed by two secondary polymerization reactions, i.e., plus and minus reactions. In the first or plus reaction, a single type of nucleotide is added, thereby, all extensions end with that base alone, whereas in the second or minus reaction, three nucleotides are added resulting in the synthesis of sequences up to the position before the next missing

Table 2.2 Publically available genome sequences of some important plant species

Species	Genome size	Reference
Arabidopsis thaliana (Thale cress)	125 Mbp	Arabidopsis Genome Initiative (2000)
Oryza sativa (rice)	430 Mbp	International Rice Genome Sequencing Project (2005)
Sorghum bicolor (sorghum)	818 Mbp	Paterson et al. (2009)
Zea mays (maize)	2300 Mbp	Schnable et al. (2009)
Glycine max (soybean)	1115 Mbp	Schmutz et al. (2010)
Azadirachta indica (neem tree)	364 Mbp	Krishnan et al. (2012)
Prunus persica (peach)	265 Mbp	International Peach Genome Initiative (2013)
Ananas comosus (pineapple)	526 Mbp	Ming et al. (2015)
Brassica juncea (Indian mustard)	922 Mbp	Yang et al. (2016)
Arachis duranensis (A-genome peanut)	1250 Mbp	Bertioli et al. (2016)
Arachis ipaensis (B-genome peanut)	2700 Mbp	Bertioli et al. (2016)
Citrus grandis (Pummelo)	381 Mbp	Wang et al. (2017)
Helianthus annuus (common sunflower)	3600 Mbp	Badouin et al. (2017)
Triticum turgidum (emmer wheat)	12,000 Mbp	Avni et al. (2017)
Pennisetum glaucum (pearl millet)	1.79 Gb	Varshney et al. (2017)
Dioscorea rotundata (white Guinea yam)	580 Mbp	Tamiru et al. (2017)
Piper nigrum (black pepper)	762 Mbp	Hu et al. (2019)
Cinnamomum kanehirae (stout camphor tree)	830 Mbp	Chaw et al. (2019)
Persea americana (avocado)	980 Mbp	Rendón-Anaya et al. (2019)
Cocos nucifera (coconut palm)	2420 Mbp	Lantican et al. (2019)
Actinidia chinensis (kiwifruit)	758 Mbp	WuH et al. (2019)
Camellia sinensis (tea tree)	3000 Mbp	Xia et al. (2019)

nucleotide. The products are run on a polyacrylamide gel and compared to decipher the position of nucleotides at each position.

2.5.1.3 Chain-Termination or Dideoxy Technique

Sanger's method was further modified to the Sanger's chain-termination or dideoxy technique (Sanger and Nicklen 1977). Here, radiolabelled dideoxynucleotides (ddNTPs), analogues of deoxyribonucleotides (dNTPs) with a missing 3′ hydroxyl group, are mixed with the DNA extension reaction mix. Lacking a 3′ hydroxyl group, these ddNTPs are unable to bond with the 5′ phosphate of the next dNTP, thus terminating the DNA chain extension at the location where incorporated in the nucleotide chain. Four parallel reactions are performed, each with all the reactions components for DNA extension along with all the four dNTPs and an individual ddNTP base at lower concentration. As the nucleotide chain extends, the ddNTPs get incorporated in some chains in place of the normal dNTP and the extension of that particular chain gets terminated as no further nucleotides can be added. This process is repeated for a number of cycles. The reaction products are resolved in four different lanes of a polyacrylamide gel. The reaction products get separated based on their size, each fragment ending with a labelled ddNTP. Based on the autoradiography, the sequence of the nucleotide can be read directly, starting from the smallest fragment that has moved farthest in the gel. This method is based on *sequencing-by-synthesis* and was used for sequencing the first DNA genome that of bacteriophage ϕX174 with a genome size of 5386 bases (Sanger et al. 1977).

Together these techniques formed the ***'First Generation Sequencing'*** techniques. The Sanger's sequencing technique has been improved further with the replacement of the radiolabelling with fluorometric based detection, enabling the reactions to be run in a single tube and resolution by capillary electrophoresis.

2.5.2 Second Generation DNA Sequencing

The second generation of sequencing technologies is generally based on the luminescent method of measuring pyrophosphate synthesis. The principle revolves around the conversion of pyrophosphate to adenosine triphosphate (ATP) using the enzyme ATP sulfurylase, which is then acted upon by luciferase to produce light proportionate to the amount of pyrophosphate utilized (Nyrén and Lundin 1985). This principle was exploited to decipher the sequence of a growing nucleotide chain by measuring the amount of pyrophosphate production for every nucleotide added corresponding to a template DNA fixed onto a solid surface (Hyman 1988). Using this method, sequences could be read in real-time without the need of electrophoresis.

2.5.2.1 Roche (454) Sequencing

The method of pyrosequencing evolved to become into the first major commercially successful next generation sequencing (NGS) technology. This method was taken up by a biotechnology company by the name of 454 Life Technologies and later launched by Roche in 2005. This method employed emulsion PCR in small oil beads along with massively parallel pyrosequencing of bead colonies. The pyrosequencing method is a *sequencing-by-synthesis* method, where DNA fragments are hybridized to an array of beads. These beads are so designed that amplification of individually bound templates by polymerase chain reaction (PCR) can be carried out on them. Along with this, these beads also contain enzymes that generate fluorescence on utilizing inorganic phosphates detected as a light signal (Ronaghi et al. 1996). This sequencing platform provided benefits of high efficiency and low cost for generating reads of up to 700 Mb length. A number of plant genomes, such as tomato, melon, and potato have been sequenced using this technique (Bevan and Uauy 2013; Mardis 2013).

2.5.2.2 Illumina Sequencing

This high-throughput sequencing platform, again based on the *sequencing-by-synthesis* concept was developed by Solexa, which was later acquired by Illumina (Turcatti et al. 2008). This technology introduced the concept of bridge amplification. A DNA library is prepared by genomic DNA digestion and adapter ligation. However, adapters at the 5′ and 3′ ends are different. DNA fragments are attached onto a solid surface and then amplified to form local bridge amplification DNA clusters. DNA sequence is then deciphered by adding cleavable fluorescent chain terminator nucleotides. Once a nucleotide is bound to the complementary stand, it emits a light that is different in colour for each of the four bases. The dye and the 3′ blocker are then chemically removed before the next cycle. This sequencing platform generates up to 3 billion reads per run, and a read length of 50–300 bp. Numerous genomes have been sequenced by Illumina sequencing (Balasubramanian 2011; Cao et al. 2011; Greenleaf and Sidow 2014).

2.5.2.3 SOLid Sequencing

This sequencing platform known as Sequencing by the Oligo Ligation and Detection (SOLiD) developed by Applied Biosystems utilizes the principle of primer extension by ligation reactions. This method is known as *sequencing-by-ligation*. First DNA is amplified using emulsion PCR and resulting beads containing copies of a single DNA molecule are fixed onto a slide. The DNA templates are then bound to an adapter to which a primer is annealed. Then four different fluorescently labelled nucleotides are added, which on matching to the template sequence ligate to the primer by DNA ligase. Ligation of sequence releases a signal indicating the nucleotide at that position. Once the fluorescence is captured, the tags are removed and fresh set of oligonucleotides is added for the next cycle of ligation (Shendure et al. 2008). This sequencing platform gives up to 1410 million paired-end reads per run.

2.5.2.4 Ion Torrent Semiconductor Sequencing

The fourth NGS platform, developed by Ion Torrent Systems Incorporated, was based on a semi-conductor-based detection system. Instead of using fluorescent dyes for detection, this system measures the change in pH that occurs on the release of an H^+ ion on nucleotide incorporation (Rothberg et al. 2011). Here, the DNA template is mixed with single nucleotides. If the nucleotide is complementary to the template strand nucleotide, it gets incorporated into the growing nucleotide strand, releasing one H^+ ion, which is detected by an ion sensor. This method can generate 80 million reads per run with an average length of 400 bp.

The ***Second Generation Sequencers*** were based on generation of DNA clusters by amplification of target molecules, followed by sequencing. However, simultaneous sequencing of several copies of each cluster restricts the overall read length. But still, these technologies have ushered in a genomics revolution, enabling the sequencing of a vast array of organisms in a relatively short span of time.

2.5.3 Third Generation DNA Sequencing

The third generation of sequencing technologies is powered by the progress made in single molecule imaging techniques. These set of techniques are able to sequence a single DNA molecule, without its amplification or modification in any way (Ozsolak 2012). These techniques generate longer reads, with an average read length of up to 10 kb.

2.5.3.1 Heliscope Single Molecule Sequencing

The first technique from this generation of sequencing was developed by Helicos Biosciences. This single molecule sequencing system follows an extension-based sequencing system (Harris et al. 2008). A DNA library is prepared by DNA shearing. The sheared DNA fragments are poly-A tailed and attached to flow cells by poly-T anchors. A universal primer is then annealed to the DNA templates, followed by chain extension using fluorescently labelled reversible terminator nucleotides (Bowers et al. 2009). Here, no amplification of the target sequences is required. The sequencer can generate up to 600 Mb read length in one run.

2.5.3.2 Single Molecule Real-Time (SMRT) Sequencing

This method is based on the *sequencing-by-synthesis* approach and was developed by Pacific Biosciences. This method utilizes the enzymatic activity of DNA polymerase enzyme. DNA is synthesized in small wells called zero-mode wave-guides (ZMWs; Levene et al. 2003). Target DNA molecules are ligated to adapters, denatured and annealed to a primer that is complementary to the adapter sequence. These primed DNA molecules are loaded onto the ZMWs. Each ZMW has a DNA polymerase enzyme attached at its bottom along fluorescently labelled dNTPs, each dNTP with a different fluorescent tag. On incorporation into the growing chain, the nucleotide emits fluorescent signal, and then the fluorescent tag gets cleaved away, making place for incorporation of the next nucleotide molecule (Eid et al. 2009). This method gives long read lengths of up to 20 kb.

2.5.3.3 Nanopore Sequencing

The nanopore sequencing approach employs the sequencing of single DNA molecules by passing them through nano scale pores. This technique, like other third-generation techniques, does not require DNA amplification for library preparation. A single stranded DNA molecule is passed through nanopores embedded in membrane under an electric field. As the DNA strands slowly passes through the nanopore, each base produces a characteristic change in membrane conductance. These changes in membrane conductance are then used to determine the nucleotide sequence (Healy 2007). This technology requires very low DNA quantity for sequencing and very long reads can be directly read. Another advantage offered by this technique is that the original DNA sequence is read directly, instead of a copy of the template as is done in other techniques. The nanopore sequence platform is commercially available as MinION platform by Oxford Nanopore Technologies (Jain et al. 2016). This technology has the ability to generate reads of nearly 98 kb length.

Over the last few years, a great deal of time, energy, and resources have been devoted to develop various genome sequencing technologies. An overview of sequencing strategies is summarized in Fig. 2.11. This is still a forward focused discipline and is forming the basis of research in several frontline areas of molecular biology and crop improvement.

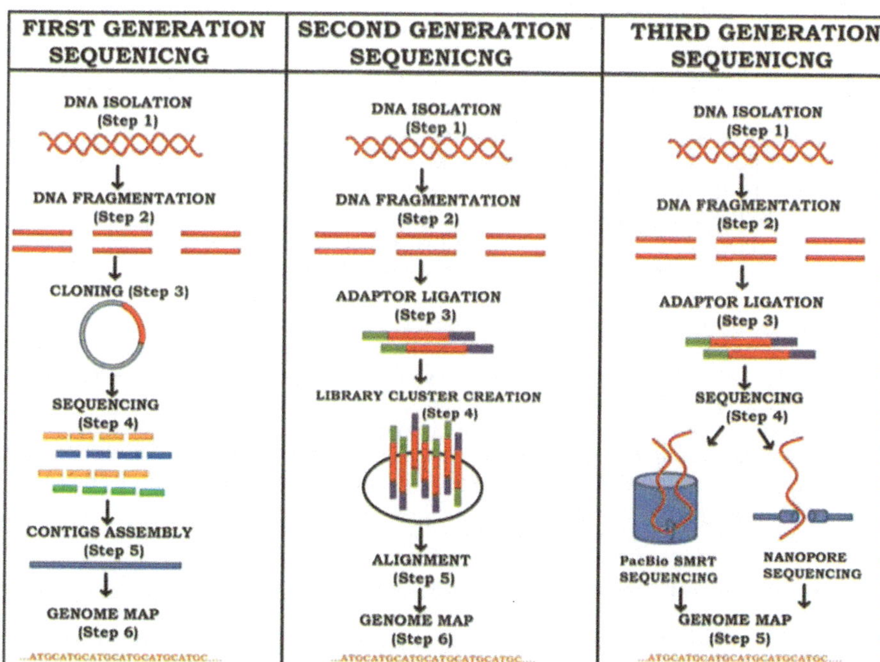

Fig. 2.11 Layout of working principle of DNA sequencing methods (Source: https://hackmd.io/@Chang/algorithms-for-DNA-sequencing#Analysis-for-sequenced-nucleic-acid-DNA-RNA)

2.6 Concluding Remarks

Plant genomes can be well-defined in terms of genome size, gene content, extent of repetitive sequences, and polyploidy/duplication events. Plants possess organellar as well as nuclear genomes and nuclear genome is the largest and most complex. The study of structure and organization of plant nuclear genome has revolutionized the study of molecular biology in plants. The underlying mechanisms involving nuclear genome have been explored and better understood, which are being integrated into crop improvement programmes. Further, whole genome sequencing has opened paths to further explore potential of nuclear genome in crop improvement. Genomics represents an interdisciplinary approach, as it covers the broad spectrum from DNA sequencing to field-based research through the integrated potential of genetics, biology, bioinformatics, molecular biology, genetic engineering, microbiology and related fields to draw extensive benefits to mankind.

Note: Figures 2.6–2.9 have been drawn using BioRender.com

References

Arabidopsis Genome Initiative (2000) Analysis of the genome sequence of the flowering plant *Arabidopsis thaliana*. Nature 408(6814):796–815

Arumuganathan K, Earle ED (1991) Nuclear DNA content of some important plant species. Plant Mol Biol Rep 9:208–218

Avni R, Nave M, Barad O, Baruch K, Twardziok SO, Gundlach H, Hale I, Mascher M, Spannagl M, Wiebe K, Jordan KW, Golan G, Deek J, Ben-Zvi B, Ben-Zvi G, Himmelbach A, MacLachlan RP, Sharpe AG, Fritz A, Ben-David R, Budak H, Fahima T, Korol A, Faris JD, Hernandez A, Mikel MA, Levy AA, Steffenson B, Maccaferri M, Tuberosa R, Cattivelli L, Faccioli P, Ceriotti A, Kashkush K, Pourkheirandish M, Komatsuda T, Eilam T, Sela H, Sharon A, Ohad N, Chamovitz DA, Mayer KFX, Stein N, Ronen G, Peleg Z, Pozniak CJ, Akhunov ED, Distelfeld A (2017) Wild emmer genome architecture and diversity elucidate wheat evolution and domestication. Science 357(6346): 93–97

Badouin H, Gouzy J, Grassa CJ, Murat F, Staton SE, Cottret L, Lelandais-Briere C, Owens GL, Carrere S, Mayjonade B, Legrand L, Gill N, Kane NC, Bowers JE, Hubner S, Bellec A, Berard A, Berges H, Blanchet N, Boniface MC, Brunel D, Catrice O, Chaidir N, Claudel C, Donnadieu C, Faraut T, Fievet G, Helmstetter N, King M, Knapp SJ, Lai Z, Le Paslier MC, Lippi Y, Lorenzon L, Mandel JR, Marage G, Marchand G, Marquand E, Bret-Mestries E, Morien E, Nambeesan S, Nguyen T, Pegot-Espagnet P, Pouilly N, Raftis F, Sallet E, Schiex T, Thomas J, Vandecasteele C, Vares D, Vear F, Vautrin S, Crespi M, Mangin B, Burke JM, Salse J, Munos S, Vincourt P, Rieseberg LH, Langlade NB (2017) The sunflower genome provides insights into oil metabolism, flowering and Asterid evolution. Nature 546(7656): 148–152

Balasubramanian S (2011) Sequencing nucleic acids: from chemistry to medicine. Chem Commun 47:7281–7286

Bennett MD, Leitch IJ (2011) Nuclear DNA amounts in angiosperms: targets, trends and tomorrow. Ann Bot 107(3):467–590

Bertioli DJ, Cannon SB, Froenicke L, Huang G, Farmer AD, Cannon EK, Liu X, Gao D, Clevenger J, Dash S, Ren L, Moretzsohn MC, Shirasawa K, Huang W, Vidigal B, Abernathy B, Chu Y, Niederhuth CE, Umale P, Araujo AC, Kozik A, Kim KD, Burow MD, Varshney RK, Wang X, Zhang X, Barkley N, Guimaraes PM, Isobe S, Guo B, Liao B, Stalker

HT, Schmitz RJ, Scheffler BE, Leal-Bertioli SC, Xun X, Jackson SA, Michelmore R, Ozias-Akins P (2016) The genome sequences of *Arachis duranensis* and *Arachis ipaensis*, the diploid ancestors of cultivated peanut. Nat Genet 48(4):438–446

Bevan MW, Uauy C (2013) Genomics reveals new landscapes for crop improvement. Genome Biol 14(6):206

Bowers J, Mitchell J, Beer E, Buzby PR, Causey M, Efcavitch JW, Jarosz M, Krzymanska-Olejnik E, Kung L, Lipson D, Lowman GM, Marappan S, McInerney P, Platt A, Roy A, Siddiqi SM, Steinmann K, Thompson JF (2009) Virtual terminator nucleotides for next-generation DNA sequencing. Nat Methods 6(8):593–595

Cao J, Schneeberger K, Ossowski S, Günther T, Bender S, Fitz J, Koenig D, Lanz C, Stegle O, Lippert C, Wang X, Ott F, Müller J, Alonso-Blanco C, Borgwardt K, Schmid KJ, Weigel D (2011) Whole-genome sequencing of multiple *Arabidopsis thaliana* populations. Nat Genet 43(10):956–963

Chaw S, Liu Y, Wu Y et al (2019) Stout camphor tree genome fills gaps in understanding of flowering plant genome evolution. Nat Plants 5:63–73

Chen ZJ (2010) Molecular mechanisms of polyploidy and hybrid vigor. Trends Plant Sci 15(2): 57–71

Clegg MT, Cummings MP, Durbin ML (1997) The evolution of plant nuclear genes. Proc Natl Acad Sci U S A 15:7791–7798

Doyle JJ, Flagel LE, Paterson AH, Rapp RA, Soltis DE, Soltis PS, Wendel JF (2008) Evolutionary genetics of genome merger and doubling in plants. Annu Rev Genet 42:443–461

Eid J, Fehr A, Gray J, Luong K, Lyle J, Otto G, Peluso P, Rank D, Baybayan P, Bettman B, Bibillo A, Bjornson K, Chaudhuri B, Christians F, Cicero R, Clark S, Dalal R, Dewinter A, Dixon J, Foquet M, Gaertner A, Hardenbol P, Heiner C, Hester K, Holden D, Kearns G, Kong X, Kuse R, Lacroix Y, Lin S, Lundquist P, Ma C, Marks P, Maxham M, Murphy D, Park I, Pham T, Phillips M, Roy J, Sebra R, Shen G, Sorenson J, Tomaney A, Travers K, Trulson M, Vieceli J, Wegener J, Wu D, Yang A, Zaccarin D, Zhao P, Zhong F, Korlach J, Turner S (2009) Real-time DNA sequencing from single polymerase molecules. Science 323: 133–138

Flemming W (1878) BeiträgezurKenntniss der Zelle und ihrerLebenserscheinungen. Arch f mikr Anat 16:302–436

Greenleaf WJ, Sidow A (2014) The future of sequencing: convergence of intelligent design and market Darwinism. Genome Biol 15(3):303

Grunennvaldt RL, Degenhardt-Goldbach J, Gerhardt IR, Quoirin M (2015) Promoters used in genetic transformation of plants. Res J Biol Sci 10:1–9

Harris TD, Buzby PR, Babcock H, Beer E, Bowers J, Braslavsky I, Causey M, Colonell J, Dimeo J, Efcavitch JW, Giladi E, Gill J, Healy J, Jarosz M, Lapen D, Moulton K, Quake SR, Steinmann K, Thayer E, Tyurina A, Ward R, Weiss H, Xie Z (2008) Single-molecule DNA sequencing of a viral genome. Science 320(5872):106–109

Healy K (2007) Nanopore-based single-molecule DNA analysis. Future Med 2(4):459–481

Hollister JD, Gaut BS (2009) Epigenetic silencing of transposable elements: a trade-off between reduced transposition and deleterious effects on neighbouring gene expression. Genome Res 19: 1419–1428

Hu L, Xu Z, Wang M et al (2019) The chromosome-scale reference genome of black pepper provides insight into piperine biosynthesis. Nat Commun 10:4702

Hyman ED (1988) A new method for sequencing DNA. Anal Biochem 74:423–436

International Peach Genome Initiative, Verde I, Abbott AG, Scalabrin S, Jung S, Shu S, Marroni F, Zhebentyayeva T, Dettori MT, Grimwood J, Cattonaro F, Zuccolo A, Rossini L, Jenkins J, Vendramin E, Meisel LA, Decroocq V, Sosinski B, Prochnik S, Mitros T, Policriti A, Cipriani G, Dondini L, Ficklin S, Goodstein DM, Xuan P, Del Fabbro C, Aramini V, Copetti D, Gonzalez S, Horner DS, Falchi R, Lucas S, Mica E, Maldonado J, Lazzari B, Bielenberg D, Pirona R, Miculan M, Barakat A, Testolin R, Stella A, Tartarini S, Tonutti P, Arus P, Orellana A, Wells C, Main D, Vizzotto G, Silva H, Salamini F, Schmutz J, Morgante M,

Rokhsar DS (2013) The high-quality draft genome of peach (*Prunus persica*) identifies unique patterns of genetic diversity, domestication and genome evolution. Nat Genet 45(5):487–494

International Rice Genome Sequencing Project (2005) The map-based sequence of the rice genome. Nature 436(7052):793–800

Jain M, Olsen HE, Paten B, Akeson M (2016) The Oxford nanopore MinION: delivery of nanopore sequencing to the genomics community. Genome Biol 17(1):239

Juven-Gershon T, Kadonaga JT (2010) Regulation of gene expression via the core promoter and the basal transcriptional machinery. Dev Biol 339(2):225–229

Krishnan NM, Pattnaik S, Jain P et al (2012) A draft of the genome and four transcriptomes of a medicinal and pesticidal angiosperm *Azadirachta indica*. BMC Genomics 13:464

Kubis S, Schmidt T, Heslop-Harrison SJP (1998) Repetitive DNA elements as a major component of plant genomes. Ann Bot 82(Suppl. 1):45–55

Lantican DV, Strickler SR, Canama AO, Gardoce RR, Mueller LA, Galvez HF (2019) *De novo* genome sequence assembly of dwarf coconut (*Cocos nucifera* L. 'Catigan Green Dwarf') provides insights into genomic variation between coconut types and related palm species. G3: Gene Genom Genet 9 (8): 2377–2393

Latchman DS (1998) Eukaryotic transcription factors, 3rd edn. Academic Press, San Diego

Levene MJ, Korlach J, Turner SW, Foquet M, Craighead HG, Webb WW (2003) Zero-mode waveguides for single-molecule analysis at high concentrations. Science 299:682–686

Lewin B (2001) Genes VII. ArtmedEditora, Porto Alegre

Mardis ER (2013) Next-generation sequencing platforms. Ann Rev Anal Chem 6:287–303

Maxam AM, Gilbert W (1977) A new method for sequencing DNA. Proc Natl Acad Sci U S A 74: 560–564

Ming R, VanBuren R, Wai CM, Tang H, Schatz MC, Bowers JE, Lyons E, Wang ML, Chen J, Biggers E, Zhang J, Huang L, Zhang L, Miao W, Zhang J, Ye Z, Miao C, Lin Z, Wang H, Zhou H, Yim WC, Priest HD, Zheng C, Woodhouse M, Edger PP, Guyot R, Guo HB, Guo H, Zheng G, Singh R, Sharma A, Min X, Zheng Y, Lee H, Gurtowski J, Sedlazeck FJ, Harkess A, McKain MR, Liao Z, Fang J, Liu J, Zhang X, Zhang Q, Hu W, Qin Y, Wang K, Chen LY, Shirley N, Lin YR, Liu LY, Hernandez AG, Wright CL, Bulone V, Tuskan GA, Heath K, Zee F, Moore PH, Sunkar R, Leebens-Mack JH, Mockler T, Bennetzen JL, Freeling M, Sankoff D, Paterson AH, Zhu X, Yang X, Smith JA, Cushman JC, Paull RE, Yu Q (2015) The pineapple genome and the evolution of CAM photosynthesis. Nat Genet 47(12):1435–1442

Monden Y, Tahara M (2015) Plant transposable elements and their application to genetic analysis via high-throughput sequencing platform. Hort J 84(4):283–294

Nyrén PL, Lundin A (1985) Enzymatic method for continuous monitoring of inorganic pyrophosphate synthesis. Anal Biochem 509:504–509

Ozsolak F (2012) Third-generation sequencing techniques and applications to drug discovery. Expert Opin Drug Discov 7(3):231–243

Paterson AH, Bowers JE, Bruggmann R, Dubchak I, Grimwood J, Gundlach H, Haberer G, Hellsten U, Mitros T, Poliakov A, Schmutz J, Spannagl M, Tang H, Wang X, Wicker T, Bharti AK, Chapman J, Feltus FA, Gowik U, Grigoriev IV, Lyons E, Maher CA, Martis M, Narechania A, Otillar RP, Penning BW, Salamov AA, Wang Y, Zhang L, Carpita NC, Freeling M, Gingle AR, Hash CT, Keller B, Klein P, Kresovich S, McCann MC, Ming R, Peterson DG, Mehboob-ur-Rahman WD, Westhoff P, Mayer KF, Messing J, Rokhsar DS (2009) The *Sorghum bicolor* genome and the diversification of grasses. Nature 457(7229): 551–556

Quispe-Huamanquispe DG, Gheysen G, Kreuze JF (2017) Horizontal gene transfer contributes to plant evolution: the case of *agrobacterium* T-DNAs. Front Plant Sci 8:Article 2015

Rendón-Anaya M, Ibarra-Laclette E, Méndez-Bravo A, Lan T, Zheng C, Carretero-Paulet L, Perez-Torres CA, Chacón-López A, Hernandez-Guzmán G, Chang TH, Farr KM, Barbazuk WB, Chamala S, Mutwil M, Shivhare D, Alvarez-Ponce D, Mitter N, Hayward A, Fletcher S, Rozas J, Gracia AS, Kuhn D, Barrientos-Priego AF, Salojärvi J, Librado P, Sankoff D, Herrera-Estrella A, Albert VA, Herrera-Estrella L (2019) The avocado genome informs deep angiosperm phylogeny, highlights introgressive hybridization, and reveals pathogen-influenced gene space adaptation. Proc Natl Acad Sci U S A 116(34):17081–17089

Ronaghi M, Karamohamed S, Pettersson B, Uhlen M, Nyrén PL (1996) Real-time DNA sequencing using detection of pyrophosphate release. Anal Biochem 242:84–89

Rothberg JM, Hinz W, Rearick TM et al (2011) An integrated semiconductor device enabling nonoptical genome sequencing. Nature 475(7356):348–352

Sanger F, Coulson A (1975) A rapid method for determining sequences in DNA by primed synthesis with DNA polymerase. J Mol Biol 94:441–448

Sanger SF, Nicklen ARC (1977) DNA sequencing with chain-terminating. Proc Natl Acad Sci USA 74:5463–5467

Sanger F, Air GM, Barrell BG, Brown NL, Coulson AR, Fiddes CA, Hutchison CA, Slocombe PM, Smith M (1977) Nucleotide sequence of bacteriophage phi X174 DNA. Nature 265:687–695

Schmidt T, Heslop-Harrison JS (1998) Genomes, genes and junk: the large-scale organization of plant chromosomes. Trend Plant Sci 3(5):195–199

Schmutz J, Cannon SB, Schlueter J, Ma J, Mitros T, Nelson W, Hyten DL, Song Q, Thelen JJ, Cheng J, Xu D, Hellsten U, May GD, Yu Y, Sakurai T, Umezawa T, Bhattacharyya MK, Sandhu D, Valliyodan B, Lindquist E, Peto M, Grant D, Shu S, Goodstein D, Barry K, Futrell-Griggs M, Abernathy B, Du J, Tian Z, Zhu L, Gill N, Joshi T, Libault M, Sethuraman A, Zhang XC, Shinozaki K, Nguyen HT, Wing RA, Cregan P, Specht J, Grimwood J, Rokhsar D, Stacey G, Shoemaker RC, Jackson SA (2010) Genome sequence of the palaeopolyploid soybean. Nature 463(7278):178–183

Schnable PS, Ware D, Fulton RS, Stein JC, Wei F, Pasternak S, Liang C, Zhang J, Fulton L, Graves TA, Minx P, Reily AD, Courtney L, Kruchowski SS, Tomlinson C, Strong C, Delehaunty K, Fronick C, Courtney B, Rock SM, Belter E, Du F, Kim K, Abbott RM, Cotton M, Levy A, Marchetto P, Ochoa K, Jackson SM, Gillam B, Chen W, Yan L, Higginbotham J, Cardenas M, Waligorski J, Applebaum E, Phelps L, Falcone J, Kanchi K, Thane T, Scimone A, Thane N, Henke J, Wang T, Ruppert J, Shah N, Rotter K, Hodges J, Ingenthron E, Cordes M, Kohlberg S, Sgro J, Delgado B, Mead K, Chinwalla A, Leonard S, Crouse K, Collura K, Kudrna D, Currie J, He R, Angelova A, Rajasekar S, Mueller T, Lomeli R, Scara G, Ko A, Delaney K, Wissotski M, Lopez G, Campos D, Braidotti M, Ashley E, Golser W, Kim H, Lee S, Lin J, Dujmic Z, Kim W, Talag J, Zuccolo A, Fan C, Sebastian A, Kramer M, Spiegel L, Nascimento L, Zutavern T, Miller B, Ambroise C, Muller S, Spooner W, Narechania A, Ren L, Wei S, Kumari S, Faga B, Levy MJ, McMahan L, Van Buren P, Vaughn MW, Ying K, Yeh CT, Emrich SJ, Jia Y, Kalyanaraman A, Hsia AP, Barbazuk WB, Baucom RS, Brutnell TP, Carpita NC, Chaparro C, Chia JM, Deragon JM, Estill JC, Fu Y, Jeddeloh JA, Han Y, Lee H, Li P, Lisch DR, Liu S, Liu Z, Nagel DH, McCann MC, SanMiguel P, Myers AM, Nettleton D, Nguyen J, Penning BW, Ponnala L, Schneider KL, Schwartz DC, Sharma A, Soderlund C, Springer NM, Sun Q, Wang H, Waterman M, Westerman R, Wolfgruber TK, Yang L, Yu Y, Zhang L, Zhou S, Zhu Q, Bennetzen JL, Dawe RK, Jiang J, Jiang N, Presting GG, Wessler SR, Aluru S, Martienssen RA, Clifton SW, McCombie WR, Wing RA, Wilson RK (2009) The B73 maize genome: complexity, diversity, and dynamics. Science 326(5956):1112–1115

Shendure JA, Porreca GJ, Church GM (2008) Overview of DNA sequencing strategies. In: John Wiley & Sons (ed.). Current protocols in molecular biology. John Wiley & Sons, Bognor Regis, West Sussex. pp. 7.1.1-7.1.11

Talbert PB, Masuelli R, Tyagi AP, Comai L, Henikoff S (2002) Centromeric localization and adaptive evolution of an Arabidopsis histone H3 variant. Plant Cell 14:1053–1066

Tamiru M, Natsume S, Takagi H et al (2017) Genome sequencing of the staple food crop white Guinea yam enables the development of a molecular marker for sex determination. BMC Biol 15:86

Turcatti G, Romieu A, Fedurco M, Tairi AP (2008) A new class of cleavable fluorescent nucleotides: synthesis and optimization as reversible terminators for DNA sequencing by synthesis. Nucl Acid Res 36:e25 vanBeneden E (1875) La maturation de l'oeuf, la fécondation et les premières de développementembryonnaire des Mammiferesd'après les recherchesfaites chez le lapin. Bull Acad Bel Cl Sci 40(2 sèr):686–736

Varshney RK, Shi C, Thudi M, Mariac C, Wallace J, Qi P, Zhang H, Zhao Y, Wang X, Rathore A, Srivastava RK, Chitikineni A, Fan G, Bajaj P, Punnuri S, Gupta SK, Wang H, Jiang Y, Couderc M, Katta M, Paudel DR, Mungra KD, Chen W, Harris-Shultz KR, Garg V, Desai N, Doddamani D, Kane NA, Conner JA, Ghatak A, Chaturvedi P, Subramaniam S, Yadav OP, Berthouly-Salazar C, Hamidou F, Wang J, Liang X, Clotault J, Upadhyaya HD, Cubry P, Rhone B, Gueye MC, Sunkar R, Dupuy C, Sparvoli F, Cheng S, Mahala RS, Singh B, Yadav RS, Lyons E, Datta SK, Hash CT, Devos KM, Buckler E, Bennetzen JL, Paterson AH, Ozias-Akins P, Grando S, Wang J, Mohapatra T, Weckwerth W, Reif JC, Liu X, Vigouroux Y, Xu X (2017) Pearl millet genome sequence provides a resource to improve agronomic traits in arid environments. Nat Biotechnol 35(10):969–976

Wang J, Wang Y, Wang Z, Liu L, Zhu XG, Ma X (2011) Synchronization of cytoplasmic and transferred mitochondrial ribosomal protein gene expression in land plants is linked to Telo-box motif enrichment. BMC Evol Biol 11:161

Wang X, Xu Y, Zhang S, Cao L, Huang Y, Cheng J, Wu G, Tian S, Chen C, Liu Y, Yu H, Yang X, Lan H, Wang N, Wang L, Xu J, Jiang X, Xie Z, Tan M, Larkin RM, Chen LL, Ma BG, Ruan Y, Deng X, Xu Q (2017) Genomic analyses of primitive, wild and cultivated citrus provide insights into asexual reproduction. Nat Genet 49(5):765–772

Watson JD, Crick FHC (1953) A structure for deoxyribose nucleic acid. Nature 171:737–738

Wicker T, Gundlach H, Spannagl M, Uauy C, Borrill P, Ramírez-González RH, De Oliveira R, International Wheat Genome Sequencing Consortium, Mayer KFX, Paux E, Choulet F (2018) Impact of transposable elements on genome structure and evolution in bread wheat. Genome Biol 19:103

WuH MT, Kang M, Ai F, Zhang J, Dong G, Liu J (2019) A high-quality *Actinidia chinensis* (kiwifruit) genome. Hortic Res 6:117

Xia E, Li F, Tong W, Yang H, Wang S, Zhao J, Liu C, Gao L, Tai Y, She G, Sun J, Cao H, Gao Q, Li Y, Deng W, Jiang X, Wang W, Chen Q, Zhang S, Li H, Wu J, Wang P, Li P, Shi C, Zheng F, Jian J, Huang B, Shan D, Shi M, Fang C, Yue Y, Wu Q, Ge R, Zhao H, Li D, Wei S, Han B, Jiang C, Yin Y, Xia T, Zhang Z, Zhao S, Bennetzen JL, Wei C, Wan X (2019) The tea plant reference genome and improved gene annotation using long-read and paired-end sequencing data. Sci Data 6:122

Yang J, Liu D, Wang X, Ji C, Cheng F, Liu B, Hu Z, Chen S, Pental D, Ju Y, Yao P, Li X, Xie K, Zhang J, Wang J, Liu F, Ma W, Shopan J, Zheng H, Mackenzie SA, Zhang M (2016) The genome sequence of allopolyploid *Brassica juncea* and analysis of differential homoeolog gene expression influencing selection. Nat Genet 48(10):1225–1232

Zhang C, Yang H, Yang H (2016) Evolutionary character of alternative splicing in plants. Bioinform Biol Insight 9(Suppl. 1):47–52

Transgenesis in Plants: Principle and Methods

3

Pradeep Kumar Singh, Pankaj Singh, Rajat Pratap Singh, and Ram Lakhan Singh

Abstract

The human population has reached 7.8 billion by 2020 and is estimated to reach 9.9 billion by the end of 2050 which is nearly 25% increase in total world population. With tremendous increase in world population, plants which are the main food source must be produced at higher rate to fulfil the food demand. The agricultural lands are limited, so, there is need to find out some other alternative route to fulfil the food requirement. In the past, traditional plant breeders use only individuals of the same or closely related species to propagate plants with desirable traits. Through the classical breeding techniques, breeders were not able to obtain desired traits which were not present within the gene pool of their target plants. However, with recent use of genetic techniques such as vector mediated transgenesis and direct gene transfer methods, we can now able to produce plants with desired traits from a completely different species. Vector mediated gene transfer methods are based on *Agrobacterium* plasmid and plant

P. K. Singh
Department of Biochemistry, Dr. Rammanohar Lohia Avadh University, Ayodhya, Uttar Pradesh, India

P. Singh
Department of Biotechnology, Dr. Rammanohar Lohia Avadh University, Ayodhya, Uttar Pradesh, India

R. P. Singh
Department of Biotechnology, Guru Ghasidas Vishwavidyalaya, Koni, Bilaspur, Chhattisgarh, India

R. L. Singh (✉)
Department of Biochemistry, Dr. Rammanohar Lohia Avadh University, Ayodhya, Uttar Pradesh, India

Nilamber-Pitamber University, Medininagar, Jharkhand, India

R. L. Singh et al. (eds.), *Plant Genomics for Sustainable Agriculture*,
https://doi.org/10.1007/978-981-16-6974-3_3

virus, whereas direct gene transfer methods are based on physical (electroporation, microinjection, particle gun bombardment methods) and chemical methods (lipofection, polyethylene glycol (PEG)-mediated, etc.). In the development of a transgenic plant, there are certain steps such as isolation of desired gene, vector constructions, transgenesis methods, transgene integration and inheritance of transgene, need to be carefully monitored during transgenesis. Production of transgenic plant will allow us to fulfil the food requirement of growing population but future of GM crops remains a point vital debate.

Keywords

Agrobacterium · Transgene · Gene gun · Transgenic plant · Gene cloning

3.1 Introduction

Through classical plant breeding techniques, high yielding strains of many different crop plants have been successfully developed in the past decades. But in recent years, with the use of genetic manipulation, a number of transgenic plants have been produced with insecticidal activity, herbicide resistance, protection against pathogenic virus, fungi and bacteria, improved nutritional quality of the plant seeds, increased post-harvest shelf life, tolerance against environmental stresses, self-incompatibility, etc. Currently, transgenic plants are cultivated in more than 175.2 million hectares in the world (James 2013; Tohidfar and Khosravi 2015). Recombinant DNA technology which has been used extensively in microbial system is also an important tool for the direct manipulation of plants bringing advancement in human life (Khan et al. 2016). Most plant cells are totipotent in nature which means entire plant can be regenerated from a single plant cell. Due to totipotent nature of plant cell, it is easier to create transgenic plants rather than the transgenic animals. A gene or genes, which are transferred to plant's cell through genetic engineering, usually from an unrelated species or kingdom, are called transgenes, and plants containing the transgenes are known as transgenic plants. The main aim to develop transgenic plant is addition of a gene to improve the agricultural, horticultural, or ornamental value of crop plant and to generate the plants as living bioreactor for the inexpensive production of economically important proteins or metabolites which are not possible to produce in bacterial as well as in animal system. Plant bioreactors have the advantages over bacterial and animals expression system of having post-translational modifications and lacking of contamination by animal pathogens (Ma et al. 2003a, b; Lienard et al. 2007). Several strategies have been developed and tested in plants to produce plant-derived pharmaceutical molecules (Ma et al. 2003a, b). Genetic transformation provides a powerful means for studying the action of genes during development or other biological processes. The first transgenic plant was produced in 1983 when a tobacco line expressing kanamycin resistance was produced (Peterson et al. 2011). Barta et al. (1986) reported that human growth hormone plant was the first plant-produced pharmaceutical product from transgenic tobacco. Later on, a number of recombinant molecules such as human biopharmaceuticals, antibodies, vaccines subunit and several other proteins have been produced in diverse plant species including cereals, legumes, fruits and vegetables. In transgenic plants formation, the aim is to introduce a new trait to the plant to produce the desired product which does not occur naturally in the species. Till now, more than 140 different plant species of crops and forest have been genetically modified to produce the new product. The important agronomic transgenic crops such as wheat, rice, *phaseolus*, barley, maize, oat, peanut, poplar, cotton, soybean and others have been produced to improve the quality and quantity of crops with application of recombinant DNA methodology (Christou 1995). New trait in food crops includes resistance to diseases, pests, chemical treatments; environmental conditions, reduction of spoilage and improving the nutrient profile of the crops, whereas new traits of non-food crops include production of biopharmaceuticals, bio-fuels and other industrially important product. Ma et al. (2003a, b) have been reported that six types of antibodies and two vaccines derived from transgenic plants have been produced and are at clinical trials stage.

Present cultivated transgenic crops have become significantly different from their wild crops in terms of production and nutrition. However in the early 1970s, numerous potential concerns have been raised in the development of GM crops but no human health problems have been identified specifically with the ingestion of transgenic crops or their products. The concerns issue was mainly focused on the production of toxic compounds and allergic reactions from transgenic food products. About the use of antibiotic resistance genes in the development of transgenic crops, no definitive evidence exists that shows these cause harm to humans, but because of public concerns, scientists are eliminating these markers who are involved in the development of transgenic plants.

In development of transgenic plants through genetic engineering involves introduction of one or more well defined genes rather than the introduction of whole genomes or parts of chromosomes. Addition of well-known defined gene in to wild plant makes toxicity testing more straightforward in transgenic plants because it is easier to identify what the new features are added in the modified plant. The advantageous feature of GM technology is that we can introduce genes from diverse plants which have little history in the food supply. Other advantages include that transgenic plants can work as a bioreactor to produce number valuable products related pharmaceuticals, therapeutic proteins, vaccines, many industrial enzymes, antibodies and many other secondary metabolites due to genetic manipulations (Sharma et al. 2004; Miao et al. 2008; Sharma and Sharma 2009; Saveleva et al. 2015, 2016). Transgenic plants also have advantages over wild plant in terms of higher yield, short time duration of production, low cost of biomass production, lack of potential contamination and conservation of eukaryotic cell, machinery mediating protein modification. Plants and prokaryotes have different nature in codon usage which leads to inefficient expression of prokaryotic protein in plants. Plants are also becoming economically important systems for producing heterologous proteins (Goddijn and Jan 1995). Due to these advantages, nowadays, transgenic plants and their products are produced at commercial level.

3.2 Development of Transgenic Crops

With recent development in the field of genetic engineering, it is now possible to transfer the desired gene/genes into target plant cell from different species. To develop a transgenic plant, parameters such as isolation of desired gene, gene cloning, gene design and package in to constructed vector, transgenic methods, integration of transgene in genome and their inheritance are carefully monitored to increase the success of the transformation event. Transgenic plants are produced by manipulation in DNA of plant cells by stable introduction of DNA sequences usually into the nuclear genome of cells which are capable of developing into a whole plant (Sharma et al. 2005). There are many diverse and complex techniques now available to produce a transformed plant cell and to develop a new improved transgenic plant variety (Potrykus 1990; Hooykaas and Shilperoort 1992; Zupan and Zambryski 1995; Sharma and Ortiz 2000). Two classes of delivery systems are used in production of transgenic plant, i.e., non-biological systems which include physical and chemical methods (Electroporation, particle gun bombardment method, microinjection, lipofection, polyethylene glycol mediated, etc.) and biological systems which includes *Agrobacterium* and virus vector mediated transformation. But before the utilization of gene product, it is very important to know the biochemical, physiological mechanisms of action and regulation of gene expression.

The process for the formation of transgenic plant requires mainly a series of following steps.

Step 1: Isolation of desired gene.
Isolation of desired gene is the first step in the transgenesis process. Reliable methods are now available for isolating the gene components from the cell nuclear genome. In isolation procedure, the initial step is the disruption of the desired organism, which may be viral, bacterial or plant cells in order to isolate the desired gene. After a series of chemical and biochemical steps, the desired gene can be obtained from the nuclear genome.

Step 2: Gene cloning.
Gene cloning is the second step in the formation of transgenic plants through genetic engineering. Upon gene isolation, the desired gene/s is cloned to make thousands of copies of the desired gene. For the gene cloning, in vivo and in vitro methods are available to clone the desired gene. Under in vivo methods, four stages are involved to clone the gene, i.e., generation of DNA fragments, joining to a vector, propagation in a host cell and selection of the transformed cells, whereas in vitro, gene cloning is carried out by PCR method.

Step 3: Gene design and packaging.
A simple functional gene consists of a promoter, coding region and terminator/stop region. But in some genes in addition to these components they also contain special sequences such as an enhancer, silencer or reporter sequences. A typical plant gene consists of the regulatory sequences and structural genes (Debnath et al. 2010). Regulatory sequences are usually located at the 5′ upstream of a gene, with its own promoter, enhancer or silencer region. Structural genes begin

with a catabolite activator protein (cap) site, followed by a leader sequence, initiation codon, introns, exons, stop codons and a polyadenylation site (poly-A tail). Once the gene of interest has been cloned, it has to be linked with regulatory DNA sequences that will control the expression of gene of interest inside the plant genome. Gene designing/packaging is done by incorporating a selectable marker gene and replacing an existing promoter according to host expression system. Selectable marker genes are also usually linked to the gene of interest to screen the transformed cell.

Step 4: Construction of suitable vector.

A vector acts as a carrier that transports the gene of interest into a target cell for replication and expression.

Step 5: Transformation.

Once the gene has constructed and cloned, it can then be introduced into the plant cells by using the transformation methods. The most common methods used in plant cell transformation are biolistic transformation using the gene gun or *Agrobacterium*-mediated transformation. The main aim in any transformation methods is to introduce the gene of interest into the nucleus of the cell without affecting the cell function. If the introduced gene gets inserted in to plant genome and retains their functions, the gene product is synthesized in to plant cell. Once the transgene integrates with the nuclear genome, it will inherit and expressed in subsequent generations. The confirmation of transgene integration with nuclear genome will be carried out through sequence of transgene constructs, selectable marker and reporter gene.

Step 6: Backcross Breeding.

It is the final step in producing transgenic crops. This is done by using conventional plant breeding methods in which transgenic plants cross with elite lines. The offspring obtained in F1 generation is repeatedly crossed back to the elite line to obtain complete transgenic line. The time required for commercial release of a new transgenic hybrid varies from 6 to 15 years. It also depends upon the available resources, crop species, nature of the gene and regulatory approval to generate the transgenic crops.

3.3 Transformation Techniques

3.3.1 Vector Mediated Gene Transfer

3.3.1.1 *Agrobacterium* Mediated Gene Transfer

There are no any naturally occurring plasmids which are known in angiosperms, whereas presence of Ti plasmid of soil bacterium *Agrobacterium tumefaciens* present in rhizospheric region of plants has great potential in plant transformation experiment. The crown gall disease of many dicotyledonous plants is the result of *A. tumefaciens* infection. Crown gall arises when a wound on the stem allows bacterium *A. tumefaciens* to infect the plant. After bacterial infection, tumourous proliferation of the stem tissue in the region of the crown arises. The crown gall tissue represents true oncogenic transformation process. The Ti (tumour inducing) plasmid of bacterial cell is responsible for crown gall disease. Ti plasmid is more than 200 kb large DNA segment which carries several genes involved in the infective process. The attractive feature of this plasmid is that after infection a segment of it is integrated into the host plant genome. This segment is called the T-DNA with size of 15–30 kb depending on the type of strain. T-DNA is integrated in the plant cell genome and is inherited on to the daughter cells as an integral part of the chromosomes. The most remarkable feature of the Ti plasmid T-DNA is that it contains important genes which are expressed in the plant cell and are accountable for the cancerous properties of the transformed cells. Along with uncontrolled growth, transformed plant cells are also capable of synthesis of unusual amino acid derivatives opines like octopine and nopaline which are not the part of normal plant metabolism. Normally bacterium provokes the synthesis of opine and uses it as sole carbon and nitrogen source. Thus *A. tumefaciens* is considered as natural genetic engineer of the plant cell for its own purposes.

3.3.1.1.1 Ti Plasmid

Two strains of *Agrobacterium* have been known for its important role in natural plant cell transformation, i.e., *A. tumefaciens* and *A. rhizogenes. A. tumefaciens* has Ti plasmid, whereas *A. rhizogenes* has Ri plasmid. Both plasmids are of about 200 kb and are interchangeable between the two interkingdom species. Ti is a tumour inducing plasmid of *A. tumefaciens,* whereas Ri is a root inducing plasmid of *A. rhizogenes*. Both Ti and Ri plasmid contains a T-DNA region having genes for production of phytohormone and opine metabolism. The T-DNA regions are efficiently transferred from Ti and Ri plasmid to plant cell and are able to integrate itself into plant genome. The Vir region of Ti and Ri plasmid facilitates the transfer and integration of T-DNA from plasmid to plant genome. Vir region includes several genes which codes for the proteins that are required for excision, movement and integration of T-DNA.

3.3.1.1.2 Induction of Tumour

Since continued occurrence of *Agrobacterium* is not necessary to maintain plant cells in their transformed state. It is found that some tumour inducing factors are transferred from the bacterium to the plant at the wounded site. Zaenen et al. (1974) first observed that virulent strains of *A. tumefaciens* possess large plasmid that are transferred between various octopine and nopaline utilizing strains. Plasmids in the octopine group are closely related to each other while those in the nopaline group are considerably more diverse. Between the groups, there are four regions of homology, including the genes directly responsible for tumour formation (Drummond and Chilton 1978; Engler et al. 1981). The presence of a plasmid in *A. tumefaciens* does not indicate the virulent nature of strain.

3.3.1.1.3 The Biology of *Agrobacterium* Infection

Infection initiates with interaction between host and parasite. The process of infection terminates with the transfer of T-DNA into plant cell genome. A Chv (chromosomal virulence) gene facilitates the host–parasite interaction. Most of these genes, for example, Chv-B, exogenes, CEL genes are associated with the biosynthesis of the polysaccharide. These polysaccharides are required for bacterial cell to firmly attach to the plant cell. Chv-D and Chv-E genes are required for maximum expression of pTi vir genes.

3.3.1.1.4 The T-DNA

24 bp imperfect direct repeat border sequences are present at T-DNA on its both sides. Three genes of pTi are involved in tumour formation. The two of these genes (iaaM and iaaH) encode enzymes which together form IAA (Indole-3-acetic acid) from tryptophan. Deletion in iaaM and iaaH produces shooty crown gall. Third gene, ipt encodes an enzyme which produces zeatin. Ipt gene encodes for isopentenyl transferase which catalyses formation of isopentenyl adenine (zeatin). Deletion in ipt gene produces root crown gall disease. The function of T-DNA genes playing important role in plant transformation is given in Table 3.1.

3.3.1.1.5 The Vir Region

The vir-A, vir-B, vir-C, vir-D, vir-E, vir-F, vir-G and vir-H are the 8 operon of Vir region. All these operons together cover 40 kb of DNA and 25 genes. Products of these genes are vital for transfer of T-DNA into plant genome. These vir genes interact only with left and right border of T-DNA. Four operons including vir-A, vir-B, vir-D and vir-G are crucial for virulence and remaining other play accessory roles. Expression of vir-A and vir-G is constitutive in nature and associated with regulation of all vir operon. Other vir operon express protein required for T-DNA transfer.

3.3.1.1.6 Genes Necessary for Transfer of T-DNA

The vir (virulence) region is located in a separate part of the Ti plasmid and is responsible for transfer of T-DNA. Vir-A is a kinase present in inner bacterial membrane and sense certain phenolic compound released by wounded plant cells.

Table 3.1 Roles of genes on T-DNA

Genes	Functions
Nos	Nopaline biosynthesis, encodes an enzyme nopaline synthetase, which produces nopaline from arginine and pyruvic acid
iaaH (aux2, tms2)	Auxin biosynthesis, encodes enzyme indole-3-acetamide hydrolase which converts indole-3-acetamide to indole –3-acetic acid
iaaM (aux1 or tms1)	Auxin biosynthesis, encodes an enzyme tryptophan-2-mono oxygenase which converts tryptophan indole-3-acetamide
Ipt (tmr or cyt)	Cytokinin biosynthesis, encodes enzymes isopentenyl transferase, which catalyse formation of isopentenyl adenine
Border sequences	For its transfer
4 bp left and right	Site of endonuclease action during transfer

Acetosyringone is a compound that has been the most extensively used to bring vir gene expression in the laboratory (Stachel et al. 1986). It is believe that the phenolic compounds acetosyringone do not attract bacteria to wounded plant cells but the bacteria simply respond to molecules like sugars and amino acids to induce the vir genes after attachment (Parke et al. 1987; Loake et al. 1988). Activated vir-A phosphorylates the vir-G protein, a transcriptional activator that activates remaining other vir genes. Besides vir-G, other genes of the bacterial chromosome also encode transcription factors which regulate vir gene expression. The induction of vir gene expression required for synthesis of proteins to form a conjugative pilus through which easy transfer of T-DNA to the plant cell takes place. The genes in the vir-B operon are involved in synthesis of pilus components. An endonuclease, products of the vir-D1 and vir-D2 genes initiates the transfer of T-DNA. These endonucleases bring in either single strand nicks or a double strand break at the 25 bp borders of the T-DNA. Vir-C12 and Vir-C2 proteins enhanced nick formation after recognition and binding to the overdrive enhancer element. The Vir-D2 protein remains covalently attached to the single stranded T-DNA. A single stranded DNA binding protein Vir-E2 coats the T-DNA strands. It was believed that the Vir-D2 protein has multiple role, for example, it protects the T-DNA from nucleases, transfers the DNA to the nucleus of plant cell and finally integrates it into the plant DNA. The Vir-D2 protein has two separate nuclear localization signals out of which C-terminal signal plays the major role in targeting the T-DNA (Tinland et al. 1992).

3.3.1.1.7 The Process of T-DNA Transfer and Integration

A single stranded form of T-DNA enters into plant cell. After their entry it immediately converted to double stranded form in the plant nuclei. The double stranded T-DNA integrates at random sites by process of illegitimate recombination due to homology in short segment of DNA in host plant genome.

The single stranded T-DNA complex is moved to the nucleus by crossing the nuclear membrane. Two Vir proteins, Vir D2 and Vir E2 are important in this step and most likely Vir F is also required in this step with a minor contribution (Hooykaas and Shilperoort 1992). The nuclear localization signals (NLS) of Vir D2 and Vir E2 play central role in nuclear targeting of single stranded T-DNA complex. The single stranded T-DNA complex is a large nucleoprotein complex with only one 5′end covalently attached Vir D2 protein per complex. Similarly nucleoprotein complex is coated by a large number of Vir E2 molecules. Integration of ssT-DNA into the plant genome is the final step of T-DNA transfer. Although the mechanism engaged in the T-DNA integration is not characterized but it is believed that the integration occurs by illegitimate recombination (Lehman et al. 1994; Puchta 1998). According to this model, paring of a small number of bases (microhomologies) is essential for a pre-annealing step between Vir D2 coupled T-DNA strand and plant DNA.

3.3.1.2 The co-Integrative Vector

The co-integrative vector is the result of homologous recombination between disarmed Ti plasmid and an intermediate vector. The intermediate vector is usually the plasmid of *E. coli* having the gene of interest. Both vectors contain some common sequences that allow the homologous recombination between these two plasmids. A large co-integrative vector containing the fused *E. coli* plasmid and disarmed Ti plasmid is formed after recombination. This co-integrative vector is introduced into the *Agrobacterium* for transformation process. The major disadvantage of this vector is its huge size which may prove an ominous challenge for manipulation. Thus, the use of this vector is replaced with the binary vector system.

3.3.1.3 The Binary Vector

Binary vectors are combination of the two separate plasmids in which T-DNA is present in one plasmid and vir gene present in another plasmid. Binary vector consists of a pair of plasmids contained disarmed T-DNA sequence (at least the left border or right border of T-DNA must be present), while the other contains vir gene and lacks entire T-DNA including border. Disarmed T-DNA plasmid is called MICRO Ti or MINI Ti, for example, Bin 19. Bin 19 possess ori for both host and *Agrobacterium*. Segment of DNA is ligated into T-region of MINI Ti and recombinant MINI Ti is cloned in *E. coli*. Conjugational transfer mechanism is used to transfer recombinant MINI Ti into *E. coli*. MINI Ti has kanamycin resistance genes for selection of *Agrobacterium* cells containing Bin 19 and neomycin resistance gene for selection of transformed plant cells. A binary vector avoids the transfer of unnecessary sequences into plant cell which occurs in case co-integrative vectors.

3.3.1.4 Plant Viral Vectors

Since long time plant viruses have been exploited as vectors for several purposes, for example, production of useful commercial proteins (Rybicki 2009). Viral genomes are excellent choice to be used as vectors because they have competent machinery and broad genome structure. Autonomously replicating virus based vectors offer different route to deliver gene of interest to plant cells. Wheat streak mosaic virus (WSMV), Barley stripe mosaic virus (BSMV) and Tobacco rattle virus (TRV) are used for transformations of monocot and dicot plants, respectively (Lee et al. 2012). Single stranded DNA viruses such as geminiviruses have been also extensively adopted as vectors for various crops. These viruses can be prepared to hold exogenous coding sequences for protein expression in important crops such as corn, oat, wheat, barley and rye (Choi et al. 2000). Different viral vector used to transfer foreign genes into plant to generate transgenic plants is described as follows:

3.3.1.4.1 Geminiviruses

Geminiviruses have the capability to infect a huge range of plant species such as wheat, cucurbits, maize, cotton, tomato, beans, ornamental plants, legumes, fruits and common weeds (Nawaz-Ul-Rehman and Fauquet 2009). Geminiviruses have a small genome of 2.8 Kb containing 4–6 overlapping open reading frames (ORFs). They are transmitted through insect vectors such as leaf hoppers and whitefly *Bemisia tabaci*. Geminiviruses are successfully designed as vectors for the production of exogenous proteins in plants (Lozano et al. 2016).

3.3.1.4.2 Tobacco Rattle Virus (TRV)

Tobacco rattle virus is a positive single stranded RNA virus. They are pathogenic plant viruses which infects over many plant species from different families. They are naturally transmitted via nematodes. TRV1 (or RNA1) and TRV2 (or RNA2) are the two genome components of TRV. The TRV1 is important for viral movement. The TRV2 genome component has genes encoding the coat protein and non-structural proteins. These non-structural proteins are not essential for experimental infection. Thus for use as a vector, few genes in TRV2 component are replaced with multiple cloning sites for inserting gene of interest (Senthil and Mysore 2014).

3.3.1.4.3 Cucumovirus

Cucumber mosaic virus (CMV) is a RNA virus that has also been extensively utilized for the production of plant made biologics. CMV has highly broad host range. CMV has trimeric RNAs which are required for virus infection and successful packaging into icosahedral capsids. Recently, Hwang et al. (2012) designed the complete tripartite genome on a binary plasmid by the replacement of the coat protein gene with α-1-antitrypsin (AAT). Cucumber mosaic virus has also been utilized as an antigen presentation system to express epitopes of porcine circovirus type 2 (PCV2) capsid protein (Gellert et al. 2012).

3.3.1.4.4 Cowpea Mosaic Virus

Cowpea mosaic virus (CPMV) is the icosahedral, positive sense RNA virus designed for the generation of vaccine and other therapeutic proteins in plants. After construction of epitope presentation system, CPMV recombinant particles have been shown to produce a strong immune response for a number of diseases (Peyret and Lomonossoff 2013).

3.3.2 Direct Gene Transfer

Direct gene transfer method is used for the transfer of foreign DNA directly into the plant genome without involvement of a biological agent. The spontaneous uptake of DNA by plant cell is quite low. Therefore, different and physical and chemical methods are used to facilitate the entry of DNA into plant cell. Direct DNA transfer methods rely on the delivery of naked DNA into plant cell in contrast to the *Agrobacterium* which is based on the vector mediated DNA transfer. Majority of direct DNA transfer methods are simple and effective and in fact several transgenic plants have been developed by this approach. But the limitations of direct DNA transfer method are its higher frequency of transgene rearrangement which will result in higher transgene copy number and lower stable transformation efficiency.

3.3.2.1 Physical Methods

Gene transfer approaches in plant system provide both qualitative and quantitative improvements in crop production. Physical methods are based on the transformation of protoplasts or intact cells either by mechanical introduction of DNA or non-mechanical DNA uptake by the protoplast. However, these methods require specialized equipment facilities and skill for the stable plant cell transformation (Singh and Shaw 1992).

3.3.2.1.1 Electroporation

Electroporation has become a favoured technique for the transfer of DNA in to plant cell (Shillito et al. 1985). Introduction of naked DNA into plant cell by exposing them to high voltage electric pulse which induces transient pores in the plasma membrane is called electroporation. Electric field pulses cause structural rearrangement in plasma membrane resulting in a temporary increase in porosity. There are two strategies to carry out the electroporation, first low voltage with long pulses (300–400 V cm^{-1} for 10–15 ms) and second high voltage with short pulses (1000–1500 V cm^{-1} for 10μs) (Shillito et al. 1985; Fromm et al. 1985). The optimum voltage and time will depend on the plant species, source of protoplast and resistance of the medium. This method is not only used for protoplasts transformation but also used for walled plant cells transformation. Electric field provides a local driving force for ionic and molecular transport through these pores. But in several cases, to facilitate gene transfer, target cells are treated with enzymes in order to increase the transformation frequency (D'Halluin et al. 1992; Laursen et al. 1994). However, gene transfer in immature rice, wheat and maize embryos can be performed without enzyme pre-treatment by using electroporation (Kloti et al. 1993; Xu et al. 1995). Several reports showed stable protoplasts transformation with electroporation (Fromm et al. 1986; Toriyama et al. 1988), whereas low voltage with long pulse produces high rate of transient transformation. The advantageous features of this technique are that it is a very simple, convenient, rapid, cost effective and also have high transformation efficiency. But there are certain limitations of electroporation method such as the amount of DNA delivery into the plant cell is very low and efficiency of electroporation is highly variable depending upon the plant material. Several physical parameters such as transmembrane potential created by the electric field, extent of porosity, duration of the permeated state, mode and duration of molecular flow, concentrations of DNA and the heterogeneity of the cell population may affect the electrotransfection efficiency (Hui 1995; Weaver 1995; Weaver and Chizmadzhev 1996).

3.3.2.1.2 Microinjection

In microinjection method, solutions containing DNA are injected directly inside the cell nuclei by using fine capillary tubes. During microinjection, there is need to immobilize the protoplasts for high rate of transformation. Immobilization of protoplast is carried out by embedding the protoplast in low melting agarose or by application of micromanipulator holding pipettes. Tobacco, alfalfa and rape protoplasts showed stable transformation with the use of microinjection method (Crossway et al. 1986; Reich et al. 1986; Neuhaus et al. 1987) and transformation frequencies ranging up to 66%. However, this procedure is very tedious, requires specialized instrumentation and recovery systems for regeneration of protoplasts. Literature also showed that plant cells with intact walls can also be transformed with the help of microinjection method. Neuhaus et al. (1987) reported that *Brassica napus* embryoids showed 80% transformation rate and about 50% of which were stable transformants with microinjection method.

3.3.2.1.3 Particle Bombardment

In particle bombardment technique, tungsten or gold particles coated with the foreign DNA is used to introduce into plant cells (Gan 1989; Birch and Franks 1991; Christou 1992; Takeuchi et al. 1992; Yao et al. 2006). This method is also known as biolistic or ballistic method. DNA coated gold or tungsten particles are loaded into a particle gun device which is involved in DNA transfer. Loaded particles are released at high speed either by using pressurized helium gas or electrostatic energy released from a droplet of water exposed to high voltage. The projected particle penetrates the plant cell walls and membranes. As the DNA coated particle enter in to cells, transgenes are released from the particle and get attached to the plant's chromosomal DNA. Kikkert (1993) result showed that higher number of stable transformants in suspension cultured cells gets increased with the use of mannitol or sorbitol as bombardment medium. The major applications of biolistic method include transient transformation studies, formation of transgenic plants and viral pathogenesis during inoculation of plants (Southgate et al. 1995; Sanford 2000; Taylor and Fauquet 2002). Biolistics method is also used for plastid transformation as well as to study of transient gene expression after 24–48 h of bombardment (Norris et al. 1993). Biolistics method has some disadvantages over *Agrobacterium* mediated cell transformation such as lower transformation efficiency, more costly, random integration of DNA to nuclear genome. It also causes multiple copy insertions resulting in to gene silencing and variation of transgene expression (Dai et al. 2001; Darbani et al. 2008). But some laboratories use linear cassettes or lower quantity of DNA to overcome this problem (Fu et al. 2000).

3.3.2.1.4 Fibre Mediated DNA Delivery

Plant materials such as embryo or embryo derived callus or cells in suspension medium are introduced in to a buffer containing DNA and silicon carbide fibres of 0.3–0.6μm in diameter and 10–100μm in length. Culture medium containing all components was vertexed which leads penetration of cell wall and plasma membrane by silicon fibres allowing the transfer of DNA inside the cells. This method is recently used for successful transformation of wheat, barley and maize cells. However, this method has several disadvantages such as DNA damage by silicon carbide fibres, unable to penetrate the hard embryonic callus in many cereals and lower transformation frequency.

3.3.2.1.5 Laser Induced DNA Delivery

Laser beams have been used to deliver DNA in to plant cell. Laser creates transient holes in plasma membrane through which DNA may enter into cell. Transfection of animal cells has been successfully transformed with laser induced DNA deliver method. High intensity laser creates high pressure on cell membrane to enhance the permeability of cell membrane (Noack and Vogel 1995). This method does not provide information about transient or stable transformation.

3.3.2.2 Chemical Methods

3.3.2.2.1 Polyethylene Glycol (PEG) Mediated

Gene transfer across the protoplast membrane is promoted by a number of chemicals such as polyethylene glycol, polyvinyl alcohol and calcium phosphate. Polyethylene glycol has become the most widely used chemical for the DNA transfer because it stimulates endocytosis and thereby causing the uptake of DNA (Negrutiu et al. 1987). However, polyethylene glycol gene transfer technique is applicable for protoplast only. Before gene transfer, protoplasts are kept in 15% polyethylene glycol (PEG) solution. Protoplasts are now exposed to exogenous DNA for the transfer of DNA in presence of PEG. After transformation, PEG is removed and intact protoplasts are cultured to form cells with walls and callus (Jogdand 2006). Calcium alginate micro beads coated with DNA along with polyethylene glycol are also used for protoplast transformation (Liu et al. 2004).

3.3.2.2.2 Liposome Fusion

Transfer of DNA into cells through liposomes is known as lipofection. Lipofection method is most preferably used in case of animal cell to transfer the DNA. However, it was first method used for plant protoplast transformation. Direct delivery of naked DNA into protoplasts may be affected by the presence of nucleases in the medium. Tobacco protoplasts have been documented for stable transformation with the use of DNA encapsulated into liposomes (Deshayes et al. 1985). DNA in liposome provides a protective cover for the DNA against nuclease attack. However, liposome mediated transformation has lower frequency as compared to electroporation or by chemical agents. Pea plant yielded segregating transformants when DNA is delivered at the time of pollination (Ahokas 1987).

3.3.2.2.3 Diethylaminoethyl (DEAE) Dextran Mediated

Plant cell transformation frequency increases when transformation is carried out with high molecular weight polymer diethylaminoethyl dextran. The efficiency can be increased up to 80% when dimethyl sulfoxide shock is given to the sample but it does not produce stable transformation. DEAE dextran mediated transformation is used to obtain the efficient transient expression. The advantages of this technique are very simple and cheap technique to transform the protoplast. The disadvantage of this technique is that it showed unpredictable pattern of foreign DNA integration and DNA faces nucleolytic cleavage during the passage into the nucleus.

3.4 Analysis and Confirmation of Transgene Integration

Analysis and confirmation of transgene (gene of interest) incorporation are confirmed through an appropriate mean relied on the transgene constructs, reporter gene and selectable marker used. Transgenic plant cells integrated with antibiotic or anti-herbicidal resistance genes are screened by the supplementation of antibiotics or herbicides to the used growing media. These supplements are allowed us to discriminate between transformed and the non-transformed plant cells. However, these methods have certain disadvantages, for example, it requires a large quantity of herbicides and antibiotics that make it expensive. Additionally, the risk of horizontal gene transfer to other bacteria is also associated with these methods. In view of this, other screening methods like reporter gene expression screening and polymerase chain reaction (PCR) are used for better accuracy as an optional screening method for transgenic plants. Some reporter genes, for example, the green fluorescent protein (GFP), β-glucuronidase (GUS) and luciferase (Luc) expression could be observed visually or directly under microscopy for confirmation of transgene expression (Naylor 1999). Quantifications of expression of the reporter gene are feasible with the use of a spectrophotometer. In addition, few more reporter gene expressions, for example, chloramphenicol acetyltransferase (CAT) and β-galactosidase (LacZ) activity are screened through enzyme assays.

A molecular technique, southern blotting is utilized for the detection of specific DNA sequences within DNA samples. Southern blotting is normally employed to identify the number of foreign gene integrated into the host genome as well as for the detection of transgene rearrangement and transgene integrity (Dai et al. 2001). In this method DNA fragments obtained after restriction enzymes digestion. Now these fragments are separated through electrophoresis, and finally transferred onto a nylon or nitrocellulose membrane. Membranes having bound DNA will be incubated in a solution with labelled probes and autoradiography is used for the detection of pattern of hybridization. The number of bands observed is equal to transgene copy number.

PCR method is one of the most sensitive methods between all the known molecular techniques used for the confirmation of the transgene integration. Successful amplification by using specific primers of the DNA fragment with expected band indicates the presence of transgene and it is further confirmed through DNA sequencing. A real time PCR (RT-PCR) is a fast, sensitive and high throughput molecular based analysis. As compared to the southern blot analysis especially of transgene copy number and zygosity detection in transgenic plants RT-PCR is a superior technique for transgene detection (Bubner and Baldwin 2004). RT-PCR is suitable as it allows for quantitative, semi-quantitative (qPCR) or qualitative (RT-qPCR) analysis of target DNA in real time. Some limitations of PCR based techniques in transgenic analysis are the generation of non-specific products and failure to amplify large exogenous DNA, multiple insertion, truncated transgene sequences and hinders precise transgene identification (Park et al. 2017). The discovery of next generation sequencing (NGS) tools and bioinformatics makes possible the study of genome and molecular characterization of complex character. The NGS data is also useful in the identification of precise genomic locations of transgene insertion which was not identified with the traditional PCR based method (Elbaidouri et al. 2013). Thus, NGS approach offers choice for high resolution analysis for transgenes insertion in transgenic crops (Lambirth et al. 2015).

3.5 Advantages of Transgenic Plants

Transgenic plants also called genetically modified (GM) plants are engineered to acquire desirable characteristics. It allows researchers to introduce new traits in the plants that are difficult or impossible through conventional methods. Transgenic plants can play a key role in crop improvement programmes. A variety of crops including soybean, cotton, canola, corn and so forth have been engineered to possess some desirable and useful features such as resistance to herbicides, insects, viruses, etc. Transgenic plants can also be employed for the production of nutraceuticals and therapeutic agents.

3.5.1 Herbicide Resistant

Chemical herbicides are widely used to control the weed population. Glyphosate is an active component of many broad-spectrum herbicides such as Round up. It is most commonly used postemergence, non-selective herbicide (Ma et al. 2016). Glyphosate does not have selectivity to distinguish between crops and weeds. Glyphosate inhibits the 5-enolpyruvylshikimate-3-phosphate synthase (EPSPS) enzyme which catalyses the critical step of the shikimate pathway in the biosynthesis of chorismate derived metabolites including the aromatic amino acids (Senseman 2007; Gong et al. 2016). Herbicide resistant plants can tolerate the effect of herbicides. The use of herbicide resistant transgenic plants simplifies the effective weed management through need-based applications of herbicides (Bonny 2008). Different herbicide resistant plants such as soybean, corn, cotton and canola have been developed (James 1999; Liu 1999; Green and Owen 2011) (Table 3.2). There are two strategies by which crops can be modified to be glyphosate tolerant. One approach is to incorporate a bacterial gene that produces glyphosate-insensitive EPSPS. The other way is to incorporate a bacterial gene that produces an enzyme which modify or degrade the glyphosate (Simoens and Van Montagu 1995). The CP4 gene of *Agrobacterium* sp. which encodes a glyphosate-insensitive EPSPS has been used as a transgene for most of glyphosate resistant crops (Cerdeira and Duke 2007). Some crops express the glyphosate degrading enzymes (glyphosate oxidoreductase) which degrade the glyphosate to aminomethylphosphonic acid (AMPA) and glyoxylate. Some glyphosate resistant crops express the enzyme glyphosate acetyltransferase (GAT) that modifies glyphosate (Green and Owen 2011). In addition, various crops have also been engineered conferring resistance to glufosinate-based herbicides. Glufosinate is also a non-selective, broad-spectrum foliar herbicide that inhibits glutamine synthetase. In the development of glufosinate resistant crops, pat or bar gene from *Streptomyces* spp. has been used as transgene.

Table 3.2 Herbicide resistant transgenic plants

Transgene	Source of transgene	Transgenic plants	Resistant to herbicide
cp4 epsps gene	*Agrobacterium tumefaciens* strain CP4	Soybean, corn, cotton, canola, sugar beet, alfalfa	Glyphosate
Zm-2mepsps gene	Maize (*Zea mays* L.)		
Glyphosate N-acetyltransferase gene	*Bacillus licheniformis*		
Glyphosate oxidoreductase gene	*Ochrobactrum anthropi* strain LBAA		
Bar gene	*Streptomyces hygroscopicus*	Canola, corn, cotton, Rice	Glufosinate
Pat gene	*Streptomyces viridochromogenes*		
Bxn gene	*Klebsiella ozaenae*	Tobacco, cotton, canola	Bromoxynil

These genes encode phosphinothricin N-acetyltransferase (PAT) or Basta N-acetyltransferase (BAR) enzyme, which catalyse the metabolic inactivation through acetylation of glufosinate (Herouet et al. 2005; Zhang et al. 2008; Green and Owen 2011).

3.5.2 Insect and Pest Resistant

Insect pests cause significant damage to crops all over the world. Although a wide variety of chemical pesticides are used for pest management, they are not effective against the wide variety of pests and contaminate the environment. In addition, some pests are resistance to these pesticides. Transgenic plants can have impacts on pests in an effective manner (Kennedy and Gould 2007). Insect-resistant transgenic plant designed for increased levels of resistance to insect pests has the potential to offer large benefits to agriculture in terms of enhanced crop protection against the pests and reduce the dependency on pesticides and their harmful effects (Ferry et al. 2003; James 2005). The development of insect-resistant transgenic plants involves the incorporation of transgene (cry) from the *Bacillus thuringiensis* (Bt) bacterium to the plant by the genetic engineering methods that make the plants resistant to pests. *B. thuringiensis* is a gram-positive, spore forming soil bacterium. During sporulation, *B. thuringiensis* produces crystalline inclusion bodies with unique insecticidal property that contain insecticidal delta endotoxins, also known as insecticidal crystal proteins (Cry proteins) (Estela et al. 2004; Flannagan et al. 2005; George 2013). These proteins are encoded by cry genes. Different strains of *B. thuringiensis* produce different Cry proteins and their toxicity varies towards different insects. The CryI proteins are specific for Lepidopteran pest species having size of 138 kDa; CryII proteins are specific for Lepidopteran and Dipteran pest species having size of 70 kDa; Cry III proteins are specific for Coleopteran pest species having size of

Table 3.3 Insect and pest resistant transgenic plants

Transgene	Source of transgene	Transgenic plant	Resistant to Pest species
cry1Ab	*Bacillus thuringiensis*	Cotton	Lepidoptera
cry1Ac and cry2Ab			
cry1Ab	*Bacillus thuringiensis*	Corn	Lepidoptera
cry9c			
cry1Aa	*Bacillus thuringiensis*	Tobacco	Lepidoptera
cry1Ab			
cry1Ac			
cry1C			
cry2A			
cry1Ab	*Bacillus thuringiensis*	Rice	Lepidoptera
cry1B			
cry1Ac, cry2A and GNA			Homoptera
cry1Ab	*Bacillus thuringiensis*	Egg plant	Lepidoptera
cry3A			Coleoptera

70 kDa and Cry IV are specific for Dipteran pest species (Hofte and Whiteley 1989; Tailor et al. 1992). The insecticidal crystal protein expressed by transgenic plants has to be ingested to affect the insects. After ingestion of this protein by the insects, the pro-toxin breaks down to active form inside the gut in presence of the high pH and gut enzymes. The active toxin binds to specific glycoprotein receptors (cadherin and APN) on the surface of gut membrane where it creates pores leading to leakage of mid-gut content, imbalance in ion concentration and death of the insect (Choma et al. 1990; Pigott and Ellar 2007; Bravo et al. 2007). Different variety of root crops, vegetables, forage crops and cereals are now being transformed with Bt genes for protection against insects (Shelton et al. 2002; Fontes et al. 2002) (Table 3.3). In addition to Bt genes, transgenic plants expressing genes for protease inhibitors, Alpha-Amylase inhibitors, enzymes and plant lectins have also been demonstrated for their resistance to pests (Fontes et al. 2002).

Virus infestation is a serious problem in agricultural cultivation (Waterhouse et al. 2001). It negatively affects the quality of crops and reduced the crop yields (Tolin 1991). Different methods have been used to inhibit the viral transmission to the plants but these strategies are not feasible. The virus resistant transgenic plants (VRTPs) confer effective resistance to viral infection. Most of the virus resistant transgenic plants are based on pathogen derived resistance (PDR) (Saharan et al. 2016). PDR refers the resistance obtained from a pathogenic virus through introduction of pathogenic viral genes either wild type or mutant into the plant genome. The expression of viral sequences in plant cells confers effective resistance to the susceptible plants (Prins et al. 2008). Many different types of transgenes have been used to develop virus resistant transgenic plants including viral coat protein (CP) genes, viral replicase genes, genes encoding movement proteins and nuclear inclusion genes as well as non-viral sequences from a variety of species (Tepfer

Table 3.4 Virus resistant transgenic plants

Transgene	Source of transgene	Transgenic plant	Resistant to Virus
Coat protein gene (CP)	Cowpea chlorotic mottle virus (CCMV)	Tobacco	CCMV
	Tobacco mosaic virus (TMV)		TMV
	Plum pox virus (PPV)		PPV
	Tobacco mosaic virus (TMV)	Tomato	TMV
	Cucumber mosaic virus (CMV)		CMV
	Zucchini yellow mosaic potyvirus (ZYMV)	Squash	ZYMV
	Watermelon mosaic 2 potyvirus (WMY2)		WMY2
	Potato virus – X (PVX)	Potato	PVX
	Potato virus – Y (PVY)		PVY
	Papaya ringspot virus (PRSV)	Papaya	PRSV
Replicase gene	Rice yellow mottle virus (RYMV)	Rice	RYMV
	Potato leaf roll virus (PLRV)	Potato	PLRV
Movement protein encoding transgene	Red clover necrotic mosaic virus (RCNMV)	Tobacco	RCNMV
	Tobacco mosaic virus (TMV)	Tobacco	TMV

2002; Saharan et al. 2016) (Table 3.4). Among the different transgene, most common transgene used for development of VRTPs is viral coat protein gene (White et al. 1995). The resistance mechanism induced by coat protein gene is mediated by protein encoded by the transgene (protein mediated) or by the transcript of the transgene (RNA mediated) (Lomonossoff 1995; Reimann-Philipp 1998). The expression of coat protein transgene is resulting in high level of protein which inhibits the disassembly of infecting virus and preventing the viral reproduction in the host cell because the plant cell blocks the viral coat protein gene in response to the high level of protein. Coat protein mediated resistance (CPMR) can provide either narrow or broad protection against the closely related viruses (Tepfer 2002). Post-transcriptional gene silencing (PTGS) is another approach to develop virus resistant transgenic plants. It involves the degradation of mRNA produced by the transgene and pathogenic virus by endogenous RNA degrading enzymes (Waterhouse et al. 2001; Galvez et al. 2014).

3.5.3 Therapeutic Proteins from Transgenic Plants

Therapeutic proteins are used in treatment of several diseases. These therapeutic proteins can be obtained from their natural sources only in limited amounts. Therefore, alternative sources are required for mass production of therapeutic proteins (Warzecha 2008). Plants can be engineered with the help of recombinant DNA

Table 3.5 Therapeutics produced by transgenic plants

Therapeutics	Potential application	Transgenic plant
Antibodies		
ScFvT84.66 (ScFv)	Cancer treatment	Rice
ScFvT84.66 (ScFv)	Cancer treatment	Wheat
Anti-HSV-2 (IgG)	Herpes simplex virus 2	Soybean
38C13	B-cell lymphoma treatment	Tobacco
Guy's 13 (secretory IgA)	Dental caries	Tobacco
RhinoRX	Common cold	Tobacco
Vaccines		
Vibrio cholera	Cholera	Rice and potato
Antiviral (griffithsin)	Severe acute respiratory syndrome (SARS)	Tobacco
PA83 and DIV	Anthrax	Brown mustard and tobacco
DPP4-fc	Middle east respiratory syndrome corona virus (MERS-CoV)	Tobacco
LTB	*Enterotoxigenic* E. coli	Potato
Other biopharmaceuticals		
Human protein C (serum protease)	Anticoagulants (protein C pathway)	Tobacco
Human hirudin variant 2	Anticoagulants (indirect thrombin inhibitors)	Ethiopian mustard
Human haemoglobin	Blood substitute	Tobacco
Human serum albumin	Liver cirrhosis	Potato, tobacco
Human aprotinin	Trypsin inhibitor for transplantation Surgery	Maize
Human interferon-α	Hepatitis B and C treatment	Rice, turnip
Glucocerebrosidase	Gaucher's disease	Tobacco

technology for production of different therapeutic proteins such as drugs, antibodies, vaccines, hormones and a variety of other biologically active pharmaceutical proteins (Fischer and Emans 2000; Ma et al. 2005). The production of human therapeutic proteins by using genetically engineered plants for pharmaceutical and other commercial interests is called as 'molecular farming' (Franken et al. 1997; Ma et al. 2003a, 2003b). Different variety of plants such as leafy crops, cereals, legumes, oilseeds, fruits, vegetables, etc. have been used for the production of therapeutic proteins (Twyman et al. 2003; Fischer et al. 2004; Daniell et al. 2001; Goldstein and Thomas 2004; Iyappan et al. 2017) (Table 3.5). Plant systems have become an alternative source for the production of therapeutic proteins to traditional production systems such as microbial cells or mammalian cell culture (Daniell et al. 2001; Goldstein and Thomas 2004). The advantages of using transgenic plants for production of therapeutic proteins include low production costs for large scale production, ability to produce complex proteins and a low risk of human pathogen contamination (Hellwig et al. 2004; Moustafa et al. 2016). The transgenic plants can produce glycosylated protein with correct folding that is an advantage over microbial systems

(Ma et al. 2003a, b; Marsian and Lomonossoff 2016). The production of biopharmaceuticals by plant system involves the transfer of gene that encodes therapeutic protein to the plant. The transgenic plant is cultivated and harvested for desired protein. The therapeutic protein is extracted, purified and possibly modified. In some instances, a vaccine can be delivered through the direct consumption of plant parts as an 'edible vaccine (Daniell et al. 2001). The plant parts such as leaves, fruits or other parts in a processed form which express a therapeutic protein could be consumed directly. Many plants such as corn, banana, tomatoes, carrots and lettuce have been used as possible oral delivery mechanisms for vaccines because all of these plants can be eaten raw which avoids protein denaturation occurring during the cooking (Sala et al. 2003).

3.5.4 Nutritional Benefits

The increase in production of food is necessary to feed the increasing world population. Plant derived foods are the major source of nutrition in human diet. The nutrient value of different food crops varies significantly (Sun 2008). Nutritional deficiency or malnutrition is the major challenge of developing countries because people rely on a single staple food crop such as rice, maize, etc. as source of nutrients (Christou and Twyman 2004). The staple foods are poor sources of nutrients and deficient in essential amino acids, minerals and vitamins (Bakshi and Dewan 2013). Most of the population in developing countries uses rice as staple food which is deficient in vitamin A. Vitamin A deficiency is quite possibly the most common in children of developing countries which is the leading cause of blindness (Ye et al. 2000; WHO 2009). The transgenic technology offers a way to improve the nutritional quality of food crops that can combat malnutrition. One of the important examples of this technology is the 'Golden Rice'. Golden rice is transgenic rice which exhibits an increased production of β-Carotene. β-Carotene is a precursor of vitamin A which is not present in endosperm of rice naturally (Xudong et al. 2000).

Golden rice was developed by incorporation of *psy* gene from daffodil and the *crtI* gene from bacterium *Erwinia uredovora* in rice endosperm (Ye et al. 2000). The transgenic Golden rice may be useful to treat the vitamin A deficiency in children of developing countries. Some other food crops have also been engineered to improve the nutritional levels by enhancing the synthesis of other vitamins and minerals (Perez-Massot et al. 2013) (Table 3.6).

3.5.5 Salt Tolerance

Abiotic stresses such as salt, drought and oxidative stresses are the environmental constraints that greatly affect plant growth and productivity (Acquaah 2007; Ahmad et al. 2012). Agricultural productivity might be increased if crops were genetically engineered to better cope with these stresses (Ahmad et al. 2012). A number of candidate genes expressed under different types of abiotic stresses have been

Table 3.6 Transgenic Plants with Enhanced Mineral and Vitamin Content

Nutrients	Transgene	Transgenic plant
Vitamin A	*Nppsy1, EucrtI*	Rice
	Zmpsy1, PacrtI	Corn
	Zmpsy1, PacrtI	Wheat
	PacrtB	Potato
Vitamin C	*Acggp*	Tomato
	Osdhar	Corn
Folic acid	*Atgtpchi, Atadcs*	Rice
Zinc	*Atzip*	Cassava
	Osnas2	Rice
Iron	*Gm ferritin and Af phytase*	Corn
	Osnas2	Rice
Calcium	*scax1*	Lettuce
	scax1	Carrot
Selenium	*Ataps1*	Indian mustard

Table 3.7 Transgenic plants with enhanced abiotic stress tolerance

Transgene	Source of transgene	Transgenic plant	Abiotic stress tolerance
Transcription factors			
TabZIP60	*Triticum aestivum*	*A. thaliana*	Drought, salt and freezing tolerance
TaMYB3R1	*Triticum aestivum*	*A. thaliana*	Drought and salt tolerance
ZmWRKY58	*Zea mays*	*Oryza sativa*	Drought and salt tolerance
ONAC063	*Oryza sativa*	*A. thaliana*	Salinity and osmotic tolerance
SlDREB1	*Solanum tuberosum*	*Solanum tuberosum*	Salt tolerance
Signaling molecules			
MAPK5	*Oryza sativa*	*Oryza sativa*	Drought, salt and cold tolerance
NDPK2	*A. thaliana*	*A. thaliana*	Salt, cold and oxidative Stress tolerance
OsMSR2	*Oryza sativa*	*A. thaliana*	Salt and drought tolerance
AtCPK6	*A. thaliana*	*A. thaliana*	Salt and drought tolerance
Ion transporters			
AtNHX1	*A. thaliana*	*Solanum lycopersicum*	Salt tolerance
TaNHX1 and TVP1	*Triticum aestivum*	*A. thaliana*	Drought and salt tolerance
NHX2	*Hordeum vulgare*	*Solanum tuberosum*	Salt tolerance
AVP1	*A. thaliana*	*A. thaliana*	Drought and salt tolerance

focused for development of transgenic plants which combat against the abiotic stresses (Jha 2019) (Table 3.7). These genes can be categorized in three groups (Li et al. 2013; Paul and Roychoudhury 2018; Jha 2019):

1. Genes involved in direct protection by synthesis of functional proteins or stress-adaptive compounds such as osmolytes, ROS scavengers or antioxidant enzymes, polyamines, late embryogenesis abundant proteins (LEA) and heat shock proteins (HSPs).
2. Genes involved in synthesis of regulatory proteins and signaling pathway intermediates such as transcription factors.
3. Genes involved in ion homeostasis such as sodium/proton transporters.

3.6 Global Status of Transgenic Crops

The global area of transgenic crops increased with an increasing proportion grown by developing countries (James 2004). It was estimated that one-third (30%) of the world transgenic crop area were grown in developing countries. The major transgenic crops which are being commercialized comprise soybean (60%), corn (23%), cotton (12%), canola (5%) and potato (~1%). The properties for which transgenic crop have been developed are herbicide tolerance, insect resistance and qualitative traits. Moreover, scientific efforts are continued to develop genetically modify plants having high economic value like cereals, floriculture, fruits, vegetables and horticulture species. In recent times, it was reported that there are 14 countries growing about 50,000 hectares or more of transgenic crops (James 2004). Thus, despite the several concerns about the transgenic crops, the hectare and number of farmers growing these crops have continued to cultivate with high rate every year after their introduction in 1996. More than eight million farmers are gaining profit from this technology (James 2004). About 90% of the beneficiaries farmers from developing countries are poor farmers in resources, whose increased revenue from these crops contribute to the lessening of poverty. In 2014, 19 out of 28 countries planted transgenic crops on more than 50,000 hectares area. These countries included USA, Brazil, China, Argentina, India, Canada, South Africa, Sudan, Uruguay, Bolivia, Australia, Myanmar, Mexico, Spain, Philippines, Burkina Faso, Paraguay, Pakistan and Colombia. There are 100 fold increases in transgenic crops cultivation while 1996 when GM commercialization started. The area for transgenic crops cultivation for the last 19 years has grown remarkably each year. The top country with highest per hectare growth in 2014 was US beating Brazil with total of ~30 million hectares (James 2014). Depending on the area of land covered by each crop in different transgenic planting countries, 4 crops, for example, soybean, maize, oilseed rape and cotton are commonly genetically modified. The global soybean plantation in year 2014 was 111 million hectares with 90.7 million hectares being genetically modified. The global cotton cultivation area was 37 million with 25.1 million hectares being genetically modified. Similarly the total area covered by transgenic maize and transgenic oilseed rape plantation was 55.2 and nine million hectares.

3.7 Major Concerns of Transgenic Crops

The use of advanced recombinant tools for changing the plant genome has created infinite opportunities of crop improvement. The use of transgenic crops has several concerns from many years. The major concerns arise due to commercialization of transgenic plants are as follows:

3.7.1 Environmental Concerns

Engineered crop plants are better suited to the environment by integration of genes for tolerance to biotic and abiotic stresses which leads to improvement in crop production (Ortiz 1998). Transgenic crops for reduced fertilizer requirement have reduced the negative impact on the soil and the subsequent effects of run-off into rivers and seepage into ground water. Disease and pest resistance are an additional area of transgenic technology that has gain criticism because virus resistance leads the evolution of new and harmful viruses (Borja et al. 1999; Rubio et al. 1999). Similarly crops engineered to produce toxins might poison non-target hosts (Concar 1999). Several evidence suggest that many targeted pest species have developed resistance to engineered genes in the similar approach as they have done to naturally occurring resistance genes (Holmes 1997). The concerns associated with transgenic plants with regard to their adverse effect on the environment, biodiversity, wildlife populations and gene transferring in non-transgenic wild plants.

3.7.2 Transgenesis and Human Health

The harmful health effects of transgenic crops increase fear and concern about the safety of feeding population (Godfrey 2000). The point of concerns associated with potential adverse health effects of consuming transgenic crops are as follows:

3.7.2.1 Allergenicity

There is no actual proof so far that genetically modified foods cause allergic reactions than conventional food. Genetic modification itself does not create any allergens but the nature of genes used for introduction into host plant may induce allergic reaction. The reports of some allergies to nuts are very common. The transgenic soybean was tested and it was found that the people who had allergic reactions to Brazil nuts also had allergic reactions to the GM soybean (Nordlee et al. 1996).

3.7.2.2 Horizontal Transfer and Antibiotic Resistance

The use of antibiotic resistance markers in transgenic crops has raised concerns to lose the ability to treat infections with antibiotic drugs. It is a fact that successful DNA transfer from transgenic plants would not arise enough to source of health problems in humans. Another important concern related to transgenic plants is the horizontal gene transfer which involved in transfer of DNA from one organism to another outside the parent to offspring route. Transfer of a resistance gene from transgenic food to microorganisms that normally resides in our mouth, stomach and intestines can gradually become competent to survive an oral dose of antibiotic medicine.

3.8 Conclusions and Future Prospective

The transgenic crops are important for hunger management, pest resistance and nutrient fulfilment. Transgenic technology may also useful in increasing crop yields, reducing risk of pest and weed attacks and increasing income of farmers. Additionally several developing countries lack the basic knowledge and regulatory infrastructures to monitor transgenic crops, and therefore are guarded about the benefits of transgenic technology. In order to remove these limitations they encourage investments in regulatory infrastructures as well as knowledge sharing among developing and developed countries. Even though these crops surely not be assumed as a magic bullet towards poverty and food demand, but they are promising answer to increase farm yields, reduce risk and increase wealth of small farmers. The worldwide release of transgenic crop and food in environment and markets has resulted in public debate in various part of the globe. Presently, there are 14 countries including 9 developing countries and 5 developed countries, growing transgenic crops. The need for recognition and detection of transgenic crops and their products has improved with the fast growth in the cultivation of transgenic crops. Labelling and traceability of transgenic material are way forward to deal with the concerns of customers and regulators. A large number of diverse strategies and protocols are available for testing of transgenic material. Thus, it is necessary to know about the methods and their applications for detection of transgenic crops and their products. Presently, existing methods for detecting transgenic crops and products are almost completely based on polymerase chain reaction, because of their high sensitivity, specificity and need for only a small amount of DNA. Particularly, real time PCR has been believed as the most potent tool for the detection and quantification of transgenic despite its high expense. In the future, scientist hopes to be able to supply vaccine and medicines through transgenic food which is useful for population of developing countries. Medications integrated into food are easier to store and transport than conventional medicine. Transgenic plants recommend a new possible approach for production and administration of human antibodies. Future generations of transgenic plants are wished for suitable towards harsh environments and for the enrichment of nutrient content, production of bio-energy and bio-fuels and production of pharmaceutical agents. Thus the potential of transgenic crops is bright and hopeful.

References

Acquaah G (2007) Principles of plant genetics and breeding. Blackwell, Oxford, UK

Ahmad P, Ashraf M, Younis M et al (2012) Role of transgenic plants in agriculture and biopharming. Biotechnol Adv 30:524–540

Ahokas H (1987) Transfection by DNA associated liposomes evidenced at pea pollination. Heriditas 106:129–138

Bakshi S, Dewan D (2013) Status of transgenic cereal crops: a review. Clon Transgen 3:119

Barta A, Sommergruber K, Thompson D et al (1986) The expression of a nopaline synthase-human growth hormone chimaeric gene in transformed tobacco and sunflower callus tissue. Plant Mol Biol 6:347–357

Birch RG, Franks T (1991) Development and optimization of microprojectile systems for plant genetic transformation. Aust J Plant Physiol 18:453–469

Bonny S (2008) Genetically modified glyphosate-tolerant soybean in the USA: adoption factors, impacts and prospects: a review. Agron Sustain Dev 28:21–32

Borja M, Rubio T, Scholthof HB, Jackson AO (1999) Restoration of wild-type virus by double recombination of tombusvirus mutants with a host transgene. Mole Plant Microbe Interact 12(2):153–162

Bravo A, Gill SS, Soberón M (2007) Mode of action of *Bacillus thuringiensis* cry and cyttoxins and their potential for insect control. Toxicon 49(4):423–435

Bubner B, Baldwin IT (2004) Use of real-time PCR for determining copy number and zygosity in transgenic plants. Plant Cell Rep 23(5):263–271

Cerdeira AL, Duke SO (2007) Environmental impacts of transgenic herbicide-resistant crops. CAB Rev: Perspect Agri Vet Sci Nutri Nat Res 2(033):14

Choi IR, Stenger DC, Morris TJ, French R (2000) A plant virus vector for systemic expression of foreign genes in cereals. Plant J 23:547–555

Choma CT, Surewicz WK, Carey PR et al (1990) Unusual proteolysis of the protoxin and toxin of *bacillus thuringiensis*- structural implications. Eur J Biochem 189:523–527

Christou P (1992) Genetic transformation of crop plants using microprojectile bombardment. Plant J 2:275–281

Christou P (1995) Strategies for variety-independent genetic transformation of important cereals, legumes and woody species utilizing particle bombardment. Euphytica 85:13–27

Christou P, Twyman RM (2004) The potential of genetically enhanced plants to address food insecurity. Nut Res 17:23–42

Concar D (1999). Dispatches from the killing fields. New Scientist 27th February p 5

Crossway A, Oakes JV, Irvine JM (1986) Integration of foreign DNA following microinjection of tobacco mesophyll protoplasts. Mol Gen Genet 202:179–185

Dai S, Zheng P, Marmey P, Zhang S, Tian W, Chen S et al (2001) Comparative analysis of transgenic rice plants obtained by *agrobacterium* mediated transformation and particle bombardment. Mol Breed 7(1):25–33

Daniell H, Streatfield SJ, Wycoff K (2001) Medical molecular farming: production of antibodies, biopharmaceuticals and edible vaccines in plants. Trends Plant Sci 6(5):219–226

Darbani B, Farajnia S, Noeparvar SH (2008) Plant transformation: needs and futurity of the transgenes. Biotechnology 7(3):403–412

Debnath M, Prasad GBKS, Bisen PS (2010) Molecular diagnostics: promises and possibilities. Springer, Netherlands, Dordrecht

Deshayes A, Herrera-Estrella L, Caboche M (1985) Liposome-mediated transformation of tobacco mesophyll protoplasts by an *Escherichia coli* plasmid. EMBO J 4:2731–2737

D'Halluin K, Bonne E, Bossut M et al (1992) Transgenic maize plants by tissue electroporation. Plant Cell 4(12):1495–1505

Drummond MH, Chilton MD (1978) Tumor inducing (Ti) plasmids of *agrobacterium* share extensive regions of DNA homology. J Bacterial 136:1178–1183

Elbaidouri M, Chaparro C, Panaud O (2013) Use of next generation sequencing (NGS) technologies for the genome-wide detection of transposition. Methods Mol Biol 1057:265–274
Engler G, Depicker A, Maenhout R, Villarroel R et al (1981) Physical mapping of DNA base sequence homologies between an octopine and a nopaline Ti plasmid of *Agrobacterium tumefaciens*. J Mol Biol 152:183–208
Estela A, Escriche B, Ferré J (2004) Interaction of *bacillus thuringiensis* toxins with larval midgut binding sites of *Helicoverpa armigera* (Lepidoptera: Noctuidae). Appl Environ Microbiol 70(3):1378–1384
Ferry N, Raemaekers RJM, Majerus MEN et al (2003) Impact of oilseed rape expressing the insecticidal cysteine protease inhibitor oryzacystatin on the beneficial predator *Harmonia axyridis* (multicoloured Asian ladybeetle). Mol Ecol 12(2):493–504
Fischer R, Emans N (2000) Molecular farming of pharmaceutical proteins. Transgenic Res 9:279–299
Fischer R, Stoger E, Schillberg S (2004) Plant-based production of biopharmaceuticals. Curr Opin Plant Biol 7:152–158
Flannagan R, Cao-Guo Y, Mathis JP et al (2005) Identification, cloning and expression of a cry1Ab cadherin receptor from European corn borer, *Ostrinia nubilalis* (Hübner) (lepidoptera: Crambidae). Insect Biochem Mol Biol 35:33–40
Fontes EMG, Pires CSS, Sujii ER et al (2002) The environmental effects of genetically modified crops resistant to insects. Neotrop Entomol 31(4):497–513
Franken E, Teuschel U, Hain R (1997) Recombinant proteins from transgenic plants. Curr Opin Biotechnol 8:411–416
Fromm ME, Taylor LP, Walbot V (1985) Expression of genes transferred into monocot and dicot plant cells by electroporation. Proc Natl Acad Sci U S A 82:5824–5828
Fromm ME, Taylor LP, Walbot V (1986) Stable transformation of maize after gene transfer by electroporation. Nature (London) 319:791–793
Fu X, Duc LT, Fontana CT et al (2000) Linear transgene constructs lacking vector backbone sequences generate low-copy-number transgenic plants with simple integration patterns. Transgenic Res 9:11–19
Galvez LC, Banerjee J, Pinar H et al (2014) Engineered plant virus resistance. Plant Sci 228:11–25
Gan C (1989) Gene gun accelerates DNA-coated particles to transform intact cells. Scientist 3(18): 25
Gellert A, Salanki K, Tombacz K, Tuboly T, Balazs E (2012) A cucumber mosaic virus based expression system for the production of porcine circovirus specific vaccines. PLoS One 7(12): e52688. https://doi.org/10.1371/journal.pone.0052688
George DGM (2013) Engineering insect-resistant crops: a review. Afr J Biotechnol 12(23): 3600–3608
Goddijn OJM, Jan P (1995) Plants as bioreactors. Trends Biotechnol 13(9):379–387
Godfrey J (2000) Do genetically modified foods affect human health? Lancet 355(9201):414
Goldstein DA, Thomas JA (2004) Biopharmaceuticals derived from genetically modified plants. Q J Med 97:705–716
Gong YY, Xu ZZ, Guo SQ et al (2016) Genome-wide analysis of the *EPSPS* genes in upland cotton. Acta Agric Boreali-Sin 31:15–21
Green JM, Owen MDK (2011) Herbicide-resistant crops: utilities and limitations for herbicide-resistant Weed Management. J Agric Food Chem 59:5819–5829
Hellwig S, Drossard J, Twyman RM et al (2004) Plant cell cultures for the production of recombinant proteins. Nat Biotechnol 22:1415–1422
Herouet C, Esdaile DJ, Mallyon BA et al (2005) Safety evaluation of the phosphinothricin acetyltransferase proteins encoded by the pat and bar sequences that confer tolerance to glufosinate ammonium herbicide in transgenic plants. Regul Toxicol Pharmacol 41:134–149
Hofte H, Whiteley HR (1989) Insecticidal crystal proteins of *bacillus thuringiensis*. Microbiol Rev 53:242–255
Holmes B (1997) Caterpillar's revenge. New Scientist:7

Hooykaas PJJ, Shilperoort RA (1992) *Agrobacterium* and plant genetic engineering. Plant Mol Biol 19:15–38
Hui SW (1995) Effects of pulse length and strength on electroporation efficiency. In: Nickoloff JA (ed) Methods in molecular biology. Plant cell electroporation and electrofusion protocols. Humana Press Inc., pp 29–40
Hwang MS, Lindenmuth BE, McDonald KA, Bryce W, Falk B (2012) Bipartite and tripartite cucumber mosaic virus-based vectors for producing the *Acidothermus cellulolyticus* endo-1,4-β-glucanase and other proteins in non-transgenic plants. BMC Biotechnol 12:66. https://doi.org/10.1186/1472-6750-12-66
Iyappan G, Hari Priya S, Kadirvelu K et al (2017) Rapid production of therapeutic proteins using plant system. Def Life Sci J 2(2):95–102
James C (1999) Global review of commercialized transgenic crops in 1999. Int Service Acquisition Agric Biotechnol Appl 12:1–7
James C (2004) Preview: Global status of commercialized biotech/GM crops. www.transgen.de/pdf/dokumente/ISAAA2004_summary.pdf
James C (2005) Global status of commercialized transgenic crops. ISAAA Briefs 34
James C (2013) Global status of commercialized biotech/GM crops. Brief no. 46. ISAAA, Ithaca, NY, USA
James C (2014) Global status of commercialized biotech/GM crops: 2014, ISAAA brief no. 49. ISAAA, Ithaca
Jha S (2019) Transgenic approaches for enhancement of salinity stress tolerance in plants. In: Singh SP et al (eds) Molecular approaches in plant biology and environmental challenges, energy, environment, and sustainability. Springer, Cham, Switzerland, pp 265–322
Jogdand SN (2006) Gene biotechnology, 2nd edn. Himalaya Publishing House, Mumbai, India, pp 237–249
Kennedy GG, Gould F (2007) Ecology of natural enemies and genetically engineered host plants. In: Kogan M, Jepson P (eds) Perspectives in ecological theory and integrated Pest management. Cambridge University Press, Cambridge, UK, pp 269–300
Khan SU, Siddique M, Nabi R et al (2016) Role of recombinant DNA technology to improve life. Int J Genom:1–14
Kikkert JR (1993) The biolistic® PDS-1000/he device. Plant Cell Tissue Organ Cult 33:221–226
Kloti A, Iglesias VA, Wunn J et al (1993) Gene transfer by electroporation into intact scutellum cells of wheat embryos. Plant Cell Rep 12:671–675
Lambirth KC, Whaley AM, Schlueter JA, Bost KL, Piller KJ (2015) Contrails: a tool for rapid identification of transgene integration sites in complex, repetitive genomes using low coverage paired-end sequencing. Genomics Data 6:175–181
Laursen CM, Krzyzek RA, Flick CE et al (1994) Production of fertile transgenic maize by electroporation of suspension culture cells. Plant Mol Biol 24:51–61
Lee WS, Hammond-Kosack KE, Kanyuka K (2012) Barley stripe mosaic virus-mediated tools for investigating gene function in cereal plants and their pathogens: virus-induced gene silencing, host-mediated gene silencing, and virus mediated overexpression of heterologous protein. Plant Physiol 160:582–590
Lehman CW, Trautman JK, Carroll D (1994) Illegitimate recombination in *Xenopus*: characterization of end joined junctions. Nucleic Acid Res 22:434–442
Li Q, Liu J, Tan D et al (2013) A genome-wide expression profile of salt-responsive genes in the apple rootstock *Malus zumi*. Int J Mol Sci 14:21053–21070
Lienard D, Sourrouille C, Gomord V et al (2007) Pharming and transgenic plants. Biotech Annu Rev 13:115–147
Liu K (1999) Biotech crops: products, properties, and prospects. Food Technol 53:42–48
Liu H, Kawabe A, Matsunaga S et al (2004) Obtaining transgenic plants using the bio-active beads method. J Plant Res 117(2):95–99
Loake GJ, Ashby AM, Shaw CH (1988) Attraction of *agrobacterium tumefaciens* C58C 1 towards sugars involves a highly sensitive chemotaxis system. Microbiology 134(6):1427–1432

Lomonossoff GP (1995) Pathogen derived resistance to plant viruses. Annu Rev Phytopathol 33: 323–343

Lozano G, Trenado HP, Fiallo-Olive E, Chirinos D, Geraud-Pouey F, Briddon RW (2016) Characterization of non-coding DNA satellites associated with sweepoviruses (genus Begomovirus, Geminiviridae)-definition of a distinct class of begomovirus-associated satellites. Front Microbiol 7:162. https://doi.org/10.3389/fmicb.2016.00162

Ma JK, Drake PM, Christou P (2003a) The production of recombinant pharmaceutical proteins in plants. Nat Rev Genet 4:794–805

Ma K-CJ, Drake PMW, Christou P (2003b) The production of recombinant pharmaceutical proteins in plants. Nat Rev Gene 4:794–805

Ma K-CJ, Barros E, Bock R (2005) Molecular farming for new drugs and vaccines. EMBO Report 6:593–599

Ma XY, Wu HW, Jiang WL et al (2016) Weed and insect control affected by mixing insecticides with glyphosate in cotton. J Integrative Agri 15:373–380

Marsian J, Lomonossoff GP (2016) Molecular pharming – VLPs made in plants. Curr Opin Biotechnol 37:201–206

Miao Y, Ding Y, Sun QY et al (2008) Plant bioreactors for pharmaceuticals. Biotechnol Genet Eng 25(1):363–380

Moustafa K, Makzhoum A, Trémouillaux-Guiller J (2016) Molecular farming on rescue of pharma industry for next generations. Crit Rev Biotechnol 36(5):840–850

Nawaz-Ul-Rehman MS, Fauquet CM (2009) Evolution of geminiviruses and their satellites. FEBS Lett 583:1825–1832

Naylor LH (1999) Reporter gene technology: the future looks bright. Biochem Pharmacol 58(5): 749–757

Negrutiu I, Shillito R, Potrykus I et al (1987) Hybrid genes in the analysis of transformation conditions. I. Setting up a simple method for direct gene transfer in plant protoplasts. Plant Mol Biol 8:363–373

Neuhaus G, Spangenberg G, Mittelsten-Scheid O (1987) Transgenic rapeseed plants obtained by the microinjection of DNA into microspore-derived embryoids. Theor Appl Genet 75:30–36

Noack J, Vogel A (1995) Streak-photographic investigation of shock wave emission after laser-induced plasma formation in water. In: Proc. SPIE 2391, laser-tissue interaction VI

Nordlee JA, Taylor SL, Townsend JA, Thomas LA, Bush RK (1996) Identification of a Brazil-nut allergen in transgenic soybeans. N Engl J Med 334:688–692

Norris SR, Meyer SE, Callis J (1993) The intron of *Arabidopsis thaliana* polyubiquitin genes is conserved in location and is a quantitative determinant of chimeric gene expression. Plant Mol Biol 21:895–906

Ortiz R (1998) Critical role of plant biotechnology for the genetic improvement of food crops: perspectives for the next millennium. Electron J Biotechnol 1(3) http://www.ejb.org

Park D, Park SH, Ban YW, Kim YS, Park KC, Kim NS, Kim JK, Choi IY (2017) A bioinformatics approach for identifying transgene insertion sites using whole genome sequencing data. BMC Biotechnol 17(1):67. https://doi.org/10.1186/s12896-017-0386-x

Parke D, Ornston LN, Nester EW (1987) Chemotaxis to plant phenolic inducers of virulence genes is constitutively expressed in the absence of the Ti plasmid in *agrobacterium tumefaciens*. J Bacteriot 169:5336–5338

Paul S, Roychoudhury A (2018) Transgenic plants for improved salinity and drought tolerance. In: Gosal SS, Wani SH (eds) Biotechnologies of crop improvement, Transgenic approaches, vol 2. Springer, Cham, Switzerland, pp 141–181

Perez-Massot E, Banakar R, Gomez-Galera S et al (2013) The contribution of transgenic plants to better health through improved nutrition: opportunities and constraints. Genes Nutr 8:29–41

Peterson MA, Shan G, Walsh TA et al (2011) Utility of Aryloxyalkanoate dioxygenase transgenes for development of new herbicide resistant crop technologies. Inform Syst Biotechnol. (http://www.isb.vt.edu/news/2011/May/Aryloxyalkanoate-Dioxygenase-Transgenes.pdf)

Peyret H, Lomonossoff GP (2013) The pEAQ vector series: the easy and quick way to produce recombinant proteins in plants. Plant Mol Biol 83(1–2):51–58
Pigott CR, Ellar DJ (2007) Role of receptor in *bacillus thuringiensis* crystal toxin activity. Microbiol Mol Biol Rev 71(2):255–281
Potrykus I (1990) Gene transfer to plants: assessment and perspectives. Physiol Plant 79:125–134
Prins M, Laimer M, Noris E et al (2008) Strategies for antiviral resistance in transgenic plants. Mol Plant Pathol 9:73–83
Puchta H (1998) Repair of genomic double-strand breaks in somatic cells by one-side invasion of homologous sequences. Plant J 13(3):331–339
Reich TJ, Iver VN, Miki BL (1986) Efficient transformation of alfalfa protoplasts by the intranuclear microinjection of Ti plasmid. J Biotechnol 4:1001–1004
Reimann-Philipp U (1998) Mechanism of resistance: expression of coat protein. Methods Mol Biol 81:521–532
Rubio T, Borja M, Scholthof HB, Jackson AO (1999) Recombination with host transgenes and effects on virus evolution: an overview and opinion. Mol Plant-Microbe Interact 12:87–92
Rybicki EP (2009) Plant-produced vaccines: promise and reality. Drug Discov Today 14:16–24
Saharan V, Jain D, Pareek S et al (2016) Viral, fungal and bacterial disease resistance in transgenic plants. In: Al-Khayri JM, Jain SM, Johnson DV (eds) Advances in plant breeding strategies: agronomic, abiotic and biotic stress traits. Springer, Cham (Switzerland), pp 627–656
Sala F, Rigano MM, Barbante A (2003) Vaccine antigen production in transgenic plants: strategies, gene constructs and perspectives. Vaccine 21:803–808
Sanford JC (2000) The development of the biolistic process. In Vitro Cell Dev Biol Plant 36:303–308
Saveleva NV, Burlakovskiy MS, Yemelyanov VV et al (2015) Transgenic plants as bioreactors for the production of substances of medicinal and veterinary importance. Ekol Genet 13(2):77–99
Saveleva NV, Burlakovskiy MS, Yemelyanov VV et al (2016) Transgenic plants as bioreactors to produce substances for medical and veterinary uses. Russ J Genet Appl Res 6:712–724
Senseman SA (2007) Herbicide handbook, 9th edn. Weed Science Society of America, Lawrence, KS, p 458
Senthil KM, Mysore KS (2014) Tobacco rattle virus-based virus induced gene silencing in *Nicotiana benthamiana*. Nat Protoc 9:1549–1562. https://doi.org/10.1038/nprot.2014.092
Sharma KK, Ortiz R (2000) Programme for the application of genetic transformation for crop improvement in the semi-arid tropics. In Vitro Cell Dev Biol Plant 36:83–92
Sharma AK, Sharma MK (2009) Plants as bioreactors: recent developments and emerging opportunities. Biotechnol Adv 27(6):811–832
Sharma AK, Jani D, Raghunath C et al (2004) Transgenic plant as bioreactors. Indian J Biotechnol 3:274–290
Sharma KK, Bhatnagar-Mathur P, Thorpe TA (2005) Genetic transformation technology: status and problems. In Vitro Cell Dev Biol Plant 41:102–112
Shelton AM, Zhao JZ, Roush RT (2002) Economic, ecological, food safety, and social consequences of the deployment of *Bt* transgenic plants. Annu Rev Entomol 47:845–881
Shillito RD, Saul MW, Paszkowski MW (1985) High efficiency direct gene transfer to plants. J Biotechnol 3:1099–1103
Simoens C, Van Montagu M (1995) Genetic engineering in plants. Hum Reprod Update 1:523–542
Singh NK, Shaw JJ (1992) Physical methods for gene transfer in plants. In: Thottappilly G, Monti LM, Mohan Raj DR, Ibadan MAW (eds) Biotechnology: enhancing research on tropical crops in Africa. IITA, Nigeria
Southgate EM, Davey MR, Power JB (1995) Factors affecting the genetic engineering of plants by microprojectile bombardment. Biotechnol Adv 13:631–651
Stachel SE, Nester EW, Zambryski PC (1986) A plant cell factor induces *agrobacterium tumefaciens* vir gene expression. Proc Natl Acad Sci U S A 83:379–383
Sun SSM (2008) Application of agricultural biotechnology to improve food nutrition and healthcare products. Asia Pac J Clin Nutr 17:87–90

Tailor R, Tippett J, Gibb G et al (1992) Identification and characterisation of a novel *bacillus thuringiensis*-endotoxin entomocidal to coleopteran and lepidopteran larvae. Mol Microbiol 7: 1211–1217
Takeuchi Y, Dotson M, Keen NT (1992) Plant transformation: a simple particle bombardment device based on flowing helium. Plant Mol Biol 18(4):835–839
Taylor NJ, Fauquet CM (2002) Microparticle bombardment as a tool in plant science and agricultural biotechnology. DNA Cell Biol 21:963–977
Tepfer M (2002) Risk assessment of virus-resistant transgenic plants. Annu Rev Phytopathol 40: 467–491
Tinland B, Koukolikova NZ, Hall MN, Hohn B (1992) The T-DNA linked VirD2 contains two distinct functional nuclear localization signals. Proc Natl Acad Sci U S A 89:7442–7446
Tohidfar M, Khosravi S (2015) Transgenic crops with an improved resistance to biotic stresses. A review. Biotechnol Agron Soc Environ 19(1):62–70
Tolin SA (1991) Persistence, establishment, and mitigation of phytopathogenic viruses. In: Levin MA, Strauss HS (eds) Risk assessment in genetic engineering. McGraw Hill, Inc., New York, pp 114–139
Toriyama K, Arimoto Y, Uchimiya H et al (1988) Transgenic rice plants after direct gene transfer into protoplasts. J Biotechnol 6:1072–1074
Twyman RM, Stoger E, Schillberg S et al (2003) Molecular farming in plants: host systems and expression technology. Trends Biotechnol 21:570–578
Warzecha H (2008) Biopharmaceuticals from plants: a multitude of options for posttranslational modifications. Biotechnol Genetic Eng Rev 25:315–330
Waterhouse PM, Wang MB, Lough T (2001) Gene silencing as an adaptive defence against viruses. Nature 411:834–842
Weaver JC (1995) Electroporation theory. In: Nickoloff JA (ed) Methods in molecular biology. Plant cell electroporation and electrofusion protocols. Humana Press Inc., Totowa NJ, pp 3–28
Weaver JC, Chizmadzhev YA (1996) Theory of electroporation: a review. Bioelectrochem Bioenerg 41(2):135–160
White PS, Morales F, Roossink M (1995) Interspecific reassortment of genomic segments in the evolution of cucumoviruses. Virology 207:334–337
WHO (2009) Global prevalence of vitamin A deficiency in populations at risk 1995–2005. In: WHO global database on vitamin A deficiency. World Health Organization, Geneva
Xu X, Shi H, Li B (1995) Transgenic Indica Rice plants by electroporation of seed embryo cells. In: Spangenberg G (ed) Potrykus I. Gene transfer to plants. Springer lab manual, Springer, Berlin, Heidelberg
Xudong Y, Al-Babili S, Kloti A et al (2000) Engineering the provitamin a (β-carotene) biosynthetic pathway into (carotenoid-free) rice endosperm. Science 287:303–305
Yao Q, Cong L, Chang JL et al (2006) Low copy number gene transfer and stable expression in a commercial wheat cultivar via particle bombardment. J Exp Bot 57(14):3737–3746
Ye X, Al-Babili S, Klöti A, Zhang J, Lucca P et al (2000) Engineering the provitamin A (beta-carotene) biosynthetic pathway into (carotenoid-free) rice endosperm. Science 287:303–305
Zaenen I, Larebeke N, Teuchy H, Schell J (1974) Supercoiled circular DNA in crown gall inducing *agrobacterium* strains. J Mol Biol 86:109–127
Zhang W, Chen J, Zhang H et al (2008) Overexpression of a rice diacylglycerol kinase gene OsBIDK1 enhances disease resistance in transgenic tobacco. Mol Cells 26:258–264
Zupan JR, Zambryski P (1995) Transfer of T-DNA from *agrobacterium* to the plant cell. Plant Physiol 107:1041–1047

Genetically Modified Crops and Their Applications

4

Minal Garg

Abstract

Continual rise in global population and, therefore, demand for food (grains and animal proteins/ fats) present massive challenge to agriculture to generate the steady supply of improved cultivars. Advancements in modern breeding programs and the genome level technologies exploit the use of genes from all the sources to develop and produce genetically modified plants for improved crop yield, nutrition potential/ quality, and lower yield losses from disease and environmental factors such as changing climate and soil depletion. Over the past few decades, advancements in biotechnological tools have revolutionized the world market with the production of genetically modified (GM) crops that possess novel combination of genetic material. To promote the global acceptance of GM crops, it is highly imperative to spread the awareness about the social gains, economic benefits, and health improvements obtained through GM crop adoption to the farmers, consumers, policy makers/ regulators, and science/ media communicators.

Keywords

Crop yield · GM crops · Stable expression · Transgenic methods · Transformation

4.1 Introduction

According to the U.S. Census Bureau population clock (https://www.census.gov/popclock/), currently the global world population is 7.6 billion and by the year 2050, it is expected to be somewhere between 8.3 billion and 10.9 billion. The

M. Garg (✉)
Department of Biochemistry, University of Lucknow, Lucknow, Uttar Pradesh, India

R. L. Singh et al. (eds.), *Plant Genomics for Sustainable Agriculture*,
https://doi.org/10.1007/978-981-16-6974-3_4

unprecedented growth in population is considered to be majorly responsible for death toll due to suffering from hunger and malnutrition. Two-third of such people live in Asia and the Pacific and a quarter in Sub-Saharan Africa. Unlike deadly diseases, hunger and malnutrition are curable conditions and require all the necessary measures to cope up with the increased global demand for staple crops, feed, and livestock. Continual rise in global population and, therefore, demand for food (grains and animal proteins) present massive challenge to agriculture to generate the steady supply of improved cultivars.

The U.N. Food and Agricultural Organization (FAO) estimates the requirement of increase in agricultural production by at least 60% globally and by 70% in developing countries by the year 2050 (Ray et al. 2013). Special emphasis is required on the annual increase of 2.4% in the production of major global crops including maize, wheat, rice, and soybean which constitute around 66% of total calories in the form of "global" diet. Urbanization, desertification, salinization, soil degradation, and demand for pastureland are continuously minimizing the availability of finite amount of arable land for food production per person. Besides these factors, climate change and limited agricultural resources including fertilizer, water, and procedures to manage pest/ weed further compound the problem of low crop productivity. Contradicting the very fact of continual requirement of increased crop productivity, the current trends in yield gains are either maintaining status quo or are declining under present agricultural practices (Lobell et al. 2011; Ray et al. 2012).

It has been possible to obtain sustainable and high yields of staple crops in the developed countries of North America and Europe by means of conventional breeding. However, many parts of Asia in the middle of twentieth century faced acute famine. In 1940s, advancements in new mechanized agricultural technologies and cropping systems helped Norman Borlaug, an eminent plant breeder (awarded with noble prize in 1970), to develop a breeding program for high-yielding and disease resistant wheats. This helped in a version of the famine in Mexico to great extent (Phillips 2013). The success story of wheat crop was followed by the development of high-yielding disease resistant cultivars of rice, IR8. It is possible to introduce new sexually compatible desirable genes/ traits into daughter progenies after sexual crossing by means of conventional breeding. However, the resultant progenies with desired traits could also inherit the multitude of undesirable traits from their parents. Thereby, such crosses could affect and reduce the final crop yield, also known as "yield drag." Use of chemical mutagens also could not introduce selective traits into the breeding lines. It was soon realized that altering genetic material by the conventional mating or natural recombination is almost impossible. Therefore, it is highly unlikely to achieve dramatic gain in crop yields as noted in green revolution by pursuing conventional breeding practices.

Improvements in the genetics and the production of stable heritable transgenic lines of major crops as per regional/ local needs and environment may reverse the existing trends of poor productivity and thereby achieve the goal of yield gains. Advancements in modern breeding programs based on molecular and genomics driven technologies, such as marker assisted breeding and genotyping-by-sequencing, could not only dramatically increase the crop yield but also reduce the breeding timelines. Modern breeding programs employ the application of novel genetic modification strategies and offer the use of genes from all sources regardless of their origin in order to improve crop yield, nutrition potential/ quality, and lower yield losses from disease and environmental factors such as changing climate and soil depletion (Oliver 2014).

The cross between wheat and rye led to the development of stable hybrid for the first time in laboratory in 1884 but unfortunately it could not survive in nature. This was followed by the experiments on chromosome doubling (where daughter progeny receives the copy of chromosome from each of its parents) to successfully produce the stable and fertile polyploid plant (Ammar et al. 2004). In 1930s, polyploid plant cells were generated by treating the embryos with colchicine (a chemical that interferes with the normal process of cell division or mitosis) under in vitro culture conditions. Polyploid daughter cells thus produced were although had improved properties compared to their parents but they were not designated as GMOs/GM crops.

Genetically modified organisms (GMO) and genetically modified (GM) crops are man-made and have long been fully accepted in the medical arena. Over many decades, advancements in biotechnological tools, use of restriction enzymes as molecular scissors, and tremendous growth in recombinant DNA technology led to the production of GMOs/GM crops. Modern biotechnology tools allowed the successful delivery of genes obtained from distant sources into the target/ host cells via transformation followed by their regeneration in tissue culture in order to produce genetically modified cells/organisms that possess novel combination of genetic material. The first GM crop, known as FLAVR SAVR tomato, modified to delay premature fruit softening (by manipulating the biosynthesis of the ripening-promoting hormone ethylene and increasing levels of the antiripening polyamines) entered into the commercial market in USA in 1994. GM tobacco, modified to withstand the harmful effects of herbicide, bromoxynil, became the first commercialized crop approved by European Union in 1994. FDA and Environmental Protection Agency (EPA) approved the first pesticide resistant Bt potato in USA in 1995. Insect-resistant Bt maize and Bt cotton, herbicide tolerant soybeans, and oilseed rape were produced later in 1996. GMOs/GM crops are being widely developed to effectively treat variety of ailments and diseases. Their use as biopharmaceuticals (products obtained from genetically modified bacteria/GMO), for example, production of human insulin for treating diabetes is well established (Aggarwal 2012). Another prime example is the golden rice whose production has been envisioned to deliver a cheap and effective dietary source of vitamin A in the form of its precursor, β-carotene (an essential component of rhodopsin which is the fundamental light absorbing pigment in the human eye), to cure blindness and a compromised immune system (Gosse and Manochia 1996). Tremendous applications/ benefits of GM crops necessitate the efforts towards unmatched partnership between public and private sectors to fund for their sustainable production.

Despite an overwhelming consensus received from the scientists and experts worldwide on the safety of foods produced from genetically modified crops, a group of anti-GM crops activists have been successfully sowing doubt about GM crops in the general population. Despite the negative propaganda about the GM crops by small group of activists, the benefits associated with GM crops, their safe and wide applications among the mankind are well understood.

4.2 Transgenic Methods to Produce Genetically Modified Crops

Large number of agronomically and horticulturally important monocot and dicot plant species are being transformed to raise transgenics. Genes/ pathways derived from bacteria and other organisms, plant species belonging to different taxonomic groups and even the synthetic genes are being exploited for their transfer and expression to engineer plant metabolism for their improved yield and quality (Fig. 4.1).

4.2.1 *Agrobacterium Tumefaciens* Mediated Transformation

Agrobacterium tumefaciens, gram negative soil bacterium, and a causative agent of crown gall disease, is capable of transferring T-DNA of its large Ti (tumor-inducing) plasmid into the nucleus of infected target cells. Target gene is placed between the left and right border repeats of T-DNA after replacing the originally located genes involved in auxin, cytokinin, and opine synthesis. Ti plasmid virulence vir region, three chromosomal virulence loci, chvA (*A. tumefaciens* chromosomal virulence A), chvB (*A. tumefaciens* chromosomal virulence B), and pscA/exoC locus, and

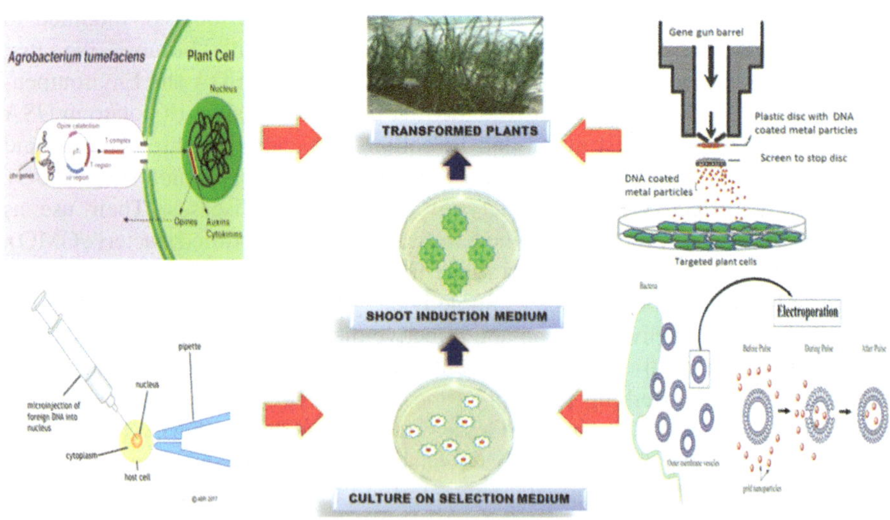

Fig. 4.1 Biological and physical methods of gene transfer to raise transgenic plants

25-bp T-DNA border sequences regulate the transfer and stable integration of transgene into plant genome. Certain plant proteins including BTI1 (virB2-interacting protein), VIP1 (vegetative insecticidal proteins), Ku80 (ATP-dependent DNA helicase 2 subunit KU80), CAK2Ms (plant ortholog of cyclin-dependent kinase-activating kinases), histones-H2A, H3–11, H4, SGA1 (silencing group A), UDP (uridine diphosphate) glucosyltransferase, and GALLS interacting proteins actively participate in T-DNA and virulence protein transfer, cytoplasm trafficking, nuclear targeting, T-DNA integration, stability and expression, and defense responses and thereby significantly regulate *Agrobacterium*-mediated transformation. Following the removal of *Agrobacterium*, the transformed cells are grown into plants under selection by standard tissue culture. Factors like genotype of the plant, explant, vector/ plasmid, bacterial strain, composition of culture medium (salt concentration, sugars, growth regulators), temperature, time and pH of co-cultivation medium, antibiotics (cefotaxime, carbenicillin, kanamycin, timentin), chemicals (acetosyringe, L-cysteine, dithiothreitol, and sodium thiosulfate), surfactants (silwet L77, pluronic acid F68, Tween20), selectable markers (neomycin phosphotransferase, phosphinothricin acetyltransferase, and hygromycin phosphotransferase), tissue damage, suppression, and elimination of *Agrobacterium* infection after co-cultivation influence *Agrobacterium*-mediated transformation (Mehrotra and Goyal 2012). Organism's host specificity, limited host range, and adaptability of the culture/ regeneration systems of certain plant species to the mechanisms of gene transfer are some of the potential limitations associated with *Agrobacterium*-mediated transformation.

Biolistic transformation or gene bombardment is based on coating of DNA to be transferred on the microscopic beads (made of inert metal particles of gold/tungsten), its attachment to the plastic gun and loading in gene gun. Helium gas at high pressure provides the force for the gun and upon gun firing, ruptures disk breaks, drives the plastic bullet down a shaft, accelerates the microprojectiles, and allows the DNA/metal particles to emerge from the gun with great speed and force. This results in the penetration of DNA coated beads to the cytoplasm of target cells or tissues of diverse types in vacuum. Two principal helium-driven gene gun devices, the Accell gene gun by Agracetus, Inc. and the Helios gene gun by Bio-Rad Laboratories are extensively used. The major concern of stable transformation by biolistic methods is random, multi-copy transgene integration in inverted repeat arrangement, which can lead to transgene silencing. For efficient production of single copy events, the strategy of using minimal transgene cassettes (2.50–2.73 ng of DNA cassette per shot) instead of bombarding with whole plasmid DNA is recommended and has been applied in the transformation of many important monocotyledonous species including wheat, maize, pearl millet (*Pennisetum glaucum*), and sugarcane (*Saccharum officinarum*) (Ismagul et al. 2018). The competence of gene transfer by bombardment method depends on the type of gene gun; bombardment force; circulation of carrier particles at target site; type of targeted cells/ tissues; amount of DNA; ratio, dimension, and material of carrier molecules; and density of carrier particles and macromolecules.

4.2.2 Microinjection

Microinjection is a physical method of transformation for direct gene transfer in a host independent manner. It is based on the injection of DNA carried into a very fine metal microinjection needles or glass micropipettes with 0.5–10 μm diameter, under the microscopical control, into the defined cells without damaging them. Immobilization of the protoplasts/ target cells in a thin layer of alginate eliminates the need of a holding capillary and greatly facilitates the process of microinjection. Southern analysis of genomic DNA isolated from regenerated plants confirms the stable integration of full length microinjected genes into high molecular weight DNA. Microinjection is possible in 50–100 cells in 1 h with the transformation efficiency of up to 20%. Stable integration of foreign gene and its transmission to the next generation in a Mendelian fashion were identified in crossing experiments (Schnorf et al. 1991). Microinjection has been not only a slow process and expensive method but also requires trained/ skilled personnel. Nevertheless, reliability, reproducibility, and affordability make it an attractive technology for transferring genetic material into living cell.

4.2.3 Electroporation

Electroporation, an electrical transformation method, is based on the formation of transient pores in the plasma membranes of cells upon applying short high voltage pulses. The field strength and pulse duration may be chosen within the range of 100–5000 V/cm and 0.01–100 ms, respectively. These microscopic aqueous pores/ electropores in cell membrane facilitate the uptake of plasmid DNA for its stable integration into the target genome. Overheating due to rigorous electrical conditions is the primary cause of cell deaths. Plasmid DNA concentration, up to about 100 μg/ mL; linearized plasmid DNA; heat shock; and addition of polyethylene glycol (PEG) stimulate stable transformation. Besides, protoplast size, pulse type, culture medium, and temperature are also known to affect transformation efficiency. Stable transformation frequency is reported to lie within the range of 0.0001–0.1% of the electroporated protoplasts. Exogenous gene transferred in the target protoplasts during electroporation appears to be inherited as a single dominant character in a Mendelian fashion (Ozyigit 2020).

4.2.4 Chemical Methods

Chemical methods for genetic transformation are based on the use of calcium phosphate, polyethylene glycol, DEAE-Dextran, and liposomes. Calcium phosphate-mediated transfer is based on the formation of calcium phosphate precipitate by mixing the desired quantity of DNA with calcium chloride and potassium phosphate solutions. This is followed by the incubation of cells/ tissues with precipitated DNA which can later be taken inside via endocytosis. PEG- mediated transfer is mainly used to transform the protoplasts. Soaking of protoplasts in PEG-containing solution facilitates the endocytosis and uptake of DNA. DEAE (diethyl aminoethyl)-Dextran mediated transfer employs the use of commercially available, low cost, and simple DEAE-Dextran as a transfection medium, for the transformation of target cells. Liposome-mediated transfer is based on the application of artificial lipid vesicles known as liposomes. Liposomes are made of positively charged cationic lipids which can readily interact with negatively charged cell membranes. They surround the delivery molecule and enable its transfer via fusion with cell membrane (Ozyigit 2020).

4.3 Benefits of GM Crops

4.3.1 Herbicide and Insect-Tolerant Transgenic Plants

Unwanted and useless plants grown as weeds along with main crops not only harbor insects but also compete for light and nutrients and thereby reduce the global crop yield by 10–15%. Inevitable use of herbicides to kill the weeds is known to have environmental and ecological consequences even for main crops derived by conventional breeding programs. Herbicide tolerant GM crops are engineered or stacked to express multiple traits to tolerate the specific broad-spectrum herbicides that are sprayed to kill the surrounding weeds. Successful release of glyphosate-resistant soybean and corn in 1996 and 1998, respectively, and their widespread adoptability flooded the American agriculture with 93% of soybeans, 82% of cotton, and 85% of corn glyphosate tolerant plants. Over 63% of GM crops growing globally have herbicide tolerant traits. The most common are glyphosate and glufosinate tolerant varieties and occupy the largest planting area of biotech crops from 1996 to 2018.

Overexpression of EPSPS gene [5-enolpyruvylshikimate 3-phosphate synthase, an enzyme catalyzes shikimate pathway for the synthesis of aromatic amino acids and a competitive inhibitor of glyphosate] or expression of insensitive form of EPSPS gene in the GM plants make them resistant to the killing effects of glyphosate (Funke et al. 2006). Genetically modified crops resistant to glufosinate/ phosphinothricin (glutamine synthetase inhibitor and disrupts nitrogen metabolism) is created by overexpressing the bar or pat genes (phosphinothricin acetyltransferase that detoxify phosphinothricin) isolated from *Streptomyces* (Wehrmann et al. 1996). The first glufosinate-resistant crop, canola, was brought to market in 1995, followed by corn in 1997, cotton in 2004, and soybeans in 2011. Alfalfa, argentine canola, carnation, chicory, cotton, creeping bentgrass, flax, linseed, maize, potato, polish canola, rice, soybean, sugar beet, tobacco, and wheat are the major approved herbicide tolerant with single and or stacked genes crops for food, feed, and/or cultivation globally.

Owing to the beneficial impact on the environment, herbicide tolerant crops are gaining widespread adaptability (Dillen et al. 2013). Use of GM herbicide tolerant crops has accelerated the adoption of conservative tillage (no-till and reduced-till) practices, reduction in the land use for agriculture, enhancement of soil quality, reduction in water runoff, conservation of nutrients, increase in water infiltration, reduction in greenhouse gases, and increase in the use of less toxic, more environmentally friendly chemicals. Crop rotation, herbicide tolerance gene-stacking technologies and field management technologies as a part of a broader integrated weed management program take care of the problems associated with the evolution of herbicide tolerance in weeds (Oliver 2014).

Fourteen percent of crop productivity of high-yielding genotypes of agronomically important crops is estimated to be lost globally due to pest insects. Application of chemical pesticides although has significantly improved the pest management practices but they are reported to impose adverse effects on human health, other biological organisms and environment. Therefore, it is imperative to drastically discourage their application in agriculture and environment for practicing safe and sustainable farming.

Insect-tolerant GM crops are rapidly growing in acreage globally and as effective alternatives, they are well integrated in agriculture ecosystems. The primary transgene used to produce insect-resistant crops is isolated from the bacteria *Bacillus thuringiensis* (Bt) and codes for CRY protein toxin/ insecticidal crystal protein. These toxins when engulfed affect the gut cells of specific agronomic caterpillar and beetle pests that feed on the Bt-GM crop plants. Thus, kill the insects by preventing digestion but are innocuous to vertebrates, including humans, and have no impact on the Bt-GM plants. Novel family of insecticidal proteins called vegetative insecticidal proteins (Vip) produced by B. *thuringiensis* during its vegetative stage. Vip toxins do not form crystals and following its stable integration into the plant genome, it confers resistance against coleopteran species and wide range of lepidopteran insects. These proteins are biodegradable and do not persist in environment (Abbas 2018). Besides being stably integrated into the genome of GM crops, these toxin proteins are also sprayed over the crop in the form of formulations and or bacterial preparations as a microbial pest control agent. Cotton, cowpea, eggplant, maize, poplar, potato, rice, soybean, sugarcane, and tomato are being developed and produced as Bt-GM crops. Genes code for protein toxins with different mode of mechanisms and are isolated from various sources to produce insect-resistant transgenic plants (Table 4.1).

Improved pest management program, reduction in insecticide use, greater sustainable crop productivity, safe environment for the growth of non-target organisms, and less mycotoxin production in food obtained from GM crops are the potential benefits associated with insect-tolerant GM crops. Transfer of transgene (gene flow) from herbicide/ insect-tolerant GM crops to wild relatives or non-GMO crops close by and environmental/ecological concerns that relate to biodiversity are the issues that need to be addressed.

4.3.2 Abiotic Stress Tolerant GM Crops

Food security, growth, and yield of plants are greatly influenced by global climate changes and environmental stresses. Around less than 10% of the soils used in

Table 4.1 Insecticidal proteins isolated from various sources and their mode of mechanisms to produce insect-tolerant GM crops

Source	Insecticidal protein toxin	Insecticidal activity/mode of action	Transgenic plants
Bacillus thuringiensis	Insecticidal crystal proteins	Bt protein gets activated in the gut's alkaline condition and punctures the midgut leaving the insect unable to feed	Cotton, cowpea, eggplant, maize, poplar, potato, Rice, soybean, sugarcane, tomato
Bacillus thuringiensis	Vegetative insecticidal proteins	Active form of Vip toxins binds to the midgut epithelial membrane, forms the pore and prevents digestion in insects	Cotton, maize, tobacco, poplar, Rice
Plants and animals	Protease inhibitors	Inhibit the proteolytic activity of enzymes specific to insects and phytopathogens	Alfalfa, Arabidopsis, oilseed rape, potato, Rice, sugarcane, tobacco, wheat, tomato
Plants	α-Amylase inhibitors	Inhibit the activity of amylase of specific plant pests	Azuki bean, common bean, pea, Chick pea
Plants	Lectins	Reduced larval weight, mortality, feeding inhibition, delays in total developmental duration, adult emergence and fecundity on the first and second generation	Tomato, Rice, sugarcane, tobacco, maize, mustard, Arabidopsis, rapeseed, wheat
Streptomyces species	Cholesterol oxidase	Catalyzes the oxidation of 3-hydroxysteroids to ketosteroids and hydrogen peroxide. Disrupts the midgut epithelial membrane and marked stunting of growth of several lepidopteran pests	Tobacco
Plants and insects	Chitinases	Digests chitin in the peritrophic matrix or exoskeleton of the pests	Maize, tobacco
Plants	Tryptophan decarboxylase	Catalyzes the decarboxylation of tryptophan to tryptamine; accumulated levels of tryptamine adversely affect feeding behavior and physiology of Malacosoma disstria and Manduca sexta and results in poor larvae growth, apparently due to a postingestive mechanism	Tobacco, poplar, canola, petunia
Agrobacterium	Isopentenyl transferase	Involved in cytokinin biosynthesis. Products of secondary metabolic pathways exhibit antifeedant effects, and reduced larval weight of specific pests including *M. sexta* and *M. persicae* nymphs	Potato, cucumber, tobacco, tomato
Plants	Peroxidase (POD) and polyphenol oxidase (PPO)	Increase the inhibitory effect of 5CQA (5-Caffeoylquinic acid) and cholorogenic acid by oxidizing the dihydroxy groups to ubiquinones that covalently bind to nucleophilic (-SH2 and NH2) groups of proteins, peptides, and amino acids. Contribute to lignification, consume oxygen, produce quinones and thus reduce plant digestibility for the insects	Tobacco, tomato, potato, Faba bean, Rice, turf grass, alfalfa

agriculture are free from environmental stresses and therefore it necessitates the development and production of genetically modified crops with improved tolerance to abiotic stresses. Among the abiotic stresses, salinity and drought remain the greatest constraint to crop production and is responsible for 70% reduction in crop yield.

Variety of transgenics are developed to express transgenes involved in signaling cascades and in transcription control for the improved yield under stress conditions. These transgenes include the expression of DREB (dehydration responsive element binding protein) genes involved in drought, salinity, and freezing to increase drought tolerance in rice, maize, and wheat; SNAC1, a transcription factor [derived from the names of proteins with DNA binding domain, namely NAM (No Apical Meristem), ATAF1 (Arabidopsis thaliana activating factor1), and CUC2 (Cup-Shaped Cotyledon)] for enhanced tolerance to severe drought and salt stress during reproductive and vegetative growth without any negative phenotypic effects or yield loss in rice; and OsWRKY11 (member of WEKY group of transcription factor) under control of heat shock protein promoter (HSP101) to prolong the survival and retain water under severe drought in rice (Bakshi and Dewan 2013). Transgenic rice with the overexpression of HSP101 and HVA1 (*Hordeum vulgaris* abundant protein) genes exhibits significant improvement in plant growth recovery after heat stress and dehydration tolerance, respectively. Overexpression of abscisic acid-responsive element binding protein (AREB) /ABF (ABRE binding factor) regulon in the transgenic plants makes them insensitive to salt, drought, heat, and oxidative stress (Singh and Laxmi 2015).

4.3.3 GM Crops Expressing Therapeutic Molecules

Constant threat of disease-causing microorganisms prompts the pharmaceutical and biotechnological industries to exploit the plants as biofactories for the rapid development and consistent production of plant-derived biopharmaceutical agents. High yields of proteins/ therapeutic molecules; lower production, processing, storage and transportation cost; increased shelf life; and elimination of pathogen contamination are the benefits associated with plant-derived molecules of therapeutic significance.

Stable nuclear and plastid expression of recombinant pharmaceutical proteins in plants dates back to 1989 which mark the development of transgenic-tobacco-derived functional immunoglobulins. Since then, series of experiments were conducted to produce edible vaccines in transgenic rice, carrot, soybean, tomato, potato, tobacco, sugar beet, lettuce, cotton, cauliflower, wheat, *Brassica* and *Arabidopsis*. Stable/ transient nuclear and chloroplast expression of varieties of pharmaceutical molecules (antigens, vaccine, antibodies, drugs) extends the production potential of the transgenic plants. Development and production of biopharmaceutical agents include human lysozyme and Cholera toxin-B (CTB) in rice; hemagglutinin surface protein, heat labile enterotoxin B, pE2 polypeptide (envelop glycoprotein), HPV 16-L1 (human papilloma virus type 16-L1) capsomere, HCV (hepatitis c virus) core protein, CMV VLPs (cytomegalovirus virus-like particles),

HPV-16 L2 (human papilloma virus type 16-L2), influenza virus M2E (matrix protein 2), GFP/HFBI (green fluorescent protein/ hydrophobin fusions for high-level transient protein), HIV-1 pr55GAG (human immunodeficiency virus type 1 polyprotein), HIV monoclonal antibody 2G12, COPV L1-protein (canine oral papilloma virus major capsid protein), GFP HA (hemagglutinin) peptide, GFP, GUS (β-glucuronidase), HBc (VLPs) (hepatitis B core antigen virus-like particles), active dust mite allergens, Hepatitis B core antigen, SAG1 protein (*Toxoplasma gondii* surface antigen), Ce16a/aadA (aminoglycoside resistance protein), βC1/GFP (betasatellite of tomato yellow leaf curl China virus/GFP), pyMSP1$_{19}$ (*Plasmodium yoelii* merozoite surface protein 1), LTB (lymphotoxin β), immune-dominant antigens, TB (tuberculosis) vaccine protein, viral coat B5 candidate pB5 [Variola virus (human)], RV VLPs (rotavirus-like particles), CTB-ESTA6 (cholera toxin B-subunit 6 kDa early secretory antigenic target), MV-H protein (measles virus strain Halle), HA1-protein, (Influenza hemagglutinin) HA1-5 (VLPs) (Influenza hemagglutinin 1-5 virus-like particles), HA7-7 (Influenza hemagglutinin 7-7), HA1–5/1 (Influenza hemagglutinin 1-5/1), Chimeric CMVs, pE2, HPV16-L2 epitope, HPV16-L1mAbs, and HPV11-L1-NLS (human papilloma virus type 11-L1 with nuclear localization signal) proteins in tobacco; FMDV 2A (Foot-and-mouth disease virus 2A) catalytic peptide, sDPT polypeptide (synthetic gene encoding the recombinant polypeptide against diphtheria, pertussis, and tetanus), and rabies nucleoprotein in tomato; heat labile enterotoxin B, Seed-specific LTB in soybean; heat labile enterotoxin B in carrot; GFP in sugar beet; anti-spectinomycin in *Brassica*; GUS in cotton; CTB-MSP1 AMA-1 (cholera toxin-B subunit merozoite surface protein-1 and apical membrane antigen-1), CTB-ESTA6, S-HBsAg (surface antigen of the hepatitis B virus), and dengue virus tetraepitope peptide (cE-DI/IIp) in lettuce; SARS-Cov (severe acute respiratory syndrome coronavirus) in cauliflower; and HPR (By2) (hepatoglobin-related protein), and HPV11-L1-NLS proteins in *Arabidopsis* (Fahad et al. 2015).

4.3.4 Biofortification of Crops with Nutrients by Transgenic Approaches

Agricultural system has been so far designed to focus on increased grain yield and crop productivity but not to promote human health. This approach has resulted in the production of nutrient-deficient food crops. Thus, conventional approaches of crop production do not support healthy lives of consumers but instead result in their impaired development, stunted mental and physical growth, sickness, poor health, diminished livelihoods, and reduced national socioeconomic development especially in developing countries. About 30% of world's population are anemic and suffer from hidden hunger. The prevalence of anemia is more acute in developing countries with Africa and South-East Asian countries are the most affected ones. Transgenic approaches have made it possible to shift the agriculture from producing the larger quantities of food crops to producing safe and nutrient-rich food crops in sufficient quantities (Chizuru et al. 2003).

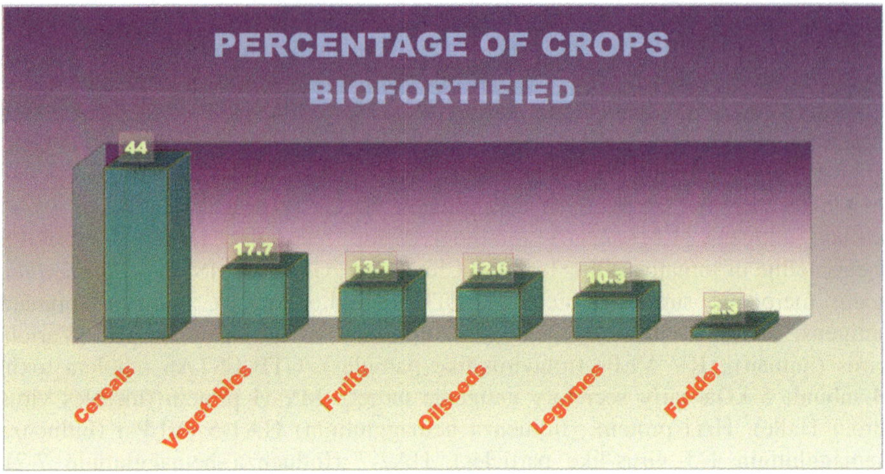

Fig. 4.2 Percentage of crops biofortified using transgenic approaches

Most of the crops including staple crops like rice, wheat, maize, sorghum, cassava, barley, soybean, lupine, common bean, potato, sweet potato, tomato, cauliflower, banana, lettuce, carrot, canola, and mustard are targeted by transgenic approaches (Garg et al. 2018) (Fig. 4.2).

Poor heritability, linkage drag, and limited availability of genetic diversity for the targeted component especially in oilseed crops allow the practice of cost-effective and sustainable approach of transgenics for the production of high-yielding biofortified crops. Around 40 known nutrients in adequate amounts including essential micro and macronutrients, amino acids, fatty acids, improved vitamin content, and higher levels of antioxidants are the important nutritional targets for the biofortification of crops. The crops with improved nutrient content not only provide enough calories to meet the energy needs but also all the vitamins and essential minerals/nutrients which are required for the healthy and productive lives of the consumers (Table 4.2).

Genetic modifications or reconstruction of selected pathways in the host plant species are targeted in order to increase the bioavailability of nutrients with enhanced concentration and redistribute them to the edible tissues/ parts of commercial crops. Reports on successful experimentations for biofortification of cereals, legumes, vegetables, oilseeds, fruits, and fodder crops are recently published (Table 4.3).

(continued)(continued)(continued)(continued)

4.3.4.1 Cereals and Oil Rich Crops

Expression of genes that code for PSY (phytoene synthase) and carotene desaturase result in the enhanced synthesis of phytoene (precursor of β-carotene) and β-carotene in golden rice is considered as an important breakthrough to tackle the problem of provitamin A deficiency in the consumers (Burkhardt et al. 1997). Overexpression of genes encoding Arabidopsis GTP-cyclohydrolase I (GTPCHI) and

Table 4.2 Nutritional targets for biofortification of crops

NUTRITIONAL TARGETS	
MICRONUTRIENTS	Minerals: Zinc, Iron, Copper, Manganese, Iodine, Nickel, Cobalt, Molybdenum, selenium
	Vitamins: A: Retinol; B_1: Thiamin; B_2: Riboflavin; B_3: Niacin; B_5: Pantothenic acid; B_6: Pyridoxine; B_7: Biotin; B_9: Folic acid; B_{12}: Cobalamin
MACRONUTRIENTS	Minerals: Calcium; Magnesium; Potassium; Sulphur; Sodium; Chlorine, Phosphorus
	Amino acids: Leucine; Isoleucine; Lysine; Histidine; Methionine; Phenylalanine; Threonine; Tryptophan
	Fatty acids: Linoleic acid; Linolenic acid
ANTIOXIDANTS	Flavonoids

aminodeoxychorismate synthase (ADCS) in rice results in the 150 fold increase in folate levels (Storozhenko et al. 2007; Blancquaert et al. 2015). One hundred gram of such rice is sufficient to meet the daily requirements of folate in an adult individual. The genes code for nicotianamine aminotransferase, OsIRT1 (*Oryza sativa* iron transport protein 1), OsNAS1 (*Oryza sativa* nicotianamine synthase 1), OsNAS2 (*Oryza sativa* nicotianamine synthase 2), soybean ferritin, and common bean ferritin are overexpressed to produce genetically modified rice for higher levels of iron to address the challenge of iron deficiency anemia (Goto et al. 1999; Lee et al. 2012; Trijatmiko et al. 2016). Overexpressing OsIRT1 gene and mugineic acid synthesis genes from barley HvNAS1 (*Hordeum vulgare* nicotianamine synthase 1), HvNAAT-A (*Hordeum vulgare* nicotianamine aminotransferase A), HvNAAT-B (*Hordeum vulgare* nicotianamine aminotransferase B), IDS3 (iron-deficiency-specific clone 3) in GM rice either allow an increase in levels of zinc or make the plants to absorb more Zn from the soil (Masuda et al. 2008; Lee and An 2009). Quality of protein in rice can be improved by targeting essential amino acids by allowing the expression of seed-specific genes of bean β-phaseolin, pea legumin; soybean glycinin, Sesame 2S Albumin, bacterial aspartate kinase, dihydrodipicolinate synthase (DHPS), maize DHPS, rice anthranilate synthase α-subunit, and E. coli

Table 4.3 Genetic modification of crops for their biofortification with various nutrients

Nutrients	GM Crops	Genes modified and transferred into host plants	References
Vitamins			
β-Carotene and phytoene	Rice	β-Carotene desaturase and phytoene synthase (PSY)	Burkhardt et al. (1997)
	Wheat	Bacterial PSY and carotene desaturase genes	Cong et al. (2009)
	Maize	Bacterial crtB (phytoene synthase) gene	Aluru et al. (2008)
	Soybean	PSY and carotene desaturase	Kim et al. (2012a, b)
	Potato	PSY, phytoene desaturase, and lycopene β-cyclase RNAi to silence the beta-carotene hydroxylase gene (bch)	Diretto et al. (2006); Van Eck et al. (2007)
	Tomato	1-deoxy-d-xylulose-5-phosphate synthase (dxs) and lycopene beta-cyclase gene (beta-Lcy)	Wurbs et al. (2007)
β-Carotene	Soybean	PSY and carotene desaturase	Kim et al. (2012a, b)
	Cauliflower	Insertion of a copia-like LTR (long terminal repeats) retrotransponson	Lu et al. (2006)
	Banana	PSY gene (PSY2a) of Asupina banana	Waltz (2014)
Carotene, lutein, and total carotenoids	Sweet potato	Sweet potato orange (IbOr-ins) gene	Kim et al. (2013)
Folate	Rice	GTP-cyclohydrolase I (GTPCHI) and aminodeoxychorismate synthase (ADCS)	Storozhenko et al. (2007); Blancquaert et al. (2015)

(continued)

Table 4.3 (continued)

Ascorbic acid	Maize	Dehydroascorbate reductase (DHAR)	Chen et al. (2003)
	Potato	Galactouronate reductase (GalUR)	Upadhyaya et al. (2009)
	Tomato	GDP-mannose-3' 5'-epimerase [SlGME1, SlGME2], DHAR, and co-expression of three genes GDP-mannose pyrophosphorylase, arabinonol,4-lactone oxidase, and myo-inositol oxygenase 2	Zhang et al. (2011); Haroldsen et al. (2011); Cronje et al. (2012)
Tocotrienol and tocopherol	Maize	Homogentisic acid geranylgeranyl transferase (HGGT)	Cahoon et al. (2003)
Increased content of δ-tocopherol and decreased content of γ-tocopherol	Soybean	2-methyl-6-phytyl benzoquinol methyltransferase genes (at-VTE3; at-VTE4)	Van Eenennaam et al. (2003)
Minerals			
Iron	Rice	Nicotianamine aminotransferase, OsIRT1 (*Oryza sativa* iron transport protein 1), OsNAS1 (*Oryza sativa* nicotianamine synthase 1), OsNAS2 (*Oryza sativa* nicotianamine synthase 2), soybean ferritin and common bean ferritin	Goto et al. (1999); Takahashi et al. (2001); Lee et al. (2012); Trijatmiko et al. (2016)
	Wheat	Ferritin gene from soybean and wheat (TaFer1-A)	Xiaoyan et al. (2012); Borg et al. (2012)
	Maize	Soybean ferritin and aspergillus phytase, *aspergillus Niger* phyA2, soybean ferritin, and silencing the expression of ATP-binding cassette transporter and multidrug resistance associated protein	Drakakaki et al. (2005); Shi et al. (2007); Chen et al. (2008); Aluru et al. (2011)
	Lettuce	Soybean ferritin gene	Goto et al. (2000)
Zinc	Rice	OsIRT1 gene and mugineic acid synthesis genes from barley HvNAS1	Masuda et al. (2008), Lee and

(continued)

Table 4.3 (continued)

		(*Hordeum vulgare* nicotianamine synthase I), HvNAAT-A (*Hordeum vulgare* nicotianamine aminotransferase A), HvNAAT-B (*Hordeum vulgare* nicotianamine aminotransferase B), IDS3 (iron-deficiency-specific clone 3)	An (2009)
	Barley	Phytase gene (HvPAPhy_a)	Ramesh et al. (2004)
Calcium	Carrot	Arabidopsis H+/Ca2+ transporter (CAX1)	Park et al. (2004)
Proteins and amino acids			
Proteins and amino acids (methionine, cysteine, lysine, tryptophan, aspartate)	Rice	Bean β-phaseolin, pea legumin; soybean glycinin, sesame 2S albumin, bacterial aspartate kinase, dihydrodipicolinate synthase (DHPS), maize DHPS, rice anthranilate synthase α-subunit, and E. coli aspartate aminotransferase	Zheng et al. (1995); Sindhu et al. (1997); Katsube et al. (1999); lee et al. (2003); Wakasa et al. (2006); Zhou et al. (2009); Yang et al. (2016)
Lysine, methionine, cysteine, and tyrosine)	Wheat	Amaranthus albumin gene (ama1)	Tamas et al. (2009)
Lysine and tryptophan	Maize	Antisense dsRNA targeting alpha-zeins	Huang et al. (2006)
Lysine	Barley	DHPS gene (dapA)	Ohnoutkova et al. (2012)
	Sorghum	High lysine protein (HT12)	(Zhao et al. 2003
Cysteine and methionine	Soybean	O-acetylserine sulfhydrylase, maize zein protein and cystathionine γ-synthase	Dinkins et al. 2001; Kim et al. 2012a, b; Song et al. 2013; Hanafy et al. 2013)
Methionine	Common	Methionine-rich storage albumin	Aragao et al.

(continued)

Table 4.3 (continued)

	bean	from Brazil nut	(1999)
	Lupines	Sunflower seed albumin gene	Molvig et al. (1997)
	Potato	Cystathionine γ-synthase (CgSΔ90) and methionine-rich storage protein and seed storage protein from Perilla (PrLeg polypeptide	(Dancs et al. 2008; Goo et al. 2013)
Total protein	Potato	Amaranth albumin (ama1)	Chakraborty et al. (2010)
Carbohydrates			
Dietary fiber cyclodextrins from starch	Potato	Cyclodextrin glycosyltransferases (CGT) gene	Oakes et al. (1991)
Fatty acids			
α-Linolenic acid	Rice	Soybean omega-3 fatty acid desaturase (FAD3) gene (GmFAD3)	Anai et al. (2003)
Polyunsaturated fatty acids, γ-linolenic acid, and stearidonic acid (STA)	Barley	Δ6 -desaturase (D6D) gene	Mihalik et al. (2014)
Lowering the levels of α-linolenic acids	Soybean	siRNA-mediated gene silencing of ω-3 FAD3	Flores et al. (2008); Sato et al. (2004)
Increasing the levels of γ-linolenic acid (GLA) and STA (ω-3 fatty acids)		Δ6 -desaturase gene	
Unsaturated fatty acids	Mustard	Enzyme Δ6 FAD3 that led to the production of gamma linoleic acid	Hong et al. (2002)
Antioxidants			
Flavonoids	Rice	Maize C1 and R-S regulatory genes [Myb-type and basic helix-loop-helix-type transcription factors; and phenylalanine ammonia lyase and chalcone synthase (CHS) genes	Shin et al. (2006); Ogo et al. (2013)
	Tomato	RNAi technology to suppress photomorphogenesis regulatory gene known as light mediated	Davuluri et al. (2005)

(continued)

Table 4.3 (continued)

		development protein (DET1)	
Anthocyanin	Wheat	Maize regulatory genes (C1, B-Peru)	Doshi et al. (2006)
	Potato	Chalcone synthase (CHS), chalcone isomerase (CHI) and dihydroflavonol reductase	Lukaszewicz et al. (2004)
	Sweet potato	IbMYB1 (*Ipomoea batatas* R2R3 MYB related transcription factor)	Park et al. (2015)
	Tomato	CHI gene	Muir et al. (2001)
Isoflavone	Soybean	Maize C1 and R transcription factor-driven gene activation	Yu et al. (2003)
Sterol	Tomato	Hydroxymethylglutaryl CoA (hmgr-1)	Enfissi et al. (2005)
Chlorogenic acid	Tomato	Gene silencing of hydroxycinnamoyl-CoA quinate transferase (HQT)	Niggeweg et al. (2004)
Trans-resveratrol	Tomato	Stilbene synthase	Giovinazzo et al. (2005)
Resveratrol	Apple	Stilbene synthase gene from the grapevine	Szankowski et al. (2003)
Genistin	Tomato	Isoflavone synthase (IFS)	Shih et al. (2008)

aspartate aminotransferase (Zheng et al. 1995; Sindhu et al. 1997; Katsube et al. 1999; Lee et al. 2003; Wakasa et al. 2006; Zhou et al. 2009; Yang et al. 2016). An essential fatty acid α-linolenic acid is made to express in rice by transferring the gene that codes for soybean omega-3 fatty acid desaturase (FAD3) gene (GmFAD3) (Anai et al. 2003). Levels of flavonoids with antioxidant activity can be increased by overexpressing maize C1 and R-S regulatory genes [Myb-type and basic helix-loop-helix-type transcription factors; and phenylalanine ammonia lyase and chalcone synthase (CHS) genes (Shin et al. 2006; Ogo et al. 2013). Rice grains with the expression of functional human milk protein (lactoferrin) have resulted in the development of value-added cereal-based ingredients that can be added into baby food (Nandi et al. 2002).

One of the most widely grown staple food crop, wheat has been modified for the overexpression of bacterial PSY and carotene desaturase genes for enhanced provitamin A content; ferritin gene from soybean and wheat (TaFer1-A) for higher levels of iron; Amaranthus albumin gene (ama1) to increase the levels of proteins and amino acids; and maize regulatory genes (C1, B-peru) involved in anthocyanin production for improving the antioxidant activity (Doshi et al. 2006; Cong et al. 2009; Tamas et al. 2009; Xiaoyan et al. 2012; Borg et al. 2012).

Transgenic maize is produced to enrich it with provitamin A by expressing bacterial phytoene synthase (crtB) gene; to increase the levels of tocotrienol and tocopherol by expressing homogentisic acid geranylgeranyl transferase (HGGT); to enhance the levels of Vitamin C (l-ascorbic acid) a water-soluble antioxidant to nearly 100 times by recycling oxidized ascorbic acid to reduced form by overexpressing dehydroascorbate reductase (DHAR); to enhance the bioavailability of iron by expressing soybean ferritin and Aspergillus phytase, *Aspergillus niger* phyA2, soybean ferritin, and silencing the expression of ATP-binding cassette transporter and multidrug resistance associated protein (Cahoon et al. 2003; Chen et al. 2003; Drakakaki et al. 2005; Shi et al. 2007; Aluru et al. 2008; Chen et al. 2008; Aluru et al. 2011). Transgenic maize with higher content of lysine is marketed under the trade name of Mavrea™ Yield Gard Maize by Monsanto in Japan and Mexico; Mavera™ Maize (LY038) by Renessen LLC (Netherland) in Australia, Columbia, Canada, Japan, Mexico, New Zealand, Taiwan, USA. These maize varieties are produced by antisense dsRNA targeting alpha-zeins (maize seed storage proteins with poor nutritional quality) (Huang et al. 2006).

Phytase gene (HvPAPhy_a) has been overexpressed in barley seeds to increase the bioavailability of iron and zinc (Ramesh et al. 2004). Lysine content has been enhanced in barley by overexpressing DHPS gene (dapA) (Ohnoutkova et al. 2012). Overexpression of cellulose synthase-like gene (HvCslF) results in the higher levels of dietary fibers in GM barley which significantly reduce the risk of contracting cardiovascular disease and type II diabetes in humans (Burton et al. 2011). Levels of polyunsaturated fatty acids, γ-linolenic acid, and stearidonic acid (STA) can be enhanced by expressing Δ6-desaturase (D6D) gene in transgenic barley (Mihalik et al. 2014).

4.3.4.2 Legumes and Pulses

Transgenic soybean has been modified for enhanced levels of beta-carotene through overexpression of PSY and carotene desaturase; increased content of δ-tocopherol and decreased γ-tocopherol by co-expressing 2-methyl-6-phytyl benzoquinol methyltransferase genes (At-VTE3; At-VTE4); higher levels of sulfur-containing amino acids, cysteine and methionine by expressing sulfur assimilatory enzyme, O-acetylserine sulfhydrylase, maize zein protein and cystathionine γ-synthase; and increased levels of isoflavone which is associated with reduced risk of heart disease, reduced menopausal symptoms, and reduced risk of some hormone-related cancers by overexpressing the combination of maize C1 and R transcription factor-driven gene activation and suppression of a competing pathway (Dinkins et al. 2001; Van Eenennaam et al. 2003; Yu et al. 2003; Kim et al. 2012a, b; Song et al. 2013; Hanafy et al. 2013).

Of the 20% oil in soybean, 7–10% of oil is rich in unstable fatty acid α-linolenic acids which reduce its seed oil quality due to the formation of undesirable trans-fatty acid as a result of hydrogenation. To enhance its agronomic value, the levels of α-linolenic acids are reduced by siRNA-mediated gene silencing of ω-3 FAD3 (Flores et al. 2008). However, to increase the content of γ-linolenic acid (GLA) and stearidonic acid (STA) (ω-3 fatty acids), soybean has been genetically modified for the overexpression of Δ6 -desaturase gene (Sato et al. 2004). Mustard, an economically significant crop has been modified for enhanced levels of unsaturated fatty acids by overexpressing the gene that codes for enzyme Δ6 FAD3 that led to the production of gamma linoleic acid (Hong et al. 2002).

4.3.4.3 Vegetables and Fruits

Potatoes being the world's fourth most important source of calories have been extensively researched to enhance its nutritional content. Potato tubers are enriched with beta-carotene by incorporating the genes for PSY, phytoene desaturase, and lycopene β-cyclase or by using RNAi to silence the beta-carotene hydroxylase gene (bch) which converts beta-carotene to zeaxanthin (Diretto et al. 2006; Van Eck et al. 2007). Potatoes are targeted to enhance the content of vitamin C (ascorbic acid) by overexpressing strawberry GalUR (galacturonate reductase); methionine by co-expressing cystathionine γ-synthase (CgSΔ90) and methionine-rich storage protein and seed storage protein from Perilla (PrLeg polypeptide); total protein content by expressing Amaranth albumin (ama1); carbohydrate content by expressing cyclodextrin glycosyltransferases (CGT) gene to produce dietary fiber cyclodextrins from starch; anthocyanins by simultaneous expression of CHS, chalcone isomerase (CHI), and dihydroflavonol reductase (Oakes et al. 1991; Lukaszewicz et al. 2004; Dancs et al. 2008; Upadhyaya et al. 2009; Chakraborty et al. 2010; Goo et al. 2013). Sweet potato, an alternative source of bioenergy and natural antioxidants are genetically modified to increase the contents of carotene, lutein, and total carotenoids by overexpressing sweet potato orange IbOr-Ins gene in white fleshed sweet potato and anthocyanins by overexpression of IbMYB1 (*Ipomoea batatas* R2R3 MYB related transcription factor), a key regulator of its biosynthesis in storage roots (Kim et al. 2013; Park et al. 2015).

Cassava, another staple crop, although is resistant to different types of stresses but deficient in many nutrients. Transgenic varieties of cassava biofortified for increased levels of iron, beta-carotene, and zinc are under development and field trials. Levels of beta-carotene, vitamins, and minerals are high in carrots but it is deficient in calcium content. Transgenic carrots are being developed by over expressing the Arabidopsis H^+/Ca^{2+} transporter (CAX1) to enrich it with high calcium concentration (Park et al. 2004). Transgenic lettuce, the most popular leafy vegetable, is developed to enrich it with iron levels by expressing a soybean ferritin gene (Goto et al. 2000). Nutritional values of cauliflower which is rich in antioxidant phytonutrients are further enhanced with beta-carotene by the insertion of a copia-like LTR retrotransponson in the Or (mutant orange cauliflower) (Lu et al. 2006).

Tomato is the most popular worldwide and an important source of vitamin C, micronutrients, and other phytonutrients. Higher concentration of isoprenoid lycopene contributes to its red color. Efforts are being made to increase the levels of other isoprenoids like sterol by overexpressing 3-hydroxymethylglutaryl CoA (hmgr-1); phytoene and beta-carotene by expressing 1-deoxy-d-xylulose-5-phosphate synthase (dxs) and lycopene beta-cyclase gene (beta-Lcy); flavonoid contents by using RNAi technology to suppress photomorphogenesis regulatory gene known as light mediated development protein (DET1); ascorbic acid by overexpressing

GDP-mannose-3′,5′-epimerase [SlGME1, SlGME2], DHAR, and co-expression of three genes GDP-mannose pyrophosphorylase, arabinono1,4-lactone oxidase, and myo-inositol oxygenase 2; anthocyanin by expressing CHI gene; chlorogenic acid by gene silencing of Hydroxycinnamoyl-CoA quinate transferase (HQT); trans-resveratrol by expressing stilbene synthase; and genistin by the overexpression of isoflavone synthase (IFS) gene (Muir et al. 2001; Niggeweg et al. 2004; Giovinazzo et al. 2005; Enfissi et al. 2005; Davuluri et al. 2005; Wurbs et al. 2007; Shih et al. 2008; Zhang et al. 2011; Haroldsen et al. 2011; Cronje et al. 2012). Transgenic apple has been targeted with stilbene synthase gene from the grapevine for its enhanced production (Szankowski et al. 2003). Banana, an important food crop has been bioengineered with PSY gene (PSY2a) of Asupina banana to increase the expression of beta-carotene (Waltz 2014).

4.4 Public Acceptance of GM Crops

Farming of GM crops has been massively increased since the time witnessed their development on laboratory scale. GM crops have become the fastest adopted crop technology in the history of modern agriculture. In 2018, GM crops are being planted by approximately 17 million farmers over 191.7 million hectares in 26 countries. Currently, total of 10% of world's arable surface area is being utilized for GM crops plantation and this corresponds to an increase of 113 fold since 1996. According to the International Service for the Acquisition of Agri-biotech Applications, five industrial countries plant GM crops as a major share of 46% on 88.6 million hectares of land (https://www.isaaa.org/resources/publications/pocketk/16/). Soybean, maize, cotton, and canola are the majorly grown GM crops worldwide.

4.4.1 Global Acceptance of GM Crops Among Farmers

Import and or cultivation authorizations have been issued by more than 65 countries since 1994. Adoption rate of GM crops like maize, cotton, and soybean in the USA; soybeans in Brazil and Argentina; cotton in India and China; and oilseed rape in Canada skyrocketed to more than 90% within no time of their inception (James 2014). One of the interesting examples citing the significance of herbicide tolerant GM soybean comes from Romania. Owing to high adoption rate of 68% of GM soy varieties that were grown on 137,000 ha land, its production soared and the surplus soybeans were exported to other European countries in 2006. In 2007, when Romania joined the European Union, farmers were declined cultivation authorization of GM soy and they were forced to grow conventional seed varieties. This marked the heavy shrinkage in the GM soy planting area by 70% and strong reduction in profitability. As a result of which, Romania became dependent on expensive soybean imports like the rest of Europe (Otiman et al. 2008).

More than 95% of food-producing animals (100 billion animals during the last decade) consume GM feed in USA alone. No detrimental effects of the GM feed versus conventional feed on the health and performance of these animals were observed (Van Eenennaam and Young 2014). Transgenic, papaya ringspot-virus (PRSV) resistant papaya trees were introduced in Hawaii in 1998, but their cultivation gained momentum following the devastating outbreak of PRSV infections. Growing transgenic varieties greatly helped Hawaii papaya industry was saved from extinction (Gonsalves and Gonsalves 2014). Besides saving crop from extinction, increase in farmer's average profits by 68%; increase in crop yields by 22%; decline in the expense for pesticides by 39%; saving of time; ease of use and more flexibility in planning for crop plantation are some of the pronounced benefits associated with GM varieties compared to the conventional crops. Hence, despite higher seed cost for genetically modified varieties, farmers profit financially by planting them (Qaim 2009; Carpenter 2013; Fernandez-Cornejo et al. 2014; Brookes and Barfoot 2014).

4.4.2 Consumers' Attitudes Towards GM Crops

Despite superficial awareness and knowledge about GM foods, majority of consumers in USA expressed little or no concerns about food and agricultural biotechnology and were likely to buy GM food products. Permissive approval policies for GM food products are adopted by the United States authorities where the finished products are not required for GM labeling. Nevertheless, contrary to the situation in USA, stringent approval policies for GM crop cultivation and strict labeling regulations in the finished food and feed products (if it contains genetic modifications) for GM food are being adopted by the European Union since 1990 (Bernauer and Meins 2003; Du 2014).

Anti-GMO activists belonging to non-government organizations are although less successful in USA but in Europe, their often-sensationalistic campaigns highlighted GMOs as a threat to biodiversity, farmer autonomy, and food safety (Ansell et al. 2006; Doh and Guay 2006). Government policies and campaigns against GMOs by the activists together leave negative perception. As a result, a large percentage of consumers in Europe do not intend to purchase GM food compared to the consumers in USA. Owing to the potential benefits of GM crops, the percent of GM food supporters in Europe are being increased year by year. Remote availability of GM food in the market and negative voice of administration do not deter the European consumers to avoid purchase of GM foods (Aerni 2011). Spain and Portugal are among the countries which grow Bt maize and have the highest GM food approval rate. Countries like Austria, Germany, or France ban GM cultivation and have a very low approval rate (Gaskell et al. 2010).

In other geographical regions like China, government is investing huge sums of money towards the development and cultivation of GM crops to meet the food demands of its growing population. GM insect-resistant cotton and Bt rice have fetched special attention for their commercial planting by the Chinese government.

Development and cultivation of GM crops on large scale have brought significant global welfare gains. The principal countries over the past few years that gained economically the most from the GM crops are USA (US$7.3 billion), Brazil (US$3.8 billion), India (US$1.5 billion), Argentina (US$2.1 billion), China (US$1 billion), Canada (US$0.82 billion), and others (US1.8 billion) for a total of US$18.2 billion. Global GM crops and seeds market revenue are expected to increase by 8.3% in 2022 to 10.5% in 2025 (Brookes and Barfoot 2018). Of the global economic profit, 48% is shared by the farmers in the industrial countries while 52% is shared by the farmers in the developing countries. Other than farmers, seed and technology providers and consumers also reap a significant fraction of the total welfare gain. Alleviating pressure to expand agricultural surfaces because of increased GM crop productivity; reduced environmental footprint of production methods; reduced emission of greenhouse gases; and safe environment are the other significant benefits associated with GM crops (Carpenter 2011; Barfoot and Brookes 2014).

4.5 Conclusion

The future of GM technologies seems to be very promising to meet the global needs for food, feed, and fiber in a sustainable manner. Advancements in the genome level technologies and conventional breeding methods pave the way to generate the effective variants/ genes and exploit their inheritable functions in the genetically modified plants for improved quality and yield. In today's era, it is highly imperative to spread the awareness about the social gains, economic benefits, and health improvements obtained through GM crop adoption to the (i) farmers and consumers in order to make informed-choice about the crops to be grown and consumed, respectively; (ii) policy makers and regulators to design and enable the biosafety guidelines to adopt and commercialize GM crops; and (iii) science and media communicators to disseminate the potential benefits of the GM technology among the masses.

Conflict of Interest The author discloses no potential conflicts of interest in this work.

References

Abbas MST (2018) Genetically engineered (modified) crops (*bacillus thuringiensis* crops) and the world controversy on their safety. Egypt J Biol Pest Control 28:52

Aerni P (2011) Do political attitudes affect consumer choice? Evidence from a large-scale field study with genetically modified bread in Switzerland. Sustainability 3:1555–1572

Aggarwal S (2012) What's fueling the biotech engine – 2011 to 2012. Nat Biotech 30:1191–11974

Aluru M, Xu Y, Guo R et al (2008) Generation of transgenic maize with enhanced provitamin a content. J Exp Bot 59(13):3551–3562

Aluru MR, Rodermel SR, Reddy MB (2011) Genetic modification of low phytic acid 1-1 maize to enhance iron content and bioavailability. J Agric Food Chem 59(24):12954–12962

Ammar K, Mergoum M, Rajaram S (2004) The history and evolution of triticale. In: Mergoum M, Gómez-Macpherson H (eds) Triticale improvement and production. Food and Agriculture Organization of the United Nations Rome, pp 1–10
Anai T, Koga M, Tanaka H et al (2003) Improvement of rice (Oryza sativa L.) seed oil quality through introduction of a soybean microsomal omega-3 fatty acid desaturase gene. Plant Cell Rep 21(10):988–992
Ansell C, Maxwell R, Sicurelli D (2006) Protesting food: NGOs and political mobilization in Europe. In: Ansell C, Vogel D (eds) What's the beef. The MIT Press, Cambridge, MA, USA, pp 97–122
Aragao FJL, Barros LMG, De Sousa MV et al (1999) Expression of a methionine-rich storage albumin from the Brazil nut (Bertholletia excelsa H.B.K., Lecythidaceae) in transgenic bean plants (Phaseolus vulgaris L., Fabaceae). Genet Mol Biol 22(3):445–449
Bakshi S, Dewan D (2013) Status of transgenic cereal crops: a review. Clon Transgen 3:1
Barfoot P, Brookes G (2014) Key global environmental impacts of genetically modified (GM) crop use 1996–2012. GM Crops Food 5:149–160
Bernauer T, Meins E (2003) Technological revolution meets policy and the market: explaining cross-national differences in agricultural biotechnology regulation. Eur J Polit Res 42:643–683
Blancquaert D, Van daele J, Strobbe S et al (2015) Improving folate (vitamin B9) stability in biofortified rice through metabolic engineering. Nat Biotechnol 33:1076–1078
Borg S, Brinch-Pedersen H, Tauris B et al (2012) Wheat ferritins: improving the iron content of the wheat grain. J Cereal Sci 56:204–213
Brookes G, Barfoot P (2014) Economic impact of GM crops. GM Crops Food 5:65–75
Brookes G, Barfoot P (2018) Farm income and production impacts of using GM crop technology 1996–2016. GM Crops Food 9(2):59–89
Burkhardt PK, Beyer P, Wuenn J et al (1997) Transgenic rice (Oryza sativa) endosperm expressing daffodil (Narcissus pseudonarcissus) phytoene synthase accumulates phytoene, a key intermediate of provitamin a biosynthesis. Plant J 11:1071–1078
Burton RA, Collins HM, Kibble NA et al (2011) Over-expression of specific HvCslF cellulose synthase-like genes in transgenic barley increases the levels of cell wall (1,3;1,4)-β-d-glucans and alters their fine structure. Plant Biotechnol J 9(2):117–135
Cahoon EB, Hall SE, Ripp KG (2003) Metabolic redesign of vitamin E biosynthesis in plants for tocotrienol production and increased antioxidant content. Nat Biotechnol 21:1082–1087
Carpenter JE (2011) Impact of GM crops on biodiversity. GM Crops 2:7–23
Carpenter JE (2013) The socio-economic impacts of currently commercialised genetically engineered crops. Int J Biotechnol 12:249–268
Chakraborty S, Chakraborty N, Agrawal L et al (2010) Next-generation protein-rich potato expressing the seed protein gene AmA1 is a result of proteome rebalancing in transgenic tuber. Proc Natl Acad Sci U S A 107(41):17533–17538
Chen Z, Young TE, Ling J et al (2003) Increasing vitamin C content of plants through enhanced ascorbate recycling. Proc Natl Acad Sci U S A 100:3525–3530
Chen R, Xue G, Chen P et al (2008) Transgenic maize plants expressing a fungal phytase gene. Transgenic Res 17(4):633–643
Chizuru N, Ricardo U, Shiriki K et al (2003) The joint WHO/FAO expert consultation on diet, nutrition and the prevention of chronic diseases: process, product and policy implications. Public Health Nutr 7(1a):245–250
Cong L, Wang C, Chen L et al (2009) Expression of phytoene synthase1 and carotene desaturase crtI genes result in an increase in the total carotenoids content in transgenic elite wheat (*Triticum aestivum* L.). J Agric Food Chem 57(18):8652–8660
Cronje C, George GM, Fernie A et al (2012) Manipulation of L-ascorbic acid biosynthesis pathways in Solanum lycopersicum elevated GDP-mannose pyrophosphorylase activity enhances L-ascorbate levels in red fruit. Planta 235:553–564
Dancs G, Kondrak M, Banfalvi Z (2008) The effects of enhanced methionine synthesis on amino acid and anthocyanin content of potato tubers. BMC Plant Biol 8:65

Davuluri GR, Van Tuinen A, Fraser PD et al (2005) Fruit-specific RNAi-mediated suppression of DET1 enhances carotenoid and flavonoid content in tomatoes. Nat Biotechnol 23:890–895
Dillen K, Demont M, Tillie P et al (2013) Bred for Europe but grown in America: the case of GM sugar beet. New Biotechnol 30:131–135
Dinkins RD, Reddy MSS, Meurer CA et al (2001) Increased sulfur amino acids in soybean plants overexpressing the maize 15 kDa zein protein. In Vitro Cell Dev Biol Plant 37:742–747
Diretto G, Tavazza R, Welsch R et al (2006) Metabolic engineering of potato tuber carotenoids through tuber-specific silencing of lycopene epsilon cyclase. BMC Plant Biol 6:13
Doh JP, Guay TR (2006) Corporate social responsibility, public policy, and NGO activism in Europe and the United States: An institutional-stakeholder perspective. J Manag Stud 43:47–73
Doshi KM, Eudes F, Laroche A et al (2006) Transient embryo specific expression of anthocyanin in wheat. In Vitro Cell Dev Biol Plant 42:432–438
Drakakaki G, Marcel S, Glahn RP et al (2005) Endosperm-specific co-expression of recombinant soybean ferritin and aspergillus phytase in maize results in significant increases in the levels of bioavailable iron. Plant Mol Biol 59(6):869–880
Du L (2014) GMO labelling and the Consumer's right to know: a comparative review of the legal bases for the consumer's right to genetically modified food labelling. McGill J Law Health 8:1–42
Enfissi E, Fraser PD, Lois LM et al (2005) Metabolic engineering of the mevalonate and nonmevalonate isopentenyl diphosphate-forming pathways for the production of health-promoting isoprenoids in tomato. Plant Biotechnol J 3:17–27
Fahad S, Khan FA, Pandupuspitasari NS et al (2015) Recent developments in therapeutic protein expression technologies in plants. Biotechnol Lett 37:265–279
Fernandez-Cornejo J, Wechsler S, Livingston M (2014) Mitchell genetically engineered crops in the United States. USDA Economic Research Service, Washington, DC, USA
Flores T, Karpova O, Su X et al (2008) Silencing of GmFAD3 gene by siRNA leads to low alpha-linolenic acids (18:3) of fad3-mutant phenotype in soybean [Glycine max (Merr.)]. Transgenic Res 17(5):839–850
Funke T, Han H, Healy-Fried ML et al (2006) Molecular basis for the herbicide resistance of roundup ready crops. PNAS 103(35):13010–13015
Garg M, Sharma N, Sharma S et al (2018) Biofortified crops generated by breeding, agronomy, and transgenic approaches are improving lives of millions of people around the world. Front Nutr 5:12
Gaskell G, Stares S, Allansdottir A et al (2010) Europeans and biotechnology in 2010: winds of change? Brussels, Belgium, European Commission
Giovinazzo G, D'Amico L, Paradiso A et al (2005) Antioxidant metabolite profiles in tomato fruit constitutively expressing the grapevine stilbene synthase gene. Plant Biotechnol J 3(1):57–69
Gonsalves CV, Gonsalves D (2014) The Hawaii papaya story. In: Smyth SJ, Phillips PWB, Castle D (eds) Handbook on agriculture, biotechnology and development. Edward Elgar Publishers, Northhampton, MA, USA, pp 642–660
Goo YM, Kim TW, Lee MK, Lee SW (2013) Accumulation of PrLeg, a Perilla legumin protein in potato tuber results in enhanced level of Sulphur-containing amino acids. C R Biol 336(9):433–439
Gosse ME, Manochia M (1996) The first biopharmaceuticals approved in the United States: 1980–1994. Drug Inf J 30:991–1001
Goto F, Yoshihara T, Shigemoto N et al (1999) Iron fortification of rice seed by the soybean ferritin gene. Nat Biotechnol 17:282–286
Goto F, Yoshihara T, Saiki H (2000) Iron accumulation and enhanced growth in transgenic lettuce plants expressing the iron-binding protein ferritin. Theor Appl Genet 100:658–664
Hanafy MS, Rahman SM, Nakamoto Y et al (2013) Differential response of methionine metabolism in two grain legumes, soybean and azuki bean, expressing a mutated form of Arabidopsis cystathionine γ-synthase. J Plant Physiol 170(3):338–345

Haroldsen VM, Chi-Ham CL, Kulkarni S et al (2011) Constitutively expressed DHAR and MDHAR influence fruit, but not foliar ascorbate levels in tomato. Plant Physiol Biochem 49: 1244–1249

Hong H, Datla N, Reed DW et al (2002) High-level production of γ-linolenic acid in Brassica juncea using a Δ6 desaturase from pythium irregular. Plant Physiol 129(1):354–362

Huang S, Frizzi A, Florida CA (2006) High lysine and high tryptophan transgenic maize resulting from the reduction of both 19- and 22-kD alpha-zeins. Plant Mol Biol 61(3):525–535

Ismagul A, Yang N, Maltseva E et al (2018) A biolistic method for high-throughput production of transgenic wheat plants with single gene insertions. BMC Plant Biol 18:135

James C (2014) Global status of commercialized biotech/GM crops: 2014; ISAAA brief. ISAAA, Ithaca, NY, USA

Katsube T, Kurisaka N, Ogawa M et al (1999) Accumulation of soybean glycinin and its assembly with the glutelins in rice. Plant Physiol 120:1063–1073

Kim MJ, Kim JK, Kim HJ et al (2012a) Genetic modification of the soybean to enhance the β-carotene content through seed-specific expression. PLoS One 7(10):e48287

Kim WS, Chronis D, Juergens M et al (2012b) Transgenic soybean plants overexpressing O-acetylserine sulfhydrylase accumulate enhanced levels of cysteine and Bowman-Birk protease inhibitor in seeds. Planta 235(1):13–23

Kim SH, Kim YH, Ahn YO et al (2013) Downregulation of the lycopene ε-cyclase gene increases carotenoid synthesis via the β-branch-specific pathway and enhances salt-stress tolerance in sweetpotato transgenic calli. Physiol Plant 147(4):432–442

Lee S, An G (2009) Over-expression of OsIRT1 leads to increased iron and zinc accumulations in rice. Plant Cell Environ 32:408–416

Lee TTT, Wang MMC, Hou RCW et al (2003) Enhanced methionine and cysteine levels in transgenic rice seeds by the accumulation of sesame 2S albumin. Biosci Biotechnol Biochem 67:1699–1705

Lee S, Kim YS, Jeon US (2012) Activation of rice nicotinamine synthase 2 (OsNAS2) enhances iron availability for biofortification. Mol Cell 33:269–275

Lobell DB, Schlenker W, Costa-Roberts J (2011) Climate trends and global crop production since 1980. Science 333(6042):616–620

Lu S, Van Eck J, Zhou X et al (2006) The cauliflower or gene encodes a DnaJ cysteine-rich domain-containing protein that mediates high levels of β-carotene accumulation. Plant Cell 18:3594–3605

Lukaszewicz M, Matysiak-Kata I, Skala J et al (2004) Antioxidant capacity manipulation in transgenic potato tuber by changes in phenolic compounds content. J Agric Food Chem 52: 1526–1533

Masuda H, Suzuki M, Morikawa KC et al (2008) Increase in iron and zinc concentrations in rice grains via the introduction of barley genes involved in phytosiderophore synthesis. Rice 1:100–108

Mehrotra S, Goyal V (2012) Agrobacterium-mediated gene transfer in plants and biosafety considerations. Appl Biochem Biotechnol 168:1953–1975

Mihalik D, Gubisova M, Klempova T et al (2014) Transgenic barley producing essential polyunsaturated fatty acids. Biol Plant 58(2):348–354

Molvig L, Tabe LM, Eggum BO et al (1997) Enhanced methionine levels and increased nutritive value of seeds of transgenic lupins (Lupinus angustifolius L.) expressing a sunflower seed albumin gene. Proc Natl Acad Sci U S A 94(16):8393–8398

Muir SR, Collins GJ, Robinson S et al (2001) Overexpression of petunia chalcone isomerase in tomato results in fruit containing increased levels of flavonols. Nature 19:470–474

Nandi S, Suzuki YA, Huang J et al (2002) Expression of human lactoferrin in transgenic rice grains for the application in infant formula. Plant Sci 163:713–722

Niggeweg R, Michael AJ, Martin C (2004) Engineering plants with increased levels of the antioxidant chlorogenic acid. Nat Biotechnol 22:746–754

Oakes JV, Shewmaker CK, Stalker DM (1991) Production of cyclodextrins, a novel carbohydrate, in the tubers of transgenic potato plants. Biotechnology (N Y) 9(10):982–986

Ogo Y, Ozawa K, Ishimaru T et al (2013) Transgenic rice seed synthesizing diverse flavonoids at high levels: a new platform for flavonoid production with associated health benefits. Plant Biotechnol J 11:734–746

Ohnoutkova L, Zitka O, Mrizova K et al (2012) Electrophoretic and chromatographic evaluation of transgenic barley expressing a bacterial dihydrodipicolinate synthase. Electrophoresis 33(15): 2365–2373

Oliver MJ (2014) Why do we need GMO crops in agriculture? Mo Med 111(6):492–507

Otiman IP, Badea EM, Buzdugan L (2008) Roundup ready soybean, a Romanian story. Bull Univ Agric Sci Vet Med Cluj-Napoca Anim Sci Biotechnol 65:352–357

Ozyigit II (2020) Gene transfer to plants by electroporation: methods and applications. Mol Biol Rep 47:3195–3210

Park S, Kim C-K, Pike LM et al (2004) Increased calcium in carrots by expression of an Arabidopsis H^+/Ca^{2+} transporter. Mol Breed 4:275–282

Park SC, Kim YH, Kim SH et al (2015) Overexpression of the IbMYB1 gene in an orange-fleshed sweet potato cultivar produces a dual-pigmented transgenic sweet potato with improved antioxidant activity. Physiol Plant 153(4):525–537

Phillips RL (2013) Norman Ernest Borlaug. 25 March 1914–12 September 2009. Biogr Mems Fell R Soc:59

Qaim M (2009) The economics of genetically modified crops. Annu Rev Resour Econ 1:665–694

Ramesh SA, Choimes S, Schachtman DP (2004) Over-expression of an Arabidopsis zinc transporter in Hordeum vulgare increases short-term zinc uptake after zinc deprivation and seed zinc content. Plant Mol Biol 54(3):373–385

Ray DK, Ramankutty N, Mueller ND et al (2012) Recent patterns of crop yield growth and stagnation. Nat Commun 3:1293

Ray DK, Mueller ND, West PC et al (2013) Yield trends are in sufficient to double global crop production by 2050. PLoS One 8(6):e66428

Sato S, Xing A, Ye X, Schweiger B et al (2004) Production of γ-linolenic acid and stearidonic acid in seeds of marker-free transgenic soybean. Crop Sci 44(2):646–652

Schnorf M, Neuhaus-Url G, Galli A et al (1991) An improved approach for transformation of plant cells by microinjection: molecular and genetic analysis. Transgenic Res 1(1):23–30

Shi J, Wang H, Schellin K et al (2007) Embryo-specific silencing of a transporter reduces phytic acid content of maize and soybean seeds. Nat Biotechnol 25(8):930–937

Shih CH, Chen Y, Wang M (2008) Accumulation of isoflavone genistin in transgenic tomato plants overexpressing a soybean isoflavone synthase gene. J Agric Food Chem 56(14):5655–5661

Shin YM, Park HJ, Yim SD et al (2006) Transgenic rice lines expressing maize C1 and R-S regulatory genes produce various flavonoids in the endosperm. Plant Biotechnol J 4:303–315

Sindhu AS, Zheng Z, Murai N (1997) The pea seed storage protein legumin was synthesized, processed, and accumulated stably in transgenic rice endosperm. Plant Sci 130:189–196

Singh D, Laxmi A (2015) Transcriptional regulation of drought response: a tortuous network of transcriptional factors. Front Plant Sci 6:895

Song S, Hou W, Godo I et al (2013) Soybean seeds expressing feedback-insensitive cystathionine γ-synthase exhibit a higher content of methionine. J Exp Bot 64(7):1917–1926

Storozhenko S, De Brouwer V, Volckaert M et al (2007) Folate fortification of rice by metabolic engineering. Nat Biotechnol 25(11):1277–1279

Szankowski I, Briviba K, Fleschhut J et al (2003) Transformation of apple (Malus domestica Borkh.) with the stilbene synthase gene from grapevine (*Vitis vinifera* L.) and a PGIP gene from kiwi (*Actinidia deliciosa*). Plant Cell Rep 22:141–149

Takahashi M, Nakanishi H, Kawasaki S et al (2001) Enhanced tolerance of rice to low iron availability in alkaline soils using barley nicotianamine aminotransferase genes. Nat Biotechnol 19:466–469

Tamas C, Kisgyorgy BN, Rakszegi M et al (2009) Transgenic approach to improve wheat (*Triticum aestivum* L.) nutritional quality. Plant Cell Rep 28(7):1085–1094
Trijatmiko K, Duenas C, Tsakirpaloglou N et al (2016) Biofortified indica rice attains iron and zinc nutrition dietary targets in the field. Sci Rep 6:19792
Upadhyaya CP, Young KE, Akula N et al (2009) Over-expression of strawberry d-galacturonic acid reductase in potato leads to accumulation of vitamin C with enhanced abiotic stress tolerance. Plant Sci 177(6):659–667
Van Eck J, Conlin B, Garvin DF et al (2007) Enhancing beta-carotene content in potato by rnai-mediated silencing of the beta-carotene hydroxylase gene. Am J Potato Res 84:331–342
Van Eenennaam AL, Young AE (2014) Prevalence and impacts of genetically engineered feedstuffs on livestock populations. J Anim Sci 92:4255–4278
Van Eenennaam AL, Lincoln K, Durrett TP et al (2003) Engineering vitamin E content: from Arabidopsis mutant to soy oil. Plant Cell 5:3007–3019
Wakasa K, Hasegawa H, Nemoto H et al (2006) High-level tryptophan accumulation in seeds of transgenic rice and its limited effects on agronomic traits and seed metabolite profile. J Exp Bot 57:3069–3078
Waltz E (2014) Vitamin A super banana in human trials. Nat Biotechnol 32:857
Wehrmann A, Van Vliet A, Opsomer C et al (1996) The similarities of bar and pat gene products make them equally applicable for plant engineers. Nat Biotechnol 14(10):1274–1278
Wurbs D, Rup S, Bock R (2007) Contained metabolic engineering in tomatoes by expression of carotenoid biosynthesis genes from the plastid genome. Plant J 49(2):276–288
Xiaoyan S, Yan Z, Shubin W (2012) Improvement Fe content of wheat (Triticum aestivum) grain by soybean ferritin expression cassette without vector backbone sequence. J Agric Biotechnol 20:766–773
Yang QQ, Zhang CQ, Chan ML et al (2016) Biofortification of rice with the essential amino acid lysine: molecular characterization, nutritional evaluation, and field performance. J Exp Bot 67(14):4285–4296
Yu O, Shi J, Hession AO et al (2003) Metabolic engineering to increase isoflavone biosynthesis in soybean seed. Phytochemistry 63:753–763
Zhang C, Liu J, Zhang Y et al (2011) Overexpression of SlGMEs leads to ascorbate accumulation with enhanced oxidative stress, cold, and salt tolerance in tomato. Plant Cell Rep 30:389–398
Zhao ZY, Glassman K, Sewalt V et al (2003) Nutritionally improved transgenic sorghum. In: Plant biotechnology 2002 and beyond. Springer, Netherlands, pp 413–416
Zheng A, Sumi K, Tanaka K et al (1995) The bean seed storage protein β-phaseolin is synthesized, processed and accumulated in the vacuolar type-II protein bodies of transgenic rice endosperm. Plant Physiol 109:777–786
Zhou Y, Cai H, Xiao J et al (2009) Over-expression of aspartate aminotransferase genes in rice resulted in altered nitrogen metabolism and increased amino acid content in seeds. Theor Appl Genet 118:1381–1390

Transcriptomics in Plant

5

Pratik Satya, Sougata Bhattacharjee, Debabrata Sarkar, Suman Roy, Laxmi Sharma, and Nur Alam Mandal

Abstract

Within a span of about 20 years, transcriptomics has established itself as an indispensible tool in almost all the areas of plant research. This chapter provides information on the rapid development of this important research area over a short period. Here, we present an overview of plant transcriptomics with an outline of the basic processes and tools including study design, RNA isolation, library preparation, sequencing platforms and bioinformatics analyses for annotation, pathway mapping and differential gene expression. A brief overview of the current status of transcriptomics in plants is presented followed by examples from a fibre producing plant, jute (*Corchorus* spp., Malvaceae), where transcriptomic researches have been proved very useful to understand biology and genetics of economically important traits.

Keywords

Transcriptomics · Transcriptome · NGS · Sequencing · Library · Gene expression · Jute

P. Satya (✉) · D. Sarkar · S. Roy · L. Sharma · N. A. Mandal
ICAR-Central Research Institute for Jute and Allied Fibres, Barrackpore, Kolkata, West Bengal, India

S. Bhattacharjee
ICAR-Vivekananda Parvatiya Krishi Anusandhan Sansthan, Almora, Uttarakhand, India

R. L. Singh et al. (eds.), *Plant Genomics for Sustainable Agriculture*,
https://doi.org/10.1007/978-981-16-6974-3_5

5.1 Introduction

Analysis of genes and genomes is a major research arena for understanding life process. A genome encompasses the whole set of inherited genetic material, carrying genes, regulatory sequences, repetitive elements and other components. Only a small part of it carries active genes, many of which are differentially expressed in tissues during various developmental phases or/and in response to external stimulations. If any of these conditions (time, cell or tissue type or environment) change, the cell adjusts to the new condition by changing the pattern and degree of gene expression. The term transcriptome, first used by Charles Auffray in 1996 (Piétu et al. 1999), refers to the total set of expressed mRNA molecules in a particular cell/cell type at a given physiological state in a specific environment. Such dynamic nature of transcriptome, which is not observed at genome level, provides opportunities to study the response of the organism to the change in environment or to study its growth and development. By analogy, transcriptomics is a collection of tools and techniques used to study the transcriptome. Over time, the use of the term 'transcriptomics' has expanded to include other coding and non-coding RNAs expressed in the cell, such as long non-coding RNA, as the same techniques can be used for mRNA or other RNA characterization by tweaking crucial steps. In plant science, transcriptomics is employed in various research arenas, such as to study environmental responses of plants under biotic or abiotic stresses, to understand basic biological processes like germination or fertilization, to identify genes and metabolic pathways, to decipher biological basis of crop productivity or to mine for novel phytochemicals from plant sources. Transcriptomics also helps to identify potential targets for a disease (e.g. ssRNA virus can be targeted using CRISPR/Cas12a/DNA ternary complex) and discovery of gene regulatory proteins (e.g. ChIP-Seq helps in identification of transcription factor and their exact binding sites on DNA). Its application has been extended to the areas of genomic manipulation, such as DNA free genome editing, which relies on RNA rather than DNA for making transgenic using CRISPR-Ribonucleoprotein complexes. Thus, transcriptomics has applications beyond identification of genes and characterization of their functionality and is an indispensable tool for solving fundamental biological questions.

While the sequencing technologies are same for both the genome and the transcriptome, the output sequence information has some basic differences. First of all, genome sequencing captures all the coding and non-coding sequences by first sequencing raw reads (contigs), and then stitches these contigs to a full length genome sequence. Transcriptome sequencing, on the other hand, captures all the mRNA sequences that are synthesized in a specific tissue/cell, where multiple copies of one particular gene are captured, each transcript being a 'raw read'. These are then aligned and matched with the gene sequences present in the genome (reference based) or de novo (based on robust gene identification algorithms). This allows a quantitative evaluation of transcript abundance by comparing the relative copy number of a read. Thus, relative expression of a gene in two or more transcriptomes, ideally from same plant and sequenced using same platform can be done to understand the biological role of the gene. This ability to interpret gene function is the most important strength of transcriptomics, which cannot be obtained from genome sequencing. Second, genome sequence is individual-level information; thus it is fixed for a genotype. Transcriptome sequence, on the other hand, exhibits variations in different cells within a genotype. Even the same cell exhibits difference in transcriptome sequence under different environmental stimuli, making the transcriptome much more variable and informative than the genome. This inherent variability allows a wider application of transcriptomics in biological sciences, particularly to study development and responses to environmental changes. Third, gene expression is controlled by a variety of regulators, including small molecules, other genes (transcriptional factors), methylation (epigenomic modification), and external factors. This multi-dimensional cross-talk makes interpretation of transcriptome data more complex. Obviously, such dimensional complexity of transcriptome data requires specific robust mathematical analysis, big data analysis platforms and trained human resources to find the desired 'needle' from the transcriptomic haystack. To provide a better overview of transcriptomics, some specific terminologies are presented in Table 5.1.

5.2 Historical Development

Although expression analysis has long been utilized as a technique to establish gene functionality, large scale cDNA analysis was first undertaken under the Human Genome Project. In a seminal work published in the journal 'Science', Adams et al. (1991) generated 600 expressed sequence tags (EST) after cloning randomly selected cDNA from human brain tissue and showed that 337 of these coded for novel genes. They predicted that their approach would allow mapping most of the human genes within a few years. Within 5 years, the first human transcript map carrying 16,000 genes was generated (Schuler et al. 1996). Since then, this approach of large-scale gene characterization through cloning and sequencing of ESTs has been proved to be extremely useful for gene identification and characterization. A variety of subsequent methods were developed, ultimately bypassing the cloning step (direct sequencing of cDNA fragments). Development of DNA microarray technique (Schena et al. 1995) was the first milestone for large-scale gene expression analysis. Several large-scale platforms of microarray based gene expression systems emerged rapidly, including serial analysis of gene expression (Velculescu et al. 1995) and cDNA fingerprinting (Clark et al. 1999). At the same time, new clustering and multivariate algorithms for robust statistical analysis of large-scale gene expression data started to appear. One such clustering technique (Eisen et al. 1998) came from David Botstein's group, who is well-known as a pioneer researcher in DNA marker development. The research in plant transcriptomics gained momentum when Zhu and Wang (2000) designed the first large-scale expression array containing 8835 probes for *Arabidopsis* genes and generated over 500 transcriptome profiles. Since then, almost all the branches of plant science have resorted to transcriptome analysis for solving research problems, which can be envisaged from the sharp rise

Table 5.1 Terminologies associated with transcriptomics

Terminology	Explanation
Adapter	Short oligonucleotide sequences that are ligated to the 5′ and 3′ ends of DNA fragment during library preparation for sequencing. They match to the sequences present on the surface of the flow cells
Alignment	The process of matching two sequences. Two types of alignment strategies, local alignment and global alignment are used for aligning two nucleotide sequences
Barcode (tag)	A unique DNA sequence attached to template sequence before sequencing. Useful for multiplex sequencing, pooling of libraries, post-sequencing analyses, etc.
Cluster	Multiple copies of a sequence around a template, formed by bridge amplification. Each cluster grows in size as sequencing proceeds until a desired size of about 1000 copies are reached, and represents a single template sequence
Contig	A stretch of continuous nucleotide sequence
Coverage level	The average number of sequenced nucleotides that match with the reference nucleotide
De novo Assembly	Assembly of a set of RNA sequences without the support of a reference sequence
FASTQ file	A text output file of NGS sequencing containing the sequence and quality information of every sequenced base
Flow cell	A specially designed glass slide containing lanes for sequencing. The templates are fixed (immobilized) on the flow cell surface, so that enzymes can synthesize multiple copies using the template as source
Indels	Insertions and deletions in DNA sequences. Indels identified from a transcriptome analysis may be due to sequencing error or due to true mutations
Kmer length	A sequence can be broken down into small sequences (words) that can be overlapping or non-overlapping. These are used for rapid matching of sequences during matching with reference genome or matching between multiple sequences. The length of the word is the kmer length
Paired-end sequencing	Sequencing a fragment of DNA from both end
Q-score	A measure for error in base calling during sequencing. A Phred score is a quality score defined by the negative logarithm of the error probability
Reference-based assembly	Assembly of a set of RNA sequences based on a reference sequence
RNA-seq	An abbreviation of 'RNA-sequencing', a technique for sequence analysis of RNA from a sample. The sample can contain full spectrum of the RNA of a cell, tissue or organism (transcriptome), specific components of RNA (mRNA, snRNA, etc.), or partial sequences
Variant discovery	Identification in variation in genetic material between two cell, tissue or individual. Detects single nucleotide polymorphism (SNP), InDel (insertion-deletion) and variation in RNA secondary structure

in research publications. A search in Pubmed Central of the US National Library of Medicine (https://www.ncbi.nlm.nih.gov/pmc) with keyword 'plant transcriptome' retrieved only 22 hit in 2000, which increased rapidly to 10,408 in 2020. By 2019, the One Thousand Plant Transcriptome Initiative sequenced the transcriptomes of 1124 species of Viridiplantae, and reconstructed the phylogeny of the major clades. To date, this is the most exhaustive documentation of plant transcriptomics. Analysis of major gene families revealed the role of gene and genome duplications in evolution although some components of the species tree remain still unresolved. The same group is now working on transcriptomics of ten thousand plants to develop a more robust phylogenetic species tree.

Advances in next-generation sequencing as well as biocomputing technologies during the past two decades resulted in development of several approaches for sequencing of genome and transcriptome. The first generation sequencers developed by Applied Biosystem Instruments (ABI) employed Sanger sequencing with fluorescent probes and used early-generation computers to collect and analyse data. In 1982, GenBank, the first public repository for sequence data was established, and a number of genome sequences were deposited by 2000. Two very important technologies, polymerase chain reaction (PCR) and shotgun sequencing revolutionized the field of genome sequencing and analysis during this period. However, post-2000 period was dominated by various new chemistry-based sequencing technologies, collectively referred as next-generation sequencing (NGS) technologies. The first wave came with the development of sequencing by synthesis (SBS), a technology based on massively parallel signature sequencing (MPSS) on microbeads. The commercial venture of MPSS was started by Lynx Therapeutics. Shankar Balasubramanian and David Klenerman developed the SBS technology and formed Solexa. Lynx Therapeutics merged with Solexa in 2005 and Solexa was acquired by Illumina in 2007. In 2004, 454 Life Science (now acquired by Roche) offered a pyrosequencing based NGS platform, and new models based on this system came in 2005-06 (454 GS 20), 2007 (454 GS FLX) and was further improved such as 454 GS FLX+, but was discontinued in 2013. By 2005, another commercial venture, Life Technologies developed SOLiD (Sequencing by Oligonucleotide Ligation and Detection). Illumina Inc. after acquiring Solexa started its own sequencing platform in 2009 and developed three very popular sequencing technologies, Hiseq, Miseq and Novaseq, and later developed Genome Analyzer platform, which is also based on SBS. A third generation of sequencing platforms like Pac Bio, nanopore and electron microscopy-based systems are currently being developed and utilized for large-scale sequencing, filling of gaps in existing sequences and resequencing of hundreds and thousands of samples. These improvements led to drastic reduction in cost and time of sequencing whole genome and transcriptome. A timeline of various events in transcriptome analysis is presented in Table 5.2.

5.3 Pipeline for Transcriptome Analysis in Plants

A pipeline or workflow of transcriptome analysis is an outline of the sequential processes to be followed to generate a transcriptome sequence and further analyse it as per the researcher's requirement. The processes can be divided in few major steps,

Table 5.2 A timeline of different transcriptomics technologies

Technology intervened	Year of intervention	Reference	Comment(s)
Northern Blot	1977	Alwine et al.	Gene specific detection, not applicable for global gene profiling
Sanger sequencing	1977	Frederick Sanger	First sequencing platform but very slow and costly
RT-PCR	1984	After PCR discovery by Kary Mullis	For cDNA synthesis from mRNA, routine transcriptome work
Microarray/ Affymetrix gene chip	1990s	Fodor et al. (1991), Schena et al. (1995)	Gene expression profiling and differential gene expression study
RACE	1989	Frohman and Martin	For cDNA end information, not useful for global transcript profiling
ESTs	1991	Adams et al.	High throughput single pass partial cDNA sequencing; now EST-clusters (unigene) used
Competitive PCR	1992	Siebert and Larrick	For differential gene expression analysis, not used recently
Antisense/Co-Suppression	1992	Richard Jorgensen	Functional transcript knocked down, targeted approach, now become obsolete
Improved DDRT-PCR	1993	Liang et al.	Differential gene expression study, target specific approach, not useful in organism level
Microarray system	1995	P Brown and R Davis	cDNA sequences on glass slides
SAGE/CAGE	1995	Veculescu et al.	Representative partial sequencing of transcripts, tags gives useful information about cell/tissue specific transcript profile.
Two-dimensional microarray	1995–96	P Brown's group	Fluorescent detection, high speed
Initiation of the concept of 'sequencing by synthesis'	Mid 1990s	S. Balasubramanian, and D. Klenerman at Cambridge	Detected motion of DNA polymerase during synthesis by fluorescent labelling
Patent filed for nanopore sequencing	1995	Church, Deamer, Branton and colleagues	The concept of nanopore sequencing developed
SSH	1996	Diatchenko et al.	Identify novel gene, very useful tool but not amenable for whole transcriptome level

(continued)

Table 5.2 (continued)

Technology intervened	Year of intervention	Reference	Comment(s)
RNAi	1998	Fire et al.	Targeting mRNA for functional validation, gene specific approach
Clustering of microarray data	1998	Eisen et al.	Improved statistical analysis and interpretation
Oligonucleotide microarray system/ GeneChip platform	1999	Affymerix	In situ synthesis of oligos on chip
Massively Parallel Signature Sequencing (MPSS)	2000	Brenner et al.	Sequencing throughput accelerated, useful for cell level when using NGS technology
qRT-PCR/Real-Time PCR-based analysis	2001	Livak et al.	Quantification of mRNA expression
Next Generation Sequencing (NGS) platforms	2004 onwards	Various commercial ventures	See previous section for historical development
454 sequencing	2005	Life Sciences (Roche Diagnostics)	Used for transcriptome study, technology withdrawn in 2013
SOLiD (Sequencing by Oligonucleotide Ligation and Detection)	2006	Applied Biosystems Inc. (later Life Technologies)	Can generate 60 Gb data per run
Genome Analyzer	2006	Solexa	Sequenced 1 GB per run
Single molecule detected by nanopore	2008	Gundlach's group	Used MspA nanopore
Single Molecule Real-Time (SMRT) sequencing	2009	Craighead, Korlach, Turner and Webb	Sequencing is performed in a SMRT cell containing nanowells
Third generation sequencing (TGS) platforms	2009 onwards	Various commercial ventures	Pacific Biosciences, Oxford Nanopore Technology, Quantapore (CA-USA), and Stratos (WA-USA)
Single Molecule Real-Time (SMRT) sequencing commercialized	2011	Pacific Biosciences	Can sequence longer reads, base calling less accurate than Illumina short read sequencing
MinION sequencer	2014	Oxford Nanopore	Portable device, up to 30 GB
NovaSeq platforms	2017	Illumina	Up to 6 TB read capacity
HiFi (High Fidelity)	2019	Pacific Biosciences	Can generate Circular Consensus Sequences (CCSs) approximately 10–20 kbp-long
Sequel II sequencer	2019	Pacific Biosciences	Contains 8 million nanowells SMRT Cell, capacity 160 GB.
R10 Nanopore sequencing	2019	Oxford Nanopore	Double sensor for more efficient base calling

namely, design of the study, RNA isolation, prepare a sequencing library, sequencing of the library, processing of the raw reads to obtain clean reads, assembly of the sequences and annotation of the transcriptome (Fig. 5.1). Further bioinformatics or wet-lab analyses are performed based on the research need. A transcriptome analysis pipeline can be objective specific. For example, several pipelines have been developed for differential gene expression analysis.

5.3.1 Transcriptomic Study Design

Any experiment needs to be planned methodically by applying appropriate tool (s) for testing the hypothesis. Testing a biological hypothesis using transcriptomics requires selecting an appropriate sequencing platform, determining the number of replicates and use of a statistically robust design. In plant RNA-seq experiments, at least three biological replicates are recommended by the European Molecular Biology Laboratory (EMBL). Biological replicates are, however, only required when inferences are to be drawn on population rather than the individual organism itself. In case of studies with plant, we primarily draw inference on population, so determining the number of biological replicates is essential for drawing a robust conclusion. In some cases, however, biological replicates may be avoided. For example, if in some experiment two tissues from the same plant are compared, or two plants of same genotype cultured in same flask are compared, one may not use biological replicate. But usually in an experiment, particularly for differential gene expression analysis, plants are grown in different conditions, where biological replicates are required. Selecting technical replicates depends on the technical reproducibility of the sequencing platform, which is high for most of the advanced sequencing systems although several processes during library preparation can introduce bias in output. Cost is another important issues, because with each replicate sequencing cost is increased, thus the researcher has to sacrifice some accuracy in case of budget constraint. A number of statistical techniques are available for interpreting un-replicated transcriptome data. Software like NOISeq (Tarazona et al. 2011) and GFOLD are effective for expression analysis of genes that have strong biological response (Khang and Lau 2015). Use of three or more replicates improves the power of the study, allowing identification of genes with weak biological response.

Another important issue is the read depth/read coverage of the transcriptome. If the experiment is a pilot scale study, or high quality reference sequence information is available, one may select low read depth and more number of replicates. But if the RNA-seq sequencing is de novo, more read depth would be preferable. Although such benchmark studies are rare, Liu et al. (2013) observed that an increase in number of DE genes with sequencing depth has diminishing returns after 10 million reads and suggested increasing replication over read depth. Lamarre et al. (2018) observed that the optimal threshold to control the false discovery rate (FDR) is approximately 2^{-r} (r = replicate number). They showed that 20 million reads per

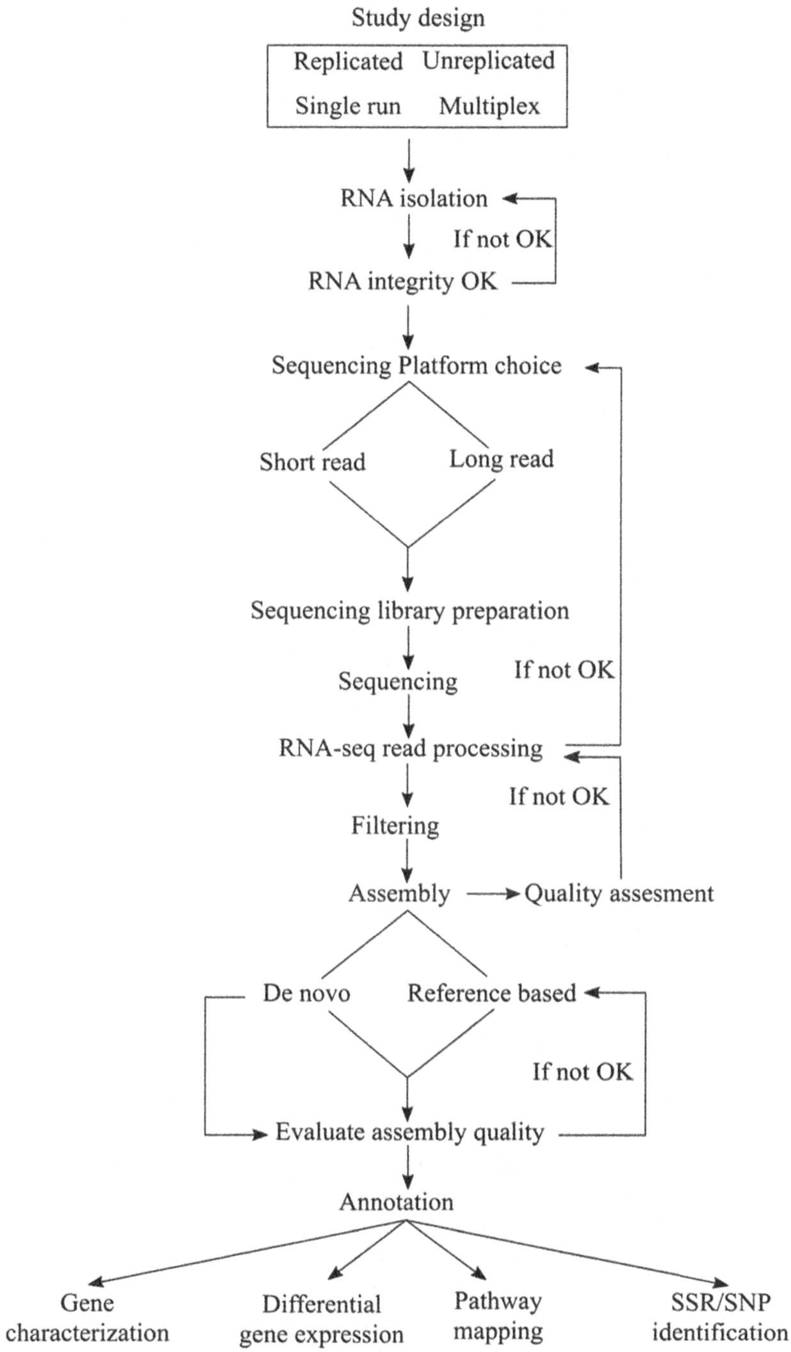

Fig. 5.1 A simplified pipeline for transcriptomics in plant

sample and four biological replicates would be required to capture 1000 differentially expressed genes in tomato.

5.3.2 RNA Isolation and Processing

Transcriptomics starts with quality RNA isolation. Since there are various kinds of RNA in cell that differ in length, the procedure for isolation of RNA will vary based on the experimental requirement. Standard RNA-seq captures only the protein mRNAs that are generally >200 bp, thus the RNA isolation procedure is standardized in such a way that short RNAs (<100 bp) are washed out, and the isolated RNA is enriched with mRNAs. To isolate short RNAs, specific silica-based membranes are used. The quality of RNA, in addition to standard spectrophotometric quality assessment, is evaluated by RNA integrity number (RIN), which is determined in an Agilent BioAnalyzer by 18s/28s rRNA electrophoresis. A RIN value of >7 (range 1–10) is well accepted for RNA-seq analysis. The next step is to remove the rRNA and tRNA, which together constitutes 96–98% of the total RNA sample, and to retain only mRNA (2–3%) in the sample. The mRNA portion is recovered from the total RNA pool by poly-dT primers that specifically bind to the poly-A tail of mRNA. Alternatively, the rRNA and tRNA can be removed by binding to probes specific to these RNAs. In case of mRNA capture, long non-coding RNAs (lncRNA) are also removed, so this method cannot be used to capture and sequence lncRNA. On the other hand, probe based methods are not full-proof and some rRNA and tRNA remnants are always present in the sample after processing. This method also requires probe information, which requires prior sequence knowledge.

5.3.3 Library Preparation

The preparation of sequencing library is the most important step of RNA-seq. It depends on the sequencing platform and sequencing strategies used. Principally, a RNA-seq library is a pool of cDNA fragments (in case of sequencing by synthesis). On an Illumina platform, the RNA pool is fragmented to a size of 50–300 bp (read length) either enzymatically, chemically or mechanically. The cDNA is synthesized either by single end sequencing or paired-end sequencing using reverse transcriptase, using a specific PCR system called bridge amplification. For the first strand synthesis, oligo-dT primer, random primer or adaptor ligated primers can be used, each of which has its own advantages and limitations. The oligo-dT primers are biased towards 3′-end, and will miss all the fragments that lack poly-A tail. Random primers capture all these fragments but suffer from drawbacks like non-random binding and loss of strand information. The ligated primers are better than the other two systems for capturing the mRNA pool. The second strand is synthesized by DNA polymerase using specially designed primers.

5.3.4 Library Sequencing

Different sequencing platforms employ different strategies for sequencing the cDNA library. The following are the major sequencing platforms widely used in plant transcriptomics. Technological advances in each of these systems have resulted in tremendous improvement in sequencing power, output quality and cost reduction.

5.3.4.1 Roche (454) FLX

454 Life Sciences (Roche Diagnostics) was first commercialized in 2005 and currently Genome Sequencer (GS) FLX System and GS FLX Titanium series platforms are available. After preparing template as discussed above, beads along with the attached DNA fragments are removed from the emulsion and loaded into the wells (PicoTiter Plate). Each well contains only one bead. Pyrosequencing principle (luciferase-based light detection on pyrophosphate release when a base is added in sequencing process) is used for sequencing (Ronaghi et al. 1996). From template preparation to data processing the FLX system (read length 450–500 bases) takes 10 h per run (generates 400-Mb sequence data). Recently developed GS FLX Titanium XL+ platform can generate 1 Gb sequence data with read length of 1000 bp.

5.3.4.2 Illumina/Solexa

The Illumina sequencing system is based on the principle of sequencing by synthesis (SBS). The solid phase PCR is then carried out inside flow cell, which is also called fold-back PCR or bridge PCR (Fedurco et al. 2006). The system works on reversible terminator technology. The templates are immobilized on a proprietary flow cell array and are ligated with adaptors carrying barcode or inline index (both are unique short sequences to discriminate reads of different pools). After the first cycle of cDNA elongation, the 5′ end of the single strand DNA bends and binds to a functional group on the flowcell, and the original templates are washed away. The bridge fragments are made double stranded and PCR is performed on these bridges to generate several millions of dense clusters. The first sequencing cycle is performed by adding four fluorescently labelled terminator nucleotides with primers and DNA polymerase. After incorporation of each dNTP, the polymerization is terminated to image the fluorescence tag, the dye is enzymatically removed and the next dNTP is incorporated to extend the chain, which allows recording of every fluorescent signal, thereby determining the sequence of the template. Few recent platforms like, Illumina Genome Analyzer 1 Gb and HiSeq 600 Gb are very popular. Illumina read length generally varies from 35 to 150 bases. IlluminaHiSeq 2000 platform yields 400 Gb of sequence data in a single run (takes 7–8 days). Another model, HiSeq X Ten can generate 1.8 Tb sequence data. In 2017, Illumina introduced Novaseq platforms which are more efficient, generating up to 6 TB sequence data and claims to complete sequencing of 48 genomes in less than two days.

5.3.4.3 ABI SOLiD

SOLiD (sequencing by oligonucleotide ligation detection) platform utilizes oligonucleotide probes (8 bp long, each having two unique nucleotides at 3′end and labelled with fluorophore at the 5′ end) ligation for detecting the base of transcripts while sequencing. It was commercialized by Applied Biosystems in 2005 as SOLiD 3.0 platform (Shendure et al. 2005). In this technology, single layer beads are immobilized in an acrylamide matrix on a glass slide along with attached DNA molecules. A set of 16 oligos (for 4 bases of nucleic acid) are required for hybridization with template cDNA while sequencing in each reaction. While encoding base in sequencing, each unique base pair of 3′ end of the probe is assigned one out of four possible colours for ease of detection and analysis. During sequencing, each base in the template is sequenced twice and hence SOLiD technology is said to be highly accurate. The SOLiD 3.0 platform yields read length of 50 bases only and can generate approx. 20 Gb sequence data per run. SOLiD 5500 and SOLiD 5500 XL systems were introduced to increase the sequence data of up to 300 Gb per run (Edwards et al. 2013).

5.3.4.4 Ion Torrent (Semiconductor-Based Life Technologies)

This technology was developed by Ion Torrent Systems Inc. and was commercialized in 2010. It utilizes a semiconductor-based device, also called ion chip, that senses the H+ ions generated during DNA extension by DNA polymerase (measures the induced pH changes by the release of hydrogen ions (Rothberg et al. 2011)). The ion chip, having wells of 3.5-μm-diameter, is located directly over the electronic sensor. The voltage signal is proportional to the number of bases incorporated in the new strand synthesized by DNA polymerase and the detection system is non-optical scanning, which eliminates use of fluorophores, thereby reducing cost and increasing speed of detection. In 2012, another high throughput technology was released, called 'Ion Proton', which increased output by an order of magnitude of 10× but the read length was drastically reduced in comparison with Ion Torrent (200 bp instead of 400 bp).

5.3.4.5 Pacific Biosciences

Single molecule real-time (SMRT) sequencing was developed by Nanofluidics, Inc. and commercialized by Pacific Biosciences, USA. In this technology, template is prepared through ligation of single-stranded hairpin structured adaptor to the cDNA ends (thereby generating a bell-shaped structure called SMRT-bell). Single molecules of DNA polymerase are immobilized at the bottom using biotin–streptavidin interaction in zepoliter-sized wells, also called zero-mode waveguides (ZMWs), and four dNTPs in high concentration with different fluorophore labelled are used for rapid DNA synthesis using strand displacing polymerase (Levene et al. 2003). One advantage is that a cDNA molecule can be sequenced multiple times. Moreover, direct sequencing instead of clonal multiplication allows the sequence to be read in real-time (Eid et al. 2009). Each SMRT cell can generate ~50 k reads and up to 1 Gb of data in 4 h.

5.3.4.6 Oxford Nanopore Technologies

In a nanopore system, a sequencing flow cell composed of hundreds of micro-wells containing a synthetic bilayer and punctured by biologic nanopores (Wang et al. 2015). Sequencing is achieved simply by precise measuring the changes in current induced as a result of incorporation of bases through the nanopores with the help of a molecular motor protein. Library is prepared by ligating adapters to cDNA ends in a manner that first adapter can bind with motor enzyme and second adapter (a hairpin oligonucleotide) can bind with another HP motor protein. Therefore, simultaneously two strands can be sequenced from a single molecule and increase the accuracy in comparison with SMRT technology. This is a highly throughput technology where a single run (18 h) can generate more than 90 Mb of sequencing data with maximum read lengths of more than 60 kb using MinION platform (USB-powered, portable sequencer) (Ashton et al. 2015).

5.3.5 Quality Control

The raw sequence data output from the system is obtained in 'FASTQ' format. A quality score, known as Phred quality score (Q) determines the quality of the sequence. Generally, Q >28 indicates good quality of the transcriptome, while Q <20 has a poor quality. Several other parameters, such as technical artefacts (adaptor, primer dimer, etc.) and biological artefacts (other sequence contamination) can interfere with the quality. To test these parameters, number of overrepresented sequence, duplicate reads and kmer count (a measure for technical artefact) are examined. Once such artefacts are determined, the contaminated sequences are removed by filtering and trimming using processing software to generate processed reads.

5.3.6 Read mapping, assembly and annotation

Once the reads are generated, they are to be assembled to identify the genes. Since the transcriptome reads are of very small length (30–100 nt) (though some platforms produce longer reads) and are to be matched ideally against genome sequence of the same organism (which is in case of plants can go up to thousands of megabases), a robust annotation system is required. Mostly, a compression algorithm is applied to reduce the computational load. Burrows–Wheeler algorithm is one such compression tool that helps in fast annotation of the sequences. Several annotation pipelines are available for de novo and reference-based annotation of transcriptomes and differential gene expression.

5.4 Bioinformatics Software for Transcriptome Analysis

Bioinformatics software is at the core of transcriptomics. These software filter raw data, assembly the filtered sequences into transcripts, annotate their biological function and mine the transcriptome for various information including SSRs, SNPs, regulatory genes, differentially expressed genes, metabolic pathways, genetic causes and responses of disease or stress, transposable elements and many more. While using software depends on use of platform and purpose of experimentation, some software are often preferred due to their high reliability and accuracy.

5.4.1 Filtering

As described earlier, filtering involves cleaning and trimming of unwanted sequences from the reads and quality assessment. FASTX-Toolkit (Gordon and Hannon 2010) is widely used for filtering of transcriptome raw reads. For quality inspection of a transcriptome, FastQC (Andrews 2010) is a good choice. During sequencing, the raw reads are stored in 'FASTQ' format by the sequencer, which merge the sequence (FASTA) with a quality score, called Phred score, which is determined by error probability of base calling. A higher Phred score indicates more confidence in base calling, i.e., sequencing quality.

5.4.2 Assembly

Errors in assembly can seriously impair transcriptome quality. A single transcript may be fragmented and scored as multiple transcripts, causing loss in information, or multiple transcripts may be erroneously joined together constructing a chimera, creating problems in annotation. Many genes exist as duplicates or gene families having high sequence similarity. Correct assembly of fragmented reads of these genes is extremely difficult, which is another source of error. Several tools are available for assembly, some of which are bundles of software or assembly pipeline. Often it is better to use more than one assembly for finding out the best one, which obviously depends on the sequence type, sequence quality and method of assembly (reference-based/de novo). Since a de novo assembly generates transcripts only based on RNA-seq data, it is more erroneous than reference-based assembly. The basis of de novo assembly is generation of a de Bruijn Graph based on *kmer* decomposition of the read. Therefore, *kmer* length is an important factor for de novo assembly. A shorter *kmer* has more coverage, but at the same time has more chance to be read from multiple transcripts. For de novo assembly, several tools are available, of which Trinity (Grabherr et al. 2011), SOAPdenovo-Trans (Xie et al. 2014), Trans-ABySS (Robertson et al. 2010) and rnaSPAdes (Bankevich et al. 2012) are more popular. Trinity is a well trusted assembly pipeline for de novo assembly and is recommended by various researchers as it has high transcript recovery and accuracy (Freedman and Weeks 2020). A software, TransRate can compare various assemblies by giving a quality score (Smith-Unna et al. 2016), which can be used for selecting appropriate assembly. Wang and Gribskov (2017) compared eight de novo assembly tools (BinPacker, Bridger, IDBA-tran, Oases-Velvet, SOAPdenovo-Trans, SSP, Trans-ABySS and Trinity) at different kmer length (25-71) and observed that SOAPdenovo-Trans had the highest base coverage, while Trans-ABySS was best in gene coverage and recovery of full-length transcripts. They recommended performing de novo assembly even when reference genome is available, as transcript fragmentation, incorrect/incomplete gene annotation and exon level differences are major reasons for difference in annotation and differential gene expression. Holzer and Marz (2019) observed that for short read sequences, Trinity, SPAdes, and Trans-ABySS, were better than other tools, but no tool was best for all data sets. These results show that evaluation of different assemblies is a critical step for good assembly construction. Another tool that can be used for de novo transcriptome analysis for gene expression is RSEM (RNA-Seq by Expectation-Maximization), which uses Bowtie/Bowtie2/STAR for read alignment and EBseq for differential gene expression (Li and Dewey 2011). New methods like principles of information theory and abundance of alternate spliced transcripts are being applied to improve the efficiency of de novo assembly (Mao et al. 2020).

For reference-based genome guided assembly, the chance of error is less, but the quality of transcriptome depends on the reference genome/transcriptome quality. Several reference based assemblers are available, such as Cufflinks (Trapnell et al. 2010), StringTie (Pertea et al. 2015), TransComb (Liu et al. 2016), Bayesembler (Maretty et al. 2014), CLASS2 (Song et al. 2016) and Scallop (Shao and Kingsford 2017). In addition, Trinity has options for genome guided de novo assembly. Comparative estimations show that StringTie produces more accurate assembly than Cufflinks or Bayesembler, but results may vary depending on sequence quality. An updated version of StringTie, StringTie2 is now available that can assemble longer reads (>200) efficiently (Kovaka et al. 2019). The RNA-seq reads are first aligned using a spliced aligner such as HISAT/HISAT2 (Kim et al. 2015) or STAR (Dobin and Davis 2013). Alignment outputs are stored as SAM (Sequence Alignment/Map) or BAM (Binary Alignment/Map format) file format, which are used as input files for differential gene expression analysis tools. New alignment-free assemblers, based on kmer matching, for example, Salmon (Patro et al. 2017) and Kallisto (Bray et al. 2016) are faster than the alignment-dependent methods like StringTie, but have lower efficiency in detecting low-abundance transcripts and novel transcripts. Another assembler, Necklace (Davidson and Oshlack 2018) is useful when the reference sequence is incomplete. It requires the RNA-seq read to be assembled, the incomplete reference genome and one or more well-annotated genome from related species, and builds a super Transcriptome merging all inputs.

5.4.3 Annotation

For de novo transcriptome assemblies, annotation is required to identify the function of the transcript, while in reference based assemblies, the transcripts are matched to annotated reference, so further annotation is not required. One may, however, improve over the previous annotation, as the databases used for annotation (for example, BLAST databases) are updated frequently. The tools of genome and transcriptome annotation are same, based on BLAST databases, which have a variety of algorithms and tools for annotation of RNA-seq data.

5.4.4 Differential Gene Expression

Perhaps the most common use of plant transcriptome analysis is study of differential expression of genes (DEG or DGE) of tissues having different treatments or stress conditions. The basic principle is to identify the number of sequenced reads mapped to a single gene, which is a measure of expression of the gene in the sample. Several other factors influence this count, such as gene length (longer transcripts have more fragments mapped), sequencing depth and expression level of other genes. Therefore, a normalized measure is required to estimate DEG. A couple of such measures are widely used in DEG analyses. The RPKM (Reads PerKilobase per Million mapped reads) is a measure where mapped reads are first normalized to reads per million (RPM) with a scaling factor of 106, which is then divided by the length of the gene. For paired-end sequencing, FPKM (Fragment Per Kilobase per Million mapped reads) is used, which follows the same normalization RPKM, with the difference that that two paired reads are considered as a single unit. Another measure is TPM (Transcript Per Million), where the mapped reads are first normalized with the length of the gene followed by with the total of the normalized reads scaled by the factor 106. Significance of gene expression can be tested by estimating mean and variance of expression of a gene over replicates, which means that replicated data should be generated for DEG. A number of other measurements and plots, such as false discovery rate, MA plot and volcano plot can be generated to understand DEG data. Software like DeSeq2 (Love et al. 2014) and NOIseq (Tarazona et al. 2011) provide these normalized read counts for comparing gene expression and perform clustering or other multivariate techniques to study relationship of samples in terms of gene expression. Most commonly, the clustering is described with a heat map showing gene expression values.

5.4.5 Pathway and Gene Ontology Mapping

Once the DEGs are identified, the next step is to understand their biological roles. While annotations using blast identify the closest homolog from the database, more meaningful biological information can be derived by DEG. For this, two approaches, pathway mapping and gene ontology (GO) mapping are very helpful. The Kyoto Encyclopaedia of Genes and Genomes (KEGG) (https://www.genome.jp/kegg/pathway.html) maintains databases and tools for mapping a gene onto metabolic pathways, which is extensively used by researchers for assigning annotated genes to metabolic pathways, and in case of DEG, helps to identify reaction paths overexpressed and underexpressed within a metabolic pathway under two or more different conditions. However, only a small fraction of the genes identified in a transcriptome or from DEG are annotated by KEGG. The GO Consortium (http://geneontology.org), on the other hand, can provide biological meaning to more number of genes, assigning them under broad categories of cellular component, molecular function and biological process, under which several sub-categories are available, which sequentially describe the 'ontology' of the gene via a GO map. Another approach, cluster of orthologous groups (COG) (Tatusov et al. 2000) classify the annotated genes into several cluster of orthologous groups. The query protein sequences can be searched using 'blastp' against the COG database. Most gene annotation pipelines have capacity for searching these databases and assign biological meaning to the RNA-seq transcripts. The European Molecular Biology Laboratory (EMBL) hosts another gene ontology search tool, EggNOG (Huerta-Cepas et al. 2019), that includes non-supervised orthologs (NOGs) for functional characterization of a gene.

5.5 Transcriptomics of Plants

5.5.1 *Arabidopsis thaliana*

A. thaliana, the mouse ear cress, is a model plant species for biological researches on plant. Consequently, *Arabidopsis* transcriptomes are the most researched transcriptomes. Before the advent of NGS technologies, large scale gene expression experiments were carried out using microarray, which still provides useful information on expression pattern of the A. thaliana genes. The Unite´ de Recherche en Ge´nomique Ve´ge´tale (URGV), France hosts a publicly available database of *Arabidopsis* transcriptomes, CATdb (Complete *Arabidopsis* Transcriptome Database) (http://urgv.evry.inra.fr/CATdb), a collection of 281 *Arabidopsis* projects mainly obtained from the microarray data resources generated by the URGV transcriptome platform. It provides access to CATMA (Complete *Arabidopsis* Transcriptome MicroArray), developed by a European consortium. The CATMA probes cover over 85% of the genes present in *Arabidopsis* providing gene sequence tags for individual genes. It has further been extended to 20 other species and presently contains data on 353 projects. The Salk Institute hosts a *Arabidopsis* Transcriptome Genomic Express Database (http://signal.salk.edu/cgi-bin/atta), containing data from the *Arabidopsis* transcriptome Tilling array, exosome, At-TAX (a whole genome tilling array) and DNA methylome. It provides a pictorial description of the expression pattern of the genes. The *Arabidopsis* Information resource (Tair) (https://www.*Arabidopsis*.org/index.jsp) also contains exhaustive functional genomics resources on *Arabidopsis*. Various other databases are publicly available to researchers for transcriptomics studies in *Arabidopsis*, making it the most researched plant species (Table 5.3).

Table 5.3 A list of *Arabidopsis* functional genomics databases

Database	Description	url
Arabidopsis Transcriptome Genomic Express Database	Contains information from tilling, methylome, expression analysis. Provides gene specific expression profile	http://signal.salk.edu/cgi-bin/atta
Arabidopsis RNA-seq Database	Gene expression levels from 20,000+ public *Arabidopsis* RNA-Seq libraries	http://ipf.sustc.edu.cn/pub/athrna/
ARTADE -- *Arabidopsis* Tiling-Array-based Detection of Exons	Annotation of genome-wide tiling-array data	http://omicspace.riken.jp/ARTADE/
Arabidopsis Gene Regulatory Information Server (AGRIS)	Contains promoter sequences, transcription factors and their target genes	http://*Arabidopsis*.med.ohio-state.edu/
Arabidopsis Small RNA Project database (ASRP)	Information on small nuclear RNA	http://asrp.danforthcenter.org/
Arabidopsis Next Gen sequence database	A part of Next Gen sequence database at Donald Danforth Plant Science Centre	https://mpss.meyerslab.org/
AthaMap	Genome-wide map of potential transcription factor and small RNA binding sites	http://www.athamap.de/
CATMA	Provides high quality Gene-specific Sequence Tags (GSTs) covering most *Arabidopsis* genes	http://www.catma.org/
ePLANT	Multiple visualization tools for gene expression	http://bar.utoronto.ca/eplant/
Expression Atlas	Contains results of 962 experiments including *Arabidopsis*, rice and maize	https://www.ebi.ac.uk/gxa/plant/experiments
SeedGenes	Information on genes with essential function during seed development	http://seedgenes.org/
TraVA	A database of gene expression profiles based on RNA-seq	http://travadb.org/

5.5.2 Current Status of Transcriptomics in Crop Plants

Use of transcriptomics in understanding the biology and cultivation of the crop plants is rising sharply in the present century. However, the sequence read archive (SRA) deposits in NCBI (https://www.ncbi.nlm.nih.gov/sra/) for the most important 25 crops of the world show a skewed pattern. Out of these, the number of SRA deposits for ten crops (maize, rice, wheat, Brassica, soybean, tomato, cotton, tea, potato and sugarcane) are about 0.35 million, which is ten times higher than the SRA deposits for the other 15 crops (0.036 million) (Fig. 5.2), indicating that more transcriptomics research is needed for harvesting the benefit of this technology in minor crops. Maize and rice, two principal food crops have received maximum attention to the transcriptomics researchers, comprising about 53% of the total SRA deposits.

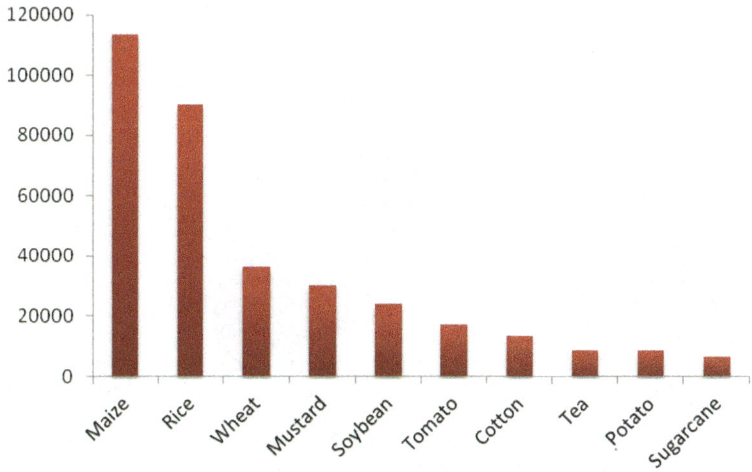

Fig. 5.2 Crop-wise top ten species with SRA (sequence read archive) deposits in NCBI SRA database (as on 11.02.2021)

Similarly, in Pubmed Central (https://www.ncbi.nlm.nih.gov/pmc/), a total of 0.156 million hits were recorded for these 25 cultivated crops (Fig. 5.3, data up to 31.12.2020). The distribution shows that over 50% of these hits are from four principal crops, rice (18.6%), wheat (11.8%), maize (13.3%) and soybean (9.1%), while another 26.4% are from tomato, potato and cotton. More emphasis on transcriptomics of other economically important crops would be required for having a better understanding of genetic basis of economically important traits in these crops. In the next section, we will give some examples of use of transcriptomics in jute (*Corchorus* spp.). Despite being the second most important fibre crop (after cotton), jute transcriptomics has received comparatively less attention and support than the food crops or even the beverage crops like tea and coffee. However, within a short time frame, transcriptomics has helped to understand a number of biological processes in jute, which is an inspiring example of the benefits of transcriptomics in crop plants.

5.6 Transcriptomics in Jute: An Overview

The jute plant represented by two species *Corchorus olitorius* L. and *Corchorus capsularis* L. (Malvaceae, subfamily Grewoideae) is cultivated for production of long, tough fibre synthesized in bast (phloem) tissue. The fibre, known as jute fibre is a lignocellulosic fibre is used for production of sacks, bags, burlaps, geotextiles, fibre composites and various other diversified products. It is valued globally as the most important non-textile fibre. While the principal producers of jute are India, Bangladesh and China, it is globally used to pack food grains and is in high demand

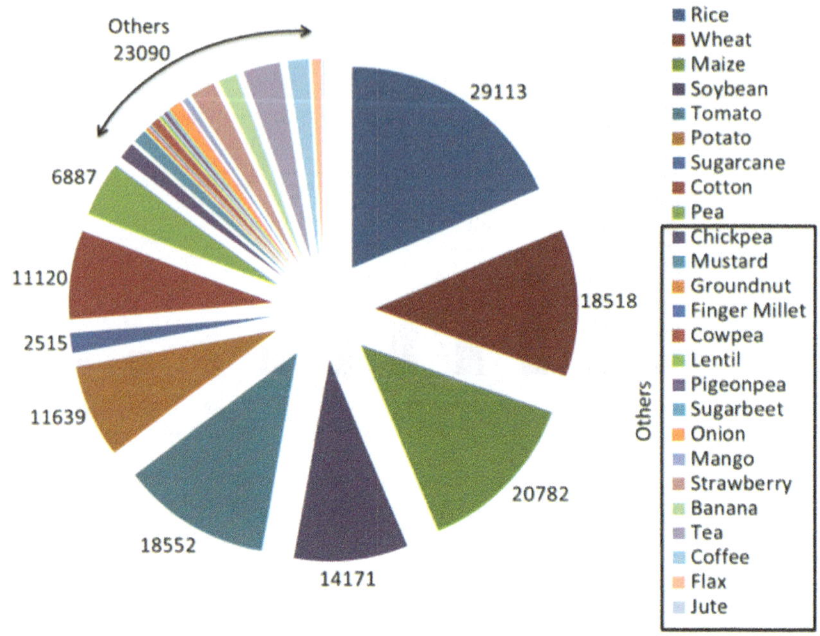

Fig. 5.3 Comparative crop-wise research focus in transcriptomics as indicated by number of hits returned by Pubmed Central on search with keywords "Transcriptome"+"respective crop name" between 2000 and 2020. Most of the researches (85%) are focused on major crops (rice, wheat, maize, soybean, tomato, potato, sugarcane, cotton and pea)

for replacement of synthetic polypropylene bags. Apart from producing natural fibres, jute plant consumes over 14 ton of CO_2/ha during its vegetative growth period of about 120 days and fixes nutrition to the soil by addition of leaf litters. It, therefore, is a climate-friendly crop that produces climate-friendly natural fibre. Consequently, research in jute genomics and transcriptomics has attracted considerable attention in recent decades in the wake of the rising concerns over climate change. Due to low genetic variability in jute at population level, researchers have concentrated more on transcriptomics to understand the genetics of economically important traits rather than using genomic tools like linkage mapping and genomic selection. This has generated a large amount of sequence information, identifying genes, regulatory sequences, metabolic pathways and genic markers. Till 11.02.2021, the number of SRA deposits for both the jute species was 714, which included full length high coverage transcriptomes of various tissues, as well as low coverage sequences from mapping experiments. The earliest reference of jute sequence was deposited to NCBI SRA archive in 2015 by the Central Research Institute for Jute and Allied Fibres, India, which were the RAD-seq (restriction-site associated DNA sequence) of jute cv. Sudan Green (SRX591273) and mutant bast fibre shy (*bfs*) and their F_2 plants. These were sequenced using IlluminaHiseq 2000 platform generating 2.2 Gb and 1.8 Gb sequences for Sudan Green and *bfs*, respectively. In this study, RAD-seq data for 330 F_2 genotypes were also deposited. The first SRA deposit for the first whole transcriptome sequence of jute was submitted by the same institute in 2015, providing transcriptome data for bast tissue of a mutant deficient in lignified phloem fibre production (*dlpf*) and its wild-type cv. JRC-212. Transcriptomes of different tissues including bast, hypocotyl, developing stem, root, fibre cell, leaf and flower have been generated in jute. In addition, expression of genes under different conditions such as salt-stressed and GA_3-treated plants has been investigated.

5.6.1 Transcriptome Assembly

A number of assemblers have been used for annotation and functional characterization of jute genes. Chakraborty et al. (2015) and Satya et al. (2018) performed de novo assembly of bast transcriptome using three assemblers CLC Genomics Workbench (v6.0; CLC bio, Aarhus), SOAPdenovo-Trans and Trinity. Islam et al. (2017) performed a reference-based assembly of the fibre cell transcriptome using Cufflinks. Yang et al. (2020) also developed reference genome based assembly using Bowtie and TopHat. The first transcriptome assembly with publicly available TSA (Transcriptome shotgun assembly) was generated by Chakraborty et al. (2015) using Illumina™Hiseq 2000 platform generating a total of 72,750,724 raw read and 67,424,930 clean reads for cv. JRC-212. After de novo assembly using CLC workbench, SOAPdenovo-Trans and Trinity, a total of 34,163 genes were identified. Among the three assemblers, Trinity was found to be the best performing with maximum percentage of unigene recovery.

5.6.2 Gene Discovery from Transcriptome Data

Wide variations have been reported for the number of genes expressed in different tissues of both the jute species. The earliest transcriptome study (Chakraborty et al. 2015) reported presence of 29,000-34,000 genes in the bast tissue of *C. capsularis*, which can be publicly accessed from the TSA database of NCBI. Some reports contain an exorbitantly high number of genes expressed (over 72,000) in jute, which probably needs to be verified. Overall, jute has an estimated number of 35,000-40,000 annotated genes (Table 5.4).

5.6.3 Orthologous Group Identification and Gene Ontology

Chakraborty et al. (2015) used four annotation databases, Nr, SwissProt, KEGG and COG for functional interpretation of the bast transcriptome assembly (Fig. 5.4), and reported that Nr-annotation was superior to the other three systems. Further, they identified gene ontology using Blast2GO (Conesa et al. 2005) and obtained the GO functional classifications using WEGO (Ye et al. 2006). Satya et al. (2018) followed

Table 5.4 Major transcriptomics studies in jute

Tissue (s)	Sequencing platform	Species	Unigenes identified	Reference
Tissues from Vegetative growth period, flowering period, bast of technical mature period, fruit	Illumina™ Hiseq 2500	*C. olitorius*	33,312 in total, 15,491 common in all the tissues	Yang et al. (2020)
Whole plant	Illumina™ HiSeq 4000	*Corchorus* sp. L.	72,674	Tao et al. (2020)
Shoot apiece	Illumina NextSeq 500	*C. olitorius*	14,050	Choudhary et al. (2019)
Hypocotyl	Illumina™ Hiseq 2000	*C. capsularis*	32,821-39,076 (annotated)	Satya et al. (2018)
Fibre cell	Illumina™ Hiseq 2500	*C. capsularis and C. olitorius*	37,031 (*C. olitorius*), 30,096 (*C. capsularis*)	Islam et al. (2017)
PEG-treated tissue	Illumina HiSeq X Ten	*C. olitorius*	45,831	Yang et al. (2017a, b)
Salinity-stressed tissue	Illumina HiSeq 4000 platform	*C. olitorius*	72,278	Yang et al. (2017a, b)
Pooled RNA from various tissues	Illumina™ Hiseq 2000	*C. capsularis*	48,914	Zhang et al. (2015)
Bast	Illumina™ Hiseq 2000	*C. capsularis*	34,163–29,463	Chakraborty et al. (2015)

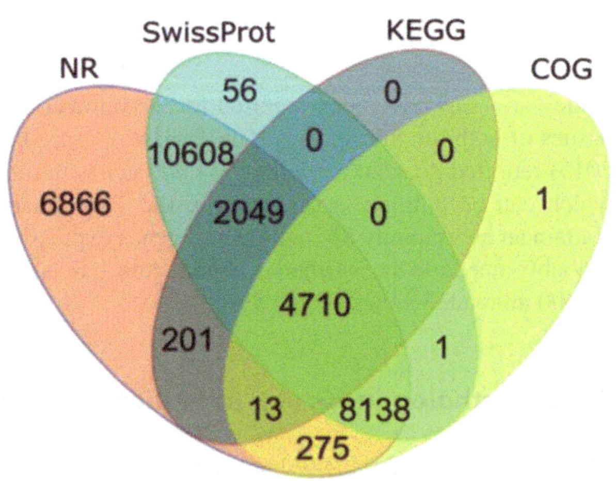

Fig. 5.4 A venn diagram representing annotations of the bast transcriptome of *Corchorus capsularis* cv. JRC-212 (Chakraborty et al. 2015) using four annotation databases. Note that SwissProt, KEGG and COG annotations did not add much information over Nr-annotations

similar methodology for annotation of hypocotyl transcriptomes and confirmed that Nr-annotation performed better. Both the studies screened the COG database to retrieve and classify COG functional categories of the genes. In case of reference-based assemblies, orthology and ontology analysis are not required, as the transcriptome is assembled based on a reference genome having genes with assigned functions.

5.6.4 Identification of Novel genes

Often, transcriptomics leads to discovery of novel genes that were unknown to exist in that species, family or even may be unknown to plant Kingdom. During examination of the role of β-galactosidases in hypocotyl development in jute, Satya et al. (2018) discovered a novel class of beta-galactosidases that are similar to prokaryotic β-galactosidase (Fig. 5.5). The prokaryotic β-galactosidase (a member of Glycosyl Hydrolase-2 or GH-2 family of enzymes) converts lactose to glucose and galactose, and was thought to be lost in higher eukaryotes. The domains of the GH-2 β-galactosidases are highly conserved in prokaryotes, consisting of three protein domains Glyco-hydro_2_N, Glyco-hydro_2 and Glyco-hydro_2_C that are linked to a Bgal_Small_N domain by another β-sandwich domain of unknown function (DUF4981). As such, plant cannot utilize lactose as a food source, which was thought to be due to absence of the prokaryotic GH-2 β-galactosidase. They, on the other hand, contain a number of β-galactosidases of GH-35 family, which function in cell wall formation by breaking galactose linked with other molecules. Satya et al. (2018) observed that a homolog of *E. coli* lacZ gene (codes for β-galactosidase) with this five-domain architecture is present not only in jute but in all the plants starting from algae to woody perennials. Phylogenetic study revealed that the plant GH-2 β-galactosidases evolved from the prokaryotic β-galactosidases. It was transferred from prokaryotes to lower plants (Marchantiophyta and Bryophyta) via Charophytic green algae and from lower plants to higher plants via Lycophyta. Protein modelling revealed remarkable similarity between the plant and prokaryotic GH-2 β-galactosidases despite having low sequence similarity.

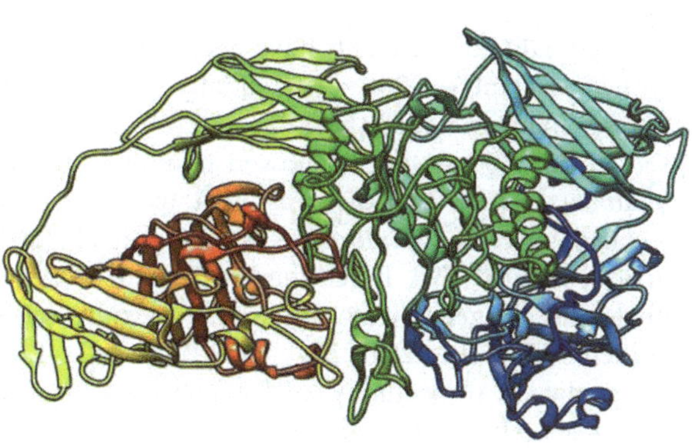

Fig. 5.5 A 3-D predicted protein structure of a novel prokaryotic β-galactosidase gene of jute discovered in plant lineage by hypocotyl transcriptomics. The structure was generated using Phyre2

Particularly, the catalytic residues that cause a nucleophile attack on the β-1,4 linkage of glucose and galactose were found to be conserved in higher plants.

5.6.5 Metabolic Pathway Identification

For metabolic pathway analysis, annotated genes are mapped to the KEGG database (Kanehisa et al. 2008) using the KEGG Automatic Annotation Server (Moriya et al. 2007). Chakraborty et al. (2015) characterized the phenylpropanoid biosynthesis pathways in jute that lead to monolignol formation and genes involved in the secondary cell wall development. They identified a total of sixteen genes with multiple isoforms which were involved in lignin biosynthesis and jute fibre formation. Islam et al. (2017) described the major genes involved in fibre formation in jute from genomic and transcriptomic datasets and observed that *C. capsularis* exhibits higher ATPase activity, oxidoreductase activity, transmembrane transport, vacuolar transport and homeostasis, suggesting that it has wider environmental adaptability. In another study, Satya et al. (2020) characterized the pectin biosynthesis pathways in jute, identifying 18 genes involved in interconversion of nucleotide-sugars, salvage biosynthesis of sugar-acids and polymerization of pectin monomers. Of these, 17 were involved in nucleotide-sugar interconversion and one, galacturonosyltransferase (GAUT) for polymerization of pectin monomers. A total of 12 *GAUT* genes were identified from both the species, which phylogenetically were distributed in seven subclades. Two of these, CcGAUT3 and CcGAUT12 were identified as the primary pectin homo-polymerizing enzymes. Both the CcGAUT3 and CcGAUT12 had an N-terminal transmembrane domain that carried a consensus motif ((R)–(X)2–(R)) for proteolytic cleavage. It was predicted that a CcGAUT3-CcGAUT12 complex may be involved in polymerization of galacturonic acid monomers in jute. The study also reported that the core pectin biosynthesis pathway is conserved in higher plants. Species that produce high mucilage, such as *Ziziphus jujube* exhibited high conservation with the jute pectin biosynthesis genes.

5.6.6 DEG Analysis

Only a few DEG experiments have been conducted in jute. Choudhary et al. (2019) identified a total of 240 differentially expressed transcripts between delayed flowering mutants under short-day (*pfr59*) in comparison with cv. JRO-524 and observed that 10 transcripts showed homology to known photoperiodic genes of *Arabidopsis*. DEG analysis was also used for identification of drought-stress associated genes in *C. olitorius* by Yang et al. (2017a, b). A drought sensitive cultivar exhibited 794 DEGs under drought stress, while in a drought tolerant cultivar only 39 genes were differentially expressed. Recently, Yang et al. (2020) identified 576/379, 291/227, 2367/255 and 1766/736 genes (upregulated/downregulated), respectively, in the stem bast, fruit, flower, and leaf compared to other tissues. They observed that 26 genes of the secondary metabolite biosynthesis pathway were consistently upregulated in the bast and the phenylpropanoid biosynthesis pathway genes were significantly upregulated in flower.

5.6.7 Marker Development

5.6.7.1 SSR

One of the major applications of plant transcriptomics is to identify EST-SSRs or genic SSRs. In jute, Zhang et al. (2015) discovered 1906 EST-SSRs with a frequency of di-, tri-, tetra-, and penta-nucleotide repeat types of 12.0%, 56.9%, 21.6% and 9.5%, respectively. They identified 113 transcription factor associated SSRs and 3 SSRs for cellulose synthase. Later, more SSRs (12,772) were identified from a bast transcriptome, with an average frequency of one SSR per 3.86 Kb (Satya et al. 2017). About 45.4% of the sequences exhibited repeat length between 10 and 15 nt. and 46.2% of the SSR loci were about 300–2000 nt. The number of repeats varied from 6 to 15 for dinucleotides, 5–8 for trinucleotides and 5–6 for tetranucleotides. Of the dinucleotide repeats, (TA/AT)6 was the most frequent (9.3%). They also identified 961 compound SSRs (7.5%). They also designed 39 phenylpropanoid biosynthesis pathway gene-specific SSR markers, seven SSR markers for peroxidase genes, and 24 SSRs for the genes involved in bast fibre formation. They also reported 4457 transcription factors (TF) and identified 2163 TF-SSRs. The study designed 1079 SSR primers and validated 120 of them using gel electrophoresis studies. Saha et al. (2017) identified 4509 SSRs and developed a set of 2079 flanking primer-pairs. They also developed a web-based SSR repository of jute (http://jutemarkerdb.icar.gov.in/). All these SSRs were found to show moderate polymorphism and were able to generate high intra-specific and inter-specific diversity.

5.6.7.2 SNP and InDel

Zhang et al. (2015) identified a total of 12,518 SNPs in jute with transition and transversion frequencies of 59.2 and 22.3%, respectively. Most of the SNPs were of the synonymous SNP type (99.37 %). Yang et al. (2018) identified 51,172 InDel sites in 18,800 unigenes of jute, which were distributed in 94 InDel types. Mononucleotide InDels were more (23,028) than bi-nucleotide (9824) or tri-nucleotide (9182) ones. The polymorphism information content of InDel markers in jute varied from 0.340 to 0.680, with an average of 0.491.

5.7 Conclusion

During the past 20 years, transcriptomics has established itself as an essential tool in plant biology. Advances in next generation sequencing have opened up new avenues for in-depth investigations of the sequence of events in a biological process at single cell level. The cost of transcriptomics studies have been reduced by several folds in recent years, allowing its wider application in plant biology and crop improvement. Moreover, publicly deposited transcriptomics studies not only benefit the researchers working for specific crops, but also enrich the public databases, allowing better precision for gene annotation in future researches. The role of transcriptomics will be invaluable for future plant research, particularly to battle various abiotic stresses escalating due to climate change, soil degradation, higher population pressure and water stress.

References

Adams MD, Kelley JM, Gocayne JD et al (1991) Complementary DNA sequencing: Expressed sequence tags and human genome project. Science 252(5013):1651–1656

Andrews S (2010) FastQC: a quality control tool for high throughput sequence data. Available online at: http://www.bioinformatics.babraham.ac.uk/projects/fastqc

Ashton PM, Nair S, Dallman T et al (2015) MinIONnanopore sequencing identifies the position and structure of a bacterial antibiotic resistance island. Nat Biotechnol 33(3):296–300

Bankevich A, Nurk S, Antipov D et al (2012) SPAdes: a new genome assembly algorithm and its applications to single-cell sequencing. J Comput Biol 19(5):455–477

Bray NL, Pimentel H, Melsted P et al (2016) Near-optimal probabilistic RNA-seq quantification. Nat Biotechnol 34(5):525–527

Brenner S, Johnson M, Bridgham J (2000) Gene expression analysis by massively parallel signature sequencing (MPSS) on microbead arrays. Nat Biotechnol 18(6):630–634

Chakraborty A, Sarkar D, Satya P et al (2015) Pathways associated with lignin biosynthesis in lignomaniac jute fibres. Mol Genet Genom 290(4):1523–1542

Choudhary SB, Saha D, Sharma HK et al (2019) Transcriptional analysis of a delayed-flowering mutant under short-day conditions reveal genes related to photoperiodic response in tossa jute (*Corchorus olitorius* L.). Ind Crop Prod 132:476–486

Clark MD, Panopoulou GD, Cahill DJ et al (1999) Construction and analysis of arrayed cDNA libraries. In: Weissman SM (ed) Methods enzymol, vol 303. Academic Press, San Diego, CA, pp 205–233

Conesa A, Gotz S, García-Gómez JM et al (2005) Blast2GO: a universal tool for annotation, visualization and analysis in functional genomics research. Bioinformatics 21(18):3674–3676

Davidson NM, Oshlack A (2018) Necklace: combining reference and assembled transcriptomes for more comprehensive RNA-Seq analysis. Gigascience 7(5):giy045

Diatchenko L, Lau YF, Campbell AP et al (1996) Suppression subtractive hybridization: a method for generating differentially regulated or tissue-specific cDNA probes and libraries. Proc Natl Acad Sci 93(12):6025–6030

Dobin A, Davis CA, Schlesinger Fet al (2013) STAR: ultrafast universal RNA-seq aligner. Bioinformatics 29(1):15–21

Edwards D, Batley J, Snowdon RJ (2013) Accessing complex crop genomes with next-generation sequencing. Theor Appl Genet 126(1):1–11

Eid J, Fehr A, Gray J et al (2009) Real-time DNA sequencing from single polymerase molecules. Science 323(5910):133–138

Eisen MB, Spellman PT, Brown PO et al (1998) Cluster analysis and display of genome-wide expression patterns. Proc Natl Acad Sci 95(25):14863–14868

Fedurco M, Romieu A, Williams S et al (2006) BTA, a novel reagent for DNA attachment on glass and efficient generation of solid-phase amplified DNA colonies. Nucleic Acids Res 34(3):e22

Fire A, Xu S, Montgomery MK et al (1998) Potent and specific genetic interference by double-stranded RNA in *Caenorhabditis elegans*. Nature 391(6669):806–811

Fodor SPA, Read JL, Pirrung MC et al (1991) Light-directed, spatially addressable parallel chemical synthesis. Science 251:767–773

Freedman A, Weeks N (2020) Best practices for *de novo* transcriptome assembly with Trinity. Available via https://informatics.fas.harvard.edu/best-practices-for-de-novo-transcriptome-assembly-with-trinity.html.

Frohman MA, Martin GR (1989) Rapid amplification of cDNA ends using nested primers. Techniques 1:165–173

Gordon A, Hannon GJ (2010) FASTX-Toolkit. FASTQ/A short-reads pre-processing tools http://hannonlab.cshl.edu/fastx_toolkit

Grabherr MG, Haas BJ, Yassour M et al (2011) Full-length transcriptome assembly from RNA-Seq data without a reference genome. Nat Biotechnol 29(7):644–652

Hölzer M, Marz M (2019) *De novo* transcriptome assembly: a comprehensive cross-species comparison of short-read RNA-Seq assemblers. Gigascience 8(5):giz039

Huerta-Cepas J, Szklarczyk D, Heller D et al (2019) eggNOG 5.0: a hierarchical, functionally and phylogenetically annotated orthology resource based on 5090 organisms and 2502 viruses. Nucleic Acids Res 47(D1):D309–D314

Islam MS, Saito JA, Emdad EM et al (2017) Comparative genomics of two jute species and insight into fibre biogenesis. Nat Plants 3(2):16223

Kanehisa M, Araki M, Goto S (2008) KEGG for linking genomes to life and the environment. Nucleic Acids Res 36:D480–D484

Khang TF, Lau CY (2015) Getting the most out of RNA-seq data analysis. PeerJ 3:e1360

Kim D, Langmead B, Salzberg SL (2015) HISAT: a fast spliced aligner with low memory requirements. Nat Methods 12(4):357–360

Kovaka S, Zimin AV, Pertea GM et al (2019) Transcriptome assembly from long-read RNA-seq alignments with StringTie2. Genome Biol 20(1):278

Lamarre S, Frasse P, Zouine M et al (2018) Optimization of an RNA-seq differential gene expression analysis depending on biological replicate number and library size. Front Plant Sci 9:108

Levene MJ, Korlach J, Turner SW et al (2003) Zero-mode waveguides for single-molecule analysis at high concentrations. Science 299(5607):682–686

Li B, Dewey CN (2011) RSEM: accurate transcript quantification from RNA-Seq data with or without a reference genome. BMC Bioinformatics 12(1):323

Liang P, Pardee A (1993) Distribution and cloning of eukaryotic mRNAs by mean of differential display: refinements and optimization. Nucleic Acids Res 21(14):3269–3275

Liu Y, Ferguson JF, Xue C et al (2013) Evaluating the impact of sequencing depth on transcriptome profiling in human adipose. PLoS One 8(6):e66883

Liu J, Yu T, Jiang T et al (2016) TransComb: genome-guided transcriptome assembly via combing junctions in splicing graphs. Genome Biol 17(1):213

Livak KJ, Schmittgen TD (2001) Analysis of relative gene expression data using real-time quantitative PCR and the 2− ΔΔCT method. Methods 25(4):402–408

Love MI, Huber W, Anders S (2014) Moderated estimation of fold change and dispersion for RNA-seq data with DESeq2. Genome Biol 15:550

Mao S, Pachter L, Tse D et al (2020) RefShannon: A genome-guided transcriptome assembler using sparse flow decomposition. PLOS One 15(6):e0232946

Maretty L, Sibbesen JA, Krogh A (2014) Bayesian transcriptome assembly. Genome Biol 15:501

Moriya Y, Itoh M, Okuda S et al (2007) KAAS: an automatic genome annotation and pathway reconstruction server. Nucleic Acids Res 35:W182–W185

Patro R, Duggal G, Love MI et al (2017) Salmon provides fast and bias-aware quantification of transcript expression. Nat Methods 14(4):417–419

Pertea M, Pertea GM, Antonescu CM et al (2015) StringTie enables improved reconstruction of a transcriptome from RNA-seq reads. Nat Biotechnol 33(3):290–295

Piétu G, Mariage-Samson R, Fayein NA et al (1999) The Genexpress IMAGE knowledge base of the human brain transcriptome: a prototype integrated resource for functional and computational genomics. Genome Res 9(2):195–209

Robertson G, Schein J, Chiu R et al (2010) *De novo* assembly and analysis of RNA-seq data. Nat Methods 7(11):909-912
Ronaghi M, Karamohamed S, Pettersson B et al (1996) Real-time DNA sequencing using detection of pyrophosphate release. Anal Biochem 242(1):84–89
Rothberg JM, Hinz W, Rearick TM et al (2011) An integrated semiconductor device enabling non-optical genome sequencing. Nature 475(7356):348–352
Saha D, Rana RS, Chakraborty S et al (2017) Development of a set of SSR markers for genetic polymorphism detection and interspecific hybrid jute breeding. Crop J 5(5):416–429
Satya P, Chakraborty A, Jana S et al (2017) Identification of genic SSRs in jute (*Corchorus capsularis*, Malvaceae) and development of markers for phenylpropanoid biosynthesis genes and regulatory genes. Plant Breed 136(5):784–797
Satya P, Chakraborty A, Sarkar D et al (2018) Transcriptome profiling uncovers β-galactosidases of diverse domain classes influencing hypocotyl development in jute (*Corchorus capsularis* L.). Phytochemistry 156:20–32
Satya P, Sarkar D, Vijayan J et al (2020) Pectin biosynthesis pathways are adapted to higher rhamnogalacturonan formation in lignocellulosic jute (*Corchorus* spp.). Plant Growth Regul 14: 1–7
Schena M, Shalon D, Davis RW et al (1995) Quantitative monitoring of gene expression patterns with a complementary DNA microarray. Science 270(5235):460–470
Schuler GD, Epstein JA, Ohkawa H et al (1996) Entrez: molecular biology database and retrieval system. Methods Enzymol 266:141–162
Shao M, Kingsford C (2017) Accurate assembly of transcripts through phase-preserving graph decomposition. Nat Biotechnol 35(12):1167–1169
Shendure J, Porreca GJ, Reppas NB et al (2005) Accurate multiplex polony sequencing of an evolved bacterial genome. Science 5741:1728–1732
Siebert PD, Larrick JW (1992) Competitive PCR. Nature 359:557–558
Smith-Unna R, Boursnell C, Patro R et al (2016) TransRate: reference-free quality assessment of *de novo* transcriptome assemblies. Genome Res 26(8):1134–1144
Song L, Sabunciyan S, Florea L (2016) CLASS2: accurate and efficient splice variant annotation from RNA-seq reads. Nucleic Acids Res 44(10):e98
Tao AF, You ZY, Xu JT et al (2020) Development and verification of CAPS markers based on SNPs from transcriptome of jute (*Corchorus* L.). Acta Agronomica Sinica 46(7):987–996
Tarazona S, García F, Ferrer A et al (2011) NOIseq: a RNA-seq differential expression method robust for sequencing depth biases. EMBnet J 17(B):18–19
Tatusov RL, Galperin MY, Natale DA et al (2000) The COG database: a tool for genome-scale analysis of protein functions and evolution. Nucleic Acids Res 28:33–36
Trapnell C, Williams BA, Pertea G et al (2010) Transcript assembly and quantification by RNA-Seq reveals unannotated transcripts and isoform switching during cell differentiation. Nat Biotechnol 28(5):511–515
Velculescu VE, Zhang L, Vogelstein B et al (1995) Serial analysis of gene expression. Science 270(5235):484–487
Wang S, Gribskov M (2017) Comprehensive evaluation of *de novo* transcriptome assembly programs and their effects on differential gene expression analysis. Bioinformatics 33(3): 327–333
Wang Y, Yang Q, Wang Z (2015) The evolution of nanopore sequencing. Front Genet 5:449
Xie YL, Wu GX, Tang JB et al (2014) SOAPdenovo-Trans: *de novo* transcriptome assembly with short RNA-Seq reads. Bioinformatics 30:1660–1666
Yang Z, Dai Z, Lu R et al (2017a) Transcriptome analysis of two species of jute in response to polyethylene glycol (PEG) induced drought stress. Sci Rep 7(1):1–11
Yang Z, Yan A, Lu R (2017b) *De novo* transcriptome sequencing of two cultivated jute species under salinity stress. PLoS One 12(10):e0185863

Yang Z, Dai Z, Xie D, Chen J, Tang Q, Cheng C, Xu Y, Wang T, Su J (2018) Development of an InDel polymorphism database for jute via comparative transcriptome analysis. Genome 61:323–327

Yang Z, Wu Y, Dai Z et al (2020) Comprehensive transcriptome analysis and tissue-specific profiling of gene expression in jute (*Corchorus olitorius* L.). Ind Crop Prod 146:112101

Ye J, Fang L, Zheng H et al (2006) WEGO: a web tool for plotting GO annotations. Nucleic Acids Res 34:W293–W297

Zhang L, Ming R, Zhang J et al (2015) *De novo* transcriptome sequence and identification of major bast-related genes involved in cellulose biosynthesis in jute (*Corchorus capsularis* L.). BMC Genomics 16:1062

Zhu T, Wang X (2000) Large-scale profiling of the *Arabidopsis* transcriptome. Plant Physiol 124(4):1472–1476

Molecular Breeding and Marker-Assisted Selection for Crop Improvement

6

Akarsh Parihar and Shiwani

Abstract

In an era of speedily increasing population and unpredictable incidence of climate changes, enhancing agricultural productivity has imposed a big challenge to the crop scientists. The situation becomes graver with the continuously shrinking land resources. To cope up with such an alarming situation, it will be quite beneficial to intervene the recent modern tools of genomics like molecular breeding and marker-assisted selection for crop improvement. Marker-assisted breeding involving trait introgression for biotic and abiotic resistance, breaking the genetic plateau and quality improvement will require a prior attention. In this chapter, we present an outline of the conventional breeding techniques, molecular breeding involving marker-assisted selection/breeding, DNA markers and mapping populations that have massive potential to perk up the effectiveness and accuracy of conventional plant breeding through marker-assisted selection (MAS), advantages of marker-assisted selection and its commonly used applications in plant breeding. Consideration is also given to genotyping methodologies and exploitation of genetic diversity. Finally, the approaches to study genotype-phenotype associations like QTL mapping, GWAS/Association mapping, transcriptomics and other techniques will also be discussed. Achieving a significant impact on crop improvement by MAS represents the great challenge as well as opportunity for agricultural scientists. The objective of this chapter is to present and describe the methods of molecular breeding and their genetic underpinnings.

A. Parihar (✉)
Department of Genetis & Plant Breeding, Anand Agricultural University, Anand, Gujarat, India

Shiwani
Centre of Excellence in Biotechnology, Department of Agricultural Biotechnology, Anand Agricultural University, Anand, Gujarat, India

R. L. Singh et al. (eds.), *Plant Genomics for Sustainable Agriculture*,
https://doi.org/10.1007/978-981-16-6974-3_6

Keywords

Molecular breeding · Marker-assisted selection · Mapping population · Foreground selection · background selection

6.1 Introduction

Plant breeders have been remarkably succeeded in constantly enhancing crop yield to meet the continuously increasing global demand for food, feed and fibre. The productivity increase in two major cereals wheat and rice in the history of humankind led by nobel laureate Norman E. Borlaug, so-called green revolution provides the most significant instance of the contribution of science towards an improved food security (Borlaug and Dowswell 2005). Despite enormous efforts made by the crop scientists in development of elite genotypes, the rate of increase in cereals productivity in the past decade has not met the global demand, especially in wheat and rice. Consequent upon, an increase in rate of gain in crop yield is required to maintain growing population that gradually seeks a nutritionally balanced and protein-enriched diet. The confront faced by the modern breeders becomes more daunting in view of (i) global warming and the consequent increased frequency of environmental constraints (e.g., drought, floods, high temperatures), (ii) the declined availability of natural resources (e.g., water, fertilizers, arable land), (iii) the rising cost of fuels, and (iv) reduction of environmental footprint of agriculture, hence improving its long-term sustainability. Accordingly, agriculture will require producing more sustainably with fewer resources. Genomics has come up like a new breeding standard which is based upon molecular approaches and already contributed to speed up the yield gain commonly achieved through conventional breeding practices. However, for an effective integration of conventional and molecular breeding approaches massive adoption of genomics-assisted selection is required (Leng et al. 2017). A better perceptive of the underlying QTLs for yield and its variability across seasons will be helpful in devising effective genomics-assisted breeding strategies for enhancing crop performance under wide range of environmental scenarios. MAB offers unequaled opportunities in comparison to conventional breeding approaches, to analyse the genetic background of traits, mainly those traits which are quantitatively inherited such as yield, biomass production, and other agronomic traits.

6.2 Conventional Plant Breeding

Conventional breeding (classical breeding or traditional breeding) is the development of new varieties (cultivars) of plants by using older tools and natural processes, as contrary to the newer, more sophisticated and sometimes radical tools of molecular plant breeding (Jain and Kharkwal 2004). In conventional breeding, desirable traits are assembled from different but usually closely related plants into the new cultivar using the techniques of crossing (hybridization) (Acquaah 2012). Consequently, the product of conventional breeding only emphasizes target traits which preexist in the genetic potential of the species, without introducing new genes.

Since 1900, Mendel's laws of inheritance gave the scientific basis for plant breeding. The genes located on the chromosomes govern all the traits and their manipulation by selection/recombination is basic underlying principle for plant breeding. There are three major breeding methods in crops, namely

(a) Selection, where some genotypes from a population are selected on the basis of desired trait and population produced from the same plant is selfed separately to produce a homozygous stable elite genotype.
(b) Hybridization includes crossing between two genetically dissimilar parents and thereby making a selection in the F_2 generation for a desired genotype followed by repeated selfing to get the desired stable genotype.
(c) Ploidy manipulation, where addition/deletion of a chromosome or a part of chromosome or a complete set of chromosome is done to manipulate the ploidy level of plant species to enhance the desired trait.

Although in crop improvement significant pace has been made by phenotypic selections for agronomically important traits but substantial difficulties are frequently found during the process. To develop a new variety through conventional plant breeding, it generally takes upto 8 to 10 years. Plant breeders are interested in to speed up the phenotypic selection and to make this process more efficient by using new technologies. Development of the molecular markers was a major breakthrough to overcome this limitation. Marker-assisted selection has become more important as large number of genes are being identified and functions and interactions between them are elucidated. Optimized strategies are needed to integrate phenotypic selection with MAS. Marker-assisted selection has been demonstrated as a very constructive practice in plant breeding. By using these techniques, breeders have been able to create agriculturally important crop varieties containing genes resistance to biotic and abiotic stress that was not possible before the introduction of DNA marker technology.

6.3 Molecular Markers and Genotyping Methodologies

6.3.1 Molecular Marker

Molecular marker technology has changed the fortune of plant breeding. Different molecular markers have been developed and their ultimate use in plant breeding has geared up crop improvement. Molecular breeding requires easy availability and accessibility of genomic resources. Technological advances in genomics era have provided an array of resources like whole genome sequences, transcriptomes, molecular markers, genetic linkage maps, etc.

6.3.2 Genetic Markers

Genetic markers are significant developments in the area of plant breeding. Marker is a nucleotide sequence with known location of chromosome which is controlling a gene or a phenotype. Markers are closely linked with the target gene in chromosome and they act as a sign or flags. Genetic markers are generally categorized into two groups: classical markers and DNA/molecular markers. Morphological, cytological and biochemical markers are types of classical markers and restriction fragment length polymorphism (RFLP), random amplified polymorphic DNA (RAPD), amplified fragment length polymorphism (AFLP), simple sequence repeats (SSRs), single-nucleotide polymorphism (SNP) and diversity arrays technology (DArT) markers are some examples of DNA markers (Jiang 2013).

Molecular markers are the nucleotide sequences which can be investigated through the polymorphism between the nucleotide sequences of diverse individuals. The basis of the polymorphism can be insertion, deletion, duplication, point mutations and translocation. Depending on application, ideal molecular markers should meet the following criteria for their efficient use in marker-assisted breeding:

- Highly reproducible and should be highly polymorphic.
- Even distribution throughout the whole genome.
- Co-dominance in nature.
- Should be free of any developmental stage specificity.
- Single copy and non epistatic.
- Simple, efficient and inexpensive.
- Automation should be easy.
- High availability and suitability to be multiplexed.
- Genome-specific in nature especially in case of polyploids.
- No harmful effect on phenotype.

6.3.3 Classification of Molecular Markers

Molecular markers are classified on the basis of:

1. Gene action (co-dominant or dominant markers).
2. Detection method (hybridization-based or polymerase chain reaction (PCR) based).
3. Mode of transmission (paternal/maternal organelle inheritance, biparental/maternal nuclear inheritance) (Semagn et al. 2006).

Various types of molecular markers have been successfully utilized in plant breeding for the improvement of various agricultural crops. Mariya et al. (2017), while working on grain protein content and genetic diversity among 28 rice genotypes at biochemical and molecular levels, encompassed phenotyping for protein and protein fractions and SSR analysis. The data analyzed for crude protein, total protein, and its fractions, viz. albumin, globulin, prolamin, and glutelin, revealed significant differences among 28 rice genotypes. In case of fractions, albumin and globulin showed positive and significant correlation between them at both genotypic and phenotypic levels, while prolamin and glutelin also exhibited positive and significant correlation at both the levels. Analysis of 13 SSR primers generated a total of 185 alleles, out of which all were polymorphic with an average of 14 bands per primer. The subsequent segment provides brief information about the molecular markers. Comparison of the important characteristics of commonly used molecular marker has been given in Table 6.1.

6.3.4 Application of Molecular Markers in Crop Science

6.3.4.1 Evolution and Phylogeny

Molecular markers are being used to reform the genetic map to acquire information about the phylogeny and evolution. Wang et al. (2017) have evaluated the genetic diversity and population structure of wild soybean by using chloroplast and nuclear gene sequences. Walunjkar et al. (2015), while working on diverse germplasm of pigeon pea, reported that both RAPD and SSR techniques may provide consistent data, and thereby can be used to study genetic diversity in pigeonpea, showing concordant values of genetic diversity. With minor fluctuations, the four wild genotypes (*Cajanus scarabaeoides, Rhynchosia rufescens, Cajanus cajanifolius,* and *Rhynchosia cana*) and the four cultivars (GTH-1, GT-100, ICPL-87, and GT-1) were present distinctly in the same subgroups both in the dendrograms obtained with RAPD and SSR analysis. Hence, it can be assumed that the results of diversity analysis with SSR markers validate the finding of RAPD analysis.

6.3.4.2 Investigation of Heterosis

Wu et al. (2013) have used SSRs markers for the investigation of diversity and heterosis in rice. SSR markers have been used for the investigation of heterotic groups and patterns in rice (Xie et al. 2014). Some studies show that transcriptome analysis has been utilized to examine the genes involved in heterosis (Guo et al. 2017; Li et al. 2017).

6.3.4.3 Identification of Haploid Plants and Cultivars Genotyping

Double haploid (DH) is very important mapping population for quantitative trait loci (QTL) mapping (Khush and Virmani 1996). DH plants allow the accurate detection of candidate genes of interest (Kunzel et al. 2000; Belicuas et al. 2007). The R1-nj (Navajo) anthocyanin colour marker has been effectively used for the identification of haploids (Melchinger et al. 2015). Similarly, SNP and SSR markers have also been used to detect double haploid and hybrids in maize (Tang et al. 2006) and rice (Shahid et al. 2013; Wu et al. 2015).

6.3.4.4 Genetic Diversity Assessment

Genome sequencing offers a great advantage to explore the genetic diversity in a very big germplasm (Nawaz et al. 2016). Molecular markers have been effectively

Table 6.1 Important characteristics of molecular markers

Characteristics	RFLP	RAPD	SSR	SSCP	CAPS	SCAR	AFLP	SNP	Dart	SRAP
Co-dominant/ dominant	Co-dominant	Dominant	Co-dominant	Co-dominant	Co-dominant	Co-dominant	Dominant	Co-dominant	Dominant	Dominant
Abundance	High	High	Medium	Low	Low	Low	High	Very high	Very high	Low
Reproducibility	Low	High	Medium	Medium	High	High	High	High	High	High
Degree of polymorphism	Medium	Medium	Medium	Low	Low	Medium	Medium	High	High	Medium
Locus specificity	Yes	No	No	Yes	Yes	Yes	No	Yes	Yes	Yes
Technical requirement	High	Low	Medium	Medium	High	Medium	Medium	High	High	Medium
Quantity of DNA required	High	Medium	Low	Low	Low	Low	Medium	Low	Low	Low
Major application	Physical mapping	Gene tagging	Genetic diversity	SNP mapping	Allelic diversity	Gene tagging and physical mapping	Gene tagging	Genetic mapping	Genotyping	Genotyping

applied for the determination of genetic diversity and genetic material classification in soybean (Wang et al. 2015), potato (Tiwari et al. 2013) and rice (Naeem et al. 2015). DArtseq and SNP are robust molecular markers and have successfully utilized for the genetic diversity analysis and genetic mapping in wheat (Akbari et al. 2006), carrot (Grzebelus et al. 2014) and common beans (Brinez et al. 2012).

6.3.4.5 Utilization of Molecular Markers in Backcrossing

Some desired traits are selected from exotic germplasm/plant genetic resources and transferred into crop plants by repetitive backcrossing (Simmonds 1993). Many genes of desired traits have been transferred from wild relatives to important cultivated varieties by using marker-assisted selection. Mostly SSR markers are being used for this purpose. Wang et al. (2005) have successfully transferred the *Lgc-1* locus by using two SSR markers which is related to low gluten level in *japonica* rice with 93–97% selection efficiency. Barley yellow mosaic virus is an important disease in barley and *rym4* and *rym5* are genes responsible for resistance to this disease and various markers have been developed to select these genes in barley population (You-Xin et al. 2012).

6.3.4.6 Linkage Map Construction

Markers like RFLP, ALP, SSR, ESTs, Dart and SNPs have been widely used for the construction of linkage maps (Semagn et al. 2006). The number of markers used depends upon the genome size as large number of markers is required for the species with larger genome size. With the advancement in new generation sequencing technologies thousands of markers have been extensively utilized for high resolution genetic mapping (Dhingani et al. 2015).

6.3.4.7 Varietal and Hybrid Identification

SSR markers have been used to check the purity of many crops like maize and rice (Li et al. 1999; Wang et al. 2003; Islam et al. 2012).When hybrids are detected using SSRs, the parents should be provided, and then primers are selected on the condition of the known parent. This method is very effective in confirming the purity and authenticity of a plant (Singh et al. 2016).

6.4 QTL Mapping and Applications

The agricultural traits of economic interest are majorly polygenic/quantitative in nature and controlled by many genes on the same or different chromosome. The chromosomal regions containing genes related with quantitative traits are referred to as QTL. The method utilizing molecular markers to locate the QTLs is known as QTL mapping.

Notably, mapping of QTL controlling the target traits facilitates plant breeders to use marker-assisted selection on the basis of polymorphic molecular markers flanking the relevant loci.

There are different types of QTLs, grouped into different categories on the basis of their effect size, effect of the environment on their expression, the type of effect produced by them, and the manner of their action.

1. ***Main effect QTLs*:** This has a direct effect on the expression of the concerned traits. A main effect QTL is described as a *major QTL* if it explains 10% or more of phenotypic variance for the trait, whereas a QTL with an effect size smaller than 10% is termed as *minor QTL*.
2. ***Epistatic QTLs*:** Epistatic QTLs are the same as modifying genes or modifiers, they interact with the main effect QTLs to influence the trait phenotype and together constitute the genetic background.
3. ***eQTLs*:** Many QTLs affecting the expression level like the level of RNA transcript of various genes produced in a tissue have been described as *expression QTLs* (*eQTLs*) or *regulatory QTLs*.
4. ***mQTLs*:** These QTLs control metabolic traits and metabolite levels. These generally show epistatic interactions and have moderate phenotypic effects. Metabolic traits have been generally found to possess much lower heritability than eQTLs and mQTLs for a specific trait.

6.4.1 QTL Mapping

There are four salient requirements for QTL mapping:

(1) An appropriate mapping population.
(2) A dense marker linkage map with good resolution for the species.
(3) Reliable phenotypic evaluation for the target trait.
(4) Appropriate software packages for QTL detection and mapping.

The general procedure for QTL mapping is briefly summarized below.

1. Two homozygous lines having contrasting phenotypes for the trait of interest are selected to cross to develop a suitable mapping population.
2. The mapping population is evaluated for the target trait in replicated trials conducted over locations and years, the practice is known as *phenotyping*.
3. The two parents of the mapping population are screened with large number of markers covering the entire genome at a sufficient density, and polymorphic markers are identified.
4. All the individuals of the mapping population are screened using these polymorphic markers; this practice is termed as *genotyping*.
5. The data of marker genotype are used to assemble a framework linkage map for the population depicting the order of markers and genetic distance between them in terms of centimorgans (cM).
6. The trait phenotype data and the marker genotype data are analysed to identify association between marker genotypes and the trait phenotype. In other words,

the plants are divided into separate groups on the basis of their marker genotype. For each of these groups, mean and variance for the trait phenotype are estimated and used for comparison. In case the genotype groups for a marker differ significantly for the trait of interest, it is concluded that the concerned marker is associated with the trait, i.e., the marker is most likely linked to a QTL controlling the trait phenotype.

6.4.2 Mapping Population

A population which is suitable for linkage mapping of genetic markers is well-known as *mapping population*. Mapping populations are developed by crossing two or more genetically diverse lines. Generally, the parents used for hybridization will be from the same species. But in some cases, where intraspecific variation is limited, related species can be used as parents. Parihar et al. (2010), while working on RAPD through bulk segregant analysis employed for the F2 population of a cross between WBPH-resistant parent, Gurjari, and highly susceptible parent, Jaya, comprised of the resistant as well as susceptible plants, reported that out of the total 50 random primers surveyed, a single linked marker OPA08 was identified to be putatively linked to resistant gene, as was evident by its presence in almost all the resistant bulks and vice versa. There are mainly two types of mapping populations, viz., primary and secondary mapping populations. *Primary mapping populations* are generated by hybridization between two homozygous lines usually having contrasting forms for the trait of interest. *Secondary mapping populations* are created by crossing two lines or individuals selected from a mapping population. The primary mapping populations are of the following different types:

(1) F_2, (2) F_2_derived F_3 (F_2:$_3$), (3) backcross (BC), (4) backcross inbred lines (BILs), (5) dou- bled haploids (DHs), (6) recombinant inbred lines (RILs), (7) near-isogenic lines(NILs), (8) chromosomal segment substitution lines (CSSLs), (9) immortalized F_2, (10) advanced intercross lines, (11) recurrent selection back-cross(RSB) populations, and (12) interconnected populations (Fig. 6.1). A summary of the characteristic features of the important mapping populations is given in Table 6.2.

6.4.3 Methods for QTL Detection and Mapping

QTL mapping methods have to resolve the three major issues:

1. The QTL genotypes of different individuals are not observed and, as a result, have to be deduced.
2. (2)Since there are potentially thousands of possible loci in the whole genome, an appropriate genetic model for QTL analysis has to be selected from among large number of possible models.

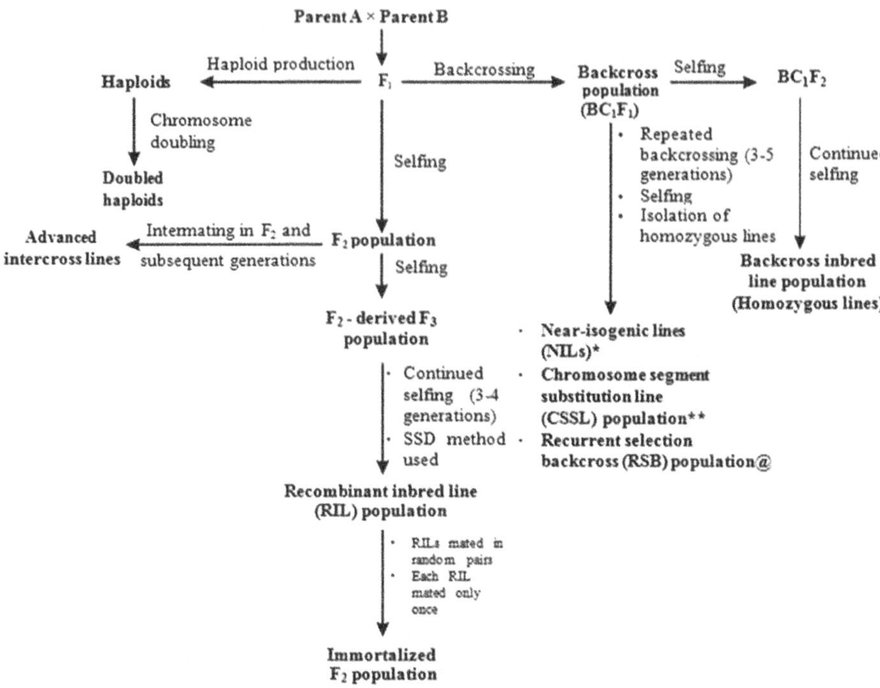

Fig. 6.1 Schematic representation of various biparental mapping populations. *Gene introgressed by repeated backcrossing followed by selection for the gene. **Repeated backcrossing without selection. @Individual with highest trait value in each backcross generation is selected and backcrossed with the recurrent parent. Source: Singh and Singh (2015)

3. The loci located in the same chromosome are correlated and, as a consequence, difficult to separate.

QTL analysis is an area of intensive research activity as it poses a variety of challenging questions that need to be resolved for obtaining reliable and reproducible results. QTL analysis approaches have been classified into the two main groups: (1) single QTL mapping and (2) multiple QTL mapping (Zou 2009). Each of these groups, in turn, comprises several approaches. Most of these approaches use maximum likelihood parameter estimation, regression analysis or Bayesian models for the detection of QTLs.

1. **Simple QTL mapping**
   These methods are able to detect a single QTL at a time. These methods do not take into account to other QTLs affecting the target trait that may be present in the genome. However, quantitative traits are considered to be governed by several polygenes, which do not locate in a single QTL. Therefore, these methods tend to be less reliable than those from multiple QTL methods. But these methods are the

Table 6.2 Comparative summary on some of the common features of mapping population

Features	Mapping population						
	F_2	DH	Backcross	RIL	NIL	CSSL	Immortalized F_2
Perpetuation	Ephemeral	Ephemeral	Ephemeral	Perpetual	Perpetual	Perpetual	Perpetual
Genetic composition	Homozygotes and heterozygotes	Homozygotes	Homozygotes and heterozygotes	Homozygotes	Homozygotes	Homozygotes	Homozygotes and heterozygotes
Each genotype represented by	One plant	One line	One plant	One line	One line	One line	One line
Number of generations required to develop	Two	Two	Two	7–8 or more	8–10	7–10	One (after RILs are developed)
Number of crosses made	One	One	Two (F1 and backcross)	One	6 or more	6 or more	Many
Number of recombination events	One	One	One	About two	One + the number of backcrosses	One + the number of backcrosses	About two
Segregation ratios for dominant and co-dominant markers	Different	Same	Different	Same	Same	Same	Different

(continued)

Table 6.2 (continued)

	Mapping population						
Features	F_2	DH	Backcross	RIL	NIL	CSSL	Immortalized F_2
Merits	Suited for oligogene mapping and mapping of heterosis loci. The degree of dominance can be estimated	Useful for mapping both qualitative and quantitative characters. Epistasis can be detected	Suitable for mapping oligogenes	Suitable for oligogene mapping, QTL mapping, fine mapping and positional cloning. Epistasis can be detected	Suitable for tagging the qualitative and quantitative traits. Quite useful in functional genomics. Epistasis can be detected	Suited for mapping of both oligogenes and QTLs. Also used for fine mapping.	Suited for genetic analysis of heterosis and detection of heterotic loci

Source: Singh and Singh (2015)

simplest and the earliest approaches for QTL mapping. The two main methods in this category are single-marker analysis and simple interval mapping.

(A) **Single-Marker Analysis**

Single-marker analysis (*SMA*), also called *single-point analysis*, is the simplest and the earliest used method of QTL detection (Soller and Brody 1976). In this method, each marker is separately tested for its association with the target trait. The phenotypic means for the plants placed in the different marker genotype groups are compared to detect a QTL at or near the site of the marker. The significance of differences between the means of the marker classes can be tested by Student's *t*-test, analysis of variance, linear regression analysis, likelihood ratio test, or maximum likelihood estimation. The *t*-test can be applied when the marker genotype has only two classes and individuals in population are classified according to genotype at a marker locus. For *t*-test the significance of difference between the traits means for the two marker genotype groups is tested. A significant difference between the trait mean indicates that the marker is to be linked to a QTL. This procedure is repeated for every marker locus evaluated in the mapping population. The magnitude of difference between the phenotypic mean and the marker genotype classes provide an estimation of the effect produced by the substitution of a single allele at QTL locus.

(B) **Simple Interval Mapping**

Interval mapping (IM) also known as *simple interval mapping (SIM)* has been developed by Lander and Botstein (1989). This method is regarded as the second level method of QTL mapping. SIM has become a standard QTL mapping procedure and has been further extended as composite interval mapping and multiple interval mapping procedures. SIM requires a marker linkage map for QTL search as it uses neighbouring marker pairs to define marker intervals and searches QTLs within these intervals. SIM makes a one-dimensional search for QTL at many locations, at every 1 or 2 cM, of each marker interval. The SIM model considers at a time a single QTL affecting the concerned trait, and each marker interval is analysed independent of the other marker intervals.

2. **Multiple QTL Mapping.**

Multiple QTL mapping (MQM) combines SIM and multiple regression analysis to incorporate all important QTLs found in the genetic model (Jansen 1994).

MQM offers the following advantages:

1. Consideration of other QTLs affecting the trait tends to reduce residual variation and increase the QTL detection power.
2. Linked QTLs can be detected as separate QTLs.
3. The estimates of QTL effects are more reliable than those with single QTL methods.
4. QTL × QTL interaction can be detected. But when too many markers are included as cofactors in the model, the QTL detection power tends to decline in comparison to SIM.

The main MQM methods include:

(a) composite interval mapping,
(b) multiple interval mapping,
(c) Bayesian multiple QTL mapping.

6.4.3.1 Composite Interval mapping

Composite Interval Mapping combines interval mapping with multiple regression analysis (Jansen 1994; Zeng 1994). The effect of QTLs present in other marker intervals of the same chromosome is controlled by CIM which increase the accuracy of QTL detection. CIM first carries out single-marker analysis then typically builds up the model as multiple QTL model using stepwise regression method. In this approach, the marker with highest LOD score is selected first then the marker with second highest LOD score is added, and the two markers are reevaluated for significance. If both the markers remain significant, the marker with next highest LOD score is added to the model, and the significance of the three markers is reevaluated.

6.4.3.2 Multiple Interval Mapping

The *multiple interval mapping (MIM)* approach is devised for simultaneous QTL mapping in multiple marker intervals (Kao et al. 1999). MIM avoids the complicated procedure used in CIM for the selection of background markers, but it uses several selection methods like forward search method and forward and backward selection methods to search for the best genetic model. MIM is able to take into account epistatic interactions, if present, among the multiple QTLs included in the model. As number of QTLs included in the model is increased, there is an exponential increase in number of parameters. As a result, the MIM implementation is computationally intensive.

6.4.3.3 Bayesian Multiple QTL Mapping.

Bayesian QTL mapping has been designed for the discovery of multiple QTLs. Number of QTLs treated as a random variable and reversible-jump Markov Chain Monte Carlo (MCMC) procedure is used further for specific modeling (Satgopan et al. 1996; Banerjee et al. 2008). In a Bayesian model, a prior distribution is selected, from which the posterior distribution is derived, and inferences are drawn from the posterior distribution. Both CIM and Bayesian methods use maximum likelihood functions. The advantages of prior distribution decline with the increase in sample size. Therefore, for most bi-parental mapping populations (population size
in hundreds), the Bayesian method may offer little advantage over the conventional mapping, particularly when high-density maps are available, and the genotype data are nearly complete. The Bayesian mapping methods are flexible in handling the ambiguity related to the QTL number, locations of the QTLs, and missing genotypes of QTLs. Bayesian models estimate the probability that a QTL exists in a given marker interval; this feature is regarded as the major advantage of these methods.

6.4.4 Factors Affecting the QTL Detection

The genetic properties of QTL, environmental factors, experimental errors in phenotype in band size of population are main factors affecting the QTL detection (Bernardo et al. 2015). There is direct effect of environment on the expression of quantitative traits and when some experiments are conducted on the same sites for various seasons, helps in the detection of environmental effect on the QTL (George et al. 2003).The population size also directly influences the QTL mapping studies, population which are large in size results in precise mapping and facilitates the detection of QTLs with less prominent effects (Tanksley 1993; Haley and Andersson 1997). Imprecise phenotyping and genotyping create experimental errors. The distance between markers can be affected by errors in phenotypic and genotypic data (Hackett 2002).

6.4.5 QTL Validation

After QTL detection, it is required to validate that particular QTL. For this reason, diverse populations will be developed by crossing different parents in order to check the presence of a particular QTL in other populations with different genetic background. NILs mapping populations are usually used for validation and confirmation of QTL (Collard et al. 2005). Confirmation of QTL provides the information that marker should be used for MAS or not (Ogbonnaya et al. 2001).

6.4.6 Software for QTL Mapping

Some methods of QTL analysis like SMA and regression interval mapping can be performed using standard statistical software. But other QTL analysis methods require special software packages for their implementation. Most of these software packages are listed at http://www.linkage.rockefeller.edu.soft. Most of these packages would give similar, if not the same, results for the same datasets, but they differ with respect to the required data format, computer platform used, user interface, graphic output, etc.

(i) MapMaker/QTL
(ii) QTLCartographer
(iii) PLABQTL
(iv) MapManagerQT/QTX
(v) R/QTL
(vi) R/QTLBIM
(vii) QTLExpress
(viii) FlexQTL
(ix) INTERQTL

(x) MCQTL
(xi) QGene

6.5 Molecular Breeding

Molecular breeding is a technique that uses molecular markers in combination with linkage maps and genomics to improve a particular trait. There are different molecular breeding strategies like:

- Marker-assisted selection (MAS).
- Marker-assisted backcrossing (MABC) and.
- Marker-assisted recurrent selection (MARC) (Ribaut et al. 2010).

MAS involves indirect phenotypic selection where individuals are selected based on marker pattern.

6.5.1 Marker-Assisted Selection or Marker-Aided Selection (MAS)

It is a selection process where a trait of interest is selected on the basis of a marker which should be tightly linked with the trait of interest.

The greater part of MAS works uses DNA-based markers in the present era. In 1923, Sax firstly reported an association of a simply inherited genetic marker with a quantitative trait in plants, he observed that segregation of seed size associated and segregation for seed coat colour marker were associated with beans (*Phaseolus vulgaris* L.). Plant breeders mostly use MAS for the identification of suitable dominant or recessive alleles across a generation and for the identification of the most favourable individuals across the segregating progeny (Francia et al. 2005). Kulshrestha et al. (2020) reported that molecular marker is a tool to study the resistance in tomato genotypes against root knot nematode. The resistance for Meloidogyne species is imparted by Mi family genes in tomato. Mi-1 was found to confer resistance towards *M. incognita, M. javanica* and *M. arenaria.*

6.5.1.1 Prerequisites for Marker-Assisted Breeding Program

Marker-assisted breeding needs more facilities as compared to conventional breeding. Pre-requisites essentials for marker-assisted breeding (MAB) in plants are listed below.

(a) Appropriate marker system and reliable markers should be there.
(b) High-throughput marker detection system.
(c) Genetic maps need to provide framework for the detection of marker and trait association so to choose markers and using it in marker-assisted selection.
(d) Efficient data processing and management system.

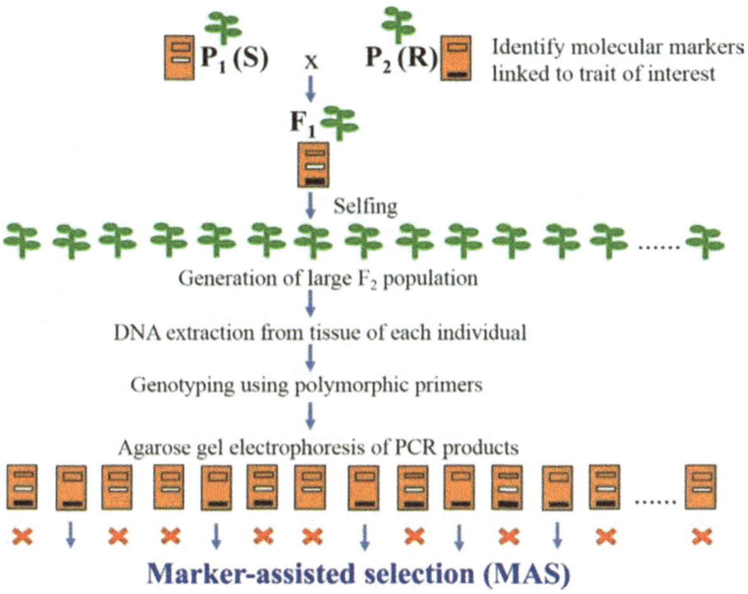

Fig. 6.2 Basic procedure for marker-assisted selection

The common procedure of MAS is given in Fig. 6.2.

Marker-assisted selection includes the following major methods:

- Molecular marker-based screening of populations for genotypes of interest.
- Marker-assisted backcross: one or more QTLs can be transferred from donor to recipient parent by repeated backcrossing so that desired trait of interest can be approved.
- Gene pyramiding schemes: where genes (two or more) identified in multiple lines/parents are accumulated into a single genotype.
- Marker-based recurrent selection: a complex scheme used for more loci involving several generations of selection and random mating of selected individuals.
- Selection: based on combining molecular and phenotypic data.

6.5.1.2 Important MAS Schemes

Important schemes used for MAS are as follows:

1. Marker-assisted backcrossing.
2. Marker-assisted recurrent selection.
3. Gene pyramiding.
4. Genomic selection.

Table 6.3 Examples of advantages of MAS over phenotypic selection

Trait/situation	Discouraging feature of phenotypic selection	Reference
Foreground selection		
Soybean cyst nematode	Time consuming, high cost	Young (1999); Bernardo (2008)
Cereal cyst nematode (wheat)	Slow speed, very high cost	Brennan and Martin (2007)
Crown rot resistance (wheat)	Slow speed, high to very high cost	Brennan and Martin (2007)
Small-scale quality tests (wheat)	Slow to moderate speed, medium to very high cost	Brennan and Martin (2007)
Amylose content (rice)	Reliable estimation is cumbersome	Gopalakrishnan et al. (2008)
"Quality protein" trait governed by o_2 mutant allele (maize)	Expensive biochemical assay, recessive inheritance	Babu et al. (2005)
Provitamin A (maize)	Time consuming, high cost	Muthusamy et al. (2014)
Background selection		
Recovery of recurrent parent genome	Slow progress (82% recovery in BC_4F_7) as compared to MAS (97% recovery in $BC_2F_{2:3}$)	Randhawa et al. (2009)
Recombinant selection	Very poor effectiveness	Young and Tanksley et al. (1989)

Source: Singh and Singh (2015)

1. **Marker-assisted backcrossing (MABC)**
   In marker-assisted backcross breeding, a useful trait is transferred from a donor parent into a recurrent parent, which is a superior variety deficient in this trait. The F_1 from cross between the donor and recurrent parent and the subsequent progenies are backcrossed to the recurrent parent. As a result, donor parent's genome is progressively replaced by the recurrent parent genome. The gene/QTL which is being transferred from the donor parent must be maintained by a rigorous phenotypic selection or else it would be rapidly replaced by the recurrent parent allele. Since the ultimate result of a backcross program is the transfer of target gene(s)/QTL(s) into the recurrent parent genotype, this process is often referred to as *gene/QTL introgression*. Some of the examples where MAS is preferable to phenotypic selection are summarized in Table 6.3.

   Backcross breeding has been extensively used because in each crop there are some varieties that are popular with the farmers. Therefore, farmers are more likely to accept a superior version of such a variety than an entirely new variety. Finally, backcross method can be continued to use for transgene introgression because in many crop species either land races or obsolete varieties have to be utilized for genetic transformation in view of technical difficulties. In such cases, backcross program must be second-handed to transfer the transgenes from the agronomically inferior transgenic lines (used as donor parent) into the elite varieties (serving as recurrent parent).

Objectives of backcross breeding are as follows:

(a) Transfer of the desired trait from a donor parent into a recurrent parent.
(b) Recurrent parent genome should be recovered.
(c) Complete exclusion of the donor genome, leaving only the target gene/QTL.

Molecular markers can be utilized to achieve all the three objectives.

- Target gene/QTL related markers enable indirect selection of the gene/QTL (foreground selection).
- Co-dominant markers distributed across the genome allow plants with the highest percentage of the recurrent parent genome to be selected (background selection).
- Co-dominant markers on either side of the target gene may be used to select unusual recombinants that do not have the genome of the donor outside these markers (recombinant selection).

(A) **Foreground Selection**

The indirect selection for the target gene/QTL on the basis of the associated marker genotype was suggested by Tanksley (1983) and was named Hospital and Charcosset's foreground selection (Hospital and Charcosset 1997). Foreground selection will be highly preferable to phenotypic selection when phenotypic evaluation for the target trait is problematic for anyone or more of the several reasons. In addition, it will greatly facilitate multiple QTL transfer and multitrait introgression and would replace disease tests during selection. Finally, the combination of oligogenic and polygenic resistance to plant diseases and insect pest would make it indispensable. The efficacy of foreground selection depends mainly on the genetic distance between the marker and the target gene/QTL; the closer is the gene/QTL marker is, the higher the foreground selection efficiency will be. The genetic distance between a marker and a gene/QTL indicates the progeny frequency at which, due to recombination, the relationship between the marker and the target gene/QTL allele is anticipated to alter.

(B) **Background Selection**

Tanksley et al. (1989) suggested the use of molecular markers to promote the recovery of recurrent parent genome and called *background selection* by Hospital and Charcosset (1997). It has been reported that different NILs developed independently from the same cross by selection for the same marker/gene usually contain different lengths of the donor genome flanking the marker/gene. It has been shown that two to four backcrosses together with background selection can recover the recurrent parent genotype to the same coverage as is achieved with six backcrosses united with phenotypic selection for recurrent parent phenotype. Background selection may also be used in a pedigree program to ensure the recovery of a specified level of the genome from one of the parents that may have more desirable features than the other parent.

(C) **Recombinant Selection**

A specific form of background selection is defined in terms of recombinant selection (Collard and Mackill 2008a, b) to remove the donor parent genome flanking the target gene/QTL (Young and Tanksley 1989). To minimize linkage drag recombinant selection guarantees the transfer of the target gene/QTL with a minimum of the donor parent genome. The adverse effect of gene linked to the target gene/QTL on the output of gene transfer lines is linkage drag. Often it is very difficult to eliminate undesirable linked genes in backcross programs. A surprisingly large amount of donor genome may remain in lines derived from several backcrosses. For example, tomato cultivars developed by transfer of *Tm2* gene from *Lycopersicon peruvianum* contained donor parent genome segment as 4 cM even after 20 backcrosses. It is remarkable that one cultivar derived after 11 backcrosses had the complete chromosome arm having the *Tm2* gene (>51 cM). The strategy of recombinant selection is based on markers located at <5 cM, preferably ~1 cM, on any side of the markers employed for foreground selection. These markers permit selection for such recombinants that have the target gene/QTL but lack the genome of donor parent beyond the markers used for foreground selection. Recombinant selection can save several generations of backcrossing without imposing a high cost.

Among the methods of molecular breeding, MABC has been used most extensively and successfully used in plant breeding. In several plants, it has been applied to various type of traits, such as rice, wheat, maize, barley, pear millet, soybean, tomato, etc. (Collard et al. 2005; Dwivedi et al. 2007; Xu 2010).

- In maize, the integration of the *Bt* transgene into various corn genetic backgrounds has been achieved by using MABC.
- Aroma in rice is regulated by a recessive gene because of an eight base-pair deletion and three single-nucleotide polymorphism in a gene which codes for betaine aldehyde dehydrogenase 2 (Bradbury et al. 2005a). This finding helps the aromatic and non-aromatic rice varieties to be classified and discriminates between homozygous recessive and dominant as well as heterozygous individuals in segregating population for the trait. MABC has been used for aroma selection in rice (Bradbury et al. 2005b).
- *opaque2* gene for high lysine has been incorporated in corn by means of MABC (Babu et al. 2005).
- Sebolt et al. (2000) used MABC for two QTL for seed protein content in soybeans.
- In tomato, Tanksley and Nelson (1996) proposed a MABC strategy, called advanced backcross- QTL (AB-QTL), to transfer resistance genes from wild relative/unadapted genotype into elite germplasm. The method has proven successful for various agronomically significant tomato traits, including fruit quality and resistance to black mould (Tanksley and Nelson 1996; Bernacchi et al. 1998; Fulton et al. 2002).
- Furthermore, AB-QTL has been used in other crop species, such as rice, barley, wheat, maize, cotton and soybean, demonstrating collectively that this method is

successful in transferring favourable alleles from the unadapted /wild germplasm to elite germplasm (Wang and Chee 2010; Concibido et al. 2003).

- A marker linked (0.7 cM) to the Yd2 gene for barley yellow dwarf virus resistance was successfully used in barley to select resistance in a backcrossing scheme (Jefferies et al. 2003).
- Marker-facilitated backcrossing has also been successfully used in maize to enhance complicated traits such as grain yield. Using MABC, by three generations of backcrossing, accompanied by two generations of selfing, six chromosomal segments each in two elite lines, Tx303 and Oh43, were transferred into two commonly used inbred lines, B73 and Mo17. Based on initial test-cross hybrid assessments, the improved lines with better performance were then picked. Zhao et al. (2012) reported that an important quantitative trait locus (named qHSR1) for maize head smut resistance was successfully incorporated into ten high-yielding inbred lines.

2. **Marker-Assisted Recurrent Selection(MARS)**

   This is the technique in which molecular markers are applied at each generation in order to target all traits of interest; it was proposed in the1990s (Bernardo and Charcosset 2006). In this technique, crossing is performed in selected individuals at every crossing and selection cycle. MARS is especially involved with the improvement of F_2 population that is achieved through one cycle of MAS (having phenotypic data with marker scores) followed by performing 2–3 cycles of marker-based selections (having marker scores only). It is a simple technique that can be easily implemented without requiring any prior knowledge of QTLs, and the selection depends entirely on the associations formed during the MARS programme between the marker and the trait (Eathington et al. 2007). MARS is a recurrent selection scheme using molecular markers to classify and select multiple genomic regions involved in the expression of complex traits to assemble the best-performing genotype within a single population or across related populations, as described by Ribaut et al. (2010).

(A) **MARS in Cross-Pollinated Crops**

Lande and Thompson (1990) proposal was soon adopted by maize breeders as marker-assisted recurrent selection (MARS), particularly in private seed companies. MARS has been mostly used for improving F_2 populations from suitable crosses before inbred isolation from them. Recurrent selection schemes were originally proposed for accumulation of desirable alleles in maize populations prior to inbred isolation from them. In these schemes, plants are selected on the basis of either their phenotype or testcross performance. In all possible combinations, the self-selected progeny of selected plants is intermingled to produce the population for the next selection period. In this way, the selection may be continued for as many cycles as desired. The testcross parent, i.e., the tester, used in the scheme may have either narrow genetic base (selection for specific combining ability) or broad genetic base (selection for general combining ability) (Allard 1960).

(B) **F_2 Enrichment and MARS in Self-Pollinated Crops**

In self-pollinated crops, recurrent selection based on phenotypic evaluation has been used to boost quantitative traits by accumulating favourable polygenic alleles regulating the target characteristics (Singh 2012). MARS scheme can be applied to segregating generations from suitable crosses of self-pollinated crops and the selected plants may be mated in pairs to generate the population for the next cycles of selection (Lande and Thompson 1990). The efficiency of MARS for one or two generations in intercrosses between RILs or DH lines derived from unique crosses was greatly impacted in computer simulations by the accuracy of the QTL position.

In the F_2 enrichment approach, MAS is used to eliminate from a F_2 population all plants that are homozygous for the unfavourable allele of one or more of the target QTLs. Therefore, only certain F_2 plants that are either homozygous or heterozygous for the beneficial alleles of all target QTLs are retained. The frequency of such plants would be$(3/4)^n$ in F_2, where n is the number of target QTLs. Thus, the F_2 enrichment approach dramatically increases the frequency of desired homozygous lines recovered from a F_2 population. Generally, F_2 enrichment is applied to the F_2 generations, but it can be used in backcrosses, three-way crosses, and double crosses as well.

3. **Gene pyramiding**
This is a technique in which several QTLs/genes are introgressed into a cultivar for a single or multiple traits which is deficient for these traits. This technique is primarily applied to increase the degree of tolerance to various diseases and insects by simultaneously selecting two or more genes (Luo et al. 2012).

Nelson (1978) introduced the idea of gene pyramiding to grow crop varieties with robust disease resistance by putting together few to many different oligogenes for resistance to the given disease. In general terms, gene pyramiding may be used to describe putting together of two or more genes in a single line/ variety regulating a single trait. Gene pyramiding is relatively straight-forward when the same donor parent contributes all the genes. But when two or more donor parents have to be used, relatively simple strategies can be used for gene pyramiding. Often, genes governing two or more different traits are introgressed into a single recurrent parent; this should be called *multitrait introgression* in the place of gene pyramiding.

There are quite a lot of examples of successful pyramiding of genes, QTLs, and both genes and QTLs related mainly to disease resistance, and QTLs for yield and yield-related traits (Table 6.4). For commercial purposes, a BB-resistant edition of Pusa Basmati 1 (PB1), known as 'Improved Pusa Basmati 1' (IPB1), containing xa13 and Xa21 genes from the donor parent IRBB55 (Gopalakrishnan et al. 2008), and a BB-resistant variant of Samba Mahsuri, known as 'Improved Samba Mahsuri', containing xa5, xa13, and Xa21 genes from the donor parent SS1113 (Sundaram et al. 2008), has been released.

Table 6.4 Examples of gene/QTL introgression in different crop plants

Crop	Trait	Gene/QTL	Reference
Rice	Amylose content	Wx	Gopalakrishnan et al. (2008)
	Root depth	Four QTLs	Shen et al. (2001)
	Root depth	QTL2, QTL7, QTL9, QTL11	Steele et al. (2006)
	Rice yellow mottle virus resistance	QTL7, QTL12	Ahmadi et al. (2001)
	Submergence tolerance	QTL SUB-1	Neeraja et al. (2007)
	Bacterial blight resistance	xa13, Xa21	Joseph et al. (2004)
		xa5, xa13, Xa21	Sanchez et al. (2000); Sundaram et al. (2008)
		Xa4, xa5, xa13, Xa21	Huang et al. (1997)
	Bacterial blight, yellow stem borer and sheath blight	*Xa21, bt,* RC7 chitinase, Bt	Datta et al. (2002)
	Blast disease	*Pil, Piz-5, Pill, Pita*	Hittalmani et al. (2000)
Maize	Protein quality	o2	Babu et al. (2004, 2005)
	Days to silking	Three QTLs	Bouchez et al. (2002)
	Yield	Two QTLs	Bouchez et al. (2002)
	Yield	Yield QTLs	Schmierer et al. (2004)
	Heterosis	2–4 genomic regions	Stuber et al. (1999)
	Northern leaf blight and head smut resistance	Oligogenes Ht1 and Ht2	Min et al. (2012)
Barley	Barley stripe rust resistance	Oligogene Recurrent parentsx and QTL4, QTL5, QTL7	Castro et al. (2003)
	Barley yellow mosaic virus	*rym1 rym5*	Okada et al. (2004)
Wheat	Powdery mildew	*Pm2, Pm4a*	Liu et al. (2000)
Tomato	Acyl sugar content	3–5 genomic regions	Lawson et al. (1997)
	Fruit quality	Five genomic regions	Lecomte et al. (2004)

Source: Singh and Singh (2015) and Collard and Mackill (2007)

Strategy for Gene Pyramiding

1. Donor parent is crossed with the recurrent parent, and the F_1 and the subsequent progeny are repeatedly backcrossed with recurrent parent.
2. When the genes to be pyramided are present in different donor parents, they can be introgressed into recurrent parent by two ways.

First Way

- Each donor parent is used in a *separate backcross program* with the recurrent parent to recover the target gene from each donor parent in the genetic background of recurrent parent either in heterozygous or homozygous state.

- These derived lines of recurrent parent are then crossed together to produce a complex hybrid.
- The pyramided version of recurrent parent having all the target genes is recovered from this hybrid by selfing coupled with selection.

Second Way

- All the donor parents are ordered into a *single backcross program* according to a suitable mating scheme. The complex hybrid obtained from either scheme is used in a backcross program with the recurrent parent to recover the pyramided version of recurrent parent (Ishii and Yonezawa 2007).

The conventional breeding schemes were considered for selection based on phenotype and are not well suited to fully exploit the marker technology. This recognition has encouraged the development of innovative breeding schemes (Table 6.5) designed to take advantage of the markers data.

6.5.2 Advantages of MAB Over Conventional Breeding

MAB advantages in comparison to conventional breeding methods:

(a) MAB allows selection of individuals at a seedling stage which proves very useful especially in case of backcrossing and recurrent selection, where crossing among selected individuals is required.
(b) MAB does not have any impact of varying environment and selection of good lines can be done under greenhouse and off season nurseries.
(c) Co-dominant markers in MAB allow selection of recessive alleles even in heterozygous condition and selfing or test crossing is not necessary.
(d) MAB allows the gene pyramiding efficiently and is more effective for the traits controlled by multiple genes/QTL.

6.5.3 Drawbacks of MAB

(a) Due to lack of marker polymorphism or selectable marker-trait association, all markers cannot be applied across populations.
(b) All markers are not very convenient to be used by the breeders due to the dominant nature of some marker types and these are required to be converted into breeder friendly markers like RAPD into SCAR (Sequence Characterized Amplified Region), etc.
(c) Use of multiple markers is required to avoid any failure in selection due to recombination between the markers and the gene/QTL.
(d) Inaccurate estimates of QTL locations and effects result in slower progress than expected. There are many factors like algorithms, mapping methods, number of

Table 6.5 A summary of various breeding methods, including innovative schemes, using MAS

S. no.	Breeding scheme	Objective	Chief features
1	Marker-assisted backcrossing (MABC)	Introgression of genes/QTLs from one or more donor parents into an recurrent parent	F_1 and subsequent generations backcrossed to recurrent parent; foreground, background, and recombinant selections, usually, based on MAS
(a)	Single gene/QTL introgression	Removal of a specific defect of recurrent parent	As above
(b)	Gene pyramiding	Accumulation into the recurrent parent of different genes/QTLs affecting a trait	Genes introgressed individually into recurrent parent by parallel MABC and brought together in the end
(c)	Multitrait introgression	Accumulation into the recurrent parent of genes/QTLs affecting several different traits	As above
(d)	Single backcross-DH scheme	Introgression of genes/QTLs from an elite donor parent into an recurrent parent	MAS in BC_1F_1 for target traits; haploids produced and subjected to MAS; DH produced and evaluated; no background MAS
(e)	Advanced backcross QTL mapping	Introgression of genes/QTLs from unadapted germplasm	Selection against deleterious traits in the backcross generations; BC_1S_1/BC_2S_1 used for QTL identification/used in breeding programs
(f)	Inbred enhancement-QTL mapping	Introgression of QTLs from donor parent into an elite recurrent parent deficient in the trait	Introgression line library constructed; lines evaluated for QTL detection and mapping; superior lines used in breeding/ as varieties
2	Breeding by design	Development of a line with the ideal genotype created, initially, in silico for high performance	The ideal genotype designed using information on marker-trait association; this genotype is constructed by combining the target genomic regions from various donor parents
3	Pedigree MAS	To ensure the presence of the desired genomic regions in the derived lines by fixing these regions	Genomic regions of interest identified from data generated in breeding activities; MAS in early segregating generations to fix these regions
4	Single large-scale MAS (SLS-MAS)	As above	Genomic regions of interest identified from appropriate crosses; MAS for fixing the target regions in F_2/F_3; subsequent generations as per pedigree scheme

(continued)

Table 6.5 (continued)

S. no.	Breeding scheme	Objective	Chief features
5	Marker-evaluated selection (MES)	Development of genotypes for adaptation and performance in specific ecosystems	Genomic regions of interest identified by changes in marker allele frequency in the target ecosystems; MAS used for these regions
6	Marker-assisted recurrent selection (MARS)	Isolation of improved inbreds/ purelines by increasing the frequency of desirable alleles in the population	Markers showing considerable association with the trait (s) used for MAS; selected plants intermated and their progeny subjected to MAS; may continue for several cycles
7	Genomic selection (GS)/ genome-wide selection (GWS)	Selection for all the QTLs affecting the trait irrespective of the significance of marker–trait associations	Genome-wide markers used for MAS based on genomic estimated breeding values; marker effects predicted from a suitable training population.
8	Heterosis breeding	Development of superior hybrid varieties	Heterotic groups identified on the basis of marker data; complementing groups crossed to produce hybrids Genomic regions involved in heterosis identified; target regions introgressed into appropriate inbreds to enhance hybrid performance.

Source: Singh and Singh (2015)

polymorphic markers, and population type and size which affect the efficiency of QTL detection.

(e) Large number of breeding programs has not been equipped with adequate facilities and conditions for a large-scale adoption of MAB in practice are not applicable to large-scale breeding programmes due to unavailability of sufficient infrastructure.
(f) MAB schemes cannot be designed for large-scale use in breeding programs unless the plant breeders are trained with procedures and implementation of these methods.
(g) Higher startup and labour costs.

6.6 Association Mapping

Association mapping (AM) also known as 'linkage disequilibrium mapping' takes advantage of linkage disequilibrium to link phenotypes to genotypes of unrelated individuals for mapping QTLs, hence discovering genetic associations. To conduct the association mapping, one must have the mapping population consisting of a diverse set of individuals chosen from natural populations. Association mapping reveals a significant association between molecular markers and a phenotypic trait. AM, at statistical level, gives the covariance between the marker polymorphism and trait of interest (Jannink and Walsh 2002; Zhang et al. 2016a). It saves time in comparison to linkage mapping and ensures better mapping resolution with a higher number of recombination events. Due to availability of more genetic variations with larger background, AM facilitates to find out higher number of alleles (Zhang et al. 2016b; Kraakman et al. 2006).

Advantage of association mapping

- Populations for AM are from existing germplasm, this saves time efforts and cost required.
- QTL associated marker identified through association mapping can be directly used for MAS as they are identified from diverse germplasm.
- QTL would be present in multiple genetic backgrounds of breeding material. Therefore, QTL discovered in such materials would be one that will be able to articulate itself in a range of genetic backgrounds and would be useful in other breeding programs.

Limitation of association mapping

- The effect of a rare and agronomically valuable allele on the target trait cannot be detected.

6.6.1 Populations Used for Association Mapping

The success of association mapping majorly depends upon the type of the population used. A natural/breeding population or a family-based population may be utilized. AM also uses biparental and multiparent populations, but single biparental populations are commonly not used for AM. Mapping populations like doubled haploid, F_3, etc., or families derived from several biparental crosses made by mating a group of inbreds in diallel scheme or in a random manner are used for AM. Among the multiparent populations, two populations, namely, multiparent advanced generation intercrosses (MAGIC) and nested association mapping (NAM) populations, have become populations of choice since they allow both AM and linkage mapping and can even be used for variety development.

1. **NAM Population**

   The *nested association mapping (NAM) population*, proposed by Yu et al. (2008), used for linkage mapping of QTLs as well as AM. The NAM scheme was initially developed for maize. RILs developed from a diverse set of parents are used to develop this population. It requires a fewer number of markers than GWAS in population-based association panels, and the resolution obtained is higher than QTL linkage mapping. The NAM strategy has higher power than AM because the controlled crosses made for generating NAM populations minimize population structure and familial relatedness. Further, being a biparental population, the frequencies of otherwise rare alleles get increased in the biparental families generating the NAM population. It facilitates cost-effective genome-wide scans and permits sharing of the NAM panel with other researchers. The NAM and similar methods have main statistical challenge related to the estimation of probability that alleles of various loci which are identical in state are also identical by descent.

2. **MAGIC Population**
   The *multiparent advanced generation intercross* (*MAGIC*) populations comprise a set of RILs produced from a complex cross or a set of crosses involving multiple parents. These populations can also be used for linkage and association mapping of multiple traits for which the parents differ, and multiple alleles at the target loci can also be detected. The development of these populations is accompanied with several rounds of recombination, which increases the precision and resolution of QTL mapping. MAGIC populations can be derived from breeding lines and germplasm lines of interest to breeders. MAGIC populations were developed in rice (using *indica* and *japonica* lines) and wheat and used in QTL mapping and variety development in rice.

6.6.2 Types of Association Mapping

1. **Genome-wide association study.**
2. **Candidate gene based AM.**

1. **Genome-wide association study**
   In *genome-wide association studies* (*GWAS*), the markers used for the genotyping are distributed evenly and densely over the whole genome. In this approach, all the loci controlling the traits, which show variation in the sample, can be evaluated in one go. The markers number required for genotyping would be much higher in cross-pollinated than in self-pollinated species because the LD decays much faster in the former than in the later. It is essential that a genome-wide linkage map of markers of the concerned species must be available to permit the selection of an appropriate set of markers. Besides, substantial resources and effort will be requisite for consistent phenotyping of the variable traits. Finally, it would be required to make thousands of independent comparisons among marker loci, when large number of markers is used. This would necessitate a large sample size (one thousand or more individuals) to permit the detection of QTLs with moderate effect size. The populations like RILs are highly suitable for genome-wide scanning of QTLs as only a few hundred markers need to be evaluated and they provide greater statistical power to evaluate the effect of a genomic region than AM. This would necessitate a large sample size (one thousand or more individuals) to permit the detection of QTLs with moderate effect size.
2. **Candidate gene based AM**

The approach where the analysis is restricted to the genomic regions having the candidate genes/QTLs for the desired trait of interest is well-known as candidate gene approach. A candidate gene is a gene expected to be involved in the control of the trait of interest. The information from different sources like comparative genomics, genome sequence annotation, transcript profiling, QTL analysis, commonly used to discover the candidate genes. This greatly reduces the target genomic region, which can be analysed with a high density of markers. Further, the total number of markers used as well as the sample size will also be considerably reduced. The candidate genes are generally discovered from loss of function mutations in laboratory strains. Therefore, sometimes it is difficult to determine as to how well these mutations relate to the variation present in the trait in natural populations. Despite these difficulties, the candidate gene based AM used to discover genes involved in the control of many morphological, phenological, and stress resistance traits (Ingvarsson and Street 2011). This approach has an advantage of ability to identify a QTL where genome-wide AM fails to detect a significant marker-trait association after false discovery rate (FDR) correction is applied. Additionally, the use of this approach with GWAS tends to increase the strength and accuracy of QTL detection (Gupta et al. 2014).

Statistical software used for association mapping in plants are summarized in Table 6.6.

6.7 Conclusion

From application point of view, although conventional selection based on phenotypic evaluation will likely remain the basis for most breeding programs, particularly in the public domain, genomics-assisted selection, and its applications are being increasingly adopted and have become prevalent as compared to conventional practices. As the twenty-first century unfolds, a multitude of genomics and post-genomics platforms are at hand to inflate our perceptive of the genetic basis of crop performance and to improve the effectiveness of selection procedures for the release of new, improved cultivars. Resequencing has revolutionized the way breeders deal with their germplasm and provides unsurpassed opportunities for a deeper mining of allelic diversity and harnessing its full potential. Nonetheless, our perceptive of the functional basis of yield and additional quantitative traits is still limited. The elusive nature of the QTLs that govern yield and yield stability across different environmental conditions is a difficult obstacle toward a more effective selection targeting specific loci and an improved understanding of quantitative traits. Notably, GS is being applied irrespective of our degree of considerate of the genetic architecture of quantitative traits. Importantly, MAS and GS should be considered as complementary rather than alternative approaches, the utilization of which should be determined on a case-by-case basis. Bioinformatics and user-friendly data bases will play an essential role for handling and managing the deluge of data produced by the molecular and phenotypic platforms. Along this line, it is significant to highlight that

Table 6.6 Statistical software packages used for association mapping in plants

S. No.	Software package	Brief description
Free packages		
1	*TASSEL*	LD statistic calculation and graphic visualization; sequence analysis
2	*EMMAX*	Fast computation, for large AM studies, corrects for population structure and kinship (http://genetics.cs.ucla.edu/emmax/)
3	*GenAMap*	Implements structured association mapping, employs various algorithms, good graphical presentation (http://sailing.cs.cmu.edu/genamap/)
4	*GenABEL*	GWAS for both quantitative and qualitative traits (http://www.genabel.org/packages/GenABEL)
5	*FaST-LMM*	AM based on large samples of up to 120,000 individuals (http://fastlmm.codeplex.com/)
6	*GAPIT*	Implements CMLM, R-based, fast computation (http://www.maizegenetics.net/gapit)
7	*STRUCTURE*	Population structure analysis; generates *Q* matrix; computation intensive (http://pritch.bsd.uchicago.edu/structure.html)
8	*SPAGeDI*	Kinship analysis; generates *K* matrix (http://www.ulb.ac.be/sciences/ecoevol/spagedi.html)
9	*EINGENSTRAT*	Association analysis; PCA to generate *P* matrix used in the place of *Q* matrix (http://genepath.med/harvard.edu/~reich/software.html)
10	*MTDFREML*	MLM analysis of animal breeding data; can be used for plants (http://aipl.arsusda.gov/curtvt/mtdfreml.html
11	*Q*	Generic package; convenient for simulation work; useful for researchers with good statistics and computer programming background (http://www.r-project.org/)
Commercial packages		
1	*ASREML*	MLM analysis for animal breeding data, can be used for plants (http://www.vsni.co.uk/products/asreml)
2	*GenStat*	Implements GLM and MLM, corrects for population structure (http://www.vsni.co.uk/software/genstat)
3	*JMP Genomics*	Computation of population structure and kinship coefficient (marker-based) (http://www.jmp.com/software/genomics/)
4	*SAS*	statistical package used for data analysis and methodology work (http://www.sas.com)

Source: Gupta et al. (2014)

any molecular approach aiming to discover genes/QTLs and test their effects should preferably be carried out in an experimental context whose results are as relevant as possible and readily applicable to the conditions prevailing in farmers' fields. Ultimately, a more effective exploitation of genomics approaches to enhance crop performance will depend on their integration with conventional breeding. Although it is not possible to predict to what extent and how quickly conventional breeding will eventually be replaced by genomics-assisted breeding, the future release of improved cultivars will be expedited through a systematic genome-based manipulation, particularly by means of genome-editing approaches, of the loci that administrate the crop performance and the desired features targeted by breeders.

6.8 Future Thrust

- Marker-assisted selection in combination with genetic engineering and *in-vitro* mutagenesis can be used to incorporate and/or manipulate novel genes to increase the genetic diversity.
- High-throughput genotyping can be utilized to improve genome-wide association studies (GWAS) in crop species, which leads to the detection of SNPs linked with economically important traits.
- Association mapping as it based upon non-random association of candidate genes or markers on a high resolution map can be used to find out new marker-trait associations or to validate associations that were found through conventional genetic mapping. It facilitates the discovery of more useful markers for wide range of genotypes.
- There is need to develop the optimum balance between conventional and molecular breeding, and the "best" balance will be unique to each situation, crop, selection scheme, environment and opportunities for different selection methods. More emphasis should be given to combined selection systems, instead of viewing MAS as a substitute for phenotypic or field selection.
- There is an absolute need to expand the infrastructure for the management, conservation, and annotation of the respective crop genomic sequences that will be produced in the near future.
- In particular, "speed breeding" that does not allow an appropriate phenotypic assessment required by conventional selection will be ideal to further enhance the benefits of genome-assisted selection.

References

Acquaah G (2012) Principles of plant genetics and breeding. Wiley-Blackwell, Oxford

Ahmadi N, Albar L, Pressoir G, Pinel A, Fargette D, Ghesquière A (2001) Genetic basis and mapping of the resistance to rice yellow mottle virus. III. Analysis of QTL efficiency in introgressed progenies confirmed the hypothesis of complementary epistasis between two resistance QTLs. Theor Appl Genet 103:1084–1092

Akbari M, Wenzl P, Caig V, Carling J, Xia L, Yang S, Uszynski G, Mohler V, Lehmensiek A, Kuchel H, Hayden MJ (2006) Diversity arrays technology (DArt) for high-throughput profiling of the hexaploid wheat genome. Theor Appl Genet 113:1409–1420

Allard RW (1960) Principles of plant breeding. Wiley, New York

Babu R, Nair S, Prasanna BM, Gupta HS (2004) Integrating marker-assisted selection in crop breeding – prospects and challenges. Curr Sci 87:607–619

Babu RSK, Nair A, Kumar S, Venkatesh JC, Sekhar NN, Singh G, Srinivasan HS, Gupta (2005). Two-generation marker-aided backcrossing for rapid conversion of normal maize lines to quality protein maize (QPM). Theor Appl Genet 111:888-897

Banerjee S, Yandell BS, Yi N (2008) Bayesian quantitative trait loci mapping for multiple traits. Genetics 179:2275–2289

Belicuas PR, Guimaraes CT, Paiva LV, Duarte JM, Maluf WR, Paiva E (2007) Androgenetic Haploids and SSR markers as tools for the development of tropical maize hybrids. Euphytica 156(1–2):95–102

Bernacchi D-T, Beck-Bunn D, Emmatty Y, Eshed S, Inai J, Lopez V, Petiard H, Sayama J, Uhlig D, Zamir SD, Tanksley S (1998) Advanced backcross QTL analysis of tomato: II. Evaluation of near-isogenic lines carrying single-donor introgressions for desirable wild Qtl-alleles derived from *Lycopersicon hirsutum* and *L.pimpinellifolium*. Theor Appl Genet 97:170–180

Bernardo R (2008) Molecular markers and selection for complex traits in plants: learning from the last twenty years. Crop Sci 48:1649–1664

Bernardo R, Charcosset A (2006) Usefulness of gene information in marker assisted recurrent selection: a simulation appraisal. Crop Sci 46(2):614–621

Bernardo A, Wang SST, Amand P, Bai G (2015) Using next generation sequencing for multiplexed trait-linked markers in wheat. PLoS One 10(12):e0143890

Borlaug NE, Dowswell CR (2005) Feeding a world of ten billion people a 21st century challenge. In: Tuberosa R, Phillips RL, Gale M (eds) Proceedings of the international congress in the wake of the double helix: from the green revolution to the gene revolution. 27–31 May 2003. Bologna Italy Avenue Media, Bologna, pp 3–23

Bouchez A, Hospital F, Causse M et al (2002) Marker assisted introgression of favorable alleles at quantitative trait loci between maize elite lines. Genetics 162:1945–1959

Bradbury LMT, Fitzgerald TL, Henry RJ, Jin Q, Waters DLE (2005a) The gene for fragrance in rice. Plant Biotech J 3:363–370

Bradbury LMT, Henry RJ, Jin Q, Reinke RF, Waters DLE (2005b) A perfect marker for fragrance genotyping in rice. Mol Breeding 16:279–283

Brennan JP, Martin PJ (2007) Returns to investment in new breeding technologies. Euphytica 157: 337–349

Brinez B, Blair MW, Kilian A, Carbonell SAM, Chiorato AF, Rubiano LB (2012) A Whole genome dart assay to assess germplasm collection diversity in common beans. Mol Breed 30:181–193

Castro AJ, Capettini FLAVIO, Corey AE, Filichkina T, Hayes PM, Kleinhofs ANDRIS, Kudrna D, Richardson K, Sandoval-Islas S, Rossi C, Vivar H (2003) Mapping and pyramiding of qualitative and quantitative resistance to stripe rust in barley. Theor Appl Genet 107:922–930

Collard BCY, Mackill DJ (2007) Marker-assisted selection: an approach for precision plant breeding in the twenty-first century. Philos Trans R Soc Lond B Biol Sci 363(1491):557–572

Collard BC, Mackill DJ (2008a) Marker assisted selection: an approach for precision plant breeding in the twenty-first century. Phil Trans R Soc B 363:557–572

Collard BCY, Mackill DJ (2008b) Marker-assisted selection : an approach for precision plant breeding in the twenty-first century. PhilosTrans R SocLond B Biol Sci 363:557–572

Collard BCY, Jahufer MZZ, Brouwer JB, Pang ECK (2005) An introduction to markers, quantitative trait loci (QTL) mapping and marker-assisted selection for crop improvement: the basic concepts. Euphytica 142:169–196

Concibido VC, La Vallee B, Mclaird P, Pineda N, Meyer J, Hummel L, Yang J, Wu K, Delannay X (2003) Introgression of a quantitative trait locus for yield from *Glycine soja* into commercial soybean cultivars. Theor Appl Genet 106:575–582

Datta K, Baisakh TN, Tu KMJ, Datta S (2002) Pyramiding transgenes for multiple resistance in rice against bacterial blight, yellow stem borer and sheath blight. Theor Appl Genet 106:1–8

Dhingani RM, Umrania VV, Tomar RS, Parakhia MV, Golakiya B (2015) Introduction to Qtl mapping in plants. Ann Plant Sci 4(04):1072–1079

Dwivedi SL, Crouch JH, Mackill DJ, Xu Y, Blair MW, Ragot M, Upadhyaya HD, Oritiz R (2007) The molecularization of public sector crop breeding: progress, problems and prospects. Adv Agron 95:163–318

Eathington SR, Crosbie TM, Edwards MD, Reiter RS, Bull JK (2007) Molecular markers in a commercial breeding program. Crop Sci 47(Suppl. 3):S154–S163

Francia E, Tacconi G, Crosatti C, Barabaschi D, Bulgarelli D, Dall'Aglio E, Valè G (2005) Marker assisted selection in crop plants. Plant Cell Tissue Org 82(3):317–342

Fulton TM, Bucheli P, Voirol E, Lopez J, Peetiard V, Tanksley SD (2002) Quantitative trait loci (QTL) affecting sugars, organic acids and other biochemical properties possibly contributing to flavor, identified in four advanced backcross populations of tomato. Euphytica 127:163–177

George ML, Prasanna BM, Rathore RS, Setty TAS, Kasim F, Azrai M, Vasal S, Balla O, Hautea D, Canama A, Regalado E (2003) Identification of QTLs conferring resistance to downy mildews of maize in Asia. Theor Appl Genet 107(3):544–551

Gopalakrishnan S, Sharma RK, Anand Rajkumar K, Joseph M, Singh VP, Singh AK, Bhat KV, Singh NK, Mohapatra T (2008) Integrating marker assisted background analysis with foreground selection for identification of superior bacterial blight resistant recombinants in Basmati rice. Plant Breed 127:131–139

Grzebelus D, Iorizzo M, Senalik D, Ellison S, Cavagnaro P, Macko-Podgorni A, Heller-Uszynska-K, Kilian A, Nothnagel T, Allender C, Simon PW (2014) Diversity, genetic mapping and signatures of domestication in the carrot (*Daucus carota* L.) genome, As revealed by diversity arrays technology (Dart) markers. Mol Breed 33:625–637

Guo H, Mendrikahy JN, Xie L, Deng J, Lu Z, Wu J, Li X, Shahid MQ, Liu X (2017) Transcriptome analysis of neo-tetraploid rice reveals specific differential gene expressions associated with fertility and heterosis. Sci Rep 7:40139

Gupta PK, Kulwal PL, Jaiswal V (2014) Association mapping in crop plants: opportunities and challenges. In: Friedmann T, Dunlap J, Goodwin S (eds) Advances in genetics, vol 85. Academic Elsevier, Waltham, MA, USA, pp 109–148

Hackett CA (2002) Statistical methods for Qtl mapping in cereals. Plant Mol Biol 48(5–6):585–599

Haley CS, Andersson LE (1997) Linkage mapping of quantitative trait loci in plants and animals. In: Dear PH (ed) Genome mapping: a practical approach. IRL Press, Oxford, pp 49–71

Hittalmani S, Parco A, Mew TV, Zeigler RS, Huang N (2000) Fine mapping and DNA marker-assisted pyramiding of the three major genes for blast resistance in rice. Theor Appl Genet 100: 1121–1128

Hospital F, Charcosset A (1997) Marker-assisted introgression of quantitative trait loci. Genetics 147:1469–1485

Huang N, Angeles ER, Domingo J, Magpantay G, Singh S, Zhang G, Kumaravadivel N, Bennett J, Khush GS (1997) Pyramiding of bacterial blight resistance genes in rice: marker-assisted selection using RFLP and PCR. Theor Appl Genet 95(3):313–320

Ingvarsson PK, Street NR (2011) Association genetics of complex traits in plants. New Phytol 189: 909–922

Ishii T, Yonezawa K (2007) Optimization of the marker based procedures for pyramiding genes from multiple donor lines: I. Schedule of crossing between the donor lines. Crop Sci 47:537–546

Islam MN, Molla M, Rohman MM, Hasanuzzaman M, Islam SMN, Rahman L (2012) DNA fingerprinting and genotyping of cotton varieties using SSR markers. Not Bot Horti Agrobo 40:261–265

Jain HK, Kharkwal MC (2004) Plant breeding Mendelian to molecular approaches. Springer, Dordrecht

Jannink JL, Walsh B (2002) Association mapping in plant populations. In: Kang MS (ed) Quantitative genetics, genomics and plant breeding. CAB International, Oxford, pp 59–68

Jansen RC (1994) High resolution of quantitative traits into multiple loci via interval mapping. Genetics 136:1447–1455

Jefferies SP, King BJ, Barr AR, Warner P, Logue SJ, Langridge P (2003) Marker-assisted backcross introgression of the *Yd2* gene conferring resistance to barley yellow dwarf virus in barley. Plant Breed 122:52–56

Jiang GL (2013) Molecular markers and marker-assisted breeding in plants. In: Andersen SB (ed) Plant breeding from laboratories to fields. Rijeka Intech, pp 45–83

Joseph M, Sharma RK, Anand Rajkumar K et al (2004) Combining bacterial blight resistance and Basmati quality characteristics by phenotypic and marker assisted selection in rice. Mol Breed 13:377–387

Kao CH, Zeng ZB, Teasdale RD (1999) Multiple interval mapping. Genetics 152:1203–1216

Khush GS, Virmani SS (1996) Haploids in plant breeding. In: Jain SM, Sopory SK, Veilleux RE (eds) In vitro haploid production in higher plants, Current plant science and biotechnology in agriculture. Springer, Dordrecht, pp 11–33

Kraakman AT, Martinez F, Mussiraliev B, Van Eeuwijk FA, Niks RE (2006) Linkage disequilibrium mapping of morphological, resistance, and other agronomically relevant traits in modern spring barley cultivars. Mol Breed 7(1):41–58

Kulshreshtha K, Parihar A, Parihar P (2020) Next generation based transcriptome analysis for nematode resistance in three different species of tomato (*S. lycopersicum*, *S. habrochaites* & *S. arcanum*). Plant Gene 24:100255

Kunzel G, Korzun L, Meister A (2000) Cytologically integrated physical restriction fragment length polymorphism maps for the barley genome based on translocation breakpoints. Genetics 154(1): 397–412

Lande R, Thompson R (1990) Efficiency of marker assisted selection in the improvement of quantitative traits. Genetics 124:743–756

Lander ES, Botstein D (1989) Mapping Mendelian factors underlying quantitative traits using RFLP linkage maps. Genetics 121:185–199

Lawson DM, Lunde CF, Mutschler MA (1997) Marker assisted transfer of acylsugar-mediated resistance from the wild tomato, *Lycopersicon pennellii* to the cultivated tomato, *Lycopersicon esculentum*. Mol Breed 3:307–317

Lecomte L, Duffe P, Buret M, Servin B, Causse M (2004) Marker-assisted introgression of five QTLs controlling fruit quality traits into three tomato lines revealed interactions between QTLs and genetic backgrounds. Theor Appl Genet 109:658–668

Leng PF, Lübberstedt T, Xu ML (2017) Genomics assisted breeding – a revolutionary strategy for crop improvement. J Integr Agric 16(12):2674–2685

Li YH, Xiao H, Zhang CQ, Hu GC (1999) Genetic variation of main parents of hybrid rice in china was revealed with simple sequence repeat markers. Zhiwuxue Tongbao 41:1061–1066

Li X, Shahid MQ, Xia J (2017) Analysis of small RNAs revealed differential expressions during pollen and embryo sac development in auto tetraploid rice. BMC Genomics 18(1):129

Liu J, Liu D, Tao W, Li W, Wang S, Chen P, Cheng S, Gao D (2000) Molecular marker-facilitated pyramiding of different genes for powdery mildew resistance in wheat. Plant Breed 119:21–24

Luo Y, Sangha JS, Wang S, Li Z, YangJ YZ (2012) Marker assisted breeding of Xa4, Xa21 And Xa27. In the restorer lines of hybrid rice for broad-spectrum and enhanced disease resistance to bacterial blight. Mol Breed 30(4):1601–1610

Mariya ZS, Jadeja GC, Akarsh P, Ankita P (2017) Comparative analysis for seed protein profiling and SSR markers in Rice cultivars (Oryza sativa L.) Res. J Biotechnol 12(5):12–19

Melchinger AE, Winter M, Mi X, Piepho HP, Schipprack W, Mirdita V (2015) Controlling misclassification rates in identification of haploid seeds from induction crosses in maize with high-oil inducers. Crop Sci 55(3):1076–1086

Min J, Chunyu Z, Khalid H, Nan L, Quan S, Qing M, Suwen W, Feng L (2012) Pyramiding resistance genes to northern leaf blight and head smut in maize. Int J Agric Biol 14:430–434

Muthusamy V, Hossain F, Thirunavukkarasu N, Choudhary M, Saha S, Bhat JS, Prasanna BM, Gupta HS (2014) Development of β-carotene rich maize hybrids through marker assisted introgression of β-carotene hydroxylase allele. PLoS One 9:e113583. https://doi.org/10.1371/journal.pone.0113583

Naeem M, Ghouri F, Shahid MQ, Iqbal M, Baloch FS, Chen L, Allah S, Babar M, Rana M (2015) Genetic diversity in mutated and non-mutated rice varieties. Genet Mol Res 14(4):17109–17123

Nawaz MA, Baloch FS, Rehman HM, Le B, Akther F, Yang SH, Gyuhwa C (2016) Development of a competent and trouble free DNA isolation protocol or downstream genetic analysis in glycine species. Turk J Agri Food Sci Tech 4(8):700–705

Neeraja C, Maghirang-Rodriguez R, Pamplona A, Heuer S, Collard BC, Septiningsih EM, Vergara G, Sanchez D, Xu K, Ismail AM, Mackill DJ (2007) A marker assisted backcross approach for developing submergence-tolerance rice cultivars. Theor Appl Genet 115:767–776

Nelson RR (1978) Genetics of horizontal resistance to plant diseases. Annu Rev Phytopathol 16: 359–378

Ogbonnaya FC, Subrahmanyam NC, Moullet O, De Majnik J, Eagles HA, Brown JS, Eastwood RF, Kollmorgen J, Appels R, Lagudah ES (2001) Diagnostic DNA markers for cereal cyst nematode resistance in bread wheat. Crop Pasture Sci 52(12):1367–1374

Okada Y, Kanatani R, Arai S, Ito K (2004) Interaction between barley yellow mosaic disease-resistance genes Ryml And Rym5, In the response to BaYMVstrains. Breed Sci 54:319–325

Parihar A, Pathak AR, Parihar P (2010) Identification of RAPD marker for the White Backed Plant Hopper (WBPH) resistant gene in rice. Afr J Biotechnol 9(10):1423–1426

Randhawa HS, Mutti JS, Kidwell K, Morris CF, Chen X, Gill KS (2009) Rapid and targeted introgression of genes into popular wheat cultivars using marker-assisted background selection. PLoS One 4:e5752. https://doi.org/10.1371/journal.pone.0005752

Ribaut JM, de Vicente MC, Delannay X (2010) Molecular breeding in developing countries: challenges and perspectives. Curr Opin Plant Biol 13:213–218

Sanchez AC, Brar DS, Huang N, Li Z, Khush GS (2000) Sequence tagged site marker-assisted selection for three bacterial blight resistance genes in rice. Crop Sci 40:792–797

Satgopan JM, Yandell BS, Newton MA, Osborn TC (1996) A Bayesian approach to detect quantitative trait locus Markov chain Monte Carlo. Genetics 144:805–816

Schmierer DA, Kandemir N, Kudrna DA, Jones BL, Ullrich SE, Kleinhofs A (2004) Molecular marker-assisted selection for enhanced yield in malting Barley. Mol Breed 14:463–473

Sebolt AM, Shoemaker RC, Diers BW (2000) Analysis of a quantitative trait locus allele from wild soybean that increases seed protein concentration in soybean. Crop Sci 40:1438–1444

Semagn K, Bjørnstad A, Ndjiondjop MN (2006) An overview of molecular marker methods for plants. Afr J Biotechnol 2540:25–68

Shahid MQ, Chen FY, Li HY, Wang SZ, Chen PF, Lin SQ, Liu XD, Lu YG (2013) Double-neutral genes, and for pollen fertility in rice to overcome × hybrid sterility. Crop Sci 53(1):164–176

Shen LB, Courtois B, McNally KL, Robin S, Li Z (2001) Evaluation of near-isogenic lines of rice introgressed with QTLs for root depth through marker-aided selection. Theor Appl Genet 103: 75–83

Simmonds NW (1993) Introgression and incorporation. Strategies for the use of crop genetic resources. Biol Rev 68(4):539–562

Singh BD (2012) Plant breeding, principles and methods, 9th edn. Kalyani Publishers, New Delhi

Singh S (2015) Mapping population, marker-assisted plant breeding: principles and practices. Springer. https://doi.org/10.1007/978-81-322-2316-0

Singh N, Choudhury DR, Tiwari G, Singh AK, Kumar S, Srinivasan K, Tyagi RK, Sharma AD, Singh NK, Singh R (2016) Genetic diversity trend in Indian rice varieties: an analysis using SSR markers. BMC Genet 17:127

Soller M, Brody T (1976) On the power of experimental designs for the detection on linkage between marker loci and quantitative loci in crosses between inbred lines. Theor Appl Genet 47: 35–39

Steele KA, Price AH, Shahsidhar HE, Witcombe JR (2006) Marker assisted selection to introgress rice QTLs controlling root traits into an Indian upland rice variety. Theor Appl Genet 112:208–221

Stuber CW, Polacco M, Senior ML (1999) Synergy of empirical breeding, marker-assisted selection and genomics to increase crop yield potential. Crop Sci 39:1571–1583

Sundaram RM, Vishnupriya MR, Biradar SK, Laha GS, Reddy GA, Rani NS, Sarma NP, Sonti RV (2008) Marker assisted introgression of bacterial blight resistance in Samba Mahsuri, an elite Indica rice variety. Euphytica 160:411–422

Tang F, Tao Y, Zhao T, Wang G (2006) In vitro production of haploid and doubled haploid plants from pollinated ovaries of maize (*Zea Mays*). Plant Cell Tissue Organ Cult 84(2):233–237

Tanksley SD (1983) Molecular markers in plant breeding. Plant Mol Biol Rep 1:3–8

Tanksley SD (1993) Mapping polygenes. Annu Rev Genet 27(1):205–233

Tanksley SD, Nelson JC (1996) Advanced backcross QTL analysis: a method for the simultaneous discovery and transfer of valuable QTLs from unadapted germplasm into elite breeding lines. Theor Appl Genet 92:191–203

Tanksley SD, Young ND, Patterson AH, Bonierbale MW (1989) RFLP mapping in plant breeding: new tools for an old science. Biotechnology 7:257–263

Tiwari JK, SinghBP GJ, Poonam P, Patil VU (2013) Molecular characterization of the Indian andigena potato core collection using microsatellite markers. Afr J Biotechnol 12(10): 1025–1033

Walunjkar BC, Parihar A, Singh NK, Parmar LD (2015) Genetic diversity of wild and c.ultivated genotypes of pigeonpea through RAPD and SSR markers. J Environ Biol 36(2):461–466

Wang B, Chee PW (2010) Application of advanced backcross quantitative trait locus (Qtl) analysis in crop improvement. J Plant Breed Crop Sci 2:221–232

Wang FG, Zhao JR, Guo JL, Liu LZ (2003) Series of research on establishing DNA fingerprinting pool of Chinese new maize cultivars in the establishment of a standard SSR system fitting for maize cultivars' identification. Yumi Kexue 11:3–6

Wang YH, Liu SJ, Ji SL, Zhang WW, Wang CM, Jiang L, Wan JM (2005) Fine mapping and Marker Assisted Selection (MAS) of a low glutelin content genein rice. Cell Res 15(8):622–630

Wang Y, Shahid MQ, Ghouri F, Baloch FS, Wang Y, Huang H (2015) Evaluation of the geographical pattern of genetic diversity of *glycine soja* and *glycine max* based on four single copy nuclear gene loci: for conservation of soybean germplasm. Biochem Syst Ecol 62:229–235

Wang Y, Ghouri F, Shahid MQ, Naeem M, Baloch FS (2017) The genetic diversity and population structure of wild soybean evaluated by chloroplast and nuclear gene sequences. Biochem Syst Ecol 71:170–178

Wu JW, Hu CY, Shahid MQ, Guo HB, Zeng YX, Liu XD, Lu YG (2013) Analysis on genetic diversification and heterosis in autotetraploid rice. Springer Plus 2(1):439

Wu J, Shahid MQ, Chen L, Chen Z, Wang L, LiuX LY (2015) Polyploidy enhances f1pollen sterility loci interactions that increase meiosis abnormalities and pollen sterility in autotetraploid rice. Plant Physiol 169(4):2700–2717

Xie F, He Z, Esguerra MQ, Qiu F, Ramanathan V (2014) Determination of heterotic groups for tropical Indica hybrid rice germplasm. Theor Appl Genet. https://doi.org/10.1007/s00122-013-2227-1

Xu Y (2010). Molecular plant breeding CAB International

Young ND (1999) A cautiously optimistic vision for marker-assisted breeding. Mol Breed 5:505–510

Young ND, Tanksley SD (1989) RFLP analysis of the size of chromosomal segments retained around theTm-2 locus of tomato during backcross breeding. Theor Appl Genet 77:353–359

You-Xin Y, Yan-Hong L, Jing-Fei T (2012) Wide-compatibility gene exploited by functional molecular markers and its effect on fertility of intersubspecific rice hybrids. Crop Sci 52(2): 669–675

Yu J, Holland JB, McMullen MD et al (2008) Genetic design and statistical power of nested association. Genetics 178:539–551

Zeng Z-B (1994) Precision mapping of quantitative trait loci. Genetics 136:1457–1468

Zhang P, Zhong KM, Shahid Q, Tong H (2016a) Association analysis in rice: from application to utilization. Front Plant Sci 7:1202

Zhang P, Zhong KH, Tong H, Shahid MQ, Li J (2016b) Association mapping for aluminum tolerance in a core collection of rice landraces. Front Plant Sci 7:1415

Zhao X, Tan G, Xing Y, Wei L, Chao Q, Zuo W, Lübberstedt T, Xu M (2012) Marker-assisted introgression of qHSR1to improve maize resistance to head smut. Mol Breeding. https://doi.org/10.1007/s11032-011-9694-3

Zou F (2009) QTL mapping in intercross and backcross populations. In: DiPitrello K (ed) Cardiovascular genomics, methods and protocols. Humana Press (Springer Science + Business Media), LLC, New York, pp 157–173

Bioinformatics in Plant Genomics for Next-Generation Plant Breeding

7

Pratibha Parihar

Abstract

Breeding has played a significant role in the evolution of human civilizations began with the domestication of plant and animal species estimated to date back 10,000–15,000 years ago. It provides sustainability to more than 6 billion world populations. Over the past 100 years, there is a drastic variation in the landscape for plant breeding due to uncontrolled population growth, demolition of agricultural land areas, and changing environmental conditions. Thus, it imposes a tremendous challenge on the researchers to improve the production and productivity of crops. The advent of novel genomics methods including NGS (Next-Generation Sequencing) and breeding tools has massively changed traditional breeding into next-generation breeding. Genome editing is a promising technique to alter specific genes to improve trait expression. Integrating computational tools with next-generation breeding technologies can speed up the breeding process and increase the genetic gains under different production systems. This chapter emphasizes the significance of next-generation sequencing-derived information (big data) and their analysis by omics tools to revolutionize crop improvement.

Keywords

Bioinformatics · Next-generation sequencing · Marker-assisted selection · Single nucleotide polymorphism · Genotyping · Marker discovery · Allele diversity · Genomic selection · Genome editing

P. Parihar (✉)
N.V. Patel College of Pure and Applied Sciences, CVM University, Vallabh Vidyanagar, Anand, Gujarat, India
e-mail: pratibha@nvpas.edu.in

R. L. Singh et al. (eds.), *Plant Genomics for Sustainable Agriculture*,
https://doi.org/10.1007/978-981-16-6974-3_7

7.1 Introduction

The twenty-first century is an era of globalization. Great saying by Helen Fisher "Globalization requires taking a broad contextual and long-term view." The idea is to develop a global network of science and technologies without any limitations and boundaries. Traditional plant breeding techniques included testing of different breeding strategies like hybridization and selection of best elite lines or varieties with the best combination of desired characteristics, viz., yield-attributing traits, quality traits, and insect and pest resistance. The late nineteenth century and early twenty-first century witnessed the transition period from a "Mendelian era" to the "Genomics era" and the introduction of the dry lab along with a wet lab. With the innovation of the whole-genome sequencing technique, an unlimited number of markers, CNV (copy number variation), SNP (Single Nucleotide Polymorphism), and diverse germplasm based on allele mining can be identified. Bioinformatics is playing an important role in breeding by managing the big data generated from large-scale genomics information, high-throughput sequencing, and breeding experiments. It is useful for plant breeders in performing next-generation breeding techniques like phenotypic selection, genomic selection (GS), and speed breeding, simulates genotype-by-environment interactions using multiple trait approaches, and developing crop models. In the present chapter, different tools and techniques are discussed to exploit genomic data in breeding programs for the selection of best-fit breeding strategies.

7.1.1 Traditional Breeding to Next-Generation Breeding

The traditional farming practice is nowadays a challenge owing to the high expectations of regulators, consumers, food processors, and retailers. It is due to global population blasts, changing climatic conditions, and environmental pressure. It is estimated that the global population will approach 10 billion by 2050 which needed an increase in agricultural production by 60%. It is estimated that with every rise in 1 °C temperature there is a fall in agricultural productivity by 5%. The agricultural revolution has gradually transformed the traditional agricultural system to hasten the production and productivity rate, novel crops breeding and insect, drought or heat tolerance crop with less pesticide usage. Agriculture has changed course many times, and always in pursuit of higher and better production and productivity (Heywood 1992) (Table 7.1).

The different techniques have been used since ancient times to improve the value of food crops with the introduction of superior cultivars, enhancing their yield and the nutritional quality of their products, for the healthy living of humans (Hallauer 2011). The term "plant breeding" is often used synonymously with "crop improvement" in modern society and is aided by genetic engineering and genomics approaches. Modern plant breeding programs are a combination of diverse skill sets which include genetics, statistics, agronomy, biochemistry, physiology,

Table 7.1 Agricultural revolution timeline

Agricultural revolution	Period	Transformation
First	~8000–10,000 BC	Domestication of plants and animals (also known as the Neolithic revolution)
Second	~1800 AD onwards	Mechanization of agricultural production, selective breeding, crop rotation, use of fertilizer, and land reclamation
Third	1930s–1970s	Biotechnologies/genetic engineering and genetic modification of agricultural products increase in the application of the science of chemistry and the breeding of dwarf varieties of grains, launched agribusinesses to new multinational heights. **(green revolution)**
Fourth	Present	Next-generation breeding (second green revolution), OMICS techniques, SMART farm management that tracks production in real-time. Artificial intelligence that calculates optimal harvest times and informs intermediaries of yield. (**the digital revolution)** computer simulation in plant breeding, genome editing techniques, speed breeding

bioinformatics, molecular biology, economics, and information technology making it highly interdisciplinary (Hickey et al. 2017).

7.1.2 IT (Information Technology) in Plant Breeding

At the beginning of the twenty-first century, the Human Genome Project (HGP) acts as a catalyst for the rapid and crucial growth and advancements in bioinformatics (Zwart 2015). Later other projects were launched to unravel the mysteries of genome blueprint, including the ENCyclopedia of DNA Elements (ENCODE), 1000 Genomes project, International HapMap project, Roadmap Epigenomics Project, Microbial Genome Project, and data was stored in databases like NCBI (http://www.ncbi.nlm.nih.gov/genomes/leuks.cgi), ENSEMBL (http://plants.ensembl.org/index.html), PROTOZOME (http://phytozome.jgi.doe.gov/pz/portal.html), SNP (http://snpdb.appliedbioinformatics. com.au). Bioinformatics is an interdisciplinary field of life science that hybrids biological data with information technology to collect, store, retrieve, manipulate, model, and analysis of scientific research. It mainly includes omics, big data analysis, synthetic biology, system biology, biomedicine, and whole-cell modeling, etc. The emergence of "Big Data" represents a new paradigm that transforms the studies into large-scale research. Such a huge data set when coupled with algorithms and bioinformatics tools has a profound implication on the predictive efficiency and reproducibility of results (Назипова and Nazipova 2017). The introduction of computer simulation in plant breeding has reduced its complexity. Computer simulation is useful in gene mapping, gene network analysis, and in developing different crop models to simulate crop growth. Deep Learning or Artificial Intelligence is a mathematical model that relies on an artificial neural network (ANN), based on the working of the human brain, that allows a program to learn and recognize an object or a specific pattern within a set of data and provides a bridge between the genes and the morphological characteristics of the plant. It will be helpful in next-generation breeding techniques like counting the number of flowers for millions of plants, flowering time, or fruits ripening in a shorter time. Machine learning (ML) algorithms based on "learning by finding" are used to capture the characteristics of target patterns in plant genotypes and phenotypes and are capable of processing big data using digital image processing and remote sensing. Machine learning (ML) based algorithms is used in different areas of molecular biology and genomics, including detection of transcription start site (TSS), splice site junction, promoters, positioned nucleosomes, genome assembly, genome editing, functional annotation of genes, and detecting SNPs in polyploidy plants. Ma et al. (2014) reviewed the significance of machine learning approaches in determining genotypes of gene homologs, repeat regions, and genome assembly in crop plants like Hordeum vulgar and Triticum aestivum using the RHadoop tool. Crop simulation models exploit various quantitative data to predict the effect of change in environmental conditions and various factors on plant growth and development (Hodson and White 2010). APSIM (The Agricultural Production Systems sIMulator) is a crop simulation tool that predicts the effect of climate and management conditions on various qualitative and quantitative traits in the plants (e.g., grain size, biomass, or sugar yield) (Keating et al. 2003). The ADAM-plant software package uses a stochastic simulation technique to detect the genetic changes in the population under various breeding circumstances by simulating plant populations (Liu et al. 2019). It allows the breeder to simulate speed breeding, selection of parental lines, overlapping breeding cycles for cross-pollinated and self-pollinated plant crops. The next-generation plant breeder's toolbox encompasses a wide spectrum of statistical and computational methods to bring together breeding, evolution, and diversity in the plant, genetics, statistics, computer programming, and Next-Generation Sequencing, genomics, and data science.

7.2 Genomics in Plant Breeding

New advancement in NGS sequencing technologies has revolutionized traditional crop breeding and research, into the "genomics era" of crop improvement. According to the report of "Plant Genomics - Global Market Outlook (2018–2027)" the approximate global market for plant genomics is around $6.18 billion in 2018 and is predictable to reach $14.49 billion by 2027 growing at a Compound Annual Growth Rate (CAGR) of 9.9%. With the accessibility of whole-genome reference sequences and cost-efficient genotyping platforms, breeders can detect and transfer specific alleles or characters or genomic information in the superior and desirable varieties in very little time (Lee et al. 2015). The advancement in omics technology leads to the spectacular transformation in plant breeding. The biggest challenge faced by plant researchers is the need for big data storage capacity and highly efficient and proficient analysis software (Ong et al. 2016). Michael and VanBuren (2020) reviewed recent advancements in sequencing technology like

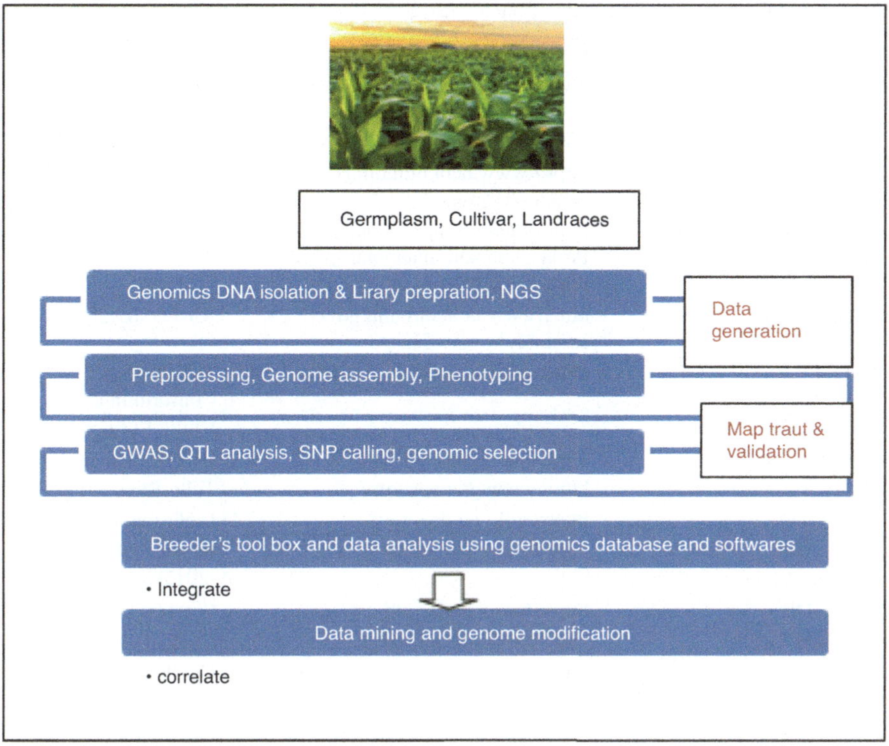

Fig. 7.1 Steps for plant genome data analysis

long-read sequencing and re-sequencing to speed up the sequencing process. In the last 20 years, more than 400 plant species have been sequenced, published and were submitted in the databases which includes both flowering and non-flowering plants (https://www.Plabipd.de/portal/web/guest/sequenced-plant-genomes). Light Speed Genomics is an optical imaging technology that uses electron microscopy in the detection of series of light patterns in the form of signals and generates high-quality reads or data (Pushpendra 2008; Xu et al. 2009). Besides genomics, plant proteomics assists breeders or crop cultivars in understanding the process and mechanism of cell division and differentiation, different stages of seed germination, flower and fruit development (Tomas et al. 2011). Further, phenomics is useful in the identification of QTLs and physiochemical responses of plants due to change in environmental conditions or genetic pressures (Robert and Mark 2011) (Fig. 7.1).

Genome research mainly includes structural genomics, functional genomics, comparative genomics, sequencing, molecular characterization, annotation, and omics data analysis which comprised of the study of molecular compositions, structures, organizations, functions, interactions/networks, and system biology of an entire plant genome. Sequencing technology in plant breeding has gained a new momentum that allows whole-genome sequencing, re-sequencing, transcriptome

Table 7.2 URLs of some important resources related to plant genome analysis for breeders

Resource	URL	Reference
Genbank	http://www.ncbi.nlm.nih.gov/genbank/	Clark et al. (2016)
NCBI reference sequence	https://www.ncbi.nlm.nih.gov/refseq	O'Leary et al. (2016)
NCBI viral genomes resource	https://www.ncbi.nlm.nih.gov/genome/viruses	Brister et al. (2015)
NCBI prokaryotic genome annotation pipeline (Prokka)	https://kbase.us/applist/apps/ProkkaAnnotation/annotate_contigs/release	Tatusova et al. (2016)
NCBI assembly	http://www.ncbi.nlm.nih.gov/assembly/	Kitts et al. (2016)
UniProt	www.uniprot.org	The UniProt Consortium and UniProt (2019)
Ensembl plants	http://plants.ensembl.org/	(Bolser et al. 2016)
Gramene	http://www.gramene.org/	Tello-Ruiz et al. (2018)
Gold	https://gold.jgi.doe.gov/	Liolios et al. (2009)
SGN	https://solgenomics.net/	Fernandez-Pozo et al. (2015)
MaizeGDB	https://www.maizegdb.org/	Portwood et al. (2019)
PlantGDB	http://www.plantgdb.org/	Dong et al. (2004)
Phytozome	https://phytozome.jgi.doe.gov/pz/	Goodstein et al. (2012)
PGDBj	http://pgdbj.jp/	Asamizu et al. (2014)
Phytozome	https://phytozome.jgi.doe.gov/pz/portal.html	Goodstein et al. (2012)

sequencing, and exome sequencing in very less time and affordable prices so that almost every species-specific genome can be sequenced at a reasonable price. Further, it offers great opportunities for targeted crop breeding, biodiversity conservation (Bevan et al. 2017) (Table 7.2).

7.3 Next-Generation Breeding

Following the green revolution and transgenic crop development, another revolutionary progress in plant breeding is the Next-Generation Breeding also known as the second green revolution. The next-generation sequencing techniques along with third-generation genotyping tools in trait discovery, genomic selection, and gene editing have revolutionize molecular breeding to next-generation plant breeding (Ray and Satya 2014). It is useful in developing low-cost, high-throughput genotyping methods for screening large populations within a small period, rapid selection of loci, and detection of flexible and polymorphic markers and creating novel combinations of alleles in a particular line. Thus "Next generations breeding techniques" are useful in reducing the breeding cycles and developing the genetic gain in terms of production and productivity. The approaches are Next-Generation

Sequencing (NGS), gene editing to allow the targeted modification of specific genes, genome-wide association studies (GWAS) for quantitative trait loci (QTLs) mapping in plants, TILLING and EcoTILLING, re-sequencing for SNPs discovery, improvement of traits by introgression QTLs into the target genotypes through marker-assisted breeding, analysis of genetic diversity, new line-breeding, genomic selection (GS), and speed breeding.

7.3.1 Next-Generation Sequencing

A flood in the biological data is being generated by high-throughput data-generating experiments like NGS (Next-Generation Sequencing), which has raised new challenges for data mining and management. The recent innovative development in sequencing technology within the market-place has reduced costs, error, and time per sequencing sample, which give rise to an exponential growth of biological data (Zhao and Grant 2011). Next-generation sequencing or second-generation sequencing is a high-throughput set of technologies based on massively parallel sequencing. It is classified into three categories: sequencing by synthesis, sequencing by sequential ligation, and single-molecule sequencing. The sequencing platforms include (1) Illuminas (Solexa) HiSeq, and MiSeq sequencing; for whole-genome sequencing based on sequencing-by-synthesis (SBS), and reversible dye-terminators, (2) Supported Oligonucleotide Ligation and Detection (SOLiD); based on sequencing by sequential ligation of oligonucleotide probes, (3) Roches 454; pyrosequencing method based on sequencing by synthesis, and(4) Helicose; based on single-molecule sequencing. All the above sequencing equipment uses the principles of the Sanger method. The basic steps in NGS involved library preparation, clonal amplification, clustering, sequencing, and data analysis. The cost of whole-genome sequencing has reduced by many folds with the advent of NGS technologies and powerful computational pipelines and thus allowing discovery, sequencing, genotyping, annotation, and detection of diversity, mutation, and polymorphic markers in a single step (Stapley et al. 2010). With the increasing usage and demand of sequencing techniques, new modification in next-generation sequencing, i.e., the third-generation sequencing or "next-next" generation sequencing (NNGS) is coming out with new insight in the sequencing. Xiao and Zhou (2020) reviewed the third-generation sequencing method including Oxford Nanopore Minion technology and Pacific Biosciences (PacBio) Single-molecule real-time (SMRT) technology. These are single sample molecule and real-time sequencing technologies. Oxford Nanopore operated on an electronics-based platform called the GridION™ system. It reads the variation in electrical signals as DNA/RNA passes through a protein nanopore. Further, these signals are decoded to specific DNA or RNA sequences. The SMRT technology allows direct RNA sequencing (Eid et al. 2009). SMRT is based on sequencing by synthesis and is a real-time technique. It has an SMRT Cell, which contains small wells called zero-mode waveguides (ZMWs). Each well immobilized with single molecules of DNA, and emits light as the polymerase incorporates nucleotide which is detected as a

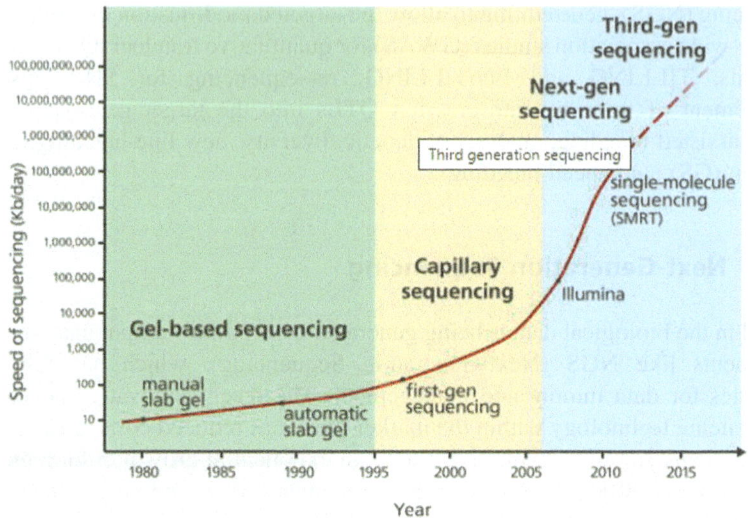

Fig. 7.2 Graphical representation of advancement in sequencing technology. (Image credit: Genome Research Limited)

fluorescent signal. Unlike next-generation sequencing these methods do not include an amplification step during sequencing and library preparation, therefore enabling single-molecule sequencing. Besides this, the expected read lengths are much higher than those of second-generation techniques, with average read lengths exceeding 250 bp–50 kbp and maximal read lengths exceeding >1 Mb (Bleidorn 2015). The two main characteristics of third-generation sequencing are PCR that does not require before sequencing, which reduces sample preparation time for sequencing, and the signals (fluorescent or electric current) are detected in real-time mode, during complementary strand synthesis. The advantage of the technique is that it generates long reads in less time and easy to perform. It is useful for genome assemblies, and also for metabarcoding and metagenomic studies. It is also suitable for the generation of high-quality assemblies of highly repetitive plant genomes (Dumschott et al. 2020). Few examples of plant species sequenced using the nanopore platform are *Juglans sigillata* (iron walnut) (Ning et al. 2020), *Eucalyptus pauciflora* (snow gum) (Wang et al. 2020), *Oryza sativa* (rice) Carolina Gold Select (Read et al. 2020), *Dioscorea dumetorum* (yam) (Siadjeu et al. 2020), and *Anthoceros agrestis* (field hornwort) (Li et al. 2020). In agricultural sciences, the NGS technique can be applicable in whole-genome sequencing (WGS), whole-genome re-sequencing (WGRS), Whole Exome Sequencing (WES), transcriptomics, Single cell sequencing, Methylation Sequencing, metagenomics, and reduced representation sequencing for high-throughput SNP genotyping. Such techniques are useful to improve breeding activities in crops of economic importance (Barba et al. 2014). Reference genome sequences combined with next-generation sequencing (NGS) led to a multitude of new approaches to detect, analyze, and visualization of genetic variation (Fig. 7.2).

7.3.1.1 RNA Sequencing

It is a genome-wide profiling of transcripts and their qualitative and quantitative measurements to expand and accelerate gene expression studies and validate gene annotation. Typically, polyadenylated RNAs enriched for subsequent cDNA synthesis and high-throughput sequencing bestow insights into the complex nature of regulatory networks. The gene expression data analysis provided a rich source of information to the breeders to understand the molecular basis of complex/diseased plant processes, and thus useful in detection of new targets for manipulating the plant process (Han et al. 2015). Schmid et al. (2005) reviewed the importance of transcriptome map in plant. The transcriptome map is a complete estimation of gene expression in all possible cells, tissues, organs, or parts of an organism during the life cycle of the organism from different developmental stages and/or environmental conditions. The first detailed transcriptome map was constructed for *A. thaliana* in 2005. Nobuta et al. (2007) constructed map from 18 samples of rice using massively parallel signature sequencing (MPSS). Beside it, researchers used different methods for gene expression analysis like Northern blot method, QRT PCR (Real-Time Quantitative Reverse Transcription PCR) (VanGuilder et al. 2008), cDNA-AFLPs (cDNA amplified fragment length polymorphisms) (Bachem et al. 1996). However, these methods are not quantitative and captured low-abundance transcripts. The Serial Analysis of Gene Expression (SAGE) (Anisimov 2008), Massively Parallel Signature Sequencing (MPSS) (Reinartz et al. 2002), and hybridization-based platforms or microarrays (Schena et al. 1995) have several advantages when compared with other methods. They can analyze thousands of different transcripts simultaneously in the same reaction and are semi-quantitative and sensitive to low-abundance transcripts. The software packages specialized for microarrays analysis are Bioconductor (http://www.bioconductor.org/help/workflows/oligo-arrays/) or MeV (http://www.tm4.org/mev/) (Saeed et al. 2003).

RNA-seq is an emerging technique to quantify gene expression alternative to microarray studies (Stiglic and BajgotM 2010). RNA deep-sequencing technologies like digital gene expression and Illumina RNA Seq (Shi et al. 2011) are both qualitative and quantitative in nature and allowed the detection of rare transcripts and splice variants. RNA sequencing involved in the conversion of RNA into cDNA, then sequencing is performed by using any NGS technology. The drawback in the method may be due to (i) the inefficient nature of reverse transcriptases (RTs), (ii) DNA-dependent DNA polymerase activity of reverse transcriptase enzyme which may cause false second strand DNA formation, and (iii) artifactual cDNA formation due to template switching.

To overcome the limitations of current RNA-Seq strategies and to reduce the costs and complications of the genome-wide transcriptome surveys, Direct RNA sequencing (DRS) technique developed by Helicos Biosciences Corporation (Ozsolak and Milos 2011). It involves sequencing of natural RNA molecules and eliminates the prior conversion to cDNA and ligation/amplification leading to improved accuracy. In plant breeding and crop improvement, RNA seq technique can be used for differential gene expression analysis, detection of novel and rare transcripts (fusion transcripts and non-coding transcripts), splicing variants, and alternative splicing. Further it is useful in annotation of gene or genome, meta-transcriptomics, and the identification of genes involved in metabolic pathways, disease response, and various developmental process (Shi et al. 2011). RNA-seq shows potential utility for next-generation breeding.

7.3.1.2 Chromatin Immunoprecipitation Sequencing (ChIP-Seq)

It is a specialized sequencing method for studying protein–DNA interactions, DNA-binding proteins, histone modifications, or nucleosomes including transcription factors, epigenetic, chromatin modifiers, and control gene expressions in plants (Xifeng et al. 2018). Barski et al. (2007) published the first large-scale profiling of histone methylations in human genome using ChIP-seq technique. The sequencing was performed on the platform of Solexa 1G genome analyzer. Robertson et al. (2007) used the ChIP-seq technique to study in vivo transcription factors in mammalian DNA sequences. The procedure for chromatin immunoprecipitation sequencing is as follows: (1) Immunoprecipitation of DNA-bound protein with a specific antibody. (2) The bound DNA is then coprecipitated and purified. (3) Library construction and sequencing are performed using techniques like Illumina and ABI SOLiD. (4) Finally computational analysis of ChIP-Seq data is done. The computational analysis involved the following steps: Mapping of all raw reads to the reference genome, the uniquely mapped reads are retained. A peak call is generated (list of enriched regions (peaks)) from mapped reads and a control profile. Peak call is analyzed to identify the binding sequence motifs which represent transcription factor binding sites (TFBSs), annotation, and functional enrichment analysis (Fig. 7.3).

ChIP-sequencing (ChIP-Seq) is used to map the transcription factors binding sites and DNA-histones binding sites (Ji 2010). R and Bioconductor packages can be used for ChIP-Seq data analysis. Four packages are used for different types of analysis. The "dada2" package is used for the trimming of high-throughput sequencing files. The alignment and quality check are done by the "QuasR" package. The most important step in ChIP-Seq is the peak calling step performed by "mosaic," "ChIPseeker," package for annotation, visualization, motif detection, and functional enrichment analysis. ChIP technique can apply to many molecular biology techniques such as cloning, microarray, real-time PCR, PCR with single-stranded conformational polymorphism, western blot analysis, and southern blot analysis. It increases the versatility of the technique (Wells 2002; Jackson and Chalkley 1981). ChIP-chip or ChIP coupled with microarrays is also a standard technique for the identification of the genome-wide DNA–proteins binding interactions (Buck and Lieb 2004). The ChIP technique is a highly efficient technique to study the chromatin structure and nuclear events during transcription. It is also used to identify target genes related to DNA-binding proteins and their regulatory enzymes. The drawback with the technique is that it is qualitative rather than quantitative and not successful with the identification of simultaneous binding of multiple proteins to one target site (Johnson and Bresnick 2002).

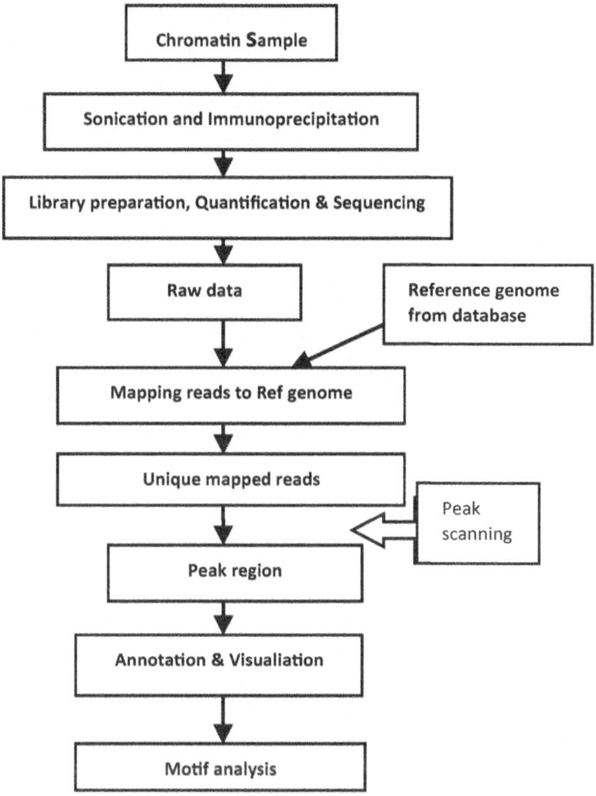

Fig. 7.3 Flow chart of the ChIP-seq procedure

7.3.2 Genome Editing

For centuries, scientists and farmers are using many techniques in crop improvement to develop new or modified varieties by changing their genetic composition. The goal of crop improvement is to develop better and improved varieties with characteristics like resistance to insects, nematodes, and disease, tolerance to salt, metal, and better yield, nutrition, and taste. The classical genetics approach relies on the methods like natural selection, hybridization, wide crossing, or induced mutation to recombine or transfer the genetic material between different germplasm within or between closely related genera or species. Thus, progress for screening phenotypes and genotypes to the development of commercial varieties is a time-consuming and labor-intensive process. Further, with the application of transgenic or genetic engineering techniques in 1980, desired genes or traits can be introduced at a random place even in the genome of the different plant species and avoided the problem of mixing the genome of two species. The successful examples of genetic engineering are bt corn, cotton, herbicide-tolerant canola, alfalfa, soybean and corn, *FLAVR SAVR* tomato, and virus-resistant squash and papaya. However, genetically modification in crops faced many health, environmental, and ethical concerns. The recent technology is gene or genome editing which uses site-specific nucleases for precise and successful genome engineering. It has revolutionized applied research in crop improvement and provided alternative ways to avoid the path of strict regulations of "genetically modified organisms or GMOs." (Andersen et al. 2015; Carroll 2011). It involves the use of engineered nucleases enzyme for the insertion, deletion, or replacement of a DNA fragment at desired locations in the genome by creating specific double-strand breaks (DSBs) and stimulates cellular DNA repair mechanisms. With the aid of gene or genome editing technique, breeders enable desirable modification in plant genome by knocking out undesirable genes such as disease vulnerability, breakdown the linkages between desirable and non-desirable traits, developing varieties with the preferred characteristics and used in target specific gene editing. It is RNA programmable method that enables the best genes to be placed in the right positions and improving beneficial traits within the organism such as drought tolerance, insect resistance, or improved nutrition.

The CRISPR–Cas (CRISPR-associated proteins) system is an adaptive or heritable anti-virus immune system in prokaryotes that memories the previous infection and uses the cas 9 enzymes to degrade the genome of the foreign molecule. The immune system captured a short DNA fragment (~ 40 bp long) from the genome of the previous attack and inserted it into the repetitive locus of the host genome via recombination called CRISPR (clustered regularly interspaced short palindromic repeat) array. These are short repetitive elements (repeats) followed by unique sequences (spacers) also known as clusters of regularly interspaced repeats. CRISPR array is flanked by a set of *cas* genes encoding the Cas proteins and preceded by an AT-rich leader sequence. When a similar foreign molecule is invaded in the host, it causes transcription and processing of the CRISPR array into long precursor CRISPR RNA (pre-cr-DNA). The transcript is recognized by a ribonuclease enzyme and digests at CRISPR spacer elements into small RNA fragments. The individual RNA fragment is known as crRNAs (for CRISPR–RNAs or **guide RNAs**). CrRNA binds with tracr RNA or trans-activating crRNA by base pairing. Tracr RNA is a component of the host immune system and has a binding site or motif for Cas protein. The CrRNA-tracr RNA complex binds to CRISPR-associated proteins (Cas proteins). This complex scans and binds to the foreign DNA at the site containing protospacer adjacent motif (PAM). Cas9 only digests the region which contains the PAM site. The gene-editing technology comprises different classes of artificially engineered site-directed nucleases (SDNs) to create site-specific double-strand breaks(DSB), such as zinc-finger nucleases (ZFNs), transcription activator-like effector nucleases (TALENs), and clustered regularly interspaced short palindromic repeats (CRISPR)-associated (Cas) nucleases RNA-guided nucleases (RGNs) and engineered meganuclease, also known as homing endonucleases (Osakabe and Osakabe 2015). The engineered CRISPR/Cas system consists of a plasmid with gene expressing cas9, gene transcribing guide RNA, and cloning site for introduction of the target specific RNA. The steps are as follows: (1) identify the target gene with adjacent PAM region, (2) amplify the target gene and ligate onto CRISPR/Cas plasmid vector, (3) transform the vector into cell lines, (4) validate the edited gene using PCR.

Table 7.3 URLs of some important resources related to detection and analysis of CRISPR systems

Resources	URL	References
CRISPRone	https://omics.informatics.indiana.edu/CRISPRone/	Zhang and Ye (2017)
CRISPRmap	http://rna.informatik.uni-freiburg.de/CRISPRmap/Input.jsp	Lange et al. (2013)
CRISPOR (guide RNA design)	http://crispor.tefor.net/	Haeussler et al. (2016)
CRISPRleader (annotates CRISPR leader boundaries)	http://www.bioinf.uni-freiburg.de/Software/CRISPRleader/	Alkhnbashi et al. (2016)
CRISPRdisco(to identify new casgenes)	https://github.com/CRISPRlab/CRISPRdisco	Crawley et al. (2018)
CRISPRstrand(to determine the orientation of CRISPR arrays)	https://www.fairshake.cloud/digital_object/2763/	Alkhnbashi et al. (2014)
CRISPRDetect (detection of CRISPR arrays)	http://crispr.otago.ac.nz/CRISPRDetect/predict_crispr_array.html	Biswas et al. (2016)
CRASS (CRISPR assembler- detecting CRISPR-Cas systems in metagenomic data)	https://ctskennerton.github.io/crass/	Skennerton et al. (2013)

The most popular tool for gene editing in CRISPR/Cas system is the CRISPR-Cas9 and CRISPR-Cas 12a that is expected to bring a revolution in the research world by opening the new era of precision plant breeding to attain desirable traits (Zhang et al. 2019). Different bioinformatics tools are used to identify the target sequences of the crRNAs, prediction of CRISPR leader sequences, characterization of PAM motifs, and CRISPR-Cas9 guide design. The protein homology search tools like Pfam/ HMMer3 can be used to predict the cas genes based on the conserved region present in them while CRISPR arrays are used for less conserved region. CRISPR Detect refines the repeat/spacer boundaries and annotates different types of sequence variations (like insertion/deletion) in near-identical repeats. PILER-CR is a useful tool for the rapid identification and classification of CRISPR repeats based on the homology search method. CRISPRdb (Grissa et al. 2007) was the first public database that contained information about CRISPR arrays from archaeal and bacterial genomes. The information in database CRISPRdb which is a part of CRISPRs web server holds about (i) Strain taxonomy browser; (ii) The CRISPRs properties page, which store information about the CRISPR's id along with its position on the genome, the number of spacers, and the consensus DR sequence. The database CRISPRCasdb combines information about CRISPR arrays and Cas annotations for more than 240 archaeal and 9242 bacterial genomes. The websites for CRISPRdb are http://crispr.i2bc.paris-saclay.fr/ CRISPR/ and CRISPRCasdb is https://crisprcas.i2bc.paris-saclay.fr/MainDb/Index (Table 7.3).

7.3.3 Restriction-Site Associated DNA Sequencing (RAD-Seq) and Genotyping-by-Sequencing (GBS)

Traditional genetic linkage mapping was based on limited markers and was costly and laborious to identify. In the age of NGS, thousands of markers can be easily added to the map even with the availability of small genomic information. High-throughput marker discovery methods using NGS technologies are restriction-site associated DNA sequencing (RAD-seq), genotyping-by-sequencing (GBS), reduced representation libraries (RRLs)/CRoPS, and multiplexed shotgun sequencing (Glaubitz et al. 2014a, b; Davey et al. 2011).

RAD-seq involved two important steps: (1) DNA fragmentation using restriction enzymes and (2) use of molecular identifiers (MID) to associate sequence reads to particular region. RAD-seq technique does not need a priori genome sequence information. In this method SNP based bulked segregation is performed in which genomic DNA is fragmented with a specific restriction enzyme followed by ligation of the barcoded adapter with molecular identifier code (Pfender et al. 2011). The processed DNA sample taken from multiple individuals (around 20) is pooled and randomly sheared (using the sonication method) so that only a subset of generated fragments contain a barcoded adapter. Further another divergent adapter is ligated. The technique uses divergent adapters which increases the chances of amplification of only those fragments containing both adapters and thus increases the accuracy of the result. The resultant amplicons are sequenced using an NGS platform. Finally, pooled samples with different identifiers are separated and SNPs are called using standard bioinformatics pipeline. If a reference sequence is available, then sequence can be processed for SNPs and insertion or deletion mutation using tools like Bowtie, Samtool, BWA, and RAD tags is used for de novo sequence. This technique is successfully used by Barchi et al. (2012) to construct a linkage map in eggplant and detection of quantitative trait loci for anthocyanin pigment of the eggplant. Yang et al. identified a resistant gene against anthracnose disease in Lupin.

Genotyping-by-sequencing (GBS) detects single nucleotide variation at the genome level, simultaneously in many individuals. GBS is a low-cost approach that uses sequence-based polymorphisms linked to the traits of interest and performs genome-wide association studies (GWAS). It has been widely applied in plant genetics and breeding. GBS uses molecular markers, such as single nucleotide polymorphism (SNP) to identify small variations in genetic sequence within populations, to reduce genome complexity, and in genotyping multiple DNA samples simultaneously. It uses a multiplex SNP genotyping platform or with high-throughput array-based technologies (e.g. Illumina Bead ArrayTM, Affymetrix GeneChipTM technology) (He et al. 2014). Initially, GBS data was generated from the restriction site-associated DNA sequencing (RAD-seq) technique (Baird et al. 2008) which uses a single restriction enzyme and sheared DNA to capture a suitable portion of the genome. In the GBS approach, genomic DNA samples are digested with restriction enzyme/enzymes and ligated to barcoded adapters in single wells, pooled, enriched by PCR, and sequencing of the amplified DNA pool. The GBS datasets are analyzed by using various bioinformatics pipelines. The coupling of high-density SNP arrays with powerful computational pipelines has allowed the fast and easy scoring of a large set of markers and has successfully applied on many crop species, e.g., rice, grapevine, peach, soybean, barley, maize, wheat, and apple (Barba et al. 2014). The basic workflow for GBS data analysis approaches has three sequential steps: (1) processing of raw data, (2) de novo assembly of the sequence tags or read alignment to a reference genome, or (3) genetic variant detection and annotation. Processing of raw data is the first key step in the processing of raw data generated from the sequencing technique. It includes demultiplexing, which involves the separation of reads into their corresponding samples based on barcode matching. Demultiplexing of Illumina reads is done using IlluminaMiSeq reporter software and CASAVA. The raw sequencing data may contain various errors or artifacts, such as poor quality bases, base-calling errors, contamination of adaptors, and duplicate reads. Correction and quality assessment is done using numerous publicly available software such as Trimmomatic, FastqMcf, FASTX-Toolkit, PRINSEQ, and cutadapt. Then De novo assembly of the sequence tags or read alignment to a reference genome is done. If a reference genome is not available for specific data, the de novo assembly method is used and the software packages are STACKS (Catchen et al. 2011); UNEAK (Lu et al. 2013); RApiD (Willing et al. 2011). It is used to produce mini-contigs which can be used as a reference for read mapping and genotyping. For short-read mapping, the reference genome is available, tools are MAQ (Li et al. 2008), STAMPY (Lunter and Goodson 2011a, b), Bowtie2 (Langmead et al. 2009), BWA (Li et al. 2009a, b, c, d, e), and SOAP2 (Li et al. 2009a, b, c, d, e). In polyploid plant species, the multi-mapped reads (align to multiple locations in the reference genome) range from 20 to 60%. Finally, the genetic variant and annotation is detected. After the completion of mapping, the next step is to identify the sequence variants (SNPs and InDels) from the processed BAM file which is the output of mapping. The tools for variant calling are SAMtools, mpileup/BCFtools (Li et al. 2009a, b, c, d, e), GATK (Van der Auwera et al. 2013), SOAP (Li et al. 2020), and GNUMAP (Clement et al. 2010). Filtering raw SNP candidates is an important step in genotyping and reducing false-positive results made from biases in the sequencing data and thus generating the data for SNP and genotype properties that fulfill specific threshold values. SAM tools are used for filtering, SNP calling, and estimation of allele frequency by applying the genotype likelihood information. The output is generated in the VCF (variant call format) format, which is a standard format for storing variant data. GBS is widely used in wheat and barley in the development of a high density-map of 20,000 and 34,000 SNPs, respectively (Poland et al. 2012) and to map QTLs for reduced plant height and spike architecture in barley (Lu et al. 2013).

The list of publicly available software and tools for GBS to identify genetic variants such as SNPs and insertions/ deletions (InDels) from NGS data generated by most major RADand GBS approaches is given in Table 7.4.

The use of NGS technologies shifted array-based genotyping assays and pre-defined SNP panels to the direct sequencing of the populations of interest, producing a genome-wide and unbiased set of markers. The restriction-site associated DNA sequencing (RAD), genotyping by sequencing (GBS), low

Table 7.4 Publicly Available Software and Tools for Genotyping-by-sequencing (GBS)

Tools	Utility	URL	References
Trimmomatic	Trimming adapter sequences and low quality regions from Illumina sequencing reads	http://www.usadellab.org/cms/?page=trimmomatic	Bolger et al. (2014)
Bowtie2	Aligning sequencing reads against a reference genome	http://bowtie-bio.sourceforge.net/bowtie2/index.shtml	Langmead et al. (2009)
SAM tools (sequence alignment/map)	Designed for manipulating alignments in the SAM Or BAM, including sorting, merging, indexing	http://samtools.sourceforge.net/	Li et al. (2009a, b, c, d, e)
BCFtools	Manipulate variant calls in the variant call format (VCF) and its binary counterpart (BCF)	http://samtools.github.io/bcftools/	Danecek et al. (2011)
GATK (genome analysis toolkit) genotyper	Provides a wide variety of tools for variant discovery and genotyping	http://www.broadinstitute.org/gatk/	Van der Auwera et al. (2013)
STACKS	Allows denovo assembly of short read GBS data and the identification of genetic variation in the absence of a reference genome	http://creskolab.uoregon.edu/stacks/	Catchen et al. (2011)
TASSEL-GBS	Implementation of a GBS analysis pipeline in the TASSEL software package	http://www.maizegenetics.net/	Glaubitz et al. (2014a, b)

coverage genome sequencing of all genotypes for a segregating population (POPSEQ) have recently employed for the development of high-density genetic maps in the large, complex, and highly repetitive barley genome and hexaploid wheat genome (Chapman et al. 2015). Golden Gate has been the most widely used platform and is 100-fold faster than gel-based methods for increasing 2–3 times maize map density.

7.3.4 TILLING and EcoTILLING

Targeting Induced Local Lesions in Genomes (TILLING) is a reverse genetic approach to facilitate mutation detection. It can detect all allelic variants of a genomics region in an artificial mutant collection (detects mutations in mutagenized populations) while Ecotype TILLING (EcoTILLING) uses natural collections to detect allelic variants for targeting genes (identifies single nucleotide polymorphisms (SNPs) within a natural population). The major steps for the TILLING are as follows: (1) mutagenize the target population, (2) DNA preparation and pooling of individuals, (3) PCR amplification and detection of heteroduplex, (4) detection of mutants, (5) sequencing the target gene segment to confirm the mutation, (6) determine the type of nucleotide change, and (7) analysis of the mutant phenotype. These methods use restriction endonucleases enzymes like CEL I or Endo I, which recognize and digest the mismatches in the double helix of DNA. This technique is important for plant breeding and crop improvement programs for the selection of the right target genes and their mutant phenotypes. The involvement of NGS technique and computation data analysis provided by gene expression studies is significantly increasing the number and quality of candidates for TILLING and EcoTILLING studies. Thus, the involvement of NGS has simplified the process of linking mutations to relevant phenotypes.

TILLING and EcoTILLING have shown successful results in many crops like rice, wheat, Arabidopsis, lotus, barley, maize, pea, and melon. The CODDLe program (Codons Optimized to Deliver Deleterious Lesions, http:// www. Proweb. org/ coddle/) facilitates gene modeling and primer designing and evaluates the probable effect of induced or natural polymorphisms on gene function (Gilchrist and Haughn 2005). CODDLe uses raw genomic and protein-coding data from public databases or the user, constructs gene models, and determines the region that has the highest density of predicted deleterious nucleotide changes. PARSESNP (Project Aligned Related Sequences and Evaluate SNPs) is a bioinformatic tool designed for displaying and analyzing nucleotide polymorphisms (http://www.proweb.org/parsesnp/parsesnp_help.html; Taylor and Greene 2003). TbyS tool is developed by the Comai lab for both TILLING and EcoTILLING data analysis (Acevedo-Garcia et al. 2017).

There are many useful and unexplored genes and alleles available in wild species, mutagenized populations, germplasm collections, cultivars, and landraces. One of the biggest challenges in agricultural genetics is to access and identify these genes and genetic variation from genomic information to phenotypic capacities. The genetic diversity within the species cannot be represented by using a single reference genome. The re-sequencing of different cultivars, landraces, and wild accessions assumes an important role to reveal domestication events, identify the gene diversity and variations, and elucidates heterosis mechanisms (Schnable and Springer 2013). Kim and Tai (2014) reported re-sequencing in Targeting Induced Local Lesions in Genomes (TILLING) populations also known as TILLING-by-Sequencing can be used for detection of induced mutations in pools of individuals and functional characterization of genes (Kurowska et al. 2011) (Table 7.5).

Innovations in the high-throughput phenotyping platforms employing new fields like remote sensing and imaging techniques (based on visible/near-infrared and far-infrared radiation, reflected and emitted by the plants, respectively) and high-performance data recording and computing evaluate plant performance in the field or controlled environments. Automated greenhouse systems (LemnaTec system, http://www.lemnatec.com/) coupled with innovative image acquisition techniques and software (Phenoscope, RootReader3D, HTPheno, RootNav, Integrated Analysis Platform) allow the continuous recording of divergent phenotypic traits over time (e.g., in barley and tomato).

Table 7.5 Steps and resources which facilitate TILLING process

	Steps	Resources
(1)	Literature mining and the selection of target gene	PubMed central and GOPubMed (ontology annotation) NCBI GenBank (nucleotide databases) *Triticum turgidum* ViroBLAST (ViroBLAST search tools) NCBI dbEST (EST databases)
(2)	Meta-analysis of gene expression data and gene ontology (GO) annotation	ArrayExpress, ArrayTrack, gene expression omnibus (GEO), Genevestigator, AgriGO, Gramene (a resource for comparative grass genomics), Genevestigator (microarray database and analysis toolbox), plant expression database (PLEXdb), Solanaceae genomics network (SGN)
(3)	Selection of gene fragments for the mutational analysis	CODDLE, SIFT (sorting intolerant from tolerant), I-Mutant3.0, PROVEAN
(4)	PCR primer Design for Mutation Screening	Primer3, PrimerQuest, primer-BLAST, OligoAnalyzer

7.3.5 SNP Discovery

Single Nucleotide Polymorphisms or Simple Nucleotide Polymorphisms (SNPs) are third-generation molecular markers and are widely used in plant genetic research and breeding (Ganal et al. 2009). NGS in SNP discovery is useful in linkage map construction, genetic diversity analyses, association mapping, identification of cloned genes, and marker-assisted selection in several species (Cortés et al. 2011; Sim et al. 2012). The different classes of SNPs are: Regulatory SNPs which are found in regulatory region of gene; Intronic SNPs present within introns; Non-synonymous SNPs or substitutions are the mutations that can alter the amino acid sequence of a protein; and synonymous SNPs present in exons that do not alter the amino acid sequence. Non-synonymous substitutions have a higher probability to bring biological change to any individual. Thus, more emphasis is given to non-synonymous SNPs as mutation at sequence level leads to structural and functional impact in terms of protein stability, which hampers the interaction with other proteins (Yates and Sternberg 2013).

In silico SNP discovery relies on the identification of variation in nucleotide position of accession read which differs from the reference genome sequence. If a reference genome is not available for a species, this is achieved by comparing reads from different genotypes using de novo assembly strategies. Novoalign (http://www.novocraft.com/main/index.php) and STAMPY (Lunter and Goodson 2011a, b) are assembly tools for mapping short reads (Nielsen et al. 2011). The mapping tool generates read assembly files and then SNP calling is performed. SNP calling software is Samtools (Li et al. 2009a, b, c, d, e), SNVer (Wei et al. 2011), GATK (Genome Analysis Toolkit) (Zhu et al. 2014), and SOAPsnp (Li et al. 2009a, b, c, d, e). Samtools and SOAPsnp tools are widely used for file conversion (SAM to BAM and vice versa), mapping statistics, variant calling, and assembly

Fig. 7.4 A simplified workflow of SNP discovery

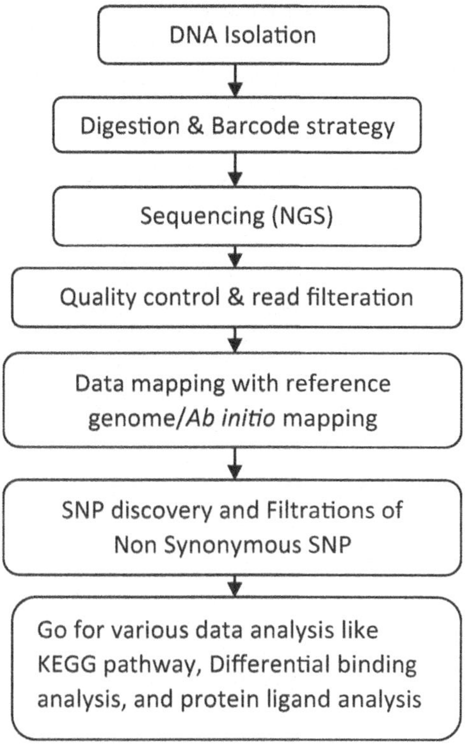

visualization. The thresholds values are adjusted based on the read length and the genome coverage achieved by the NGS data. After SNP filtering step, a list of SNP and indel coordinates is generated and graphical user interface programs (GUI) are used to visualize the result using tool Tablet (Milne et al. 2009), SNP-VISTA (Shah et al. 2005), or Savant (Fiume et al. 2010). Tablet has a user-friendly interface and it supports many file formats such as SAM, BAM, SOAP, ACE, FASTQ, and FASTA generated by different read assemblers such as Bowtie, BWA, SOAP, MAQ, and SeqManNGen. It displays contig overview, coverage information, read names, and coordinates information on scaffolds.

SNP discovery is useful for phylogenic analysis, marker-assisted selection, genetic mapping of quantitative trait loci (QTL), bulked segregant analysis, genome selection, and genome-wide association studies (GWAS). due to advantages in high throughput, high accuracy, and low-cost SNP discovery, it is a widely used method for genetic diversity analysis, breeding, and genome-wide association studies (Fig. 7.4).

7.3.6 Speed Breeding

A powerful and promising technology that allows rapid generation and advancement by shortening generation time is called "Speed breeding," (Alahmad et al. 2018; Watson et al. 2018). It developed new crop varieties faster and offering hope for food security in the world. In normal glasshouse conditions, a maximum of 2–3 generations can be achieved per year while speed breeding can speed up 4–6 generations per year. Speed breeding involved an extended photoperiod of light and controlled growth conditions such as temperature, growth media, and spacing. It accelerates the growth rate of a plant, shortens generation time, and showed genetic improvement. Speed breeding has shown a positive effect on many crops like wheat (*Triticum aestivum*), durum wheat (*T. durum*), barley (*Hordeum vulgare*), chickpea (*Cicer arietinum*), and pea (*Pisum sativum*). They have achieved up to 6 generations per year and 4 generations for canola (*Brassica napus*) successfully till yet (Watson et al. 2018). ADAM-plant tools help in the simulation of speed breeding by specifying the number of time steps or years each generation takes. Thus, for each cycle, 4–6 generations can be assumed which can be achieved within a year instead of 4 years.

7.4 Conclusion

The exploitation of big data and NGS techniques in the applied breeding methods such as MAS, MARS, and GS has accelerated effectively and efficiently the practice of next-generation breeding. Diverse shareware software and open-source platforms are available to combine genomics, data management, and breeding activities. Various tools are available on the Integrated Breeding Platform (www.integratedbreeding.net) to analyze and manage genomics and breeding data. Various complex traits such as drought tolerance, salt or metal resistance, etc. will be benefitted when genomics, novel breeding, and informatics tools are combined effectively. Thus, the amalgamation of NGS technologies, bioinformatics, and ANN based phenotyping tools in the plant sciences revolutionize precision breeding strategies to achieve genetic improvement of crops and will help breeders to obtain new cultivars with improved characteristics.

References

Acevedo-Garcia J, Spencer D, Thieron H, Reinstädler A, Hammond-Kosack K et al (2017) Mlo-based powdery mildew resistance in hexaploid bread wheat generated by a non-transgenic TILLING approach. Plant Biotechnol J 15:367

Alahmad S, Dinglasan E, Leung KM, Riaz A, Derbal N, Voss-Fels KP et al (2018) Speed breeding for multiple quantitative traits in durum wheat. Plant Methods 14:36. https://doi.org/10.1186/s13007-018-0302-y

Alkhnbashi OS, Costa F, Shah SA, Garrett RA, Saunders SJ, Backofen R (2014) CRISPRstrand: predicting repeat orientations to determine the crRNA-encoding strand at CRISPR loci. Bioinformatics 30(17):i489–i496

Alkhnbashi OS, Shah SA, Garrett RA, Saunders SJ, Costa F, Backofen R (2016) Characterizing leader sequences of CRISPR loci. Bioinformatics 32(17):i576–i585

Andersen MM, Landes X, Xiang W, Anyshchenko A, Falhof J, Østerberg JT et al (2015) Feasibility of new breeding techniques for organic farming. Trends Plant Sci 20:426–434. https://doi.org/10.1016/j.tplants.2015.04.011

Anisimov SV (2008) Serial analysis of gene expression (SAGE): 13 years of application in research. Curr Pharm Biotechnol 9:338–350

Asamizu E, Ichihara H, Nakaya A, Nakamura Y, Hirakawa H, Ishii T, Tamura T, Fukami-Kobayashi K, Nakajima Y, Tabata S (2014) Plant Genome DataBase Japan (PGDBj): a portal website for the integration of plant genome-related databases. Plant Cell Physiol 55(1):e8. https://doi.org/10.1093/pcp/pct189. Epub 2013 Dec 19. PMID: 24363285; PMCID: PMC3894704

Bachem CW, van der Hoeven RS, de Bruijn SM, Vreugdenhil D and Zabeau et al. (1996). Visualization of differential gene expression using a novel method of RNA fingerprinting based on AFLP: analysis of gene expression during potato tuber development. Plant J, 9, 745–753

Baird NA, Etter PD, Atwood TS, Currey MC, Shiver AL, Lewis ZA, Selker EU, Cresko WA, Johnson EA (2008) Rapid SNP discovery and genetic mapping using sequenced RAD markers. PLoS One 3:e3376

Barba M, Czosnek H, Hadidi A (2014) Historical perspective, development and applications of next-generation sequencing in plant virology. Viruses 6:106–136. https://doi.org/10.3390/v6010106

Barchi L, Lanteri S, Portis E, Valè G, Volante A, Pulcini L et al (2012) A RAD tag derived marker based eggplant linkage map and the location of QTLs determining anthocyanin pigmentation. PLoS One 7:43740. https://doi.org/10.1371/journal.pone.0043740

Barski A, Cuddapah S, Cui K, Roh TY, Schones DE, Wang Z, Wei G, Chepelev I, Zhao K (2007) High-resolution profiling of histone methylations in the human genome. Cell 129(4):823–837. https://doi.org/10.1016/j.cell.2007.05.009

Bevan M, Uauy C, Wulff B et al (2017) Genomic innovation for crop improvement. Nature 543: 346–354. https://doi.org/10.1038/nature22011

Biswas A, Staals RHJ, Morales SE, Fineran PC, Brown CM (2016) CRISPRdetect: a flexible algorithm to define CRISPR arrays. BMC Genomics 17:i356

Bleidorn C (2015) Third generation sequencing: technology and its potential impact on evolutionary biodiversity research. Syst Biodivers 14(1):1–8. https://doi.org/10.1080/14772000.2015.1099575

Bolger AM, Lohse M, Usadel B (2014) Trimmomatic: a flexible trimmer for Illumina sequence data. Bioinformatics. https://doi.org/10.1093/bioinformatics/btu170

Bolser D, Staines DM, Pritchard E, Kersey P (2016) Ensembl plants: integrating tools for visualizing, mining, and analyzing plant genomics data. Methods Mol Biol 1374:115–140

Brister JR, Ako-Adjei D, Bao Y, Blinkova O (2015) NCBI viral genomes resource. Nucleic Acids Res 43(Database issue):D571–D577

Buck MJ, Lieb JD (2004) ChIP-chip: considerations for the design, analysis, and application of genome-wide chromatin immuno-precipitation experiments. Genomics 83:349–360

Carroll D (2011) Genome engineering with zinc-finger nucleases. Genetics 188:773–782. https://doi.org/10.1534/genetics.111.131433

Catchen JM, Amores A, Hohenlohe P, Cresko W, Postlethwait JH (2011) Stacks: building and genotyping loci *de novo* from short-read sequences. G3: Genes| Genomes| Genetics 1(3): 171–182

Chapman JA, Mascher M, Buluc A, Barry K, Georganas E et al (2015) A whole-genome shotgun approach for assembling and anchoring the hexaploid bread wheat genome. Genome Biol 16:26

Clark K, Karsch-Mizrachi I, Lipman DJ, Ostell J, Sayers EW (2016) GenBank. Nucleic Acids Res 44(D1):D67–D72

Clement NL, Snell Q, Clement MJ, Hollenhorst PC et al (2010) The GNUMAP algorithm: unbiased probabilistic mapping of oligonucleotides from next-generation sequencing. Bioinformatics 26: 38–45

Cortés AJ, Chavarro MC, Blair MW (2011) SNP marker diversity in common bean (*Phaseolus vulgaris* L.). Theor Appl Genet 123(5):827–845

Crawley AB, Henriksen JR, Barrangou R (2018) CRISPRdisco: an automated pipeline and classes of crispr like elements. BMC Bioinf 18(1):92

Danecek P, Auton A, Abecasis G, Albers CA, Banks E, DePristo MA et al (2011) The variant call format and VCFtools. Bioinformatics (Oxford, England) 27(15):2156–2158

Davey JW, Hohenlohe PA, Etter PD, Boone JQ, Catchen JM, Blaxter ML (2011) Genome-wide genetic marker discovery and genotyping using next-generation sequencing. Nat Rev Genet 12(7):499–510. https://doi.org/10.1038/nrg3012

Dong Q, Schlueter SD, Brendel V (2004) PlantGDB, plant genome database and analysis tools. Nucleic Acids Res 32:D354–D359

Dumschott K, Schmidt MH, Chawla HS, Snowdon R, Usadel B (2020) Oxford Nanopore sequencing: new opportunities for plant genomics. J Exp Bot 71(18):5313–5322. https://doi.org/10.1093/jxb/eraa263

Eid J, Fehr A, Gray J, Luong K, Lyle J, Otto G, Peluso P, Rank D, Baybayan P, Bettman B, Bibillo A et al (2009) Real-time DNA sequencing from single polymerase molecules. Science 323(5910):133–138

Fernandez-Pozo N, Menda N, Edwards JD, Saha S, Tecle IY, Strickler SR, Bombarely A, Fisher-York T, Pujar A, Foerster H, Yan A, Mueller LA (2015) The Sol Genomics Network (SGN)—from genotype to phenotype to breeding. Nucleic Acids Res 43(D1):D1036–D1041. https://doi.org/10.1093/nar/gku1195

Fiume M, Williams V, Brook A, Brudno M (2010) Savant: genome browser for high-throughput sequencing data. Bioinformatics 26(16):1938–1944

Ganal MW, Altmann T, Roder MS (2009) SNP identification in crop plants. Curr Opin Plant Biol 12:211–217

Gilchrist EJ, Haughn GW (2005) TILLING without a plough: a new method with applications for reverse genetics. Curr Opin Plant Biol 8:211–215

Glaubitz JC, Casstevens TM, Lu F, Harriman J, Elshire RJ, Sun Q, Buckler ES (2014a) TASSELGBS: a high capacity genotyping by sequencing analysis pipeline. PLoS One 9:e90346

Glaubitz JC, Casstevens TM, Lu F, Harriman J, Elshire RJ, Sun Q et al (2014b) TASSEL-GBS: a high capacity genotyping by sequencing analysis pipeline. PLoS One 9:e90346. https://doi.org/10.1371/journal.pone.0090346

Goodstein DM, Shu S, Howson R, Neupane R, Hayes RD, Fazo J, Mitros T (2012) Phytozome: a comparative platform for green plant genomics. Nucleic Acids Res 40:D1178–D1186

Grissa I, Vergnaud G, Pourcel C (2007) The CRISPRdb database and tools to display CRISPRs and to generate dictionaries of spacers and repeats. BMC Bioinf 8:172

Haeussler M, Schönig K, Eckert H, Eschstruth A, Mianné J, Renaud JB, Schneider-Maunoury S, Shkumatava A, Teboul L, Kent J, Joly JS, Concordet JP (2016) Evaluation of off-target and on-target scoring algorithms and integration into the guide RNA selection tool CRISPOR. Genome Biol 17(1):148

Hallauer AR (2011) Evolution of plant breeding. Crop Breed Appl Biotechnol 11(3):197–206

Han Y, Gao S, Muegge K, Zhang W, Zhou B (2015) Advanced applications of RNA sequencing and challenges. Bioinformatics and Biology Insights 9:29–46

He J, Zhao X, Laroche A, Lu Z-X, Liu H, Li Z (2014) Genotyping-by-sequencing (GBS), an ultimate marker-assisted selection (MAS) tool to accelerate plant breeding. Front Plant Sci 5: 484. https://doi.org/10.3389/fpls.2014.00484

Heywood C (1992) An agricultural revolution. In: The development of the French economy, 1750–1914. Studies in economic and social history. Palgrave Macmillan, London. https://doi.org/10.1007/978-1-349-10596-0_5

Hickey JM, Chiurugwi T, Mackay I, Powell W (2017) Genomic prediction unifies animal and plant breeding programs to form platforms for biological discovery. Nat Genet 49:1297–1303

Hodson D, White J (2010) GIS and crop simulation modelling applications in climate change research. In: Reynolds MP (ed) Climate change and crop production. CABI, Oxfordshire, UK, pp 245–262

Jackson V, Chalkley R (1981) Use of whole-cell fixation to visualize replicating and maturing simian virus 40: identification of new viral gene product. Proc Natl Acad Sci U S A 78:6081–6085

Ji H (2010) Computational analysis of ChIP-seq data. In: Ladunga I (ed) Computational biology of transcription factor binding. Methods in molecular biology (methods and protocols), vol 674. Humana Press, Totowa, NJ. https://doi.org/10.1007/978-1-60761-854-6_9

Johnson KD, Bresnick EH (2002) Dissecting long-range transcriptional mechanisms by chromatin immunoprecipitation. Methods 26:27–36

Keating BA, Carberry P, Hammer G, Probert ME, Robertson M, Holzworth D et al (2003) An overview of APSIM, a model designed for farming systems simulation. Eur J Agron 18(3–4): 267–288

Kim SI, Tai HT (2014) Identification of novel rice low phytic acid mutations via TILLING by sequencing. Mol Breed 34:1717–1729

Kitts PA, Church DM, Thibaud-Nissen F, Choi J, Hem V et al (2016) Assembly: a resource for assembled genomes at NCBI. Nucleic Acids Res 44(D1):D73–D80

Kurowska M, Daszkowska-Golec A, Gruszka D, Marzec M, Szurman M, Szarejko I, Maluszynski M (2011) TILLING—a shortcut in functional genomics. J Appl Genet 52:371–390

Lange SJ, Alkhnbashi OS, Rose D, Will S, Backofen R (2013) CRISPRmap: an automated classification of repeat conservation in prokaryotic adaptive immune systems. Nucleic Acids Res 41(17):8034–8044

Langmead B, Trapnell C, Pop M, Salzberg SL (2009) Ultrafast and memory-efficient alignment of short DNA sequences to the human genome. Genome Biol 10:R25

Lee J, Chin JH, Ahn SN, Koh HJ (2015) Brief history and perspectives on plant breeding. In: Koh HJ, Kwon SY, Thomson M (eds) Current technologies in plant molecular breeding. Springer, Dordrecht

Li H, Ruan J, Durbin R (2008) Mapping short DNA sequencing reads and calling variants using mapping quality scores. Genome Res 18:1851–1858

Li H, Handsaker B, Wysoker A et al (2009a) The sequence alignment/map format and SAMtools. Bioinformatics 25(16):2078–2079

Li H, Handsaker B, Wysoker A, Fennell T, Ruan J, Homer N, Marth G, Abecasis G, Durbin R, Genome Project Data Processing Subgroup (2009b) The sequence alignment/map format and SAMtools. Bioinformatics 25:2078–2079

Li H, Handsaker B, Wysoker A, Fennell T, Ruan J, Homer N, Marth G, Abecasis G, Durbin R, Genome Project Data Processing Subgroup (2009c) The sequence alignment map format and SAMtools. Bioinformatics 25:2078–2079

Li R, Li Y, Fang X et al (2009d) SNP detection for massively parallel whole-genome resequencing. Genome Res 19(6):1124–1132

Li R, Yu C, Li Y, Lam TW, Yiu SM, Kristiansen K, Wang J (2009e) SOAP2: an improved ultrafast tool for short read alignment. Bioinformatics 25:1966–1967

Li FW, Nishiyama T, Waller M et al (2020) Anthoceros genomes illuminate the origin of land plants and the unique biology of hornworts. Nat Plants 6:259–272

Liolios K, Chen IMA, Mavrommatis K et al (2009) The genomes on line database (GOLD) in 2009: status of genomic and metagenomic projects and their associated metadata. Nucleic Acids Res 38:D346–D354

Liu H et al (2019) ADAM-plant: a software for stochastic simulations of plant breeding from molecular to phenotypic level and from simple selection to complex speed breeding programs. Front Plant Sci 9:1926. https://doi.org/10.3389/fpls.2018.01926

Lu F, Lipka AE, Glaubitz J, Elshire R, Cherney JH, Casler MD, Buckler ES, Costich DE (2013) Switchgrass genomic diversity, ploidy, and evolution: novel insights from a networkbased SNP discovery protocol. PLoS Genet 9:e1003215

Lunter G, Goodson M (2011a) Stampy: a statistical algorithm for sensitive and fast mapping of Illumina sequence reads. Genome Res 21(6):936–939

Lunter G, Goodson M (2011b) Stampy: a statistical algorithm for sensitive and fast mapping of Illumina sequence reads. Genome Res 21:936–939

Ma C, Zhang HH, Wang X (2014) Machine learning for big data analytics in plants. Trends Plant Sci 19:798–808

Michael TP, VanBuren R (2020) Building near-complete plant genomes. Curr Opin Plant Biol 54: 26–33

Milne I, Bayer M, Cardle L et al (2009) Tablet-next generation sequence assembly visualization. Bioinformatics 22(3):401–402

Nielsen R, Paul JS, Albrechtsen A, Song YS (2011) Genotype and SNP calling from next-generation sequencing data. Nat Rev Genet 12(6):443–451

Ning DL, Wu T, Xiao LJ, Ma T, Fang WL, Dong RQ, Cao FL (2020) Chromosomal-level assembly of *Juglans sigillata* genome using Nanopore, BioNano, and hi-C analysis. GigaScience 9. https://doi.org/10.1093/gigascience/giaa006

Nobuta K, Venu RC, Lu C, Beló andVemaraju et al. (2007). An expression atlas of rice mRNAs and small RNAs. Nat Biotechnol 25, 473–477

O'Leary NA, Wright MW, Brister JR, Ciufo S, Haddad D et al (2016) Reference sequence (RefSeq) database at NCBI: current status, taxonomic expansion, and functional annotation. Nucleic Acids Res 44(D1):D733–D745

Ong Q, Nguyen P, Thao NP, Le L (2016) Bioinformatics approach in plant genomic research. Curr Genomics 17(4):368–378

Osakabe Y, Osakabe K (2015) Genome editing with engineered nucleases in plants. Plant Cell Physiol 56:389–400. https://doi.org/10.1093/pcp/pcu170

Ozsolak F, Milos PM (2011) Single-molecule direct RNA sequencing without cDNA synthesis. Wiley Interdiscip Rev RNA 2(4):565–570. https://doi.org/10.1002/wrna.84

Pfender WF, Saha MC, Johnson EA, Slabaugh MB (2011) Mapping with RAD (restriction-site associated DNA) markers to rapidly identify QTL for stem rust resistance in Lolium perenne. Theor Appl Genet 122:1467–1480. https://doi.org/10.1007/s00122-011-1546-3

Poland JA, Brown JP, Sorells ME, Jannick JL (2012) Development of high-density genetic maps for barley and wheat using a novel two-enzyme genotyping-by-sequencing approach. PLoS One 7:e32253. https://doi.org/10.1371/journal.pone.0032253

Portwood JL, Woodhouse MR, Cannon EK, Gardiner JM et al (2019) The maize multi-genome genetics and genomics database. Nucleic Acids Res 47(D1):D1146–D1154

Pushpendra KG (2008) Single-molecule DNA sequencing technologies for future genomics research. Trends Biotechnol 26:602–611

Ray S, Satya P (2014) Next generation sequencing technologies for next generation plant breeding. Front Plant Sci 5:367. https://doi.org/10.3389/fpls

Read AC, Moscou MJ, Zimin AV, Pertea G, Meyer RS, Purugganan MD, Leach JE, Triplett LR, Salzberg SL, Bogdanove AJ (2020) Genome assembly and characterization of a complex zfBED-NLR gene-containing disease resistance locus in Carolina Gold select rice with Nanopore sequencing. PLoS Genet 16:e1008571

Reinartz J, Bruyns E, Lin JZ, Burcham T, Brenner S (2002) Massively parallel signature sequencing (MPSS) as a tool for in-depth quantitative gene expression profiling in all organisms. Brief Funct Genomic Proteomics 1:95–104

Robert TF, Mark T (2011) Phenomics –technologies to relieve the phenotyping bottleneck. Trends Plant Sci 16(12):635–644. Ronald DL. 2008. Population (2nd ed.)

Robertson G, Hirst M, Bainbridge M et al (2007) Genome-wide profiles of STAT1 DNA association using chromatin immunoprecipitation and massively parallel sequencing. Nat Methods 4:651–657. https://doi.org/10.1038/nmeth1068

Saeed AI, Sharov V, White J, Li J, Liang W et al (2003) A free, open-source system for microarray data management and analysis. BioTechniques 34:374–378

Schena M, Shalon D, Davis RW, Brown PO (1995) Quantitative monitoring of gene expression patterns with a complementary DNA microarray. Science 270:467–470

Schmid M, Davison TS, Henz SR, Pape UJ, Demar M et al (2005) A gene expression map of Arabidopsis thaliana development. Nat Genet 37:501–506

Schnable PS, Springer NM (2013) Progress toward understanding heterosis in crop plants. Annu Rev Plant Biol 64:71–88

Shah N, Teplitsky MV, Minovitsky S et al (2005) SNP-VISTA: an interactive SNP visualization tool. BMC Bioinformatics 6(1):292

Shi CY, Yang H, Wei CL, Yu O, Zhang ZZ (2011) Deep sequencing of the *Camellia sinensis* transcriptome revealed candidate genes for major metabolic pathways of tea-specific compounds. BMC Genomics 12:131

Siadjeu C, Pucker B, Viehöver P, Albach DC, Weisshaar B (2020) High contiguity *de novo* genome sequence assembly of trifoliate yam (*Dioscorea dumetorum*) using long read sequencing. Genes 11:274

Sim SC et al (2012) High-density SNP genotyping of tomato (*Solanum lycopersicum* L.) reveals patterns of genetic variation due to breeding. PLoS One 7:e45520. https://doi.org/10.1371/journal.pone.0045520

Skennerton CT, Imelfort M, Tyson GW (2013) Crass: identification and reconstruction of CRISPR from unassembled metagenomic data. Nucleic Acids Res 41(10):e105

Stapley J, Reger J, Feulner PGD, Smadja C, Galindo J, Ekblom R et al (2010) Adaptation genomics: the next generation. Trends Ecol Evol 25:705–712. https://doi.org/10.1016/j.tree.2010.09.002

Stiglic G, BajgotM KP (2010) Gene set enrichment metalearning analysis: next-generation sequencing versus microarrays. BMC Bioinformatics 11:176

Tatusova T, DiCuccio M, Badretdin A, Chetvernin V et al (2016) NCBI prokaryotic genome annotation pipeline. Nucleic Acids Res 44(14):6614–6624

Taylor NE, Greene EA (2003) PARSESNP: a tool for the analysis of nucleotide polymorphisms. Nucleic Acids Res 31(13):3808–3811. https://doi.org/10.1093/nar/gkg574

Tello-Ruiz MK, Naithani S, Stein JC, Gupta P, Campbell M, Olson A, Wei S, Preece J et al (2018) Unifying comparative genomics and pathway resources for plant research. Nucleic Acids Res 46 (D1):D1181–D1189

The UniProt Consortium, UniProt (2019) A worldwide hub of protein knowledge. Nucleic Acids Res 47(D1):D506–D515

Tomas T, Tibor P, Jozef S (2011) Differential proteomics of plant development. J Proteome 74(5): 577–588

Van der Auwera GA, Carneiro MO, Hartl C, Poplin R, del Angel G et al (2013) From FastQ data to high confidence variant calls: the genome analysis toolkit best practices pipeline. Curr Protoc Bioinform 43:11.10.1–11.10.33

VanGuilder HD, Vrana KE, Freeman WM (2008) Twenty-five years of quantitative PCR for gene expression analysis. BioTechniques 44:619–626

Wang W, Das A, Kainer D, Schalamun M, Morales-Suarez A, Schwessinger et al (2020) The draft nuclear genome of assembly of *Eucalyptus pauciflora*: new approaches to comparing *de novo* assemblies. GigaScience 9. https://doi.org/10.1093/gigascience/giz160

Watson A, Ghosh S, Williams MJ et al (2018) Speed breeding is a powerful tool to accelerate crop research and breeding. Nat Plants 4:23–29. https://doi.org/10.1038/s41477-017-0083-8

Wei Z, Wang W, Hu P, Lyon GJ, Hakonarson H (2011) SNVer: a statistical tool for variant calling in analysis of pooled or individual next-generation sequencing data. Nucleic Acids Res 39(19): 132

Wells J and FarnhamPJ (2002). Characterizing transcription factor binding sites using formaldehyde crosslinking and immunoprecipitation. Methods 26:48–56

Willing EM, Hoffmann M, Klein JD, Weigel D, Dreyer C (2011) Paired-end RAD-seq for de novo assembly and marker design without available reference. Bioinformatics 27:2187–2193

Xiao T, Zhou W (2020) The third generation sequencing: the advanced approach to genetic diseases. Transl Pediatr 9(2):163–173. https://doi.org/10.21037/tp.2020.03.06

Xifeng C, Vijai B, Bojun M (2018) ChIP-Seq: a powerful tool for studying protein-DNA interactions in plants. Curr Issues Mol Biol 27:171–180

Xu M, Fujita D, Hagamata N (2009) Perspectives and challenges of emerging single molecule DNA sequencing technologies. Small 5(23):2638–2649

Yates CM, Sternberg MJ (2013) Proteins and domains vary in their tolerance of non-synonymous single nucleotide polymorphisms (nsSNPs). J Mol Biol 425:1274–1286

Zhang Q, Ye Y (2017) Not all predicted CRISPR-Cas systems are equal: isolated cas genes. BMC Bioinformatics 18:92

Zhang Y, Malzahn AA, Sretenovic S, Qi Y (2019) The emerging and uncultivated potential of CRISPR technology in plant science. Nat Plants 5:778–794

Zhao J, Grant S (2011) Advances in whole genome sequencing technology. CurrPharmBiotechno 12:293–305

Zhu PY, He L, Li Y, Huang W, Xi F, Lin L, Zhi Q et al (2014) OTG-snpcaller: an optimized pipeline based on TMAP and GATK for SNP calling from ion torrent data. PLoS One 9(5): e97507

Zwart H (2015) Human Genome Project: history and assessment. H. Zwart (2015) “Human Genome Project: history and assessment”. In: International Encyclopedia of social & behavioral sciences, 2nd edn. Elsevier, Oxford. https://doi.org/10.1016/B978-0-08-097086-8.82036-X

Назипова НН, Nazipova N (2017) Big data in bioinformatics. Math Biol Bioinform 12:102–119

Whole-Genome Sequencing of Plants: Past, Present, and Future

8

Talambedu Usha, Prachurjya Panda, Arvind Kumar Goyal, Anusha A. Kukanur, Arunagiri Kamala, M. K. Prasannakumar, K. Sidhalinghamurthy, and Sushil Kumar Middha

Abstract

Advances in analytical chemistry, as well as the constant introduction of novel analytical chemistry approaches, are altering genomics at a higher level. The cost of sequencing has dropped from US$100 million for the first plant genome *Arabidopsis thaliana* to just US$1000 as a result of this rapid progress in the genomics era. Without a doubt, this will speed up and lower the cost of future sequencing technologies. This chapter highlights the major advances in agriculture. To add, this also understands past, present, and future of in era of plant genomics. This chapter highlights the genome databases, tools that are mostly used over the globe. We proposed as a futuristic approach that an altogether comprehensive genome database is to be created to host all the plant genomes to speed up the plant genomic research.

T. Usha · K. Sidhalinghamurthy
Department of Biochemistry, Bangalore University, Bengaluru, Karnataka, India

P. Panda
Faculty of Health Science, Ben-Gurion University of the Negev, Be'er Sheva, Israel

A. K. Goyal
Department of Biotechnology, Bodoland University, Kokrajhar, Assam, India

A. A. Kukanur · A. Kamala
Department of Biochemistry, Maharani Lakshmi Ammanni College for Women, Bangalore, Karnataka, India

M. K. Prasannakumar
Department of Plant Pathology, University of Agricultural Sciences, Bengaluru, Karnataka, India

S. K. Middha (✉)
DBT-BIF Facility, Department of Biotechnology, Maharani Lakshmi Ammanni College for Women, Bangalore, Karnataka, India

R. L. Singh et al. (eds.), *Plant Genomics for Sustainable Agriculture*,
https://doi.org/10.1007/978-981-16-6974-3_8

Keywords

NCBI · NGS · Plant genomics · Sequencing techniques · Plant genome databases

8.1 Introduction

The science of plant genomics encompasses the various aspects and types of sequencing, characterization, and identification of genetic composition, construction, organization, functions, and genetic interactions or biological networks of a complete plant genome. However, to draw greater insights into complete plant biology multi-omics approaches are widely employed. The science of genomics is exploited to increase the crop yield in both biotic and abiotic stress conditions, pest management, tackle food insecurity faced by the world population. It is also used to address other needs like feed, fibre, and fuel. Avalanche of scientific publications have emphasized the role of plant genomics in identifying, characterizing and enhancing the medicinal properties of plants (Gantait et al. 2014) and engineering plant species to reduce air, water and sound pollution caused by various human endeavours inclusive of industrial, farm and agriculture waste.

Foremost challenge in current scenario, genomics studies devices to engineer new plant varieties for the sustainable development of crude substances for the regular need like F (food, fuel, feed, fibre) and medicines. An ultimate green resolution to the exponentially swelling demands for these 4F and medicines using plants as a foundation is highly appreciable. Similarly, it is likely vital to diminish pollution arising from rigorous agriculture-related activities. The increasing demand for staple food and high inflation rates made human to think out of the box to meet end-users demands on time to fulfill food, nutritional and other needs. Hence, the research priorities in agriculture or plant biology require more inventions and interventions to develop next-generation agricultural products for a sustainable prospect. Hence, the regular development in next-generation sequencing (NGS) approaches in genomics technologies shows massive capacity for genetic enhancement of plants to meet twenty-first century requirements.

The major hurdle a researcher comes across while doing plant genomics includes (i) Creating a physical and genetic map for all economically useful plants or agronomic traits using NGS approach, (ii) Assigning function to a gene by functional genomics investigations under specific environmental circumstances. (iii) Creating smart and efficient computational approaches for high-throughput genotyping-phenotyping integration under specific environmental conditions for the major target traits.

8.2 Plant Genome Research

8.2.1 The History and Research During Pre-genomic Era

The proper plant genomic research began with the Mendel's work in 1866 and his contributions towards the field are well known. With the revelation of DNA structure by "Watson and Crick" in 1953, genomic research started taking its new dimension (Pray 2008). In spite of the structural elucidation, details of the DNA molecule stayed elusive for more than two decades until the nucleotide sequences of DNA came into picture. Maxam and Gilbert method was the most popular tool in 70s for the determination of the nucleotide sequences. However, because of the utilization of hazardous radioactive chemicals, it was over took by Sanger's dideoxy chain termination method. Detailed methods are explained further below in the Sect. 8.2.2.2. In 1987, Barbara McClintock explained the unexpected ability of the certain genotypes of maize which were able to give rise to both forward and reverse mutations with numerous characterized genes at extremely high speed. The transposable genes/ transposition are considered to be the exceptional phenomenon in modern biology (Muñoz-López and García-Pérez 2010). Many technologies, ranging from quantitative "Fuelgen stain analysis" to DNA renaturation studies, revealed that the bulk of crop genomes were big and full of repeating sequences. Botstein et al. (1980) proposed that DNA sequence polymorphism analysis may offer enough genetic markers for humans to allow the building of a genome-wide linkage map. Endogenous heritable markers known as "restriction fragment length polymorphisms" (RFLPs) are used to create genome-wide linkage maps. In the 1990s, it became customary to characterize quantitative trait loci or genes that influence phenotypes in a quantitative rather than qualitative Mendelian form.

Genomic sequencing became an achievable target when DNA sequencing started becoming less expensive and more efficient. Even though the cost of sequencing started declining "assembling the relatively small DNA sequences" in their appropriate order has remained a difficult undertaking (Usha et al. 2021). *Arabidopsis thaliana* was first plant to be sequenced (Feldmann and Goff 2014). Initially, it was believed that a single duplication was responsible for around 60% of the genome but further analysis on the *Arabidopsis* lineage revealed that it had undergone at least two duplications (Bevan and Walsh 2005). Previously, 25,000 genes in *Arabidopsis* and 100,000 genes in tobacco which are expressed during their lifetime were sequenced. Several attempts have been made to estimate the number of genes expressed during the plant's life time.

Later, work on plant transcriptome began. With the help of expressed sequence tag (EST) cDNA sequencing projects, databases of sequences have been developed, containing the sequences obtained from isolated mRNA from many crops, with many plants providing many sequence files that have been condensed into contigs of related sequences. For example, *Hordeum vulgare* EST library have over 440,000

EST sequences having 28,000 contigs (http://harvest.ucr.edu/), symbolizing closely related family of genes or a gene.

As a result of technological advancements, genomic research has seen the progress over the years. To name a few, "northern blot provided the semi-quantitative analysis of gene expression, quantitative RT-PCR, microarrays, Serial Analysis of Gene Expression (SAGE), Illumina and Affymetrix technologies simultaneously allows analysis of most or all of the genes expressed in a tissue."

8.2.2 Current Research in Field of Plant Genomic Studies

8.2.2.1 Introduction to Current Research

Over three decades, the brisk technological development noted in the area of genomics and proteomics is majorly in the field of consumer electronics and computation which is exemplified by the increase in Genbank size and numerous publications indexed by Medline. Deposited bases doubles every one and a half year, a concept in line with Moore's law which refers to doubling of processing power of computers every couple of years. Humongous sequencing data are lodged in private databases, Genbank data being just a tip of the iceberg. As reported in Nature journal in 2012, "10,000 human genomes" have been sequenced, which corresponds to 3×10^{13} bases sequenced. Over 10^{15} bases of genomic sequence have been generated in the past 12 months.

8.2.2.2 Sequencing, Assembly/Reassembly, Annotation

8.2.2.2.1 Sanger's Chain Termination Method

This method employs DNA primer, single stranded DNA sequence of interest as template, DNA polymerase, deoxynucleotide triphosphates (dNTPs) and a radiolabelled dideoxynucleotide (ddNTPs). The DNA polymerase affixes dNTPs to the primer using DNA sequence as a template, synthesizing complementary strand of DNA. Intermittently, when DNA polymerase incorporates ddNTPs, chain extension ceases. The oligonucleotides obtained are separated based on size by running the mixture through gel electrophoresis. The fragments separated are detected using autoradiography (Sanger et al. 1977). The Sanger's method is still considered a "Gold Standard." Though the use of autoradiography-based sequencing makes this process a tedious one.

A decade later, Leroy Hood's at CalTech and James Prober's at DuPont introduced fluorescent tags instead of radioactive labels to detect the final product. Laser-induced fluorescence detection was employed for detection (Roberts 1987). Their technology opened a new arena to various break through and paved way to the future technologies. Four fluorescent tags were labelled in Hood's technology, i.e., each of the four dideoxynucleotides were labelled with a different fluorescent dye, such as dideoxyadenosine chain terminator with red dye, dideoxycytidine chain terminator with green dye, dideoxyguanosine chain terminator with yellow dye, and dideoxythymidine chain terminator with blue dye. The advantage of this dye termination method is that DNA can be sequenced in a single reaction due to colour coding where each dye fluorescence at different wavelength. Final products are pooled and run in a single lane of polyacrylamide gel. The intensity of fluorescence was recorded using a fixed detector.

Prober created a more advanced version of Hood's method that simplified the entire reaction by performing the chain terminating reaction simultaneously with the four labelled dideoxynucleotides.

Visual interpretation of an autoradiogram produced from a sequencing gel was replaced by the established technologies, which was time-consuming and error-prone. In 1986, a bunch of visionary biochemists recognized that this new automated DNA technology could be used to solve an ambitious puzzled project called the human genome project, which was funded by the Department of Energy and other agencies such as the National Institutes of Health in the USA, as well as scientific and financial contributions from other countries.

Fluorescence-based sequencing methods have a number of drawbacks, including the fact that they are not entirely automated and require time to prepare sequencing gels, prompting scientists to seek out novel analytical methodologies. "Mass spectrometry, scanning probe microscopy and microfabricated technologies" were among the analytical technologies considered. Unfortunately, these methods lacked adequate sensitivity, resolution and speed.

Finally, ABS funded a new flexible and resilient method based on capillary electrophoresis, which was created by a small group of academic researchers and included sequence reading using laser-induced fluorescence. ABS's 3700 DNA sequencer model (1998) quickly became most popular in the sequencing business. For the first human genomes, the approach was frequently utilized to obtain maximum sequence data. Hitachi, Japan and the University of Alberta, Canada, developed the technology used in the 3700 DNA sequencer model. (Chen et al. 1991; Kambara and Takahashi 1993)

.

Celera genomics used the whole-genome shotgun sequencing fueled the human genome project. Over three billion dollars was funded to sequence the three billion bases of the human genome. The shortgun approach results in a new technique known as next-generation DNA sequencing. For tiny fragment sequencing and capillary electrophoresis remains the most popular and effective method. As a result, a slew of new, improved technologies have emerged to deliver significant advancements in sequencing performance. Das et al. 2021 tabulated the comparison of NGS techniques and table 8.1 lists a few of these technologies.

8.2.2.2.2 Pyrosequencing

The target fragment, coupled with a primer, is immobilized on a hard surface (solid), along with DNA polymerase and a single dNTP. A diphosphate group (pyrophosphate) is liberated if the dNTP connects to the strand. The sulphurylase transforms ATP into luciferase, which uses it as a substrate. This luciferase generates light while incorporating nucleotide. If no nucleotide is incorporated, then no luminescence

Table 8.1 Comparison of next-generation approaches

Sequencer	Sanger 3730xl	454 GS FLX	Illumina HiSeq 3000	Illumina HiSeq 2000	SOLiDv4	Ion torrent S5	PacBio sequel	Oxford nanopore
Sequencing mechanism	Dideoxy chain termination	Pyrosequencing	Sequencing by synthesis	Sequencing by synthesis	Ligation and two-base coding			Membrane
Read length	400–900 bp	700 bp 500–1000 bp)	50SE, 50PE, 101PE, 2 × 150bp	2 × 150bp	50 + 35 bp or 50 + 50 bp			
Accuracy	99.999%	99.9%*	98%, (100 PE)	99%, (100 PE)	99.94% *raw data	99%	99.9	85
Reads	–	1 M	3–5 G	3–5 G	1200–1400 M	10 million	500 K	1 million
Output data/ run	1.9–84 kb	0.7 gb	600 gb	600 gb	120 gb	600 bp	15 Kb	2 Mb
Time/run	20 min–3 h	24 h	3–10 Days	7 h–10 Days	7 Days for SE, 14 Days for PE	2.5–4 h	2–3 h per cell	1 min–48 h
Advantage	High quality, long-read length	Long-read length, fast	High throughput, high accuracy	High throughput, high accuracy	Accuracy	Relatively long reads; accuracy	Longest read length; no amplification errors	Long-read length; data streamed real-time
Disadvantage	High cost low throughput	Error rate with polybase more than 6, high cost,	High concentration of DNA; high	High concentration of DNA; high	Short-read assembly	High cost per Mb	Low outputs, higher costs	Low read quality;

		low throughput homopolymer errors	computation cost low throughput; short-read assembly	computation cost low throughput; short-read assembly			per Mb; error rates	low device cost
Price	Instrument $95,000, about $4 per 800 bp reaction	Instrument $500,000, $7000 per run	Instrument $690,000, $6000/(30×) human genome	Instrument $690,000, $6000/(30×) human genome	Instrument $495,000, $15,000/ 100 Gb	Instrument $495,000, $15,000/ 100 Gb	Instrument $495,000, $15,000/ 100 Gb	Instrument $1000, $15,000/ 100 Gb
CPU	Pentium IV 3.0 GHz	2* Intel Xeon X5675	2* Intel Xeon X5560	2* Intel Xeon X5560	8* processor 2.0 GHz			
Memory	1 GB	48 GB	48 GB	48 GB	16 GB			
Hard disk	280 GB	1.1 TB	3 TB	3 TB	10 TB	10 TB	10 TB	>1 TB
Automation in library preparation	No	Yes	Yes	Yes	Yes	Yes	Yes	Yes
Other required device	No	REM e system	cBot system	cBot system	EZ beads system			
Cost/million bases	$2400	$10	$0.07	$0.07	$0.13			

Key: *SE* single-end sequencing, *PE* paired-end sequencing, *bp* base pairs, *gb* gigabase, *tb* terabase happens in the process. After that, the DNA sequence is determined by treating the sample with several nucleotide triphosphates in order and recording which nucleotide produces light. Therefore, in a single run the sequence of several hundred nucleotides can be resolved (Ronaghi et al. 1998).

Wheeler and his team founded 454 Life Sciences, a Roche subsidiary, in 2008. "Fragmented genomic DNA is jumped (attached) to beads under such circumstances to ensure that not more than one fragment is connected to a bead". In this massively parallel approach of pyrosequencing, Oil emulsion is used for compartmentalization, followed by PCR reaction to amplify the fragments. This creates 107 copies of the original fragment on the bead. These beads are enriched and confined into wells on a slide. After the addition of appropriate enzymes along with all four dNTPs, finally luminescence is captured. Although every step is comparatively slow, the sequencing data is generated from "tens of millions of samples in parallel producing a gigabase of sequence per run, which is six orders of magnitude higher throughput than that produced by capillary electrophoresis instruments."

8.2.2.2.3 Reversible Terminator Sequencing

For Illumina, Mardis created yet another next-generation sequencing technology. To create a "grass of sequencing primers," sequencing templates are immobilized on a surface. To generate clonal clusters, a sophisticated surface-based variant of PCR is utilized to amplify each of the sequencing fragments and immobilize those fragments in a millimetre region on the surface. Sequencing is done by using DNA polymerase and a modified deoxynucleotide triphosphate to hybridize a primer to the immobilized fragments.

The fluorescent tag on this modified nucleotide allows it to be detected once it has been incorporated onto the expanding complementary strand. It also has a blocking group that prohibits another nucleotide from being added. Excess tagged nucleotides are rinsed out of the solution and the surface is imaged using a laser-based fluorescence detector.

The fluorescence of the templates that included the changed nucleotide is used to identify them. The label and the blocking group are then chemically removed and the process is repeated with another nucleotide to complete the sequence of each immobilized template.

This technology generates massive amounts of data (5×10^{10} bases per run), but each template only generates 100 bases of sequence, making it difficult to put the individual sequences together into a complete genomic sequence. The instrument is best used for resequencing known genomes rather than de novo sequencing of novel organisms. Illumina presently owns roughly two-thirds of the sequencing business due to its vast data output.

8.2.2.2.4 Ligation-Based Approaches

From a vast number of templates, SOLiD (Sequencing by Oligonucleotide Ligation and Detection) builds short sequences. Templates are amplified and immobilized to a solid particle using emulsion PCR, similar to the 454 method. The sequence is read by adding a fluorescently tagged oligonucleotide to tens to hundreds of millions of beads on a glass slide. The oligonucleotide and primer are covalently attached using a ligase. Only those oligonucleotides are connected to the template whose sequence matches the corresponding region. Laser-induced fluorescence is utilized to image the slide, then the fluorescence dye is removed via smart chemistry. The process is repeated, with the complimentary strand gradually extended. The 4th and 5th nucleotides in the template are evaluated in this method, which examines two nucleotide locations at a time. The synthesized strand is eliminated when the procedure is completed and a new primer with an additional base is inserted, permitting the determination of bases that were missed in the initial pass. The sequence of 25–50 nucleotides is determined using a set of five primers. The device can generate approximately 2×10^{10} bases of sequence each day and can sequence a genome in a few of days.

8.2.2.2.5 Proton Detection

Ion Torrent uses chemistry comparable to pyrosequencing and depends on the semiconductor industry's scalable manufacturing process. Ion Torrent, unlike the other detection systems, uses non-photonic detection. When a deoxynucleotide triphosphate is added to a developing oligonucleotide, a proton is formed in addition to the pyrophosphate group. The presence of this proton results in a decrease in the pH of the solution, which can be detected using a pH metre of high sensitivity. On a microfabricated device, Ion Torrent places pH sensors at the bottom of thousands of wells. The device was sequenced by amplifying a large number of DNA fragments, placing them in the wells, and then running DNA polymerase and a deoxy triphosphate through it. The pH decreases in those wells that contain a fragment that has integrated with the nucleotide, which is detected by the electronics. Despite the fact that only 20–40% of the wells in an Ion Torrent device are utilized, the massively parallel nature of the design results in significant data production. In a 2-h operation, the company believes to generate over 4×10^{11} bases of sequence with a 99.5% accuracy.

8.2.2.2.6 Sequencers Based on Single-Molecule Detection

Whole-genome sequencing for the general population as a guide to personalized medicine has sparked a lot of interest. The National Human Genome Research Institute (NHGRI) has put a lot of effort towards turning genetic sequences into therapeutic recommendations for patients at their bedside. Individual genome sequencing, on the other hand, requires a cost reduction of one to two orders of magnitude. To that purpose, the National Human Genome Research Institute is holding financial solicitations with the goal of obtaining entire genome sequences for $1000.

Harris and colleague (2008) at Helicos introduced "True Single Molecule Sequencing" with its "HeliScope technology" for sequencing a viral genome. "A short DNA strand (25 bp) is prepared and a PolyA primer added to the 3′ end. Upto a billion DNA strands are hybridized to a DNA flow cell with Oligo dT capture sites. Sequencing reactions are detected from individual molecules with the addition of fluorescent-labeled nucleotides. In its marketing, Helicos placed great emphasis on freedom from amplification, citing errors in the PCR process as a major flaw with other sequencing approaches. Unfortunately, Helicos was unable to generate sufficient cash flow to move further with this technology."

Another technology was developed by "Pacific Biosciences single molecule detection approach," termed SMRT for "Single Molecule Real Time sequencing, SMRT and PacBio."

Eid et al. (2009) used "multiple DNA polymerases and immobilized in Zero Mode Waveguides and then monitored via fluorescence for the addition of labeled nucleotides." The major let down for this approach was expensive instrumentation through using a very promising methodology. Other disadvantage was the high error rate as compared to other sequencing techniques.

8.2.2.2.7 Nanopore

Another company, Oxford Nanopore developed "their sequencing platform, based on passing single DNA strands through protein nanopores embedded in a polymer membrane (Diederichs et al. 2019). As the nucleotides pass through the protein channel, a detectable change in the membrane current occurs. Each of the four nucleotides produces a distinct signal, allowing discrimination and ultimately, a DNA sequence. If a hairpin structure is present, both the sense and antisense strands can be read for each DNA molecule. The approach can also be performed in an 'exonuclease-sequencing' fashion, where an exonuclease is positioned over the nanopore and the individual nucleotides are cleaved and translocated through the pore one at a time."

The nanopore technology works on long-read generation. Oxford Nanopore had sequenced a 48 kb virus genome in a single pass. The advantage of this technology is low cost disposable sequencing instrumentation and no space requirement for the instrument having a size equal to a portable mobile device.

8.3 Advances in Plant Genomics

8.3.1 Plant Genome Assembly from 3rd Generation Genomic Technologies

About two decades ago the genome of *Arabidopsis thaliana* was sequenced as the first plant genome using first-generation capillary sequencing. Since then, technological improvements have driven the increase in sequenced plant genomes. The automated DNA sequencing instruments of the first-generation had the potential to sequence only thousands of base pairs in a day, whereas today with latest technologies we can sequence billions of bases at much cheaper costs. Thereafter, genome sequence has been the backbone for enabling annotation of gene networks (Park et al. 2012), reveal developmental forces of evolution through comparative genomics approaches, optimization of plant breeding by cataloging genomic markers (Moose and Mumm 2008). Besides incredible advances in throughput, sequence assembly remains a major challenge, consuming more effort than just the sequence.

The introduction of second-generation sequencing technology, also known as short-read Next-Generation Sequencing (NGS) techniques, resulted in an instant increase in the quantity of plant genome sequences available, lowering sequencing costs but also lowering quality. The reads were either generated on Illumina platform alone or combined with Roche 454 second-generation sequencing platforms (Michael and VanBuren 2015). However, the assemblies were highly fragmented with a high number of contigs. This was mostly due to the short-read length, complexity and vast genome, and the occurrence of large lengths of conserved regions/repetitiveness due to transposable elements which could not be extended in the de novo assembly process (Alkan et al. 2011). Unlike vertebrate genome assemblies (Gnerre et al. 2011), large contigs and scaffolds of sequencing reads from plants are assembled into isolated gene islands among the background of high copy repeats. The large size and complexity of plant genomes along with other biological, biomolecular, and computational reasons make them particularly challenging for de novo assembly. Furthermore, gene sequences and assembly from nearly identical or same gene families are challenging and may not always be correct.

Third-generation sequencing (TGS) technologies provided a new perspective on sequencing as well as excellent accuracy in de novo assembly (Pareek et al. 2011). Pacific Biosciences (PacBio), Illumina Tru-seq Synthetic Long-Read technology, single-molecule real-time (SMRT) sequencing, BioNano Genomics (BioNano) sequencing, and Oxford Nanopore Technologies (ONT) sequencing are the key platforms commercializing TGS technology (Staňková et al. 2016). These platforms hold significant advantages when compared to the 1st and 2nd generation platforms. TGS has the potential to generate long-read lengths having higher percentage of consensus accuracy and low bias of G+C content. In addition, it also provides simultaneous epigenetic classification (Nakano et al. 2017). Likewise, in comparison to the NGS platforms, TGS platforms hold three important improvements. Firstly, increase in average read lengths up to 10,000 bp per read or more, secondly drastic decline in the sequencing time from days to hours (or to minutes for real-time applications), and thirdly decrease or removal of sequencing bias introduced by polymerase chain reaction amplification (Lu et al. 2016).

The main impediment to a high-quality genome assembly is repetitive regions. Short second-generation sequencing reads are essentially incapable of building repetitive sequences that are longer than the available read (or span) length (Berlin et al. 2015). Third-generation long sequencing reads, on the other hand, cover a greater proportion of the repetitions in a genome, making them important for generating high-quality assemblies.

Raw read mapping, read error correction, assembly of corrected reads, and assembly polishing are all steps in the de novo long-read genome assembly process. It creates lengthy read alignments first, then assembles them using overlap-based techniques such overlap–layout–consensus (OLC) algorithms. After assembly it computes the best overlap graph or string graph and finally generates the consensus sequence of the contigs from the graph. There are two ways to rectify the long-reads error. The first way encompasses self-correction, i.e., aligning the long reads against themselves to form an error corrected consensus sequence (Berlin et al. 2015) while the second approach involves hybrid error correction, i.e., the use of alignment of high-quality short reads to correct long reads (Koren et al. 2012). Hybrid correction approach is more useful when a limited amount of long-read ($<30\times$) coverage is available, on the other hand, self-correction approach is applicable to higher sequencing coverage since more consistent alignments can be made between the long reads. Nevertheless, the error correction stage forms an integral part of the assembly process, but it is practically impossible to obtain an error free assembly, specifically in long-read assemblies. However, polishing with short or long reads can help us to improve the assemblies (Sohn and Nam 2018).

At present there are a number of tools developed through TGS platforms for assembling long-read sequences. FALCON, MinHash Alignment Process (MHAP), PBJelly, HINGE, and Hierarchical Genome Assembly Process (HGAP), and HINGE are genome assemblers that utilize long reads from SMRT platforms. The genome assembler, RefAligner involves a dynamic programming algorithm to identify the best matching region in the sequence genome by aligning each molecule map to the reference maps utilizing the BioNano platform. PoreSeq and Nanocorr utilize long reads form the ONT platform for de novo sequencing analysis. Minimap/miniasm and Circlator are available for de novo assembly and circularization genome assembly analysis in both SMRT and ONT platforms.

Arabidopsis thaliana (Berlin et al. 2015) and *Oropetium thomaeum* (Vanburen et al., 2015) were the first plant genomes assembled from PacBio data alone. Being smaller in size, the genome of *A. thaliana* was assembled at chromosome-arm level, whereas the genome assembly of the *O. thomaeum* genome displayed a contig N50 of 2.4 Mb. Short-read assemblies could never achieve such contiguities, however, scaffolding (*i.e.*, ordering and orienting of contigs) through long-range read pairs is capable of generating similar contiguities (Hoshino et al. 2016).

8.3.2 Machine Learning Aided Crop Plant Genomics

The scientific and technical achievements over the last few decades have been revolutionary. The availability of plant genotypes and phenotypes at a low cost resulted in the creation of enormous, complicated datasets. However, most of the attempts to derive conclusions from these metrics and combine disparate datasets have failed. In the meantime, rapid evolution machine learning has found widespread use in science and related sectors, such as plant science and breeding.

Machine learning (ML) is a branch of Artificial Intelligence (AI), develops algorithms for predictive modelling/performing tasks from multidimensional datasets (Camacho et al. 2018). ML empowers us to analyse results from large-sized datasets, integrate complex data for meaningful interpretations without the need of mechanistic understanding. ML was used to predict possible genome crossover locations, i.e., regions of exchange of paternal and maternal genetic material in various plants (Demirci et al. 2018). It is also being used in population genetics of plants to identify advantageous mutations imposed through natural selections (Bourgeois et al. 2018; Schrider and Kern 2018). Other examples of applications of machine learning include predicting macronutrient deficiencies in tomato (Tran et al. 2019), plant stress phenotyping (Ghosal et al. 2018), sequence tagging in rice (Do et al., 2018), predicting regulatory and non-regulatory regions in the *Zea mays* genome (Mejía-Guerra and Buckler 2019), predicting mRNA expression levels (Washburn et al. 2019), polyadenylation site prediction in *Arabidopsis thaliana* (Gao et al. 2018a).

8.3.3 Accelerate the Development of New Crops via Speed Breeding

Breeding crop plants and advanced cultivars take a long time. Breeding a stable new variety typically requires 4–6 generations of inbreeding post-crossing of selected parent lines. The plant generation time, from germination of seed to the harvesting is one of the limiting factors in the process (Watson et al. 2018). For crops that are limited to 1–2 generations per year, this is time-consuming. Therefore, in order to fasten the cultivation process and ensure food availability throughout the year, it is important to minimize the generation time and accelerate the breeding programme. This can be achieved by the speed breeding procedure, by altering growth conditions, such as prolonged photoperiods and temperature (Sysoeva et al. 2010). This method can also be used to construct mapping populations, phenotype adult plant attributes, speed up backcrossing and trait pyramiding, and perform mutant studies and transformations (Watson et al. 2018).

Speed breeding has raised efficient wheat and barley production of up to six generations in a year, compared to the usual two generations per year through conventional methods. In addition, speed breeding procedures have also been successfully applied to for the production of a number of crops. In order to accelerate the rate of crop improvement, speed breeding can be integrated with several other technologies such as high-throughput phenotyping and genotyping, single seed descent (SSD), marker-assisted selection (MAS), single plant selection (SPS), genomic selection, CRISPR gene editing, etc. (Hickey et al. 2017).

8.3.4 High-Throughput Phenotyping

Plant phenotyping is a multidisciplinary research that employs a variety of techniques and methodologies to precisely assess the growth, architecture, and composition of plants at various scales (Fiorani and Schurr 2013). Individual genes, gene-by-gene interactions, and gene-by-environment interactions all influence the phenotype.

Multiple genes and their interactions with the environment govern several agronomically significant features (Mickelbart et al. 2015). However, obtaining phenotypic data remains a bottleneck, which limits crop breeding and functional genomics research (Deery et al. 2016). Plant phenomics, in general, lags considerably behind the rapidly evolving genomics technologies. Nonetheless, several high-throughput phenotyping tools for small and large plants have been developed in recent decades.

8.4 Role of Genomic Studies in Plant Research

8.4.1 Using Genomics to Improve Crop Plant Diversity and Resilience

The rapid increase in human population leads to an increase in urbanization, which threatens global food security, narrowing our access to appropriate land for agriculture (Satterthwaite et al. 2010). Concurrently, the erratic change in climatic conditions, which includes elevated temperature, changing precipitation, and upregulated levels of CO_2 and ozone, further limits our access to agricultural land and water use *via* untimed drought, floods, salinity, and frosting (Godfray et al. 2010). In addition, the rapidly rising economic growth is significantly accelerating a dietary transition, increasing consumption of meat, dairy products, and eggs, thereby demanding increased production of crops to feed more livestocks and poultry (Tilman and Clark 2014). Advancement in research and technologies, in the last century, enabled dramatic increase in food production by combining conventional crop breeding techniques and advanced agronomic practices, leading to the green revolution. However, with current trends of annual increase in yield for major crops lying merely between 0.9 and 1.6%, we are starting to saturate the benefits of green revolution (Ray et al. 2013). Furthermore, monoculture in agriculture has led to agricultural food dependence on a few major plant species, significantly reducing the genetic diversity (Khoury et al. 2014). However, irrespective of the above-mentioned challenges, the challenges of feeding the human population have to be met. Fortunately, revolution in science, conceptual, and technological innovations, and the rise of genomics are keeping our hopes alive in overcoming the challenges. Genomics enables the identification of new/additional source of genetic variation, understand principal alterations in phenotypes, identification of unique traits, and characterization of molecular pathways involved in stress tolerance (abiotic and biotic); thereby forming the core of crop improvement (Pourkheirandish et al. 2020).

8.4.2 Tracing Our Steps Back to CWRs for Genetic Diversity

Wild plants have been continuously subjected to biotic and abiotic factors of stress via nature for over a million years; natural selection has allowed plants to accumulate genes to combat stress from varied environmental sources. Genomics compared to earlier molecular technologies allow improved characterization of genetic variation available in CWR. Genomic analysis is a tool for effective utilization of existing plant biodiversity, discovery of beneficial genes in CWR and support agriculture through the transfer of these genes into crop species, aiding food security, and discovering novel genetic resources (Brozynska et al. 2014). Furthermore, genomic analysis of CWR can be translated into phenotypic analysis, increasing the exploitation of CWR in plant breeding. However, genomic/genotypic analysis must be automated to facilitate screening of larger populations and time efficiency. Nonetheless, studies have screened for drought tolerance quantitative trait loci (QTL) in wild barley introgression lines (ILs) (Honsdorf et al. 2014) and evaluate phenotypic response of *Brachypodium distachyon* (purple false brome) to nutrient deficiency (availability of nitrogen and phosphorus) using integrative digital imaging (Poiré et al. 2014).

Sequencing of wild relatives of *Arabidopsis thaliana* is essential to study the evolution and adaptation in plants. Rice (*Oryza sativa*) is utilized for cereal grasses, as a model system, to understand the biology, domestication, and improvement.

Many valuable crop plants such as wheat, peanut, sugarcane, oat, cotton, coffee, etc. have polyploid genomes. Sequencing CWRs of these complex crop plant genomes can build reliable and accurate reference genome sources for other polyploid plants. Sequences of diploid progenitors of hexaploid wheat (*Aegilops tauschii* and *Triticum urartu*), progenitors of *Saccharum officinarum* ($x = 10$) and *Saccharum spontaneum* ($x = 8$) (commonly used CWRs in sugarcane cultivars) (Berkman et al. 2014; Souza et al. 2011; Zhang et al. 2018), and diploid robusta coffee (*Coffea canephora*) (Denoeud et al. 2014) are contributing as reference genome and model plant system to understand today's commercialized species.

CWRs of barley or sourced from *Hordeum spontaneum* show resistance against various microbial plant diseases, which include *Fusarium* (crown rot) resistance (Chen et al. 2013), leaf rust and powdery mildew resistance (Schmalenbach et al. 2008), and leaf stripe resistance gene (Biselli et al. 2010). Also, they were found to possess resistant genes against harsh environmental factors like agronomic traits under post-anthesis drought (Kalladan et al. 2013), salt tolerance (Pakniyat and Namayandeh 2007), stress environments (Lakew et al. 2011). Similarly, for other chemical and grain quality factors like selenium concentration (Yan et al. 2011), sodium accumulation (Shavrukov et al. 2010), and grain hardness (Li et al. 2010).

8.4.3 De novo Crop Domestication

Incorporating wild plant resources into De novo domestication of novel crops might help us solve problems of genetic and species diversification of agricultural systems. The introduction of domestication genes into non-domesticated plants is known as De novo domestication while domestication syndrome is the attributes that are produced as a result of de novo domestication of diverse species (Cornille et al. 2014). In cereals, domestication is linked to grain dispersal in wheat, barley, and rice; apical dominance in maize; grain quality in wheat; seed dormancy in rice; grain filling in maize (Uauy et al. 2006; Dubcovsky and Dvorak 2007; Pourkheirandish et al. 2015, 2018). In fruits and vegetables, domestication genes include fruit size and seed dormancy in tomato; flowering in sunflower, etc. (Frary et al. 2000; Chen et al. 2007; Wang et al. 2018b). Interestingly, the above foreplay results from very few genes and the vegetable crops are less well characterized compared to the cereals for domestication syndrome. It is also suggested that most of it is due to loss-of-function mutations (Ramsey and Schemske 1998; Komatsuda et al. 2007; Pourkheirandish et al. 2015).

Again, genomics here has facilitated gene discovery and identification. Whole-genome sequencing enabled us to detect genes associated with non-brittle rachis in pasta wheat and seed filling in maize (Sosso et al. 2015; Avni et al. 2017); genotyping by sequencing gave us smooth awn in barley using (Milner et al. 2019); RNA sequencing revealed seed quality in soybean; and cutin responsible for water retention in barley (Li et al. 2013; Gao et al. 2018b). Genome-wide association study (GWAS) identified gene controlling seed dormancy in soybean and comparative genomics confirmed its presence in other transgenic plants (Wang et al. 2018b).

Interestingly, research and breeding on *Miscanthus*, close relatives of sorghum and sugarcane, tree tobacco, and *Jatropha curcas* (Montes and Melchinger 2016; Usade et al. 2018), etc. are booming due to their potential of being candidates for the increasing demand for biofuels.

Evidence suggests many ways to domesticate a plant. However, very little has been published on domestication experiments that are not so promising. The results advocate that artificial selection requires over 20 generations to change the phenotypes of wild or crop wild hybrids. Additionally, studies indicate that complexity in genetic control influences the time period of achievement of desired phenotype. Importantly, de novo domestication can potentially lead to higher frequency of accumulation of deleterious mutants. The phenomenon of de novo domestication must be carefully considered while planning future agricultural policies.

8.4.4 Engineering Polyploid Plants

Polyploid plants possess more than two sets of homologous chromosomes. This can be achieved either by allopolyploidy or autopolyploidy. There are many agriculturally important plants that are polyploids, such as strawberry (allo-octaploid; $8\times = 56$), potato (auto-tetraploid; $4\times = 48$), and banana (auto-triploid; $3\times = 33$) (Pourkheirandish et al. 2020). Polyploidy induction has been successful in providing plants (including crop, ornamental, and medicinal, etc.) with greater agronomic characteristics including larger seedless fruits and flowers, improved hybrid vigour, and improved pest resistance and physical stress tolerance (An et al. 2014; Tu et al. 2014). Depending on the ploidy level, different manifestations of the trait are observed (Corneillie et al. 2019).

Engineering polyploid plants could be an approach to cultivate new crop species and also reduce economic losses from existing crops (Tamayo-Ordóñez et al. 2016). Polyploid plants can be formed by two major paths, namely unreduced gametes and somatic doubling. Somatic doubling is the most common method. It occurs due to the disruption/failure of cell cycle signalling control between the second growth (G2) and the mitosis (M) phases, and thus leading to repeated DNA synthesis (Ramsey and Schemske 1998), whereas pre-meiotic genome doubling and post-meiotic genome duplication lead to the production of unreduced gametes (De Storme and Geelen 2013). Polyploidy can also be introduced via use of dinitroaniline anti-microtubule drugs such as colchicine, oryzalin, and trifluralin.

Generation of polyploid plants is mainly dependent on cell cycle phases which in turn influences their viability and vigour (Comai 2005). Specific cell cycle and meiosis genes, e.g., cyclin-dependent kinases (CDKs), cyclins (CYCs), tardy asynchronous meiosis (TAM), and omission of second division 1 (OSD1) play major role in obtaining polyploid conditions (Menges et al. 2005; Tank and Thaker 2011). However, they are known to be influenced and regulated by external factors including growth regulators, low temperature, and darkness. Thus, it is essential to recognize the molecular mechanisms and influence of internal and external factors behind cell cycle control, homologous chromosome pairing, and meiotic crossover formation. While the above task can be tedious given the vast number of crop species, discovery and characterization of orthologs of above-mentioned cell cycle genes in other species can be performed using comparative genomics approaches (Gaebelein et al. 2019).

Importantly, post-hybridization (polyploidization) in the early generations, extensive structural rearrangements of merged genomes and methylation changes are known to occur within the plant genomes (Szadkowski et al. 2010). They are known to result/influence sub-genome biases in gene content and expression via genome dominance (Bird et al. 2018). Genomic approaches can be implemented to trace post-hybridization structural rearrangements and machine learning/artificial intelligence can be used to predict the best suited grouping of diverse wild species to construct new synthetic crops which can diversify agriculture and protect food security in current changing environmental conditions (Edger et al. 2017). One of the best examples, bread wheat (*Triticum aestivum*), a major crop is an allohexaploid plant originated through multiple hybridizations. *T. aestivum* has been able to acclimatize to altered climatic zones and grow robustly due to the genes of three different genomes that it possesses.

8.4.5 Boosting Agriculture Through Better Understanding of Plant–Microbe Interactions

With the exponentially growing world population and change in earth's climatic conditions, we need to aim for alternative yet sustainable agricultural reforms in order to continue feeding the growing population. Therefore, molecular study of plant–microbe interactions becomes an equally important factor just as others. It is also projected as an alternative for sustainable agriculture (Johansson et al. 2004). Microorganisms live together with plants below the ground (rhizosphere), above ground surface (phyllosphere), within the plants as endophytes, attached to plant surface as epiphytes, and around the roots in the surrounding soil (Bennett and

Lynch 1981; Lindow and Brandl 2003). They may have significant positive, neutral, or adverse effects on the health and development of host plants (Smith and Goodman 1999). Notably, association of microbes with plants influences higher yield potential (Bhattacharyya et al. 2016). Therefore, it is becoming a priority factor/component in achieving eco-friendly and sustainable agricultural practices in this era of agrochemicals and drastic global climate change.

Plants association with microbes may be symbiotic or pathogenic. While there are some pathogenic microbes that adversely affect plant health (Pusztahelyi et al. 2015; Chagas et al. 2018), there are also bacterial communities that have been reported with potential to manipulate the plant to utilize soil resources, support efficient nutrient uptake and growth, and promote plant biotic and abiotic stress tolerance (Mendes et al. 2011; Fitzpatrick et al. 2018).

One of the strategies to curb the use of chemical pesticides is the use of Plant Growth-Promoting Rhizobacteria (PGPR) in agriculture. Extensive application of PGPR as inoculants to crops has escalated crop yield, and simultaneously there is significant reduction in the use of chemical fertilizers and pesticides, which pollute the environment and contaminate food (Adesemoye and Kloepper 2009). Crop plants expressing PGPR genes are also known to be resistant against various biotic and abiotic factors (Haggag and Habbasha 2015). Genomic platforms and tools are promising tools for better understanding and selection of beneficial strains with various improved traits for crops. Comparative genomic studies and proteome analysis have led to identification of various effector genes for traits such as nutrient uptake, imparting abiotic and biotic stress tolerance during their interaction with plants (Tshikhudo et al. 2019). For example, whole-genome sequence of *Bacillus aryabhattai* AB211 isolated from *Camellia sinensis* rhizosphere revealed signature genes for plant growth promotion, such as chemotaxis, siderophore production, phosphate solubilization, metal ion uptake, etc., making *B. aryabhattai* AB211 a probable candidate to be used as PGPR (Bhattacharyya et al., 2017). Similarly, other species with potential for PGPR include *Serratia marcescens* UENF-22GI (Matteoli et al. 2018), *Pantoea agglomerans* strain P5 (Shariati et al. 2017), *Bacillus cereus* AR156, *Bacillus subtilis* SM21, and *Serratia* sp. XY21 (Zhang et al. 2019). The advances in genomic technologies have also contributed to our understanding of the soil microbial communities through sequencing of numerous soil microorganisms (Jansson and Hofmockel 2018; das et al., 2021; Prasannakumar et al., 2021). For example, the significant increase in genomic sequences for nitrogen-fixing and phosphate-solubilizing bacteria (Ormeño-Orrillo et al. 2018).

8.4.6 Genome Editing for Nutritionally Enhanced Crop Production

One of the long-standing goals of agricultural research is to breed crops with enhanced nutritional content. Notably, widely consumed crops like wheat, maize, and rice are poor sources of many essential micro- and macro-nutrients. This by itself explains the need for the production of crops with high nutrition content. Among all, genome editing tools are successful in providing higher accuracy and efficiency in genetic modification than conventional breeding in a shorter time period. Genome editing uses sequence-specific nucleases (SSNs) to introduce targeted mutations in crops with increased proficiency and accuracy (Georges and Ray 2017). Zinc finger nucleases (ZFNs), transcriptional activator-like effector nucleases (TALENs), and clustered regularly interspaced short palindromic repeats (CRISPR)-associated endonuclease Cas9 (CRISPR/Cas9) are among the artificially engineered SSNs being widely used today (Miglani 2017; Schiml and Puchta 2016). However, no commercialization of crops from genome editing has been achieved until today. Nonetheless, research in the field carries on. Liang and coworkers (2014) reported the incidence of anti-nutritional compound phytic acid, inositol 1,2,3,4,5,6-hexakisphosphate in *Zea mays*, of which phytic acid is poorly digested in humans and is hazardous to the environment. Therefore, using genome editing tools, two gRNAs targeting the ZmIPK (Inositol Phosphate Kinase) gene that catalyses a vital step in phytic acid biosynthetic pathway were designed in order to reduce phytic acid content of maize seeds. Similarly, soybeans with increased oleic acid and reduced linoleic acid content have also been produced (Demorest et al. 2016). Let us hope the field makes greater progress in the future.

8.5 Completeness of Functional Annotation for Better Candidate Gene Identification

The completion of the human genome project enabled sequencing facilities to aim for more ambitious projects, increasing the sequencing data exponentially. However, annotation of information from these data has been lagging ever since, depriving us of a comprehensive catalog of gene models for a given species. The eukaryotic gene structure prediction is more complex and alternative splicing; a process which facilitates the synthesis of more than one protein from a single gene sequence adds on much more complexity (Schellenberg et al. 2008). A combination of *ab initio* gene prediction and homology-based methods is often used to build gene models (Campbell et al. 2014; Klasberg et al. 2016). Today, long reads sequencing technologies can generate full-length transcripts, providing new insights into the extent of alternative splicing and transcriptome diversity (Cook et al. 2019).

Genome annotations enable us to study gene functions, biochemical and regulatory pathways, or quantitative trait loci in plants. However, annotation of genes faces many hurdles and obstruction. Gene functional annotations for most of the crop plants are achieved by transmission of annotation from most similar genes in model plants like *Arabidopsis* and *Oryza sativa,* i.e., homology-based inference, without much direct experimental support. Evolutionary plant history, gene redundancy due to successive rounds of polyploidy and subsequent diploidization, and differential loss further contribute to the (Jiao and Paterson 2014; Salman-Minkov et al. 2016) complexity of annotation transfer. Notably a single reference sequence of an organism hides the true complexity of gene space variation, which leads to missing gene models. For example, a presence–absence variation profiling by Zhao and team (2018) resulted in documentation of 10,872 genes in 67 rice accessions under study that were partially absent in the *Oryza sativa* Nipponbare reference annotation. This strongly suggests the need for pan-genomics expansion.

Specialized databases through integrative genomic approaches have been developed such as SNP-Seek (rice), SoyBase (soybean), MIPSPlantsDB (*Arabidopsis*, *Medicago*, Lotus, rice, maize and tomato), and WheatGenome (wheat) (Spannagl et al. 2007a, b; Grant et al. 2009; Lai et al. 2012; Blake et al. 2016; Mansueto et al. 2017). Similarly, to classify candidate genes involved in biological processes, tools like KnetMiner and MCRiceRepGP were developed that use multicriteria decision analysis for sorting the genes (Golicz et al., 2018; Hassani-Pak and Rawlings 2017). CEGMA (Core Eukaryotic Genes Mapping Approach) relied on the evolutionary conservation of orthologous genes within in order to assess the completeness of a genome annotation (Parra et al. 2007). BUSCO (Benchmarking Universal Single-Copy Orthologs), most successful successor of CEGMA, predominantly selects single-copy genes for the formation of orthologous datasets from eukaryotic clades. However, it is not preferred for flowering plant species that are allopolyploids (Van de Peer et al. 2009). Regardless of the availability of genome annotations, functional characterization of annotated genes remains a vital challenge in molecular breeding pipelines (Scheben and Edwards 2018).

8.5.1 New Breeding Targets from Non-coding Part of Genome

The genome of most large crops encodes abundantly for non-coding genes compared to protein-coding genes (Long et al. 2017). Among all the RNAs, long non-coding RNAs (lncRNAs) are abundantly found in plants and are key regulators of the transcriptional process. lncRNAs also assist in plant development, adaptation to environmental change, and are vital molecular drivers of lineage-specific morphological evolution (Wang et al. 2018c). Interestingly, lncRNAs are suggested to be involved in plant sexual reproduction due to strong bias toward transcription in reproductive tissues (Zhang et al. 2014; Golicz et al. 2018), thereby influencing the formation of flowers, fruits, and grains. Evidence also shows that some of the lncRNAs could be functional (Huang et al. 2018).

In *Arabidopsis*, vernalization-mediated epigenetic silencing of the *FLOWERING LOCUS C (FLC)* promotes flowering in spring (Heo and Sung 2011). Similarly, rice lncRNA LDMAR was reported to regulate DNA methylation of photoperiod-sensitive male sterility (PSMS), a vital attribute responsible for the development of hybrid rice (Ding et al. 2012). Another example is that of highly conserved lncRNA *EARLY NODULIN 40 (ENOD40)* in legumes (Gultyaev and Roussis 2007), rice (Kouchi et al. 1999), and maize (Compaan et al. 2003); involved in the regulation of symbiosis between the legumes and microbes (bacteria or fungi) for organogenesis in root nodules (Gultyaev and Roussis 2007), along with mycorrhizal association in alfalfa (Van Rhijn et al. 1997). Recently, transcriptomics was to characterize the role of circRNAs in establishing resistance against *Pectobacterium carotovorum* in potato plant (Zhou et al. 2018). Several other studies focused on circRNAs response to biotic stress are being carried out in kiwifruit and tomato (Wang et al. 2017, 2018a).

The published literature makes it evident that lncRNAs play a significant role in the regulation of genetic pathways towards development of plants. They use many of the cellular machinery components to perform their diverse roles and represent as promising molecular targets for manipulation in disease responsive pathways. The studies are in nascent stages and much effort needs to be made to better understand the non-coding areas of the genome.

8.5.2 The Pan-Genome Approach

The advancing genomic technologies and our increasing knowledge on genomic variation have made us realize that single-genome reference is not sufficient as a reference to represent the intraspecific diversity (Golicz et al. 2020). The pan-genome notion was born as a result of this. The universe of genomic sequences contained in a bacterial species is referred to as the pan-genome. In 2005, many isolates of *Streptococcus agalactiae* were sequenced, representing 80% of S. agalactiae genes as core genomes. (Tettelin et al. 2005). This is very much relevant to plant species as well, exhibiting extensive phenotypic variation in agronomic characters, such as yield, metabolite biosynthesis, and response to biotic and abiotic stresses, etc.

Pan-genomes have been constructed for various model plants and key crop species, such as *Arabidopsis thaliana* (Van de Weyer et al. 2019) , *Brachypodium distachyon* (Gordon et al. 2017), *Brassica oleracea* (Golicz et al. 2016), *Oryza sativa* (Sun et al. 2017; Zhao et al. 2018), *Glycine max* (Li et al. 2014), *Triticum aestivum* (Walkowiak et al. 2020), *Brassica napus* (Song et al. 2020), *Hordeum vulgare* (Jayakodi et al. 2020), and *Solanum esculentum* (Gao et al. 2019).

Pan-genomes as references will increase the horizon of genomic analysis. For example, it improves short-read mapping accuracy in contrast to using a single reference, resulting in higher quality variant calls and more precise gene expression quantification (Eggertsson et al. 2017; Bayer et al. 2020). Increasing use of pan-genomes also accelerates the scope of incorporating transposable elements (TEs) into crop genomics, which have been neglected until recently. TEs can modify the structure and amount of gene product that is transcribed (Jiang et al. 2004).

Pan-genome investigations are aided by the increasing availability of genome sequence data, which is a result of rapid technological advancements and is also cost effective. Specifically, storing and visualizing pan-genome data are still a hurdle. There is a need to create methods and establish standards for precise and reliable functional annotation of genes and genomes. Integrative genomics approaches can be used to link various features of genes to their function in order to have a better knowledge of their possible function. While most pan-genome research has concentrated on the genic area of the genome so far, there is significant genomic and phenotypic variation outside of those regions that needs to be investigated.

While the computational power rises, someday wide pan-genomes might answer the bigger question: what genes make a plant?

8.6 The Sequenced Angiosperm Genomes

As of March 27th, 2021, the NCBI-genome browser listed completed genome of 765 angiosperm species and could be found at *https://www.ncbi.nlm.nih.gov/genome/browse/#!/overview/*. Table 8.2 indicates the list of all available sequenced angiosperm genomes. Most of these plants have great economic significance (Chen et al. 2018).

8.7 Databases for Plant Genomics/Popular Genome Databases

8.7.1 The National Center for Biotechnology Information-Genome Browser

The NCBI-Genome browser (https://www.ncbi.nlm.nih.gov/genome/) endow with genomic information (Fig. 8.1). The genome database currently comprised of 60,159 species including Eukaryotes (15,998), Prokaryotes (321,872), Viruses (42,533), Plasmids (28,557), Organelles (18,354), as accessed on 26th June, 2021. It also provide interlinking to another genome recourses like GOLD—Genomes Online Database (https://gold.jgi.doe.gov/), Bacteria Genomes at Sanger (https://www.sanger.ac.uk/resources/ downloads/bacteria/), Ensembl (http://asia.ensembl.org/index.html)

8.7.2 Plant Genome DataBase Japan (PGDBj)

TAIR10 and RAP-DB, SABRE DB, and DNA Marker DB databases are all integrated into the PGDBj database (Fig. 8.2). This database also contains information gathered from the literature, such as chromosome number and genome size, quantitative trait loci and related linkage mapping markers, and genome-specific databases (Asamizu et al. 2014).

8.7.3 EnsemblPlants

Ensemblplants database (https://plants.ensembl.org/index.html) is a genome portal for 49 plants (e.g., 49-plants.ensembl.org) (Fig. 8.3). It provides statistics of genome size and data on assembly, regulation, variation, and sequences of these plants. This database collaborated **Gramene** (http://www.gramene.org/) for its curated, open-source, integrated data source.

Table 8.2 Genome information available in NCBI- Genome recourses

Plant name	Family	Type	Size (Mb)	Chromosomes	Organelles	Plasmids	Assemblies	Reference in NCBI
Metrosideros polymorpha	Myrtaceae	flower	304.366	0	0	0	1	Metrosideros polymorpha (ID 451 78)—Genome—NCBI (nih.gov)
Micractinium conductrix	Chlorellaceae	Green algae	61.0189	0	2	0	1	Micractinium conductrix (ID 56073)—Genome—NCBI (nih.gov)
Micromonas	Mamiellaceae	Green algae	0.614822	0	0	0	5	Micromonas—Genome—NCBI (nih.gov)
Micromonas commoda	Mamiellaceae	Green algae	21.1093	17	2	0	1	Micromonas commoda (ID 44422)—Genome—NCBI (nih.gov)
Micromonas pusilla	Mamiellaceae		21.9583	0	0	0	1	Micromonas pusilla (ID 501)—Genome—NCBI (nih.gov)
Microtea debilis	Microteaceae	Plant	475.429	0	0	0	1	Microtea debilis (ID 812 80)—Genome—NCBI (nih.gov)
Microthlaspi erraticum	*Microthlaspi erraticum*	Flower	170.423	0	0	0	1	Microthlaspi erraticum (ID 87015)—Genome—NCBI (nih.gov)
Microthlaspi erraticum	Brassicaceae	Flower	1790.64	19	0	0	1	Microthlaspi erraticum (ID 87015)—Genome—NCBI (nih.gov)
Mimosa pudica	Fabaceae	Plant	557.202	0	0	0	1	Mimosa pudica (ID 701 77)—Genome—NCBI (nih.gov)

(continued)

Table 8.2 (continued)

Plant name	Family	Type	Size (Mb)	Chromosomes	Organelles	Plasmids	Assemblies	Reference in NCBI
Miscanthus lutarioriparius	Poaceae	Plant	2074.8	0	0	0	1	Miscanthus lutarioriparius (ID 95813)—Genome—NCBI (nih.gov)
Miscanthus sacchariflorus	Poaceae	Grass	2074.92	19	0	0	2	Miscanthus sacchariflorus (ID 1664 5)—Genome—NCBI (nih.gov)
Momordica charantia	Cucurbitaceae	Fruit	285.614	0	0	0	3	Momordica charantia (ID 12860)—Genome—NCBI (nih.gov)
Monoraphidium	Selenastraceae		74.6589	0	0	0	1	Monoraphidium—Genome—NCBI (nih.gov)
Monoraphidium neglectum	Selenastraceae	Algae	69.7118	0	2	0	1	Monoraphidium neglectum (ID 36372)—Genome—NCBI (nih.gov)
Monotropa hypopitys	Ericaceae	Plant	2197.49	0	0	0	1	Monotropa hypopitys (ID 44119)—Genome—NCBI (nih.gov)
Morella rubra	Myricaceae	Plant	313.009	8	0	0	1	Morella rubra (ID 12 795)—Genome—NCBI (nih.gov)
Moricandia arvensis	Brassicaceae	Flower	758.71	0	0	0	1	Moricandia arvensis (ID 98181)—Genome—NCBI (nih.gov)

Moricandia moricandioides	Brassicaceae	Flower	498.312	0	0	0	1	Moricandia moricandioides (ID 981 82)—Genome—NCBI (nih.gov)
Moringa oleifera	Moringaceae	Tree	253.894	0	0	0	1	Moringa oleifera (ID 814 77)—Genome—NCBI (nih.gov)
Morus alba	Moraceae	Tree	336.456	14	0	0	2	Morus alba (ID 86952)—Genome—NCBI (nih.gov)
Morus notabilis	Moraceae	Plant	320.379	0	0	0	1	Morus notabilis (ID 1 7692)—Genome—NCBI (nih.gov)
Mucuna pruriens	Fabaceae	Plant	397.042	0	0	0	1	Mucuna pruriens (ID 71 552)—Genome—NCBI (nih.gov)
Musa acuminata	Musaceae	Fruit	472.231	11	0	0	2	Musa acuminata (ID 10 976)—Genome—NCBI (nih.gov)
Musa balbisiana	Musaceae	Fruit	492.775	11	0	0	1	Musa balbisiana (ID 164 89)—Genome—NCBI (nih.gov)
Musa itinerans	Musaceae	Fruit	455.349	0	0	0	1	Musa itinerans (ID 44 989)—Genome—NCBI (nih.gov)
Musa schizocarpa	Musaceae	Fruit	525.283	0	0	0	1	Musa schizocarpa (ID 74 967)—Genome—NCBI (nih.gov)

(continued)

Table 8.2 (continued)

Plant name	Family	Type	Size (Mb)	Chromosomes	Organelles	Plasmids	Assemblies	Reference in NCBI
Nasturtium officinale	Brassicaceae	Plant	216.122	0	0	0	1	Nasturtium officinale (ID 8209)—Genome—NCBI (nih.gov)
Nelumbo nucifera	Nelumbonaceae	Flower	804.648	0	1	0	5	Nelumbo nucifera (ID 14095)—Genome—NCBI (nih.gov)
Nicotiana attenuata	Solanaceae	Plant	2365.68	12	0	0	2	Nicotiana attenuata (ID 13243)—Genome—NCBI (nih.gov)
Nicotiana benthamiana	Solanaceae	Leaf	61.9511	0	0	0	1	Nicotiana benthamiana (ID 10940)—Genome—NCBI (nih.gov)
Nicotiana glauca	Solanaceae	Tree	3222.83	0	0	0	1	Nicotiana glauca (ID 670 62)—Genome—NCBI (nih.gov)
Nicotiana knightiana	Solanaceae	Flower	2298.94	0	0	0	1	Nicotiana knightiana (ID 79603)—Genome—NCBI (nih.gov)
Nicotiana obtusifolia	Solanaceae	Plant	1222.77	0	0	0	1	Nicotiana obtusifolia (ID 53488)—Genome—NCBI (nih.gov)
Nicotiana otophora	Solanaceae	Plant	2689.35	0	0	0	1	Nicotiana otophora (ID 32281)—Genome—NCBI (nih.gov)
Nicotiana paniculata	Solanaceae	Flower	2190.56	0	0	0	1	Nicotiana paniculata (ID 79602)—Genome—NCBI (nih.gov)

Nicotiana rustica	Solanaceae	Plant	4231.29	0	0	0	1	Nicotiana rustica (ID 79601)—Genome—NCBI (nih.gov)
Nicotiana sylvestris	Solanaceae	Plant	2221.99	0	1	0	1	Nicotiana sylvestris (ID 13135)—Genome—NCBI (nih.gov)
Nicotiana tabacum	Solanaceae	Plant	3643.47	0	2	0	4	Nicotiana tabacum (ID 425)—Genome—NCBI (nih.gov)
Nicotiana tomentosiformis	Solanaceae	Plant	1688.47	0	1	0	1	Nicotiana tomentosiformis (ID 12239)—Genome—NCBI (nih.gov)
Nicotiana undulata	Solanaceae	Plant	1914.3	0	0	0	1	Nicotiana undulata (ID 12210)—Genome—NCBI (nih.gov)
Nissolia schottii	Fabaceae	Leaf	466.099	0	0	0	1	Nissolia schottii (ID 701 74)—Genome—NCBI (nih.gov)
Noccaea caerulescens	Brassicaceae	Plant	140.792	0	0	0	1	Noccaea caerulescens (ID 83070)—Genome—NCBI (nih.gov)
Noccaea goesingensis	Brassicaceae	Plant	150.323	0	0	0	1	Noccaea goesingensis (ID 83071)—Genome—NCBI (nih.gov)
Nothapodytes nimmoniana	Icacinaceae	Plant	1.36527	0	0	0	1	Nothapodytes nimmoniana (ID 540 72)—Genome—NCBI (nih.gov)

(continued)

Table 8.2 (continued)

Plant name	Family	Type	Size (Mb)	Chromosomes	Organelles	Plasmids	Assemblies	Reference in NCBI
Nymphaea colorata	Nymphaeaceae	Flower	409.014	14	1	0	2	Nymphaea colorata (ID 69117)—Genome—NCBI (nih.gov)
Nymphaea thermarum	Nymphaeaceae	Leaves	368.003	0	0	0	1	Nymphaea thermarum (ID 88507)—Genome—NCBI (nih.gov)
Nyssa sinensis	Nyssaceae	Plant	1001.45	22	0	0	1	Nyssa sinensis (ID 84 589)—Genome—NCBI (nih.gov)
Ochetophila trinervis	Rhamnaceae	Plant	309.116	0	0	0	1	Ochetophila trinervis (ID 70176)—Genome—NCBI (nih.gov)
Ocimum tenuiflorum	Lamiaceae	Economic	332.617	0	0	0	2	Ocimum tenuiflorum (ID 40058)—Genome—NCBI (nih.gov)
Odontarrhena argentea	Brassicaceae	Plant	183.186	0	0	0	1	Odontarrhena argentea (ID 82488)—Genome—NCBI (nih.gov)
Oenanthe javanica	Apiaceae	Plant	1278.51	0	0	0	1	Oenanthe javanica (ID 84 753)—Genome—NCBI (nih.gov)
Olea europaea	Oleaceae	economic	1141.15	23	1	0	3	Olea europaea (ID 10 724)—Genome—NCBI (nih.gov)
Ophiorrhiza pumila	Rubiaceae	Plant	440.319	0	0	0	1	Ophiorrhiza pumila (ID 97777)—Genome—NCBI (nih.gov)

Oropetium thomaeum	Poaceae	resurrection plant	243.175	0	0	0	1	Oropetium thomaeum (ID 38890)—Genome—NCBI (nih.gov)
Oryza barthii	Poaceae	Weed	347.716	12	0	0	4	Oryza barthii (ID 2 750)—Genome—NCBI (nih.gov)
Oryza brachyantha	Poaceae	Food	259.908	12	0	0	2	Oryza brachyantha (ID 10862)—Genome—NCBI (nih.gov)
Oryza coarctata	Poaceae	Food	569.986	0	0	0	1	Oryza coarctata (ID 11313)—Genome—NCBI (nih.gov)
Oryza glaberrima	Poaceae	Food	347.321	12	0	0	1	Oryza glaberrima (ID 4 58)—Genome—NCBI (nih.gov)
Oryza glumipatula	Poaceae	Food	388.593	12	0	0	1	Oryza glumipatula (ID 11318)—Genome—NCBI (nih.gov)
Oryza meridionalis	Poaceae	Food	393.639	12	0	0	2	Oryza meridionalis (ID 11319)—Genome—NCBI (nih.gov)
Oryza meyeriana var. granulata	Poaceae	Food	776.957	0	0	0	3	Oryza meyeriana var. granulata (ID 11287)—Genome—NCBI (nih.gov)
Oryza minuta	Poaceae	Food	45.1659	2	0	0	1	Oryza minuta (ID 10 965)—Genome—NCBI (nih.gov)

(continued)

Table 8.2 (continued)

Plant name	Family	Type	Size (Mb)	Chromosomes	Organelles	Plasmids	Assemblies	Reference in NCBI
Oryza nivara	Poaceae	Food	395.534	12	0	0	3	Oryza nivara (ID 2 841)—Genome—NCBI (nih.gov)
Oryza officinalis	Poaceae	Food	584.134	0	0	0	2	Oryza officinalis (ID 10 964)—Genome—NCBI (nih.gov)
Oryza punctata	Poaceae	Food	422.391	12	0	0	2	Oryza punctata (ID 10 963)—Genome—NCBI (nih.gov)
Oryza rufipogon	Poaceae	Food	384.518	0	0	0	62	Oryza rufipogon (ID 4 57)—Genome—NCBI (nih.gov)
Oryza sativa	Poaceae	Food	374.423	12	2	0	81	Oryza sativa (ID 10)—Genome—NCBI (nih.gov)
Panax notoginseng	Araliaceae	Herbs	2660.68	12	0	0	2	Panax notoginseng (ID 11326)—Genome—NCBI (nih.gov)
Panicum miliaceum	Poaceae	Millet	848.352	18	0	0	2	Panicum miliaceum (ID 44156)—Genome—NCBI (nih.gov)
Panicum virgatum	Poaceae	Grass	1130	18	1	0	1	Panicum virgatum (ID 660)—Genome—NCBI (nih.gov)
Papaver somniferum	Papaveraceae	Breadseed poppy	2715.53	11	1	0	3	Papaver somniferum (ID 12819)—Genome—NCBI (nih.gov)

Parasponia andersonii	Cannabaceae	Tree	475.834	0	0	0	1	Parasponia andersonii (ID 66935)—Genome—NCBI (nih.gov)
Passiflora edulis	Passifloraceae	passion fruit	165.657	0	0	0	1	Passiflora edulis (ID 13 910)—Genome—NCBI (nih.gov)
Penstemon barbatus	Plantaginaceae	Plant	696.306	0	0	0	1	Penstemon barbatus (ID 71049)—Genome—NCBI (nih.gov)
Penstemon centranthifolius	Plantaginaceae	Plant	4.47159	0	0	0	1	Penstemon centranthifolius (ID 32 663)—Genome—NCBI (nih.gov)
Penstemon cyananthus	Plantaginaceae	Plant	4.62226	0	0	0	1	Penstemon cyananthus (ID 13453)—Genome—NCBI (nih.gov)
Penstemon davidsonii	Plantaginaceae	Plant	2.37523	0	0	0	1	Penstemon davidsonii (ID 13464)—Genome—NCBI (nih.gov)
Penstemon dissectus	Plantaginaceae	Commercial	2.62809	0	0	0	1	Penstemon dissectus (ID 13465)—Genome—NCBI (nih.gov)
Penstemon fruticosus	Plantaginaceae	Shrub	2.31904	0	0	0	1	Penstemon fruticosus (ID 13612)—Genome—NCBI (nih.gov)
Penstemon grinnellii	Plantaginaceae	Plant	3.66352	0	0	0	1	Penstemon grinnellii (ID 32664)—Genome—NCBI (nih.gov)

(continued)

Table 8.2 (continued)

Plant name	Family	Type	Size (Mb)	Chromosomes	Organelles	Plasmids	Assemblies	Reference in NCBI
Perilla citriodora	Lamiaceae	economic	618.797	0	0	0	1	Perilla citriodora (ID 460 88)—Genome—NCBI (nih.gov)
Persea americana	Lauraceae	avocado	912.698	0	0	0	3	Persea americana (ID 11 781)—Genome—NCBI (nih.gov)
Petunia axillaris	Solanaceae	Flower	2199.86	0	0	0	1	Petunia axillaris (ID 93408)—Genome—NCBI (nih.gov)
Phalaenopsis hybrid cultivar	Orchidaceae	Flower	2687.66	0	0	0	1	Phalaenopsis hybrid cultivar (ID 34687)—Genome—NCBI (nih.gov)
Pharnaceum exiguum	Molluginaceae	Plant	287.973	0	0	0	1	Pharnaceum exiguum (ID 87254)—Genome—NCBI (nih.gov)
Phaseolus coccineus	Fabaceae	Runner bean	371.086	0	0	0	1	Phaseolus coccineus (ID 10943)—Genome—NCBI (nih.gov)
Phaseolus lunatus	Fabaceae	Lima beans	546.42	11	0	0	1	Phaseolus lunatus (ID 92 715)—Genome—NCBI (nih.gov)
Phaseolus vulgaris	Fabaceae	Bean	521.077	11	0	0	5	Phaseolus vulgaris (ID 3 80)—Genome—NCBI (nih.gov)
Phoenix dactylifera	Arecaceae	Date palm	773.189	18	2	0	4	Phoenix dactylifera (ID 2 664)—Genome—NCBI (nih.gov)

Phtheirospermum japonicum	Orobanchaceae	Plant	1226.89	0	0	0	1	Phtheirospermum japonicum (ID 95889)—Genome—NCBI (nih.gov)
Phyllostachys edulis	Poaceae	bamboo	1908.07	0	0	0	1	Phyllostachys edulis (ID 10429)—Genome—NCBI (nih.gov)
Physaria acutifolia	Brassicaceae	Plant	199.442	0	0	0	1	Physaria acutifolia (ID 83072)—Genome—NCBI (nih.gov)
Physaria fendleri	Brassicaceae	Plant	331.342	0	0	0	1	Physaria fendleri (ID 830 73)—Genome—NCBI (nih.gov)
Physaria ovalifolia	Brassicaceae	Flower	290.267	0	0	0	1	Physaria ovalifolia (ID 83078)—Genome—NCBI (nih.gov)
Physcomitrium patens	Funariaceae	Model organism	472.081	27	2	0	1	Physcomitrium patens (ID 383)—Genome—NCBI (nih.gov)
Picea abies	Pinaceae	Plant	11961.4	0	0	0	2	Picea abies (ID 11155)—Genome—NCBI (nih.gov)
Picea engelmannii	Pinaceae	Tree	24943.6	0	0	0	1	Picea engelmannii (ID 78303)—Genome—NCBI (nih.gov)
Picea glauca	Pinaceae	Plant	24621.5	0	0	0	3	Picea glauca (ID 112 58)—Genome—NCBI (nih.gov)

(continued)

Table 8.2 (continued)

Plant name	Family	Type	Size (Mb)	Chromosomes	Organelles	Plasmids	Assemblies	Reference in NCBI
Picea sitchensis	Pinaceae	Tree	18225.2	0	0	0	1	Picea sitchensis (ID 1313 7)—Genome—NCBI (nih.gov)
Picocystis	Picocystaceae	*Algae*	29.6462	0	0	0	1	Picocystis (ID 72946)—Genome—NCBI (nih. gov)
Pinus lambertiana	Pinaceae	Tree	27602.7	0	0	0	1	Pinaceae—Taxonomy—NCBI (nih.gov)
Pinus sylvestris	Pinaceae	Plant	0.985624	0	0	0	1	Pinaceae—Taxonomy—NCBI (nih.gov)
Pinus taeda	Pinaceae	Plant	22103.6	0	0	0	1	Pinus taeda (ID 1102 7)—Genome—NCBI (nih.gov)
Pistacia vera	Anacardiaceae	Tree	671.28	0	1	0	1	Pistacia vera (ID 55403)—Genome—NCBI (nih.gov)
Pisum sativum	Fabaceae	Food	3297.04	0	0	0	2	Pisum sativum (ID 120 50)—Genome—NCBI (nih.gov)
Platycodon grandiflorus	Campanulaceae	Flower	574.706	9	0	0	2	Platycodon grandiflorus (ID 56439)—Genome—NCBI (nih.gov)
Pogostemon cablin	Lamiaceae	Herb	1916.69	0	0	0	1	Pogostemon cablin (ID 73046)—Genome—NCBI (nih.gov)

Pontederia paniculata	Pontederiaceae	Plant	571.388	0	0	0	1	Pontederia paniculata (ID 44985)—Genome—NCBI (nih.gov)
Populus alba	Salicaceae	Plant	416.986	0	2	0	1	Populus alba (ID 13203)—Genome—NCBI (nih.gov)
Populus alba x Populus glandulosa	Salicaceae	Plant	415.53	0	0	0	1	Populus alba x Populus glandulosa (ID 87686)—Genome—NCBI (nih.gov)
Populus davidiana	Salicaceae	Tree	417.66	0	0	0	1	Populus davidiana (ID 51 728)—Genome—NCBI (nih.gov)
Populus deltoids	Salicaceae	Plant	428.645	19	0	0	3	Populus deltoides (ID 12 709)—Genome—NCBI (nih.gov)
Populus euphratica	Salicaceae	Tree	496.033	0	1	0	1	Populus euphratica (ID 13265)—Genome—NCBI (nih.gov)
Populus ilicifolia	Salicaceae	Plant	399.016	0	0	0	1	Populus ilicifolia (ID 4 6602)—Genome—NCBI (nih.gov)

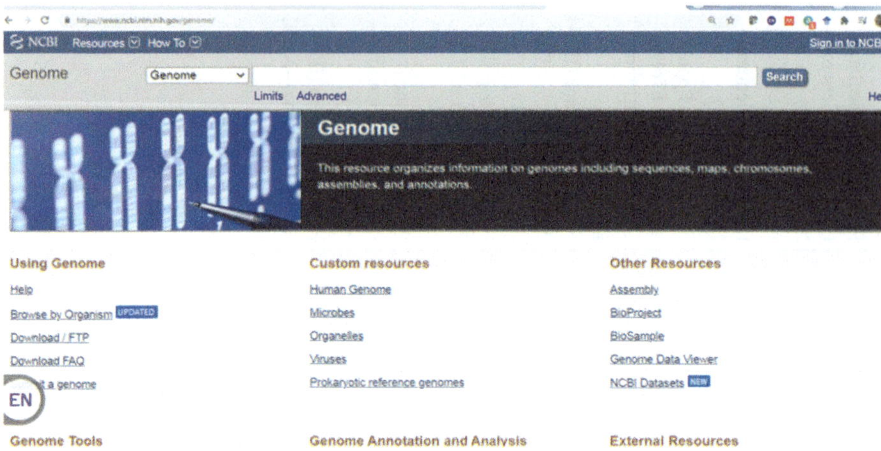

Fig. 8.1 A screenshot of NCBI-Genome database

Fig. 8.2 A screenshot of plant genome database Japan

8.7.4 Genome Size in Asteraceae Database (GSAD):

The GSAD database Release 3.0 (https://www.asteraceaegenomesize.com/) is a specific and indigenous genome size data for family Asteraceae (Fig. 8.4). The Genome sizes are currently available for 1219 species consisting 337 species (21.67%) and 46 genera (19.83%) based on 2768 records from 198 publications as per 26th June 2021 (Vitales et al. 2019).

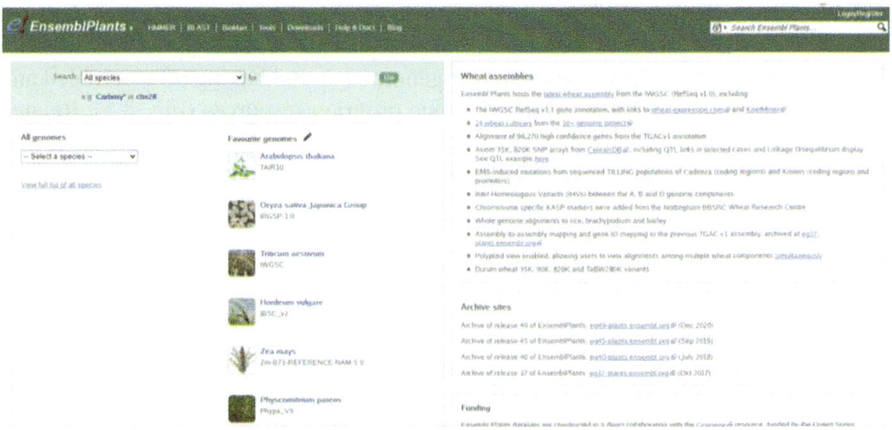

Fig. 8.3 A screenshot of EnsemblPlants database

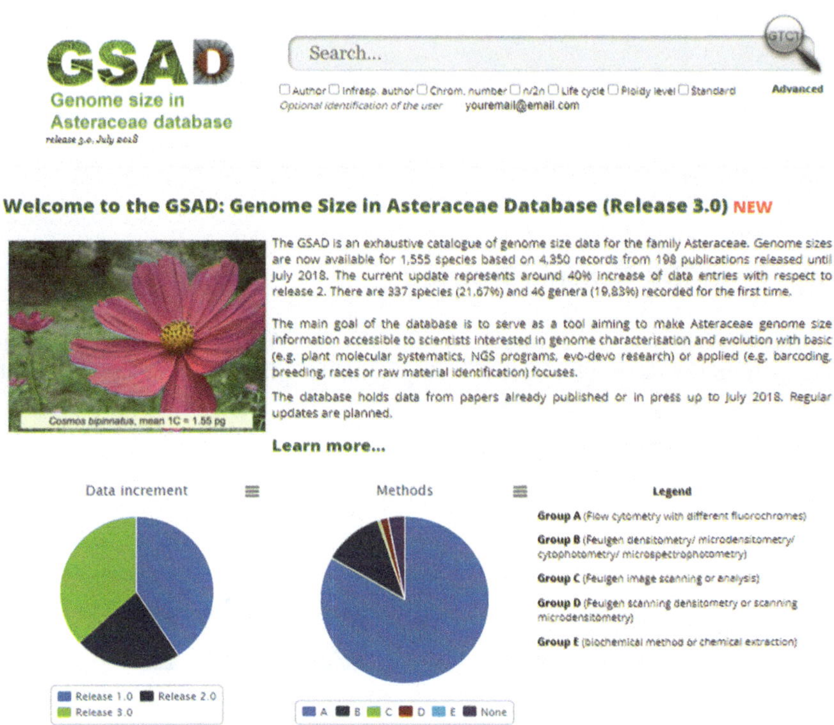

Fig. 8.4 A screenshot of GSAD database

8.7.5 Phytozome

Phytozome v12.1.6 is a plant comparative genomics portal having 93 assembled and interpreted genomes using JGI Plant Science program resources (Fig. 8.5). Related information can be found using https://phytozome.jgi.doe.gov/pz/portal.html. The genome portal is maintained by Department of Energy's Joint Genome Institute.

8.7.6 Plant DNA *C*-Values Database

The Plant DNA C-values database (release 7.1) contains *C-values* for 12,273 species including angiosperm (10,770), gymnosperm (421), 303 pteridophyte 246 ferns (monilophytes), 334 bryophyte (209 mosses, 102 liverworts, and 23 hornworts), and 445 algal species (Fig. 8.6) (Pellicer and Leitch 2019).

8.7.7 Plant rDNA Database

The Plant rDNA Database include the data from over 785 publications and approximately 2148 plant species (3783 entries) as of June 26th, 2021 (Fig. 8.7). It contains information on "chromosome number, genome size, ploidy level, telomere type and genome sequenced."

8.7.8 PlantGDB Genome Browser:

"The *Plant GDB Genome Browser* (specifically for *Viridiplantae)* provides high-quality genome browser which provides high quality spliced alignments of available

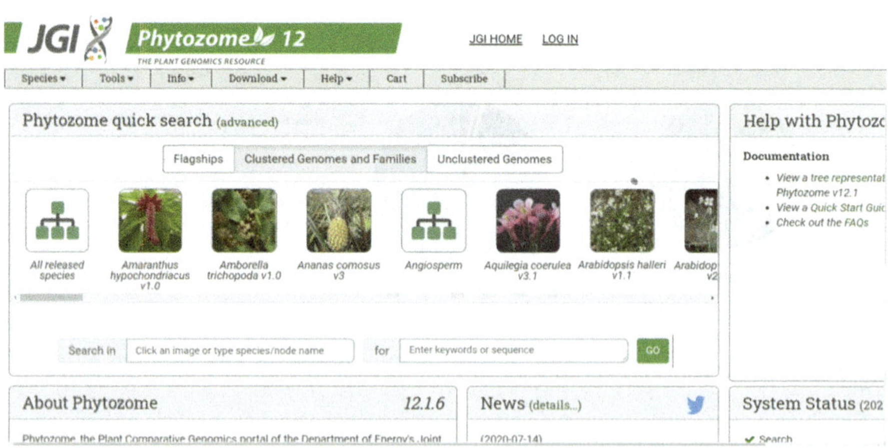

Fig. 8.5 A Screenshot of Phytozome database

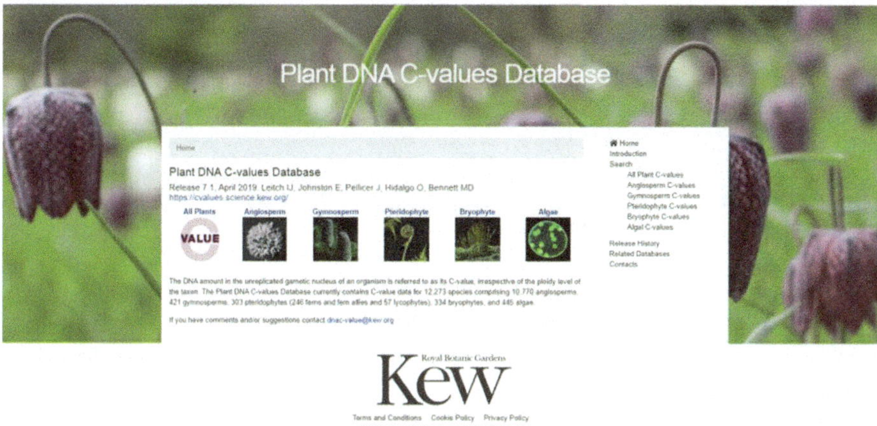

Fig. 8.6 A Screenshot of plant DNA *C*-values database

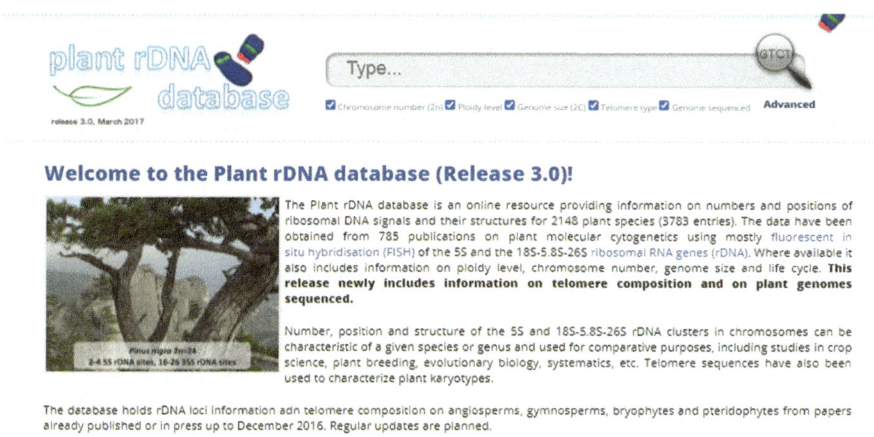

Fig. 8.7 A Screenshot of Plant rDNA database

transcripts of more than 100 plants as well as predicted proteins from related model species" (Fig. 8.8). It also provides graphical interferences for more than 14 completed genome species (Duvick et al. 2008).

8.8 Further Remarks

Flowering plant genome data is rapidly expanding in the NCBI genome. The new generation, along with big-data theories, is gaining in popularity as a means of addressing larger issues. There is currently a considerable requirement for all linked data to be integrated and maintained. More bioinformatics tools, as well as

Fig. 8.8 A Screenshot of PlantGDB database

comprehensive or huge databases, are required to attract more scientific minds to this understanding.

Remarkably, few efforts have been made to sequence the genomes of plant species connected to the evolutionary history of angiosperms such as magnoliids. There is trepidation about sharing genomic data with the scientific community as soon as the genome is made public. This increases redundancy and makes valuable collaborative cooperation between two and more groups more difficult. Even different genome databases require the integration of diverse forms of genetic data as well as the potential integration of genome analysis tools.

8.9 Closing Note?

Nearly four decades of technology advances, the genomic investigations are expected to attain maturity within a decade's time because of its drastically drop in prices. We focused on plant genomes in this chapter, but not on epigenomics. Epigenomics is the study of how individual nucleotides in the genome are modified after replication. There are a few notable examples, such as the inclusion of a methyl group in cytosine residues. This majorly implicate in gene expression. Technologies to analyse epi-genome data are desperately needed to better understand developmental biology and disease progression. This chapter's missing link also involves transcription analysis. In the discipline of transcriptomics, technologies including as reverse transcription, hybridization arrays, and whole-genome sequencing are already commonplace. Several other topics, including as metagenomics, metabolomics, and pathways associated with them, are not covered in this chapter. In the current era of genomics research, major difficulties include developing novel variety for food, nutrition, feed, fibre, pharmaceuticals, and fuel. The use of plants as a source of environmentally, commercially, and socially sustainable raw materials is the ultimate green resolution.

Acknowledgments The DBT-BIF Facility, DST-FIST Facility for PG Level O, Government of India, and the BISEP programme, run by the Government of Karnataka, India, at MLACW contributed to this research.

Conflict of Interest Statement The authors state that there were no commercial or financial relationships that may be seen as a potential conflict of interest during the research.

References

Adesemoye AO, Kloepper JW (2009) Plant-microbes interactions in enhanced fertilizer-use efficiency. Appl Microbiol Biotechnol 85:1–12

Alkan C, Sajjadian S, Eichler EE (2011) Limitations of next-generation genome sequence assembly. Nat Methods 8:61–65

An F, Fan J, Li J, Li QX, Li K, Zhu W, Wen F, Carvalho LJCB, Chen S (2014) Comparison of leaf proteomes of cassava (*Manihot esculenta* Crantz) cultivar NZ199 diploid and autotetraploid genotypes. PLoS One 9(4):e85991

https://pgdbj.jp/plantdb/plantdb.html. Accessed 26 Jun 2021

http://asia.ensembl.org/index.html. Accessed 26 Jun 2021

https://cvalues.science.kew.org/. Accessed 26 Jun 2021

https://www.plantrdnadatabase.com/. Accessed 26 Jun 2021

Asamizu E, Ichihara H, Nakaya A et al (2014) Plant Genome DataBase Japan (PGDBj): a portal website for the integration of plant genome-related databases. Plant Cell Physiol 55(1):e8

Avni R, Nave M, Barad O, Baruch K, Twardziok SO, Gundlach H et al (2017) Wild emmer genome architecture and diversity elucidate wheat evolution and domestication. Science 357(6346): 93–97

Bayer PE, Golicz AA, Scheben A, Batley J, Edwards D (2020) Plant pan-genomes are the new reference. Nat Plants 6:914–920

Bennett RA, Lynch JM (1981) Colonization potential of bacteria in the rhizosphere. Curr Microbiol 6(3):137–138

Berkman PJ, Bundock PC, Casu RE, Henry RJ, Rae AL, Aitken KS (2014) A survey sequence comparison of saccharum genotypes reveals allelic diversity differences. Trop Plant Biol 7:71–83

Berlin K, Koren S, Chin CS, Drake JP, Landolin JM, Phillippy AM (2015) Assembling large genomes with single-molecule sequencing and locality-sensitive hashing. Nat Biotechnol 33(6): 623–630

Bevan M, Walsh S (2005) The *Arabidopsis* genome: a foundation for plant research. Genome Res 15(12):1632–1642

Bhattacharyya PN, Goswami MP, Bhattacharyya LH (2016) Perspective of beneficial microbes in agriculture under changing climatic scenario: a review. J Phytology 8:26–41

Bhattacharyya C, Bakshi U, Mallick I, Mukherji S, Bera B, Ghosh A (2017) Genome-guided insights into the plant growth promotion capabilities of the physiologically versatile Bacillus aryabhattai strain AB211. Front Microbiol 8:411

Bird KA, VanBuren R, Puzey JR, Edger PP (2018) The causes and consequences of subgenome dominance in hybrids and recent polyploids. New Phytol 220(1):87–93

Biselli C, Urso S, Bernardo L, Tondelli A, Tacconi G, Martino V, Grando S, Valè G (2010) Identification and mapping of the leaf stripe resistance gene Rdg1a in Hordeum spontaneum. Theor Appl Genet 120(3):1207–1218

Blake VC, Birkett C, Matthews DE, Hane DL, Bradbury P, Jannink J (2016) The Triticeae toolbox: combining phenotype and genotype data to advance small-grains breeding. Plant Genome 9(2). https://doi.org/10.3835/plantgenome2014.12.0099

Botstein D, White RL, Skolnick M, Davis RW (1980) Construction of a genetic linkage map in man using restriction fragment length polymorphisms. Am J Hum Genet 32:314–331

Bourgeois Y, Stritt C, Walser JC, Gordon SP, Vogel JP, Roulin AC (2018) Genome-wide scans of selection highlight the impact of biotic and abiotic constraints in natural populations of the model grass *Brachypodium distachyon*. Plant J 96(2):438–451

Brozynska M, Omar ES, Furtado A, Crayn D, Simon B, Ishikawa R, Henry RJ (2014) Chloroplast genome of Novel rice germplasm identified in Northern Australia. Trop Plant Biol 7(3-4): 111–120

Camacho DM, Collins KM, Powers RK, Costello JC, Collins JJ (2018) Next-generation machine learning for biological networks. Cell 173(7):1581–1592

Campbell MS, Law MY, Holt C, Stein JC, Moghe GD, Hufnagel DE et al (2014) MAKER-P: a tool kit for the rapid creation management and quality control of plant genome annotations. Plant Physiol 164(2):513–524

Chagas FO, Pessotti RDC, Caraballo-Rodríguez AM, Pupo MT (2018) Chemical signaling involved in plant-microbe interactions. Chem Soc Rev 47(5):1652–1704

Chen DY, Swerdlow HP, Harke HR, Zhang JZ, Dovichi NJ (1991) Low-cost, high-sensitivity laserinduced fluorescence detection for DNA sequencing by capillary gel electrophoresis. J Chromatogr A 559(1-2):237–246

Chen KY, Cong B, Wing R, Vrebalov J, Tanksley SD (2007) Changes in regulation of a transcription factor lead to autogamy in cultivated tomatoes. Science 318:643645

Chen G, Liu Y, Ma J, Zheng Z, Wei Y, McIntyre CL, Zheng YL, Liu C (2013) A Novel and major quantitative trait locus for Fusarium Crown rot resistance in a genotype of Wild Barley (*Hordeum spontaneum* L.). PLoS One 8(3):e58040

Chen F, Dong W, Zhang J, Guo X, Chen J, Wang Z, Lin Z, Tang H, Zhang L (2018) The sequenced angiosperm genomes and genome databases. Front Plant Sci 9:418

Comai L (2005) The advantages and disadvantages of being polyploid. Nat Rev Genet 6(11): 836–846

Compaan B, Ruttink T, Albrecht C, Meeley R, Bisseling T, Franssen H (2003) Identification and characterization of a Zea mays line carrying a transposon-tagged ENOD40. Biochim Biophys Acta 1629:84–91

Cook DE, Valle-Inclan JE, Pajoro A, Rovenich H, Thomma BPHJ, Faino L (2019) Long-read annotation: automated eukaryotic genome annotation based on long-read cDNA sequencing. Plant Physiol 179(1):38–54

Corneillie S, De Storme N, Van Acker R, Fangel JU, De Bruyne M et al (2019) Polyploidy affects plant growth and alters cell wall composition. Plant Physiol 179(1):74–87

Cornille A, Giraud T, Smulders MJM, Roldán-Ruiz I, Gladieux P (2014) The domestication and evolutionary ecology of apples. Tends Genet 30(2):57–65

Das A, Ravinath R, Shilpa BR, Rohith BS, Goyal AK, Ramesh N et al (2021) Microbiomics and cloud-based analytics advance sustainable soil management Frontiers in Bioscience. Landmark 26:478–495

De Storme N, Geelen D (2013) Sexual polyploidization in plants: cytological mechanisms and molecular regulation. New Phytol 198(3):670–684

Deery DM, Rebetzke GJ, Jimenez-Berni JA, James RA, Condon AG et al (2016) Methodology for high-throughput field phenotyping of canopy temperature using airborne thermography. Front Plant Sci 7:1808

Demirci S, Peters SA, de Ridder D, van Dijk ADJ (2018) DNA sequence and shape are predictive for meiotic crossovers throughout the plant kingdom. Plant J. https://doi.org/10.1111/tpj.13979

Demorest ZL, Coffman A, Baltes NJ, Stoddard TJ, Clasen BM et al (2016) Direct stacking of sequence-specific nuclease-induced mutations to produce high oleic and low linolenic soybean oil. BMC Plant Biol 6(1):225

Denoeud F, Carretero-Paulet L, Dereeper A, Droc G, Guyot R et al (2014) The coffee genome provides insight into the convergent evolution of caffeine biosynthesis. Science 345(6201): 1181–1184

Diederichs T, Pugh G, Dorey A, Xing Y, Burns JR, Hung Nguyen Q, Tornow M, Tampé R, Howorka S (2019) Synthetic protein-conductive membrane nanopores built with DNA. Nat Commun 10(1):5018

Ding J, Lu Q, Ouyang Y, Mao H, Zhang P, Yao J, Xu C et al (2012) A long noncoding RNA regulates photoperiod-sensitive male sterilityan essential component of hybrid rice. Proc Natl Acad Sci U S A 109(7):2654–2659

Do H, Than K, Larmande P (2018) Evaluating named-entity recognition approaches in plant molecular biology. In: Kaenampornpan M, Malaka R, Nguyen D, Schwind N (eds) Multi-disciplinary trends in artificial intelligence. MIWAI 2018, Lecture notes in computer science, vol 11248. Springer, Cham. https://doi.org/10.1007/978-3-030-03014-8_19

Dubcovsky J, Dvorak J (2007) Genome plasticity a key factor in the success of polyploid wheat under domestication. Science 316(5833):1862–1866

Duvick J, Fu A, Muppirala U, Shabarwal M, Wilkerson MD, Lawrence CJ, Lushbourg C, Brendel V (2008) PlantGDB: a resource for comparative plant genomics. Nucleic Acids Res 36(Database issue):D959–D965

Edger PP, Smith R, McKain MR, Cooley AM, Vallejo-Marin M, Yuan Y et al (2017) Subgenome dominance in an interspecific hybrid synthetic allopolyploid and a 140-year-old naturally established neo-allopolyploid monkeyflower. Plant Cell 29(9):2150–2167

Eggertsson HP, Jonsson H, Kristmundsdottir S, Hjartarson E, Kehr B, Masson G et al (2017) Graphtyper enables population-scale genotyping using pangenome graphs. Nat Genet 49:1654–1660

Eid J, Fehr A, Gray J, Luong K, Lyle J, Otto G, Peluso P, Rank D, Baybayan P et al (2009) Real-time DNA sequencing from single polymerase molecules. Science 323(5910):133–138

Feldmann KA, Goff SA (2014) The first plant genome sequence—*Arabidopsis thaliana*. Adv Bot Res 69:91–117

Fiorani F, Schurr U (2013) Future scenarios for plant phenotyping. Annu Rev Plant Biol 64:267–291

Fitzpatrick CR, Copeland J, Wang PW, Guttman DS, Kotanen PM, Johnson MTJ (2018) Assembly and ecological function of the root microbiome across angiosperm plant species. Proc Natl Acad Sci U S A 115(6):E1157–E1165

Frary A, Nesbitt TC, Frary A, Grandillo S, Van Der Knaap E, Cong B et al (2000) fw22: a quantitative trait locus key to the evolution of tomato fruit size. Science 289(5476):85–88

Gaebelein R, Schiessl SV, Samans B, Batley J, Mason AS (2019) Inherited allelic variants and novel karyotype changes influence fertility and genome stability in Brassica allohexaploids. New Phytol 223(2):965–978

Gantait S, Debnath S, Nasim Ali M (2014) Genomic profile of the plants with pharmaceutical value. 3 Biotech 4(6):563–578

Gao X, Zhang J, Wei Z, Hakonarson H (2018a) DeepPolyA: a convolutional neural network approach for polyadenylation site prediction. IEEE Access 6:24340–24349

Gao H, Wang Y, Li W, Gu Y, Lai Y, Bi Y, He C (2018b) Transcriptomic comparison reveals genetic variation potentially underlying seed developmental evolution of soybeans. J Exp Bot 69(21):5089–5104

Gao L, Gonda I, Sun H, Ma Q, Bao K, Tieman DM, Burzynski-Chang EA et al (2019) The tomato pan-genome uncovers new genes and a rare allele regulating fruit flavor. Nat Genet 51(6): 1044–1051

Georges F, Ray H (2017) Genome editing of crops: a renewed opportunity for food security. GM Crop Food 8(1):1–12

Ghosal S, Blystone D, Singh AK, Ganapathysubramanian B, Singh A, Sarkar S (2018) An explainable deep machine vision framework for plant stress phenotyping. Proc Natl Acad Sci U S A 115(18):4613–4618

Gnerre S, MacCallum I, Przybylski D, Ribeiro FJ, Burton JN, Walker BJ et al (2011) High-quality draft assemblies of mammalian genomes from massively parallel sequence data. Proc Natl Acad Sci U S A 108(4):1513–1518

Godfray HCJ, Crute IR, Haddad L, Muir JF, Nisbett N, Lawrence D, Pretty J et al (2010) The future of the global food system. Philos Trans R Soc Lond B Biol Sci 365(1554):2941–2957

Golicz AA, Bayer PE, Barker GC, Edger PP, Kim HR, Martinez PA et al (2016) The pangenome of an agronomically important crop plant *Brassica oleracea*. Nat Commun 7:13390

Golicz AA, Bhalla PL, Singh MB (2018) MCRiceRepGP: a framework for the identification of genes associated with sexual reproduction in rice. Plant J 96(1):188–202

Golicz AA, Bayer PE, Bhalla PL, Batley J, Edwards D (2020) Pangenomics comes of age: from bacteria to plant and animal applications. Trends Genet 36(2):132–145

Gordon SP, Contreras-Moreira B, Woods DP, Des Marais DL, Burgess D et al (2017) Extensive gene content variation in the Brachypodium distachyon pan-genome correlates with population structure. Nat Commun 8(1):2184

Grant D, Nelson RT, Cannon SB, Shoemaker RC (2009) SoyBase the USDA-ARS soybean genetics and genomics database. Nucleic Acids Res 38(Database issue):D843–D846

Gultyaev AP, Roussis A (2007) Identification of conserved secondary structures and expansion segments in enod40 RNAs reveals new enod40 homologues in plants. Nucleic Acids Res 35(9): 3144–3152

Haggag WMFAH, Habbasha ES (2015) Agriculture biotechnology for management of multiple biotic and abiotic environmental stress in crops. J Chem Pharma Res 7(10):882–889

Harris TD, Buzby PR, Babcock H, Beer E, Bowers J, Braslavsky I et al (2008) Single-molecule DNA sequencing of a viral genome. Science 320(5872):106–109

Hassani-Pak K, Rawlings C (2017) Knowledge discovery in biological databases for revealing candidate genes linked to complex phenotypes. J Integr 14(1):20160002

Heo JB, Sung S (2011) Vernalization-mediated epigenetic silencing by a long intronic noncoding RNA. Science 331(6013):76–79

Hickey LT, Germán SE, Pereyra SA, Diaz JE, Ziems LA, Fowler RA et al (2017) Speed breeding for multiple disease resistance in barley. Euphytica 213:64

Honsdorf N, March TJ, Berger B, Tester M, Pillen K (2014) High-throughput phenotyping to detect drought tolerance QTL in wild barley introgression lines. PLoS One 9(5):e97047

Hoshino A, Jayakumar V, Nitasaka E, Toyoda A, Noguchi H, Itoh T et al (2016) Genome sequence and analysis of the Japanese morning glory Ipomoea nil. Nat Commun 7:13295

Huang L, Dong H, Zhou D, Li M, Liu Y, Zhang F, Feng Y et al (2018) Systematic identification of long non-coding RNAs during pollen development and fertilization in *Brassica rapa*. Plant J 96(1):203–222

Jansson JK, Hofmockel KS (2018) The soil microbiome — from metagenomics to metaphenomics. Curr Opin Microbiol 43:162–168

Jayakodi M, Padmarasu S, Haberer G, Bonthala VS, Gundlach H et al (2020) The barley pan-genome reveals the hidden legacy of mutation breeding. Nature 588(7837):284–289

Jiang N, Bao Z, Zhang X, Eddy SR, Wessler SR (2004) Pack-MULE transposable elements mediate gene evolution in plants. Nature 431(7008):569–573

Jiao Y, Paterson AH (2014) Polyploidy-associated genome modifications during land plant evolution. Philos Trans R Soc Lond B Biol Sci 369(1648):20130355

Johansson JF, Paul LR, Finlay RD (2004) Microbial interactions in the mycorrhizosphere and their significance for sustainable agriculture. FEMS Microbiol Ecol 48(1):1–13

Kalladan R, Worch S, Rolletschek H, Harshavardhan VT, Kuntze L, Seiler C et al (2013) Identification of quantitative trait loci contributing to yield and seed quality parameters under terminal drought in barley advanced backcross lines. Mol Breeding 32:71–90

Kambara H, Takahashi S (1993) Multiple-sheathflow capillary array DNA analyser. Nature 361: 565–566

Khoury CK, Bjorkman AD, Dempewolf H, Ramirez-Villegas J, Guarino L et al (2014) Increasing homogeneity in global food supplies and the implications for food security. Proc Natl Acad Sci U S A 111(11):4001–4006

Klasberg S, Bitard-Feildel T, Mallet L (2016) Computational identification of novel genes: current and future perspectives. Bioinform Biol Insights 10:121–131

Komatsuda T, Pourkheirandish M, He C, Azhaguvel P, Kanamori K, Perovic D et al (2007) Six-rowed barley originated from a mutation in a homeodomain-leucine zipper I-class homeobox gene. Proc Natl Acad Sci U S A 104(4):1424–1429

Koren S, Schatz MC, Walenz BP, Martin J, Howard JT et al (2012) Hybrid error correction and de novo assembly of single-molecule sequencing reads. Nat Biotechnol 30(7):693–700

Kouchi H, Takane KI, So RB, Ladha JK, Reddy PM (1999) Rice ENOD40: isolation and expression analysis in rice and transgenic soybean root nodules. Plant J 18(2):121–129

Lai K, Berkman PJ, Lorenc MT, Duran C, Smits L, Manoli S et al (2012) WheatGenome info: an integrated database and portal for wheat genome information. Plant Cell Physiol 53(2):e2

Lakew B, Eglinton J, Henry RJ, Baum M, Grando S, Ceccarelli S (2011) The potential contribution of wild barley (*Hordeum vulgare* ssp. *spontaneum*) germplasm to drought tolerance of cultivated barley (*H. vulgare* ssp. *vulgare*). Field Crops Res 120(1):161–168

Li WT, Huang X, Wang JR, Chen GY, Nevo E, Zheng YL, Wei YM (2010) Genetic analysis and ecological association of Hina genes based on single nucleotide polymorphisms (SNPs) in wild barley *Hordeum spontaneum*. Hereditas 147(1):18–26

Li C, Wang A, Ma X, Pourkheirandish M, Sakuma S, Wang N, Ning S, Nevo E, Nawrath C, Komatsuda T, Chen G (2013) Characterization and genetic mapping of *eceriferum-ym* (*cer-ym*), a cutin deficient barley mutant with impaired leaf water retention capacity. Breeding Sci 126(3): 637–646

Li YH, Zhou G, Ma J, Jiang W, Jin LG, Zhang Z, Guo Y et al (2014) De novo assembly of soybean wild relatives for pan-genome analysis of diversity and agronomic traits. Nat Biotechnol 32(10): 1045–1052

Liang Z, Zhang K, Chen K, Gao C (2014) Targeted mutagenesis in Zea mays using TALENs and the CRISPR/Cas system. J Genet Genomics 41(2):63–68

Lindow SE, Brandl MT (2003) Microbiology of the phyllosphere. Appl Environ Microbiol 69(4): 1875–1883

Long Y, Wang X, Youmans DT, Cech TR (2017) How do lncRNAs regulate transcription? Sci Adv 3(9):eaao2110

Lu H, Giordano F, Ning Z (2016) Oxford nanopore MinION sequencing and genome assembly. Genomics Proteomics Bioinform 14(5):265–279

Mansueto L, Fuentes RR, Borja FN, Detras J, Abrio-Santos JM, Chebotarov D et al (2017) Rice SNP-seek database update: new SNPs indels and queries. Nucleic Acids Res 45(Database issue):D1075–D1081

Matteoli FP, Passarelli-Araujo H, Reis RJA, Da Rocha LO et al (2018) Genome sequencing and assessment of plant growth-promoting properties of a Serratia marcescens strain isolated from vermicompost. BMC Genomics 19(1):750

Mejía-Guerra MK, Buckler ES (2019) A k-mer grammar analysis to uncover maize regulatory architecture. BMC Plant Biol 19(1):103

Mendes R, Kruijt M, De Bruijn I, Dekkers E, Van Der Voort M et al (2011) Deciphering the rhizosphere microbiome for disease-suppressive bacteria. Science 332(6033):1097–1100

Menges M, De Jager SM, Gruissem W, Murray JAH (2005) Global analysis of the core cell cycle regulators of Arabidopsis identifies novel genes reveals multiple and highly specific profiles of expression and provides a coherent model for plant cell cycle control. Plant J 41(4):546–566

Michael TP, VanBuren R (2015) Progress challenges and the future of crop genomes. Curr Opin Plant Biol 24:71–81

Mickelbart MV, Hasegawa PM, Bailey-Serres J (2015) Genetic mechanisms of abiotic stress tolerance that translate to crop yield stability. Nat Rev Genet 16(4):237–251

Miglani GS (2017) Genome editing in crop improvement: present scenario and future prospects. J Crop Improv 31:453–459

Milner SG, Jost M, Taketa S, Mazón ER, Himmelbach A, Oppermann M et al (2019) Genebank genomics highlights the diversity of a global barley collection. Nature Genet 51:319–326

Montes JM, Melchinger AE (2016) Domestication and breeding of *Jatropha curcas* L. Trends Plant Sci 21(12):1045–1057

Moose SP, Mumm RH (2008) Molecular plant breeding as the foundation for 21st century crop improvement. Plant Physiol 147(3):969–977

Muñoz-López M, García-Pérez JL (2010) DNA transposons: nature and applications in genomics. Curr Genomics 11(2):115–128

Nakano K, Shiroma A, Shimoji M, Tamotsu H, Ashimine N, Ohki S, Shinzato M, Minami M, Nakanishi T, Teruya K, Satou K, Hirano T (2017) Advantages of genome sequencing by long-read sequencer using SMRT technology in medical area. Hum Cell 30(3):149–161

Ormeño-Orrillo E, Aguilar-Cuba Y, Zúñiga-Dávila D (2018) Draft genome sequence of Rhizobium sophoriradicis H4 a nitrogenfixing bacterium associated with the leguminous plant Phaseolus vulgaris on the coast of Peru. Genome Announc 6(21):e00241–e00218

Pakniyat H, Namayandeh A (2007) Salt tolerance associations with RAPD markers in Hordeum vulgare L and H spontaneum C Koch. Pak J Biol Sci. https://doi.org/10.3923/pjbs200713171320

Pareek CS, Smoczynski R, Tretyn A (2011) Sequencing technologies and genome sequencing. J Appl Genet 52(4):413–435

Park SJ, Jiang K, Schatz MC, Lippman ZB (2012) Rate of meristem maturation determines inflorescence architecture in tomato. Proc Natl Acad Sci U S A 109(2):639–644

Parra G, Bradnam K, Korf I (2007) CEGMA: a pipeline to accurately annotate core genes in eukaryotic genomes. Bioinformatics 23(9):1061–1067

Pellicer J, Leitch IJ (2019) The Plant DNA C-values database (release 71): an updated online repository of plant genome size data for comparative studies. New Phytol 226(2):301–305

Poiré R, Chochois V, Sirault XRR, Vogel JP, Watt M, Furbank RT (2014) Digital imaging approaches for phenotyping whole plant nitrogen and phosphorus response in *Brachypodium distachyon*. J Integr Plant Biol 56(8):781–796

Pourkheirandish M, Hensel G, Kilian B, Senthil N, Chen G, Sameri M et al (2015) Evolution of the grain dispersal system in Barley. Cell 162(3):527–539

Pourkheirandish M, Dai F, Sakuma S, Kanamori H, Distelfeld A et al (2018) On the origin of the non-brittle rachis trait of domesticated einkorn wheat. Front Plant Sci 8:2031

Pourkheirandish M, Golicz AA, Bhalla PL, Singh MB (2020) Global role of crop genomics in the face of climate change. Front Plant Sci 11:922

Prasannakumar MK, Netravathi LM, Mahesh HB et al (2021) Comparative metagenomic analysis of rice soil samples revealed the diverse microbial population and biocontrol organisms against plant pathogenic fungus *Magnaporthe oryzae*. 3 Biotech 11:245. https://doi.org/10.1007/s13205-021-02783-y

Pray L (2008) Discovery of DNA structure and function: Watson and Crick. Nat Educ 1(1):100

Pusztahelyi T, Holb IJ, Pócsi I (2015) Secondary metabolites in fungus-plant interactions. Front Plant Sci 6:573

Ramsey J, Schemske DW (1998) Pathways mechanisms and rates of polyploid formation in flowering plants. Annu Rev Ecol Systemat 29:467–501

Ray DK, Mueller ND, West PC, Foley JA (2013) Yield trends are insufficient to double global crop production by 2050. PLoS One 8(6):e66428

Roberts L (1987) New sequencers to take on the genome. Science 238(4825):271–273

Ronaghi M, Uhlén M, Nyrén PA (1998) Sequencing method based on real-time pyrophosphate. Science 281(5375):363–365

Salman-Minkov A, Sabath N, Mayrose I (2016) Whole-genome duplication as a key factor in crop domestication. Nat Plants 2:16115

Sanger F, Nicklen S, Coulson AR (1977) DNA sequencing with chain-terminating inhibitors. Proc Natl Acad Sci USA 74(12):5463–5467

Satterthwaite D, McGranahan G, Tacoli C (2010) Urbanization and its implications for food and farming. Philos Trans R Soc Lond B Biol Sci 365:2809–2820

Scheben A, Edwards D (2018) Bottlenecks for genome-edited crops on the road from lab to farm. Genome Biol 19:178

Schellenberg MJ, Ritchie DB, MacMillan AM (2008) Pre-mRNA splicing: a complex picture in higher definition. Trends Biochem Sci 33(6):243–246

Schiml S, Puchta H (2016) Revolutionizing plant biology: multiple ways of genome engineering by CRISPR/Cas. Plant Methods 12:8

Schmalenbach I, Körber N, Pillen K (2008) Selecting a set of wild barley introgression lines and verification of QTL effects for resistance to powdery mildew and leaf rust. Theor Appl Genet 117(7):1093–1106

Schrider DR, Kern AD (2018) Supervised machine learning for population genetics: a new paradigm. Trends Genet 34(4):301–312

Shariati VJ, Malboobi MA, Tabrizi Z, Tavakol E, Owilia P, Safari M (2017) Comprehensive genomic analysis of a plant growth-promoting rhizobacterium Pantoea agglomerans strain P5. Sci Rep 7(1):15610

Shavrukov Y, Gupta NK, Miyazaki J, Baho MN, Chalmers KJ, Tester M, Langridge P, Collins NC (2010) HvNax3: a locus controlling shoot sodium exclusion derived from wild barley (Hordeum vulgare ssp. spontaneum). Funct Integr Genomics 10(2):277–291

Smith KP, Goodman RM (1999) Host variation for interactions with beneficial plant-associated microbes. Ann Rev Phytopathol 37:473–491

Sohn JI, Nam JW (2018) The present and future of de novo whole-genome assembly. Brief Bioinform 19(1):23–40

Song JM, Guan Z, Hu J, Guo C, Yang Z, Wang S et al (2020) Eight high-quality genomes reveal pan-genome architecture and ecotype differentiation of Brassica napus. Nat Plants 6(1):34–45

Sosso D, Luo D, Li QB, Sasse J, Yang J, Gendrot G, Suzuki M et al (2015) Seed filling in domesticated maize and rice depends on SWEET-mediated hexose transport. Nat Genet 47(12): 1489–1493

Souza GM, Berges H, Bocs S, Casu R, D'Hont A, Ferreira JE, Henry R, Ming R et al (2011) The sugarcane genome challenge: strategies for sequencing a highly complex genome. Tropical Plant Biol 4:145–156

Spannagl M, Noubibou O, Haase D, Yang L, Gundlach H, Hindemitt T, Klee K, Haberer G, Schoof H, Mayer KF (2007a) MIPSPlantsDB—plant database resource for integrative and comparative plant genome research. Nucleic Acids Res 35(Suppl. 1):D834–D840

Spannagl M, Noubibou O, Haase D, Yang L, Gundlach H, Hindemitt T et al (2007b) MIPSPlantsDB—plant database resource for integrative and comparative plant genome research. Nucleic Acids Res 35(Database issue):D834–D840

Staňková H, Hastie AR, Chan S, Vrána J, Tulpová Z, Kubaláková M et al (2016) BioNano genome mapping of individual chromosomes supports physical mapping and sequence assembly in complex plant genomes. Plant Biotechnol 14(7):1523–1531

Sun C, Hu Z, Zheng T, Lu K, Zhao Y, Wang W, Shi J, Wang C et al (2017) RPAN: rice pan-genome browser for ~3000 rice genomes. Nucleic Acids Res 45(2):597–605

Sysoeva M, Markovskaya E, Shibaeva T (2010) Plants under continuous light: a review. Plant Stress 4(1):5–17

Szadkowski E, Eber F, Huteau V, Lodé M, Huneau C, Belcram H, Coriton O, Manzanares-Dauleux MJ et al (2010) The first meiosis of resynthesized Brassica napus, a genome blender. New Phytol 186(1):102–112

Tamayo-Ordóñez MC, Espinosa-Barrera LA, Tamayo-Ordóñez YJ, Ayil-Gutiérrez B, Sánchez-Teyer LF (2016) Advances and perspectives in the generation of polyploid plant species. Euphytica 209:1–22

Tank JG, Thaker VS (2011) Cyclin dependent kinases and their role in regulation of plant cell cycle. Biol Plantarum 55:201–212

Tettelin H, Masignani V, Cieslewicz MJ, Donati C, Medini D, Ward NL et al (2005) Genome analysis of multiple pathogenic isolates of *Streptococcus agalactiae*: implications for the microbial "pan-genome". Proc Natl Acad Sci U S A 102(39):13950–13955. https://doi.org/10.1073/pnas.0506758102. Epub 2005 Sep 19. Erratum in: Proc Natl Acad Sci U S A. 2005 Nov 8;102(45):16530

Tilman D, Clark M (2014) Global diets link environmental sustainability and human health. Nature 515:518–522

Tran TT, Choi JW, Le TTH, Kim JW (2019) A comparative study of deep CNN in forecasting and classifying the macronutrient deficiencies on development of tomato plant. Appl Sci (Switzerland) 9(8):1601

Tshikhudo PP, Ntushelo K, Mudau FN, Salehi B, Sharifi-Rad M, Martins N et al (2019) Understanding Camellia sinensis using Omics technologies along with endophytic bacteria and environmental roles on metabolism: a review. Appl Sci (Switzerland) 9(2):281

Tu Y, Jiang A, Gan L, Hossain M, Zhang J, Peng B, Xiong Y et al (2014) Genome duplication improves rice root resistance to salt stress. Rice (N.Y). 7(1):15

Uauy C, Distelfeld A, Fahima T, Blechl A, Dubcovsky J (2006) A NAC gene regulating senescence improves grain protein, zinc, and iron content in wheat. Science 314(5803):1298–1301

Usade B, Tohge T, Scossa F, Sierro N, Schmidt M et al (2018) The genome and metabolome of the tobacco tree, Nicotiana glauca: a potential renewable feedstock for the bioeconomy. In: bioRxiv. https://doi.org/10.1101/351429

Usha T, Kumar HPP, Panda P, Goyal AK, Sukhralia S, Afreen S, Shanmugarajan D, Middha SK (2021) Deciphering the animal genomics using bioinformatics approaches. In: Modal S, Lakhan S (eds) Advances in animal genomics, pp 281–297. ISBN 9780128205952

Van de Peer Y, Fawcett JA, Proost S, Sterck L, Vandepoele K (2009) The flowering world: a tale of duplications. Trends Plant Sci 14(12):680–688

Van de Weyer AL, Monteiro F, Furzer OJ, Nishimura MT, Cevik V et al (2019) A species-wide inventory of NLR genes and alleles in *Arabidopsis thaliana*. Cell 178(5):1260–1272.e14

Van Rhijn P, Fang Y, Galili S, Shaul O, Atzmon N, Wininger S et al (1997) Expression of early nodulin genes in alfalfa mycorrhizae indicates that signal transduction pathways used in forming arbuscular mycorrhizae and Rhizobium-induced nodules may be conserved. Proc Natl Acad Sci U S A 94(10):5467–5472

Vanburen R, Bryant D, Edger PP, Tang H, Burgess D, Challabathula D et al (2015) Single-molecule sequencing of the desiccation-tolerant grass Oropetium thomaeum. Nature 527(7579):508–511

Vitales D, Fernández P, Garnatje T, Garcia S (2019) Progress in the study of genome size evolution in Asteraceae: analysis of the last update. Database 2019:baz098

Walkowiak S, Gao L, Monat C, Haberer G, Kassa MT, Brinton J et al (2020) Multiple wheat genomes reveal global variation in modern breeding. Nature 588(7837):277–283

Wang Z, Liu Y, Li D, Li L, Zhang Q, Wang S, Huang H (2017) Identification of circular rnas in kiwifruit and their species-specific response to bacterial canker pathogen invasion. Front Plant Sci 8:413

Wang J, Yang Y, Jin L, Ling X, Liu T, Chen T et al (2018a) Re-analysis of long non-coding RNAs and prediction of circRNAs reveal their novel roles in susceptible tomato following TYLCV infection. BMC Plant Biol 18(1):104

Wang M, Li W, Fang C, Xu F, Liu Y, Wang Z et al (2018b) Parallel selection on a dormancy gene during domestication of crops from multiple families. Nat Genet 50(10):1435–1441

Wang Y, Luo X, Sun F, Hu J, Zha X, Su W, Yang J (2018c) Overexpressing lncRNA LAIR increases grain yield and regulates neighbouring gene cluster expression in rice. Nat Commun 9(1):3516

Washburn JD, Mejia-Guerra MK, Ramstein G, Kremling KA, Valluru R et al (2019) Evolutionarily informed deep learning methods for predicting relative transcript abundance from DNA sequence. Proc Natl Acad Sci U S A 116(12):5542–5549

Watson A, Ghosh S, Williams MJ, Cuddy WS, Simmonds J, Rey MD et al (2018) Speed breeding is a powerful tool to accelerate crop research and breeding. Nat Plants 4(1):23–29

Wheeler DA, Srinivasan M, Egholm M, Shen Y, Chen L, McGuire A et al (2008) The complete genome of an individual by massively parallel DNA sequencing. Nature 452(7189):872–876

Yan J, Wang F, Qin H, Chen G, Eviatar N, Fahima T, Cheng J (2011) Natural variation in grain selenium concentration of Wild Barley, *Hordeum spontaneum*, populations from Israel. Biol Trace Elem Res 142(3):773–786

Zhang YC, Liao JY, Li ZY, Yu Y, Zhang JP, Li QF, Qu LH, Shu WS, Chen YQ (2014) Genome-wide screening and functional analysis identify a large number of long noncoding RNAs involved in the sexual reproduction of rice. Genome Biol 15(12):512

Zhang J, Zhang X, Tang H, Zhang Q, Hua X, Ma X et al (2018) Allele-defined genome of the autopolyploid sugarcane *Saccharum spontaneum* L. Nat Genet 50(11):1565–1573. https://doi.org/10.1038/s41588-018-0237-2. Epub 2018 Oct 8. Erratum in: Nat Genet 50(12):1754

Zhang LN, Wang DC, Hu Q, Dai XQ, Xie YS, Li Q, Liu HM, Guo JH (2019) Consortium of plant growth-promoting rhizobacteria strains suppresses sweet pepper disease by altering the rhizosphere microbiota. Front Microbiol 10:1668

Zhao Q, Feng Q, Lu H, Li Y, Wang A, Tian Q et al (2018) Pan-genome analysis highlights the extent of genomic variation in cultivated and wild rice. Nat Genet 50(8):1196

Zhou R, Zhu Y, Zhao J, Fang Z, Wang S, Yin J, Chu Z, Ma D (2018) Transcriptome-wide identification and characterization of potato circular RNAs in response to pectobacterium carotovorum subspecies brasiliense infection. Int J Mol Sci 19(1):71

Model Plants in Genomics

9

Soham Ray, Pratik Satya, Laxmi Sharma, Suman Roy, Amit Bera, Sayantan Santra, and Srinjoy Ghosh

Abstract

Basic principles of biology are generally developed, tested, and established for the first time in organisms that are easy to study, convenient to handle and have enough biological merit to generalize the derived inferences. Some of these organisms are given the status of 'model organism' provided they fulfill some basic (intrinsic, derived, and community) criteria. The fundamentals of genetics were established through Mendel's legendary work on garden pea (*Pisum sativum*) but at the early days of plant genetics maize (*Zea mays*) was the model system that got popularity. As plant science entered the genomics era and robust genetic manipulation techniques were established in some plant systems, a paradigm shift took place in the selection criteria of plants that can be promoted as 'model system'. The emergence of *Arabidopsis thaliana* and rice (*Oryza sativa*) as 'model plants' being the most prominent examples in this regard. During the past 40 years, *Arabidopsis* has overtaken all others and got established as the most preferred and frequently used model system in plant biology. Rice, on the other hand, has come up a long way to establish itself as a model monocot and it assumes paramount importance especially in the field of agriculture. However, it is also necessary to understand that a handful of model

S. Ray (✉)
ICAR-Central Research Institute for Jute & Allied Fibres, Kolkata, West Bengal, India

Institute of Applied Microbiology, Collage of Agriculture, Guangxi University, Nanning, Guangxi, People's Republic of China

P. Satya · L. Sharma · S. Roy · A. Bera · S. Ghosh
ICAR-Central Research Institute for Jute & Allied Fibres, Kolkata, West Bengal, India

S. Santra
Institute of Applied Microbiology, Collage of Agriculture, Guangxi University, Nanning, Guangxi, People's Republic of China

R. L. Singh et al. (eds.), *Plant Genomics for Sustainable Agriculture*,
https://doi.org/10.1007/978-981-16-6974-3_9

plants cannot answer every biological question. Hence, there exists the potential of expanding the horizon of 'model plants' by introducing new entries to keep pace with the ever-expanding knowledge and technology. In fact, with the recent introduction of rapid and low-cost whole-genome sequencing methodologies and precise genome editing technologies, the idea of 'model organism' is undergoing a rapid change. Under these circumstances, it seems likely that model plants in the future will be chosen based on their biological relevance rather than the operational ease and historical pedigree.

Keywords

Classical model plants · Neo-classical model plants · Maize · Arabidopsis · Rice · Evolution and phylogenetics · Genomics and genome sequencing

9.1 Introduction

About 4 billion years ago life arose on the earth (Nelson et al. 2008). Biology is the scientific study of living organisms that encompasses myriads of species extant, extinct, and yet undiscovered. Even though living organisms display enormous variation in their form, size, shape, ecological niche, locomotion, behaviour, mode of nutrition acquisition, reproduction, and all possible aspects of life; but life itself is governed by some simple set of rules which establish the foundation of biology (Nelson et al. 2008). In biological science, model systems has served a very crucial role to understand, establish and conceptualize the basic principles governing life. Biological organisms which are easy to study and convenient to handle are used for the experimental purposes to test and establish basic biological principles. While selecting an experimental organism it is also taken care that the organism should have enough biological merit to generalize the derived inferences. In other words, the selected organism should possess enough biological similarity to a broad group of organisms so that it may serve as a true representative of that group (Flavell 2009). Some of these organisms are given the status of 'model organism' provided they fulfill some basic criteria (Chang et al. 1988). Initially, the species are selected based on some desirable intrinsic criteria (primary criteria) such as small size suited for laboratory use, short lifecycle, high fecundity, ease of maintenance and culture (both in situ and ex situ), ease of genetic manipulation by crossing and mutagenesis, small genome size, etc. Next comes some derived criteria (secondary criteria) such as ease of isolating DNA, RNA, and protein to study basic molecular biology, ease of genetic transformation, regeneration and other standard laboratory operations, etc. Finally, at the later stage of adoption, when a community of user for that particular organism develops, along with it develops the community criteria (tertiary criteria) such as availability of well annotated genome and different databases, availability of reporter genes constructs optimized for the species, availability of genetic stock repositories and mutant strains, etc. In brief it can be said that study of model organism allows rapid accumulation of biologically relevant knowledge base about a broad group of species cheaply by assembling information generated by several scientists working on it over a considerably long period.

Probably the greatest contribution of plant science to modern biology is the establishment of the basic principles of genetics by Gregor Johan Mendel in 1866 through his famous experiments on pea plants (*Pisum sativum*) published as 'Versucheüber Pflanzen-Hybriden (Experiments in Plant Hybridization)' (Mendel 1996). The logic behind using pea plants in his study was the discrete phenotypic variations displayed by the species. In fact, he enumerated seven different traits which displayed contrasting characters in selected parental lines visible by the naked eye (Principles of genetics). But as a model system pea plants did not gain much popularity as Mendel's work was so ahead of his time that it was not recognized by the scientific community immediately and hence community of users did not develop for that particular plant system too. Plants like tomato (*Solanum lycopersicum*), petunia (*Petunia hybrida*), Snapdragon (*Antirrhinum majus*), barley (*Hordeum vulgare*), etc. have also been used as model plants but with limited community success. Maize (*Zea mays*), however, served as a model plant for at least a century and led to several important discoveries including the famous discovery of transposable genetic elements and also on epigenetics (Nannas and Dawe 2015). Maize as a model system gained popularity because of high phenotypic variability, easy artificial crossing, seeds of bold and non-shattering type (every seed is an individual crossover product), annual lifecycle, etc. But with the advent of Sanger's sequencing chemistry-based whole-genome sequencing technology which initiated the era of genomics, maize—as a model plant—gradually lost its *numero uno* status primarily because of its comparatively large and complex genome. From this time onwards, the selection of model plants was largely biased towards possession of small and simple genome. Loss of maize was, however, the gain for *Arabidopsis thaliana*— a miniature cruciferous weed with a rapid life cycle and small genome which gradually gained the status of the premier organism for plant science research. Although, based on its favourable intrinsic properties, Laibach proposed using *Arabidopsis* as model organism as early as in the 1940s (Laibach 1943); but the real shift of importance towards *Arabidopsis* gradually gained momentum between 1980 and 2000 during which a dedicated community of user grew up, a detailed genetic map was constructed, genetic stock centres were established and the whole-genome sequence was deciphered (Provart et al. 2016). Emergence of an enormous multinational community devoted to *Arabidopsis* research within a mere 25–30 years period reflected its worth as a model for plant biology, as well as, its suitability for addressing fundamental questions in biology common to all eukaryotes. Apart from *Arabidopsis*, rice (*Oryza sativa*) also rose to prominence as a plant model in the post-genomic era. The strengths of rice in this regard were, unlike *Arabidopsis*, it was a crop of prime importance (hence, a community of researchers was already existing), it could serve as a monocot model (*Arabidopsis* is a dicot), it possesses a relatively small diploid genome (which was completely sequenced during 2005) and it fulfilled most of the primary (intrinsic) and secondary (derived) criteria which are desirable for selection of a model plant. Another interesting difference between the two was, *Arabidopsis* was a temperate plant while rice predominantly belonged to the tropics. Hence, while the *Arabidopsis* model became more popular in Europe and the USA, the rice model gained prominence in Asia and the USA. Besides, it is also noteworthy that even though experiments in rice are relatively lengthier and costlier, as compared to that in *Arabidopsis*, owing to the lengthier life cycle and relative complex genome, but the generation and accumulation of genomic information and genetic stocks were fast-tracked because of a vast community of researchers were already actively working in rice since long back throughout the world. In effect, *Arabidopsis* and rice complimented each other perfectly representing the two lineages of angiosperm (flowering plants) –composed of ~3,50,000 reported species, which encompass virtually all the economically important crop species (Dong et al. 2009).

The plant kingdom consists of more than 4,00,000 known species which displays enormous diversity in every aspect of life (Chang et al. 1988). The use of model plants provides an amicable solution for the practical problem of handling a less convenient system in research by replacing it with more a convenient alternative. Model species can provide insight into evolutionarily conserved processes but the suitability of the model itself greatly depends upon the evolutionary relatedness between the target species and the model selected for the study. For example, *Arabidopsis*, the most well-studied model plant so far, is an annual dicotyledonous herb of the Brassicaceae family, adapted to temperate climate, having C_3 photosynthesis and very limited symbiotic ability to soil microbes (Chang et al. 1988). Hence, its suitability as a model for research in cereals (which are monocot and several of them have C_4 Photosynthesis) or pulses (which undergoes symbiosis with nitrogen fixers) or fruits (most of which are obtained from perennial trees) might be questionable. As biology entered the molecular era, biological research has greatly been driven by genomics. But, until recently, whole-genome sequencing was a very costly and time-consuming affair. Hence, molecular biology research on non-model plants was mostly chaperoned by the knowledge derived from the studies on model organisms with an additional step of validation to adopt the result into target species. In the past one decade, however, significant additions and improvements in sequencing chemistries and technologies, computational assembly and annotation techniques and development of precise genome editing tool have reduced the cost and time of molecular genetic studies in non-model plants by many folds. This has suddenly unfolded an avenue for expanding the horizon of 'model plants' by introducing new entries which will be evolutionarily more suited to represent a particular group of plants. In fact, the idea of 'model organism' itself is undergoing a rapid evolution keeping pace with recent developments in the field of biological science as a whole. This chapter intends to sketch a pen-picture of this gradual evolution which seems to culminate in blurring the boundaries between model- and non-model species.

9.2 Classical Model Plants in Biology

The story of 'Model Plants' in biology largely revolves around the 'big three'—Maize, *Arabidopsis,* and Rice. The fact can be realized quite easily by carrying out a simple analysis in PubMed (https://pubmed.ncbi.nlm.nih.gov/). Searching the database by putting the botanical names of the classical model plants as search-queries yielded at least >twofold more publications pertaining to the above-mentioned 'big three' compared to any other plants (Fig. 9.1). Hence, we would be discussing here on these three model plants.

9.2.1 Maize (*Zea mays*)

Maize (Fig. 9.2a) is a member of the grass family (Poaceae) and one of the widely cultivated cereal crops all over the world. It is suggested that around 9000-year B.P maize was originated from its ancestor teosinte (*Zea mays* ssp. *parviglumis*) in southwest Mexico. Every biology student since from school age are well familiar with Mendel's experiments with pea plant but hardly very few are aware of the fact that Mendel used maize as one of his model plants to reproduce the hybridization result that he found from pea. Till then maize has been used as a prime plant model and blessed with the attention of several legends of classical genetics and plant breeding. Maize has several advantages apart from being an economically important crop. Physically separated male and female flower (ease of controlled crossing), a large number of seeds obtained from single pollinated ear attached tightly in a cob (useful for calculating segregation ratio), larger chromosome size, and synchronized meiosis (useful for cytological analysis), high genotypic and phenotypic diversity (useful for identifying trait responsible genes), an abundance of duplication events in

Fig. 9.1 Relative contributions of different classical model plants to PubMed in terms of literature pertaining to the respective species as on 20th October, 2020

Fig. 9.2 The 'Big Three' of plant model systems. (**a**) View of a maize field and Maize cobs (shown in inset at bottom-right corner). (**b**) Mouse-ear cress plant and its flower (shown in inset at bottom-right corner). (**c**) Rice plant grown in pot, rice flower (shown in inset at top-left corner) and rice seeds (shown in inset at bottom-right corner)

genome (useful for studying molecular evolution), ease of growing in vitro and in situ, availability of genetic tool and resources, etc. are only a few of them (Strable and Scanlon 2009). During the early days, genetics studies on maize were related to map-location, mutant screening, etc. and centre of all such work was either Cornell University (Emerson School) or Harvard University (East School). People like C. Correns, R.A. Emerson, E.M. East were the founder of this research-area who influenced many notable geneticists including G. Beadle, M. Rhoades, B. McClintock whose works are regarded as each milestone of biology. Progress of plant genetics was almost synonymous with the success achieved in maize in those days. *Drosophila* and maize were two classical models for animal- and plant-biology research, respectively, in those days; and in plant science, maize research overshadowed everything else at that time (Rhoades 1984). Several important understandings of modern-day genetics are derived from research in maize. Here we will mention a few of such important contributions in brief.

After the rediscovery of Mendel's work, De Vries and Correns were focusing on xenia paradox, i.e., colour and morphological characteristics of pollen are expressed in endosperm instead of character of the female parent. Maize was the model organism used in this study. During the golden era of cytology, due to its large chromosome size and shape, maize was an attractive choice as a model. These first-generation maize geneticists developed meiotic mutant, chromosomal number variant, and translocation lines. Most of the mutational studies were related to kernel colour even Barbara McClintock used it to describe the breakage-fusion-bridge cycle and further reporting of transposable elements (McClintock 1941). Kernel pigmentation was very helpful for demonstrating transposons jumping in and out of a segment and TE has been a useful tool for mutant-based gene cloning. Two important classes of transposons—Ac/Ds and Robertson's Mutator have transposon insertion database for reverse genetics analysis (Settles et al. 2007). Maize genome size is comparatively higher (~2.3 Gb) than other plant models, *viz. Arabidopsis* (70 Mb) and Rice (0.4 Gb) but lower than few more economically important grass family members, *viz.* barley (~6 Gb) and wheat (~17 Gb) (Strable and Scanlon 2009). Hundreds of families of transposable elements constitute large sections of the maize genome which are dispersed non-uniformly. As of September, 2020 MaizeGDB (www.maizegdb.org) hosts a total of 45 maize genome sequences. B73 line was sequenced using the minimum tiling path of BAC and fosmid clones in 2009 and the current updated version is Zm-B73-REFERENCE-NAM-5.0. MaizeGDB is a comprehensive platform for researchers which provides genome sequence data and other analytical tools like BLAST, qTeller (RNA-Seq data to compare expression), CornCyc (metabolic pathway network), pedigree viewer, GenomeQC (for genome assemblies and gene structure annotations), PAST (for correlating GWAS result and metabolic pathway), SNPVersity (SNP visualization tool). With the advent of high-throughput sequencing techniques, there have been reports of extensive study on SNP variant, expressional data and gene annotation — NCBI, MaizeGDB portals are actively storing such data and making them available for public use. Taking the advantage of natural genetic diversity of maize McMullen's group developed a nested association mapping (NAM) panel. They crossed 25 inbreed lines to B73 line and captured diverse recombination events. Since then NAM population has been useful to decipher QTLs associated with complex traits like plant height, disease resistance, flowering time, etc.; not only in maize but also in other important crops including rice, wheat, sorghum (Gage et al. 2020). Another powerful technique that is mostly tested in maize is Genome-Wide Association Study (GWAS) which correlates individual marker-variations with trait-variations. There are numerous approaches in maize to use GWAS study to link specific trait including smut resistance, drought tolerance, oil biosynthesis, etc. (Wang D'hont et al. 2012; Li et al. 2013; Wang et al. 2016). Current sophisticated cytological techniques like Fluorescence In Situ Hybridization (FISH) also used maize as model platform for validation and revolutionized the field of cytogenetics (Jiang 2019). Genotype by sequencing (GBS), another high-throughput approach for precise genotyping using SNP marker, is also well-established in maize (Crossa et al. 2013). Considerable success in RNA interference (RNAi)-based gene knockout has been achieved in maize for controlling dwarf mosaic virus (Zhang et al. 2010), Sugarcane Mosaic Virus (Gan et al. 2010), developing aflatoxin-free lines (Thakare et al. 2017), increasing nutritional content (Wu and Messing 2012), etc. Targeted genome editing using the CRISPR-cas9 system has also been demonstrated successfully in maize for improvement of traits like drought tolerance (Shi et al. 2017), generating male sterile line (Chen et al. 2018a, b), imparting semi-dwarf architecture (Zhang et al. 2020). Hence, we see in the pre-genomic era, maize was the most prominent model plant available to the geneticists and in the post-genomic era, and it still continues to be one of the most important one where sophisticated tools, techniques, and databases have been established making it amenable to undertake high-end molecular biology research.

9.2.2 Mouse-Ear Cress (*Arabidopsis thaliana*)

As briefly discussed earlier, *Arabidopsis thaliana* (Fig 9.2b)—a miniature weed of the crucifer family—is currently the most preferred model plant due to its attributes like miniature size (25–35 cm in height), very short life cycle (8–12 weeks), simple laboratory growth requirements, self-fertilizing diploid plants capable of producing thousand offspring, the small genome size (120 Mb) and the ease of genetic manipulation. But things were not as rosy initially as it looks now for *Arabidopsis* researchers. Friedrich Laibach, who described the correct chromosome numbers of the plant as a Ph.D. student from Strasburger's laboratory in Bonn back in 1907, was one of the pioneer researchers lauding *Arabidopsis* as a promising plant model in 1943 (Meyerowitz 2001a, b). X-ray induced mutational studies by Laibach's student Erna Reinholz published during 1947 attracted several other groups to adopt *Arabidopsis* as an experimental model (Somerville and Koornneef 2002; Langridge 1955; Rédei 1962), etc. All these groups working on *Arabidopsis* gradually developed a community and met during XIth Genetics Congress in The Hague in 1963. As a consequence of this meeting, *Arabidopsis* Information Service (AIS) was established which later evolved to The Arabidopsis Information Resource (TAIR) database (Huala et al. 2001). Röbbelen in collaboration with other groups took the initiative to organize First International *Arabidopsis* Symposium in 1965 in Göttingen, Germany, and also established AIS seed-stock collection (which included Laibach ecotypes and several induced mutants). However, despite getting such initial success, research on *Arabidopsis* was viewed with scepticism initially, and hence to managing funds for *Arabidopsis*-research was getting difficult. To give a piece of small evidence in support of this, it would be enough to mention that a grant proposal of Re′dei, a leading *Arabidopsis*-researcher of his time, was rejected by National Science Foundation (NSF) in 1969 as the panel scrutinizing the proposal believed prokaryotes were more suited than *Arabidopsis* for genetic studies (Rédei 1992). Another counterargument for using the *Arabidopsis* model was—it is a weed with no agricultural importance. Hence, several leading groups conducting research on plant science preferred to continue with other models having direct economic importance, *viz.* maize, tomato, barley, tobacco, pea, petunia, and snapdragon. As a result of fund constraints and lack of enthusiasm *Arabidopsis* lagged far behind as a model plant than the established crops during the pre-genomic era. During the mid-1970s *Arabidopsis* research became limited to only a few laboratories, among which groups led by of Jaap van der Veen (studying chimerism in mutants) and Ian Sussex (studying embryo lethal mutants) were noteworthy (Bernier 2013).

From the late-1970s to the 1980s with the progress of classical and molecular genetics plant biologists realized the need for a single model for more collaborative work and hence search for suitable model was reinitiated. Koornneef and Meinke, in their review, entitled mentioned this period (1976–1989) as the 'period of renaissance' for *Arabidopsis* plant biology research (Koornneef and Meinke 2010). Although there was a competition between petunia, tomato, and *Arabidopsis* several influential articles were finally successful to attract promising geneticists, and eventually *Arabidopsis* research gained momentum. One of such influential papers was by György Rédei reviewing success stories of *Arabidopsis* in cytological, mutational auxotroph development and physiological studies (Rédei 1975). The next influential paper was published by Meyerowitz group reporting *Arabidopsis* haploid genome size of only ~70 MB supporting the previous claim by Bernett and Smith, i.e. *Arabidopsis* contains the smallest nuclear DNA among all known higher plants (Bennett and Smith 1976; Leutwiler et al. 1984). At the very beginning of the molecular biology era, researchers were facing difficulties to work with plants having larger genome sizes because it was tedious to clone and maintain millions of clones to cover the whole genome, whereas only a few thousand of λ-phage clones were enough to cover 99% Arabidopsis genome (Pruitt and Meyerowitz 1986). These exciting reports also attracted several molecular biologists from *Drosophila* and yeast biology including Howard Goodman, Ron Davis, and Gerry Fink. With a growing community and a more collaborative atmosphere, RFLP maps, YAC libraries, and other genetic resources were established soon. These things actually accelerated the acceptance of *Arabidopsis* as a model plant in several branches of plant biology (Chang et al. 1988; Hauge et al. 1993; Albertsen et al. 1990). Another breakthrough discovery that provided tremendous impetus to *Arabidopsis* for being appreciated as a model plant was the development of an easy and highly reproducible *Agrobacterium*-mediated transformation protocol without the use of tissue culture techniques (known as floral dip transformation) (Feldmann and Marks 1987; Zhang et al. 2006). And finally, with the release of genome sequence in the December, 2001 by *Arabidopsis* Genome Initiative (AGI)—a multi-institutional collaborative project—not only firmly established *Arabidopsis* as a model plant but made it a forerunner in plant biology research (Koornneef and Meinke 2010).

Arabidopsis on its way to becoming a model plant system also facilitated many outstanding discoveries and been used to understand fundamental processes in cell biology, developmental biology, hormonal regulation, stress biology, gene and protein networks analysis, metabolic engineering, and many more aspects. It has been used to study cellulose biosynthesis which is one of the key processes in plant morphogenesis (Arioli et al. 1998), characterization of endoglucanase in higher plants [29], identification of SPIRAL2-microtubule-associated protein (Shoji et al. 2004), plant-specific cell-cycle regulator (Churchman et al. 2006; Blilou et al. 2002; Dewitte et al. 2003), asymmetric cell division (Hemerly et al. 1993; Dong et al. 2009), vacuolar trafficking network (Sohn et al. 2003). A few classical works on understanding the developmental biology of higher plants are based on *Arabidopsis* research, *viz*. understanding the biology of flower development and its key molecular regulators (Yanofsky et al. 1990), circadian clock (Suárez-López et al. 2001; Alabadí et al. 2001), etc. Synthesis, degradation, mode of action, and genetic network of almost all the important phytohormones have been elucidated during the last century, and not surprisingly in most cases *Arabidopsis* was the model plant (Gazzarrini and McCourt 2003; Potuschak et al. 2003; Dharmasiri et al. 2005). Previously, there was a notion that *Arabidopsis* might not be suitable for studying plant–pathogen interaction because of its weedy nature imparting it high natural resistance. But later in the late 1980s, race-specific susceptibility was found in different *Arabidopsis* ecotypes against several agriculturally important pathogens like *Xanthomonas*, *Pseudomonas*, *Magnaporthe*, etc. Finally, after completing the genome sequence of Col-0 accession, *Arabidopsis* was also used as a model to study plant immune system (Nishimura and Dangl 2010). In a nutshell, the advancement of plant molecular biology almost walked hand-in-hand with the success story of *Arabidopsis* as a model plant for quite a significant time.

9.2.3 Rice (*Oryza sativa*)

Rice (Fig. 9.2c) is the most important staple food for around 4 billion people worldwide (according to IRRI 2019 annual report) and 90% of the global rice cultivation is in Asia. In the pre-genomic era, rice research was solely focused on to enhancement of its agronomic traits by classical breeding approaches, while examples of considering rice as a model system to answer the questions pertaining to basic plant biology were scanty. Even though *Arabidopsis* was a grand success as a model plant in the early days of the post-genomic era, researchers felt the need for some monocot model to answer specific questions related to that particular lineage or adopt some of the finding derived from *Arabidopsis* research to monocot lineage. And nothing was better than a member of the grass family (Poaceae). Why? The answer is simple—just three of their family members, rice, wheat, and maize, contribute to approximately 89% of the total global food production and hence are crucial for global food security (Saldivar 2016). Besides, because of its early domestication and pan-continent distribution, rice has accumulated enormous genetic diversity. In order to appreciate the fact, let us have a quick glance through the number of collections available in different genetic repositories. International Rice Genebank at International Rice Research Institute, Manila, Philippines archives more than 1,32,000 accession of rice which include cultivated, wild, and close taxonomic relatives of rice. AfricaRice is a pan-African centre for rice research and currently, 28 African nations contribute to this project. The genetic resource unit of this research station, AfricaRice Genebank Information System, also maintains dataset of ~19,000 accessions. Dale Bumpers National Rice Research Center which is part of USDA also maintains a large number of rice accessions. In our country, National Bureau of Plant Genetic Resources, New Delhi and National Gene Bank of ICAR-National Rice Research Institute possess more than 1,00,000 rice accessions each which include cultivated species as well as wild relatives.

During the fag end of last century, Izawa and Shimamoto in their publication at PNAS, first proposed rice as an ideal monocot model for molecular biology research based on four specific aspects—(i) availability of efficient transformation protocol, (ii) availability of large-scale expressed sequence tags (ESTs), (iii) highly saturate molecular map, and (iv) availability of diverse genetic stocks and resource (Izawa and Shimamoto 1996). Although maize was already available as an established monocot model, but its large genome size and relative less-responsiveness to transformation (during that time) hindered its use in upcoming molecular biology research. Here, rice fitted perfectly owing to its small genome size (compared to other grass family members) and ease of genetic transform, and it provided the required impetus for undertaking the rice genome sequencing project (Jackson 2016). The International Rice Genome Sequence Project (IRGSP) was in a true sense a multinational effort involving 11 nations initiated in 1997. Several private ventures also came forward with the independent efforts to sequence the rice genome and a private company, Monsanto, first released the draft genome using clone-by-clone shotgun approach. Other two private ventures, Syngenta (*japonica* cultivar Nipponbare) and Beijing Genomics Institute (*indica* cultivar 93-11), also took an important role in genome sequencing and their sequencing results were later incorporated in IRGSP (Vij et al. 2006).

As expected after the release of the draft sequence in 2005, research on several aspects, *viz.* uncovering the genes controlling yield contributing factors, disease resistance, domestication traits, etc. multiplied by several times. As a consequence, several databases have been developed and are now available which are dedicated towards archiving and disseminating genomic information on rice. Apart from high-throughput sequencing-based genomics and transcriptomics, proteomics and metabolomics techniques have also been standardized for rice, which has facilitated gaining a comprehensive understanding of the system as a whole. To give a brief account on a few important databases, the first one would be The Rice Annotation Project Database (RAP-db) which was conceptualized by the International Rice Genome Sequencing Project back in 2004 with a goal of providing accurate annotation of the rice genome to facilitate a comprehensive analysis of the genome structure and function (Sakai et al. 2013). With a similar objective, Michigan State University provides Rice Genome Annotation Project database (Kawahara et al. 2013). OryzaBase (Kurata and Yamazaki 2006) and RGKbase (Wang et al. 2012a, b) are two databases that provide genetic and genomic information on rice and its wild relatives. RicyerDB (Jiang et al. 2018) is an important biological database that collect information regarding yield-related traits. Rice TOGO Browser (Nagamura et al. 2011) provides an interactive platform for retrieving functional information of the rice genome. Another comprehensive database named Gramene (Ware et al. 2002), meant for genomics of grasses, has a large collection of genomic information on rice.

Let us briefly revisit some important biological aspects which required a model other than *Arabidopsis* and rice complemented perfectly to fulfill the needs. Rice contributed a lot in understanding the plant–microbe interactions. Due to the limited interaction of *Arabidopsis* with agriculturally important pathogens, rice became a potential model to understand complex plant–pathogen interaction. Cereals, as well as, other monocots possess a completely different root architecture (compared to the taproot system of eudicots like *Arabidopsis*). The dense fibrous root architecture of cereals has been in research focus because of their high efficiency in nutrient and water acquisition from the upper soil layer and their ability to bind the surface-soil together. Identification of several genes (like ARL1/CRL1), hormonal cross-talks, and QTLs regulating the differentiation and development of roots have been elucidated based on the rice model (Coudert et al. 2010). Cereal endosperm is a sophisticated storage system where photosynthetic products are converted in a stable form and stored. Understanding this process has also been an attractive research area over time because modulation of this process can reflect in grain quality. In the post-genomic era, rice has almost been an exclusive model to study endosperm development and starch biosynthesis (Nakamura 2018). Most of our understandings regarding phytohormones and microRNAs (miRNAs), which govern the overall biology of the plant, are based on the model plant *Arabidopsis*. But their validation in other crop species, especially in phylogenetically distant monocot systems, required a convenient model for study. Empowered by well-standardized molecular biology tools rice, in this case, provided an ideal platform to address those needs. Several studies in rice demonstrated and revalidated conserved networks of phytohormones and miRNAs between rice and *Arabidopsis* suggesting evolutionary conservation in basic biological aspects across plant kingdom (Narsai et al. 2010; Wang et al. 2017). Among the recent developments, CRISPR/cas9 mediated genome editing has been by far the most standardized and is being extensively used in rice among crop plants. Since the first report of editing the phytoene desaturase gene (OsPDS), this technique has been used in rice for editing several attributes like disease resistance (Os8N3 editing for bacterial leaf blight tolerance by Kim et al. (2019)), drought tolerance (OsSAPK2 editing by Lou et al. 2017), salinity tolerance (editing promoter of OsRAV2 by Duan et al. (2016)), herbicide tolerance (EPSPS gene editing by Yu et al. 2015), etc. Interested readers can see Romero and Gatica-Arias, 2019 to get an update on recent developments regarding genome editing and its applications in rice breeding. The availability of high-quality genetic- and physical- maps has made rice an excellent model for studying chromosome biology and genome evolution (Yan and Jiang 2007). In fact, rice along with its large number wild relatives, make the genus *Oryza* an ideal system to study genome evolution. National Science Foundation (NSF), United States funded 'The International *Oryza* Map Alignment Project (I-OMAP)' is being involved in sequencing genomes of all the species belonging to the *Oryza* genus (Jacquemin et al. 2013) to unlock new evolutionary secrets of the genus which will also probably identify the genetic bottlenecks imposed during domestication. Keeping in mind that *Arabidopsis* being a wild species cannot be used as a model to explain molecular genetics of domestication and issues related to it, rice probably stands as the most potent model to facilitate such studies. And now with the availability of precise genome editing techniques, some new opportunities might unfold to fix some of the inherent errors of domestication (domestication bottlenecks), not only in rice but in all the crop species in general. Hence, in terms of selecting models for higher plants rice, being a monocot, complemented well with eudicot *Arabidopsis* to cover the angiosperm clad which harbours most of the economically important plant species.

9.3 Redefining 'Model System' Concept

'Nothing makes sense except in the light of evolution' (Dobzhansky 2013). The quest of deciphering unifying principles that establish the basic foundation of life is also guided and benefited by understanding evolutionary history ached in the diverse gene pool constituting the biosphere. The traditional model plants (or any other eukaryotic model organism *per se*), however, mostly belong to recently diverged groups often referred to as 'crown eukaryotes', provide a glimpse of only the tip of the evolutionary iceberg (Goldstein and King 2016). Let us consider a situation where a gene in target species 'X' and its homolog in the model organism 'Y' govern the same function. What is the probability that the same mutation (knockout/ overexpression/ ectopic expression) in that gene and in its homolog will create similar phenotypic anomalies in the target species and in the selected model? This is a question of paramount importance for selecting a proper model organism, as the higher the probability better the model can explain gene–trait associations in the target species. Reasonable conservation between the pathways (downstream to the mutated gene in this case) responsible for trait expression, between 'X' and 'Y', is a necessary prerequisite for this to happen. So, 'Y' can serve as a befitting model organism for 'X' only if 'X' and 'Y' share significant homology (an evolutionary statement meaning '*related by descent*') between them. Ambiguities might arise at any point in time regarding selection of a proper model species as evolutionary divergence limits the relevance of substituting one species with other. But by far the best way to select a suitable model is to take a cue from the phylogenetic tree (Flavell 2009; Chang et al. 1988). Phylogenetically more closely related species are likely to serve as better models for each other and hence there should ideally be one model species identified for each key node of plant phylogeny. This understanding is not new but earlier the scientific community had technological limitations to make it happen. Fortunately, however, recent additions of high-throughput and low-cost DNA sequencing technologies, high-throughput proteome and metabolome characterization and cataloging technologies, high-end computing, and CRISPR-based precise genome editing and visualization technologies to the genomic toolkit have rationalized this dream. On the one hand, the high-throughput technologies coupled with high-end computing have made 'omics' easier, faster, and cheaper than ever before, while, on the other hand, the advent of CRISPR-based precise genome editing and visualization technologies have completely revolutionized the field of reverse genetics making gene-to-trait association easier than ever. Together these have the potential to quickly develop a near-comprehensive biological sequence database and considerably large mutant collection in a short time thus setting up the foundation for transforming any biological organism in an attractive experimental system to address a biological question (Rine 2013; Goldstein and King 2016; Borrill 2019).

9.3.1 Phylogenetics, Comparative Genomics, and Model Plant Selection

Analysing equivalent traits in several species belonging to multiple phylogenetic groups indicates what are the features (or genetic networks and pathways in terms of genomics) that are conserved, *i.e.*, which arose before the divergence of two groups under question and what are of relatively recent origin which potentially would have guided their divergence from the last common ancestor. Such studies encompassing model plants representing every important phylogenetic group of the plant kingdom will thus be able to answer some key questions regarding plant evolution as a whole and will also potentially guide crop improvement which is the pinnacle of agriculture. The plant kingdom is constituted by reported species of ~4000 chlorophytes, ~865 charophytes, ~25,100 bryophytes, ~1340 lycophytes, ~12,400 pteridophytes, ~766 gymnosperms, and ~350,000 angiosperms (Dong et al. 2009). Among these, angiosperms—the flowering plants—are of the most recent origin constituting the evolutionary history of about 167–199 million years and are most diverse containing almost all the economically important plant species (Dong et al. 2009). Angiosperms are further classified into eight extant clades, namely amborellales (represented by a single species), nymphaeales (~82 species), austrobaileyales (~100 species), monocots (~70,000), magnoliids (~9000), ceratophyllales (~6 species), chloranthales (~75 species), and eudicots (~2,62,000 species) with several sub-clades in monocots and eudicots (Zheng et al. 2014) (Fig. 9.3). More than a decade ago, Hubble (2005) in his editorial entitled "*So much more to know*" published in *Science* Journal raised some critical questions related to plant biology encompassing the aspects of plant development, floral evolution, genome size diversity, the relevance of 'junk' DNA, etc.; most of which are still unanswered. A few other longstanding questions in plant biology are regarding the evolution of differential -photosynthetic mechanism, –ecological adaptation, –mode of nutrient

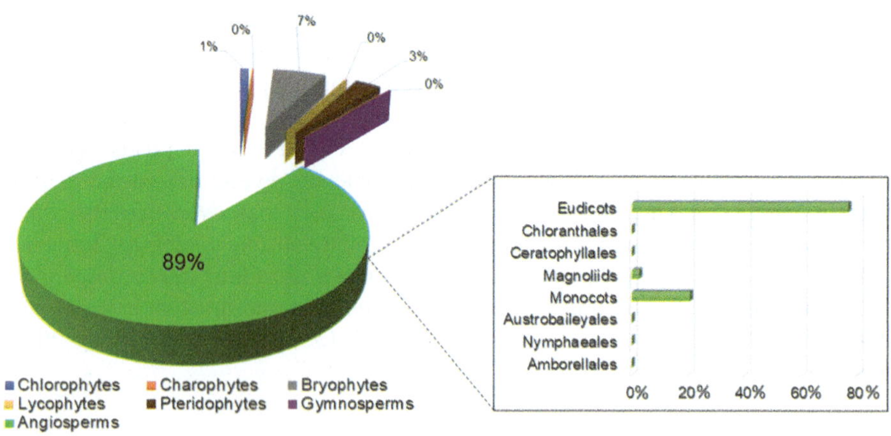

Fig. 9.3 Contribution of different groups (in terms of percentage) in constitution of the plant kingdom

acquisition, which are gradually gaining more importance under global climate change scenario. Plant phylogenetics-guided comparative genomics is expected to deliver some key answers to some of these outstanding questions. To this end inclusion of a model plant as a representative of every important taxonomic group will render relevant information not only about that particular group but regarding evolution in the plant kingdom as a whole. Efforts in this direction have already been initiated regarding which we will briefly discuss in the subsequent sections.

9.3.2 Model Species for Lower Plants

In order to study the very base of the plant kingdom, algal genome sequencing was undertaken; initially with the established models for unicellular chlorophytic algae *Chlamydomonas reinhardtii* (Merchant et al. 2007) and multicellular chlorophytic algae *Volvox carteri* (Prochnik et al. 2010) for gaining insight regarding the transition from unicellular to multicellular life. But latter another chlorophytic marine algae, *Ostreococcus tauri* having a single uninucleate cell of a mere 0.8 μm diameter containing a single mitochondrion, a single chloroplast, a single Golgi body and only 1250 ribosome (smallest free-living eukaryote) has ignited the interest of the scientific community as a 'minimalist model' for eukaryote life (Krumholz et al. 2012). Although monophyletic chlorophytes have traditionally been the more studied group in algal lineage, the other broad group represented by polyphyletic charophytes are evolutionary much closer to land plants and thus can potentially enable deeper understanding regarding terrestrial adaptation. Filamentous alga *Klebsormidium flaccidum*, belonging to one of the basal lineages of charophytes, has been proposed as a model to study this group. The reference genome sequence of *K. flaccidum* (104 Mb) has already been published (Hori et al. 2014), while efforts for sequencing genomes representing other charophyte lineages are already underway (Chang et al. 1988). Together these can provide significant insight regarding the evolution of very basic pathways common to all terrestrial plants like phytohormone regulation (Holzinger et al. 2015; Ju et al. 2015), biochemistry and physiology of plant cell wall (Domozych 2014), perception and regulation of stress responses (Holzinger et al. 2015), etc. Typical features of land plants, cumulatively classified as embryophytes, include three-dimensional body growth, complex multicellular body, and alternation of generation. Among the seven extant lineages of embryophytes, the lower three (liverwort, mosses, and hornwort) exhibit haplobiontic life cycle with the absence of vascular system and are classified together as bryophytes; while the higher four lineages (lycophytes, ferns, gymnosperms, and angiosperms) majorly exhibit diplobiontic life cycle (except some of the lycophytes and ferns) with well-developed vasculature and are grouped together as pteridophytes. *Physcomitrella patens* (moss) has been the most established model among bryophytes historically, but *Marchantia polymorpha* (liverwort) (Bowman 2016) and *Anthoceros agrestis* (hornwort) (Szövényi et al. 2015) are also coming up quickly as models to supplement useful information regarding this group which might not be accessible otherwise only by using

P. patens model. Bryophytes, as a whole, can provide some interesting evolutionary cues regarding alternation of generation, shift from two-dimensional to three-dimensional body growth, formation of reproductive organs in plants, etc. Among the vascular plants (pteridophytes), lycophytes and ferns constitute the basal lineage. They share some interesting similarities like both are non-seed forming and both of the clades contain free-living haplobiontic, as well as, diplobiontic members. But the biggest difference between the two is that, while lycophytes contain relatively small genomes, ferns are known to contain some of the largest genomes in the whole plant kingdom. *Selaginella moellendorffii* (~106 Mb genome) (Banks et al. 2011) and *Ceratopteris richardii* (~11.26 Gb genome) (Atallah and Banks 2015) are in the process to be adopted as models for lycophytes and ferns, respectively. While the former might provide insight regarding the evolution of the vascular system, the latter is supposed to shed light on the logic and mechanism of genome size explosion, in the plant kingdom.

9.3.3 Model Species for Higher Plants

The rest of the two clades in the plant kingdom, the higher (seed forming) pteridophytes—gymnosperms and angiosperm—demands exclusive attention; not only because these two contain almost all the economically important plant species, but also because these two cumulatively constitute lion-share of the diversity in the plant kingdom. Between these two, gymnosperm research traditionally lagged far behind, grossly because of their gigantic stature, unusually long lifecycle, and enormous genomes. Nevertheless, three of the coniferous species, namely, Norway spruce (*Picea abies*), white spruce (*Picea glauca*), and loblolly pine (*Pinus taeda*)—all having genome size ~20 GB, have recently been sequenced (Nystedt et al. 2013; Birol et al. 2013; Zimin et al. 2014) and are being used as a model to study wood (softwood) development, thus are valuable for timer research. Coming to the angiosperms, the classical models (*Arabidopsis*, rice, and maize) have already been discussed in detail in the previous section. Here, to put things in perspective, we will briefly mention the neo-classical models which are quickly coming up with recent technological advancements. In angiosperm lineage, the first major attempt, to establish a model which can link monocots and dicots in evolutionary terms, was probably undertaken by The Joint Genome Institute of US Department of Energy by the end of last decade when they decided to sequence *Aquilegia formosa*, belonging to the basal-most eudicot clade ranunculales, evolutionarily almost equidistant from the by far most established model plants *Arabidopsis* and rice (Flavell 2009). From the same perspective, the basal-most angiosperm, *Amborella trichopoda*, which also happens to be the lone reported extant species belonging to the amborellales, was also sequenced (*Amborella* Genome Project 2013). Within monocots, we already know that the grasses (Poales) are of utmost importance from human use perspective as it contains the major three—rice, wheat, and maize—constituting the bulk of the staple human diet globally. Among these, maize and rice were already well-established model plants, while wheat (*Triticum aestivum*) is also catching up quickly in recent times (Borrill 2019). Additional non-crop grasses like *Brachypodium distachyon* (~272 Mb genome; model for C_3 photosynthesis, plant–pathogen interaction, etc.) (Brutnell et al. 2010; Parker et al. 2008), *Setaria viridis* (~395 Mb genome; model for C_4 photosynthesis) (Brutnell et al. 2010), *Oropetium thomaeum* (~245 Mb; model for drought tolerance) (Van Buren et al. 2015), *Oryza coarctata* (~680 Mb genome; model for salinity tolerance) (Mondal et al. 2018), etc. are recently coming up as prospective model systems. Apart from grasses genomes of oil palms (*Elaeis guineensis* and *E. oleifera*, ~1.8 Gb) (Singh et al. 2013), date palm (*Phoenix dactylifera*, ~658 Mb), banana (*Musa acuminata*; ~523 Mb) (D'hont et al. 2012), and pineapple (*Ananas comosus*; ~526 Mb) (Ming et al. 2015) are now being sequenced within monocot lineage because of their economic importance and can serve as models for the respective groups/species. Other plants worth mentioning among the monocots are *Sorghum bicolor*, *Miscanthus sinensis*, *Panicum virgatum* (as terrestrial biofuel plant), and duckweeds, *viz. Lemna minor* and *Spirodela polyrhiza* (as aquatic biofuel plant) as upcoming models for research in the field of green energy (Chang et al. 1988). In dicots again, there are several new models which are coming up rapidly. Among these, apart from the already established traditional models like *Arabidopsis*, Petunia, snapdragon, tomato, etc., *Populus trichocarpa* (model for hardwood tree), *Lactuca sativa* (model for flower development), *Prunus persica* and *Fragaria vesca* (model for fruit development), *Medicago truncatula* and *Lotus japonicus* (model for biological nitrogen fixation by symbiotic association), *Linum usitatissimum* and *Gossypium hirsutum* (fibre development), etc. are gaining popularity as model systems (Chang et al. 1988). Recently, we have proposed jute (*Corchorus capsularis and C. olitorius*), a fast-growing annual bast fibre producing crop, as a model for studying hypocotyl development. Table 9.1 provides links for some important plant genome databases.

With time, publicly available plant genome databases are getting enriched both qualitatively and quantitatively at a fast pace. Let us take a look at *Genome Database for Angiosperm* (GDA) for example. It contained 236 completely sequenced angiosperm genomes representing 31 of the 64 extant angiosperm orders when last updated on 31st August 2017 (Chen et al. 2018a, b). Many of these sequenced species are either established- or upcoming- model plants. Availability of models cutting across almost all the important nodes of plant phylogeny has actually put 'Model Plant Concept' itself under rapid evolution. Traditionally being viewed as a simplified, easily traceable, and widely adopted system used to answer questions related to a larger biological theme; 'Model Plants', at present, rather means an inherently convenient system best-suited to answer a specific biological question.

Table 9.1 link for accessing genome database of some important plant species

Botanical name	Order	Type	URL
Arabidopsis thaliana	Brassicales	Wild	www.arabidopsis.org/index.jsp
Brassica oleracea	Brassicales	Vegetable	plants.ensembl.org
Camellia sinensis	Ericales	Drink	www.plantkingdomgdb.com
Lotus japonicus	Fabales	Economic	chibba.agtec.uga.edu/duplication/
Medicago truncatula	Fabales	Vegetable	www.medicagogenome.org/
Glycine max	Fabales	Vegetable, oil	plants.ensembl.org
Cajanus cajan	Fabales	Vegetable	chibba.agtec.uga.edu/duplication/
Antirrhinum majus	Lamiales	Ornamental, Economic	genomevolution.org
Populus trichocarpa	Malpighiales	Economic	phytozome.jgi.doe.gov
Linum usitatissimum	Malpighiales	Economic, Fibre	phytozome.jgi.doe.gov
Theobroma cacao	Malvales	Drink	www.cacaogenomedb.org/
Gossypium hirsutum	Malvales	Fibre	cgp.genomics.org.cn
Corchorus olitorius	Malvales	Fibre	www.ncbi.nlm.nih.gov/genome/46639
Corchorus capsularis	Malvales	Fibre	www.ncbi.nlm.nih.gov/genome/46591
Oryza sativa	Poales	Food	rice.plantbiology.msu.edu/
Zea mays	Poales	Food	plants.ensembl.org
Sorghum bicolor	Poales	Food	gramene.org/
Brachypodium stacei	Poales	wild	genome.jgi.doe.gov
Triticum aestivum	Poales	Food	phytozome.jgi.doe.gov
Citrus sinensis	Sapindales	Fruit	phytozome.jgi.doe.gov
Solanum lycopersicum	Solanales	Vegetable	phytozome.jgi.doe.gov
Solanum tuberosum	Solanales	Vegetable	phytozome.jgi.doe.gov
Petunia inflata	Solanales	Ornamental	genomevolution.org
Nicotiana benthamiana	Solanales	Economic	genomevolution.org
Vitis vinifera	Vitales	Fruit	plants.ensembl.org

9.4 Concluding Remarks

Up till now, we have tried to enumerate and illustrate, our readers, the logic behind selecting 'Model Organisms' for plant biology research; *vis-à-vis*, also made an effort to provide a brief account of changes in preferences during the selection of model plants over time. In doing so we have culminated our discussion so far by suggesting evolution in the 'Model Plant Concept' itself that is taking place in recent times. At this point, let us borrow an immortal line from A. G. Gardiner's famous essay on letter-writing in the book 'Pebbles on The Shore', 'If diamonds were as plentiful as pebbles, we shouldn't stoop to pick them up'. In the same notion let us share a very tricky question with our readers—if model plants become so plentiful, will it actually diminish the importance of model plants? We are inclined towards a negative answer to this question, at least for now and for the near future. We present our argument for such conviction. In general, the utility of model plants for testing of hypotheses in plant biology requires no defence as it has been a story of resounding success so far. However, not all inferences drawn from a model plant are directly applicable to all plant systems because of a high degree of system specializations within the plant kingdom. Hence, an in-depth understanding of evolutionary linkages to select a suitable model and a step of adaptive research to assimilate findings derived by using a model plant to the species of interest is the necessary prerequisite. It is true that the recent explosion in the field of 'omics' technologies and advances in reverse genetics tools (making targeted knock-out, knock-in, and editing of genes easier than ever) provides better opportunities to gain insight into genome-structure and -function of almost any plant species. These certainly are significant improvements as the blueprint of species-biology is written in its genome and hence these can confer almost similar standing/opportunity to multiple species to be used in understanding the molecular mechanisms underlying species-biology. But it does not ensure turning every plant species into a model species. If we enumerate the constraints in this regard, the very first one will stem from the primary criteria required for the selection of model plants. That is, some plant species due to their intrinsic properties—like dimensions (*viz.* tree species) and biology (*viz.* certain orchids) unsuitable for *ex situ* culture, unusually long life-span (*viz.* pines), or vegetative phase (*viz.* bamboo), difficulties in genome manipulation by control self- and cross-fertilization (*viz.* ramie), etc.—are not suitable for genetic studies and hence, as model will have limited success. Secondly, the success of many species to be used as model plants is constrained now and will remain as such in near future, due to the challenges in precise phenotyping. As phenomics has not attained the kind of rapid growth similar to genomics to date, the lack of enough phenotypic descriptors in many plant species limits opportunities of conducting genetic studies and, in turn, the utility, of those species as model plants. Finally, validation of gene function by reverse genetic approach still requires insertion of foreign DNA in a single cell (transformation) followed by tissue culture-based regeneration of complete plant from the single transformed cell. Non-responsiveness to tissue culture-based regeneration is a very common problem in many plant species which also limits the usability of those species as a model.

Nevertheless, empowered by the recent advances as illustrated before and also in areas like standardization of transformation and regeneration protocols in new species, *ex situ* culturing in artificial growth media and growth chambers, automated phenotyping, etc., has broadened the scope of introducing new plant species as models—thus blurring the boundaries differentiating model and non-model plants. Hence, we predict, in the future rather than selecting a model based only on its historical heritage to answer a biological question, the choice of the model plant will rather be guided by the biological question itself. We foresee a future, at least half a century down the line, where the model plant concept will still thrive but the number of usable models will significantly increase and researchers will conveniently toggle around different model systems based on the biological question to be answered.

References

Alabadí D, Oyama T, Yanovsky MJ et al (2001) Reciprocal regulation between TOC1 and LHY/CCA1 within the Arabidopsis circadian clock. Science 293(5531):880–883

Albertsen HM, Abderrahim H, Cann HM et al (1990) Construction and characterization of a yeast artificial chromosome library containing seven haploid human genome equivalents. Proc Natl Acad Sci 87(11):4256–4260

Amborella Genome Project (2013) The *Amborella* genome and the evolution of flowering plants. Science 342:1241089

Arioli T, Peng L, Betzner AS et al (1998) Molecular analysis of cellulose biosynthesis in Arabidopsis. Science 279(5351):717–720

Atallah NM, Banks JA (2015) Reproduction and the pheromonal regulation of sex type in fern gametophytes. Front Plant Sci 6:100

Banks JA, Nishiyama T, Hasebe M et al (2011) The Selaginella genome identifies genetic changes associated with the evolution of vascular plants. Science 332(6032):960–963

Bennett MD, Smith JB (1976) Nuclear DNA amounts in angiosperms. Philos Trans Royal Soc London B: Biol Sci 274(933):227–274

Bernier G (2013) My favourite flowering image: the role of cytokinin as a flowering signal. J Exp Bot 64(18):5795–5799

Birol I, Raymond A, Jackman SD et al (2013) Assembling the 20 Gb white spruce (Picea glauca) genome from whole-genome shotgun sequencing data. Bioinformatics 29(12):1492–1497

Blilou I, Frugier F, Folmer S et al (2002) The Arabidopsis HOBBIT gene encodes a CDC27 homolog that links the plant cell cycle to progression of cell differentiation. Genes Dev 16(19):2566–2575

Borrill P (2019) Blurring the boundaries between cereal crops and model plants. New Phytol 228 (6):1721–1727

Bowman JL (2016) A brief history of Marchantia from Greece to genomics. Plant Cell Physiol 57(2):210–229

Brutnell TP, Wang L, Swartwood K et al (2010) Setaria viridis: a model for C4 photosynthesis. Plant Cell 22(8):2537–2544

Chang C, Bowman JL, DeJohn AW et al (1988) Restriction fragment length polymorphism linkage map for Arabidopsis thaliana. Proc Natl Acad Sci 85(18):6856–6860

Chen F, Dong W, Zhang J et al (2018a) The sequenced angiosperm genomes and genome databases. Front Plant Sci 9:418

Chen R, Xu Q, Liu Y et al (2018b) Generation of transgene-free maize male sterile lines using the CRISPR/Cas9 system. Front Plant Sci 9:1180

Churchman ML, Brown ML, Kato N et al (2006) SIAMESE, a plant-specific cell cycle regulator, controls endoreplication onset in Arabidopsis thaliana. Plant Cell 18(11):3145–3157

Coudert Y, Périn C, Courtois B et al (2010) Genetic control of root development in rice, the model cereal. Trends Plant Sci 15(4):219–226

Crossa J, Beyene Y, Kassa S et al (2013) Genomic prediction in maize breeding populations with genotyping-by-sequencing. G3: Genes, Genomes Genetics 3(11):1903–1926

D'hont A, Denoeud F, Aury JM et al (2012) The banana (Musa acuminata) genome and the evolution of monocotyledonous plants. Nature 488(7410):213–217

Dewitte W, Riou-Khamlichi C, Scofield S et al (2003) Altered cell cycle distribution, hyperplasia, and inhibited differentiation in Arabidopsis caused by the D-type cyclin CYCD3. Plant Cell 15(1):79–92

Dharmasiri N, Dharmasiri S, Estelle M et al (2005) The F-box protein TIR1 is an auxin receptor. Nature 435(7041):441–445

Dobzhansky T (2013) Nothing in biology makes sense except in the light of evolution. Am Biol Teach 75(2):87–91

Domozych DS (2014) Penium margaritaceum: a unicellular model organism for studying plant cell wall architecture and dynamics. Plan Theory 3(4):543–558

Dong J, MacAlister CA, Bergmann DC (2009) BASL controls asymmetric cell division in Arabidopsis. Cell 137(7):1320–1330

Duan YB, Li J, Qin RY et al (2016) Identification of a regulatory element responsible for salt induction of rice OsRAV2 through ex situ and in situ promoter analysis. Plant Mol Biol 90(1-2):49–62

Feldmann KA, Marks MD (1987) Agrobacterium-mediated transformation of germinating seeds of Arabidopsis thaliana: a non-tissue culture approach. Mol Gen Genet 208(1):1–9

Flavell R (2009) Role of model plant species. Plant Genomics:1–18

Gage JL, Monier B, Giri A et al (2020) Ten years of the maize nested association mapping population: impact, limitations, and future directions. Plant Cell 32(7):2083–2093

Gan D, Zhang J, Jiang H et al (2010) Bacterially expressed dsRNA protects maize against SCMV infection. Plant Cell Rep 29(11):1261–1268

Gazzarrini S, McCourt P (2003) Cross-talk in plant hormone signalling: what Arabidopsis mutants are telling us. Ann Bot 91(6):605–612

Goldstein B, King N (2016) The future of cell biology: emerging model organisms. Trends Cell Biol 26(11):818–824

Hauge BM, Hanley SM, Cartinhour S et al (1993) An integrated genetic/RFLP map of the Arabidopsis thaliana genome. Plant J 3(5):745–754

Hemerly AS, Ferreira P, de Almeida EJ (1993) cdc2a expression in Arabidopsis is linked with competence for cell division. Plant Cell 5(12):1711–1723

Holzinger A, Becker B et al (2015) Desiccation tolerance in the streptophyte green alga Klebsormidium: the role of phytohormones. Commun Integr Biol 8(4):e1059978

Hori K, Maruyama F, Fujisawa T et al (2014) Klebsormidium flaccidum genome reveals primary factors for plant terrestrial adaptation. Nat Commun 5(1):1–9

Huala E, Dickerman AW, Garcia-Hernandez M et al (2001) The Arabidopsis Information Resource (TAIR): a comprehensive database and web-based information retrieval, analysis, and visualization system for a model plant. Nucleic Acids Res 29(1):102–105

Hubble E (2005) So much more to know. Science 309:78–102. https://doi.org/10.1126/science.309.5731.78b

Izawa T, Shimamoto K (1996) Becoming a model plant: the importance of rice to plant science. Trends Plant Sci 1(3):95–99

Jackson SA (2016) Rice: the first crop genome. Rice 9(1):1–3

Jacquemin J, Bhatia D, Singh K et al (2013) The International Oryza Map Alignment Project: development of a genus-wide comparative genomics platform to help solve the 9 billion-people question. Curr Opin Plant Biol 16(2):147–156

Jiang J (2019) Fluorescence in situ hybridization in plants: recent developments and future applications. Chromosom Res 27(3):153–165

Jiang J, Xing F, Zeng X et al (2018) RicyerDB: a database for collecting rice yield-related genes with biological analysis. Int J Biol Sci 14(8):965

Ju C, Van de Poel B, Cooper ED et al (2015) Conservation of ethylene as a plant hormone over 450 million years of evolution. Nat Plants 1(1):1–7

Kawahara Y, de la Bastide M, Hamilton JP et al (2013) Improvement of the Oryza sativa Nipponbare reference genome using next generation sequence and optical map data. Rice 6 (1):1–10

Kim YA, Moon H, Park CJ (2019) CRISPR/Cas9-targeted mutagenesis of Os8N3 in rice to confer resistance to Xanthomonas oryzae pv. oryzae. Rice 12(1):1–3

Koornneef M, Meinke D (2010) The development of Arabidopsis as a model plant. Plant J 61(6):909–921

Krumholz EW, Yang H, Weisenhorn P et al (2012) Genome-wide metabolic network reconstruction of the picoalga Ostreococcus. J Exp Bot 63(6):2353–2362

Kurata N, Yamazaki Y (2006) Oryzabase. An integrated biological and genome information database for rice. Plant Physiol 140(1):12–17

Laibach F. (1943) *Arabidopsis thaliana* (L.) Heynh. als Objekt für genetische und entwicklungsphysiologische Untersuchungen. Bot Archiv 44: 439-455

Langridge J (1955) Biochemical mutations in the crucifer Arabidopsis thaliana (L.). Heynh Nat 176(4475):260–261

Leutwiler LS, Hough-Evans BR, Meyerowitz EM (1984) The DNA of Arabidopsis thaliana. Mol Gen Genet 194(1):15–23

Li H, Peng Z, Yang X et al (2013) Genome-wide association study dissects the genetic architecture of oil biosynthesis in maize kernels. Nat Genet 45(1):43–50

Lou D, Wang H, Liang G et al (2017) OsSAPK2 confers abscisic acid sensitivity and tolerance to drought stress in rice. Front Plant Sci 8:993

McClintock B (1941) The stability of broken ends of chromosomes in Zea mays. Genetics 26(2):234

Mendel, G., 1996. Experiments in plant hybridization (1865). (*Verhandlungen des naturforschenden Vereins Brünn.)* Available online: www. mendelweb. org/Mendel. html (accessed on 1 January 2013)

Merchant SS, Prochnik SE, Vallon O et al (2007) The Chlamydomonas genome reveals the evolution of key animal and plant functions. Science 318(5848):245–250

Meyerowitz EM (2001a) Prehistory and history of arabidopsis research. Erratum 125(4):2203

Meyerowitz EM (2001b) Prehistory and history of Arabidopsis research. Plant Physiol 125(1):15–19

Ming R, VanBuren R, Wai CM et al (2015) The pineapple genome and the evolution of CAM photosynthesis. Nat Genet 47(12):1435–1442

Mondal TK, Rawal HC, Chowrasia S et al (2018) Draft genome sequence of first monocot-halophytic species Oryza coarctata reveals stress-specific genes. Sci Rep 8(1):1–3

Nannas NJ, Dawe RK (2015) Genetic and genomic toolbox of Zea mays. Genetics 199(3):655–669

Nagamura Y, Antonio BA, Sato Y et al (2011) Rice TOGO Browser: a platform to retrieve integrated information on rice functional and applied genomics. Plant Cell Physiol 52(2):230–237

Nakamura Y (2018) Rice starch biotechnology: rice endosperm as a model of cereal endosperms. Starch-Stärke 70(1-2):1600375

Narsai R, Castleden I, Whelan J (2010) Common and distinct organ and stress responsive transcriptomic patterns in Oryza sativa and Arabidopsis thaliana. BMC Plant Biol 10(1):1–25

Nelson DL, Lehninger AL, Cox MM (2008) Lehninger principles of biochemistry. Macmillan

Nishimura MT, Dangl JL (2010) Arabidopsis and the plant immune system. Plant J 61(6):1053–1066

Nystedt B, Street NR, Wetterbom A et al (2013) The Norway spruce genome sequence and conifer genome evolution. Nature 497(7451):579–584

Parker D, Beckmann M, Enot DP et al (2008) Rice blast infection of Brachypodium distachyon as a model system to study dynamic host/pathogen interactions. Nat Protoc 3(3):435
Potuschak T, Lechner E, Parmentier Y et al (2003) EIN3-dependent regulation of plant ethylene hormone signaling by two Arabidopsis F box proteins: EBF1 and EBF2. Cell 115(6):679–689
Prochnik SE, Umen J, Nedelcu AM et al (2010) Genomic analysis of organismal complexity in the multicellular green alga Volvox carteri. Science 329(5988):223–226
Provart NJ, Alonso J, Assmann SM et al (2016) 50 years of Arabidopsis research: highlights and future directions. New Phytol 209(3):921–944
Pruitt RE, Meyerowitz EM (1986) Characterization of the genome of Arabidopsis thaliana. J Mol Biol 187(2):169–183
Rédei GP (1962) Supervital mutants of Arabidopsis. Genetics 47(4):443
Rédei GP (1975) Arabidopsis as a genetic tool. Annu Rev Genet 9(1):111–127
Rédei GP (1992) A heuristic glance at the past of Arabidopsis genetics. Methods Arabidopsis Res 12:1–5
Rine J (2013) A future of the model organism model. Mol Biol Cell 25(5):549–553
Rhoades MM (1984) The early years of maize genetics. Annu Rev Genet 18(1):1–30
Romero FM, Gatica-Arias A (2019) CRISPR/Cas9: development and application in rice breeding. Rice Sci 26(5):265–281
Sakai H, Lee SS, Tanaka T et al (2013) Rice Annotation Project Database (RAP-DB): an integrative and interactive database for rice genomics. Plant Cell Physiol 54(2):e6–e6
Saldivar SS (2016) Cereals: types and composition. In: Caballero B, Finglas P, Toldrá F (eds) 2015 Encyclopedia of food and health (Edited Book). Academic Press
Settles AM, Holding DR, Tan BC et al (2007) Sequence-indexed mutations in maize using the Uniform Mu transposon-tagging population. BMC Genomics 8(1):1–2
Shi J, Gao H, Wang H et al (2017) ARGOS 8 variants generated by CRISPR-Cas9 improve maize grain yield under field drought stress conditions. Plant Biotechnol J 15(2):207–216
Shoji T, Narita NN, Hayashi K et al (2004) Plant-specific microtubule-associated protein SPIRAL2 is required for anisotropic growth in Arabidopsis. Plant Physiol 136(4):3933–3944
Singh R, Ong-Abdullah M, Low ET et al (2013) Oil palm genome sequence reveals divergence of interfertile species in old and new worlds. Nature 500(7462):335–339
Sohn EJ, Kim ES, Zhao M et al (2003) Rha1, an Arabidopsis Rab5 homolog, plays a critical role in the vacuolar trafficking of soluble cargo proteins. Plant Cell 15(5):1057–1070
Somerville C, Koornneef M (2002) A fortunate choice: the history of Arabidopsis as a model plant. Nat Rev Genet 3(11):883–889
Strable J, Scanlon MJ (2009) Maize (Zea mays): a model organism for basic and applied research in plant biology. Cold Spring Harb Protoc 10:pdb-emo132
Suárez-López P, Wheatley K, Robson F et al (2001) CONSTANS mediates between the circadian clock and the control of flowering in Arabidopsis. Nature 410(6832):1116–1120
Szövényi P, Frangedakis E, Ricca M et al (2015) Establishment of Anthoceros agrestis as a model species for studying the biology of hornworts. BMC Plant Biol 15(1):1–7
Thakare D, Zhang J, Wing RA et al (2017) Aflatoxin-free transgenic maize using host-induced gene silencing. Sci Adv 3(3):e1602382
Van Buren R, Bryant D, Edger PP et al (2015) Single-molecule sequencing of the desiccation-tolerant grass Oropetium thomaeum. Nature 527(7579):508–511
Vij S, Gupta V, Kumar D et al (2006) Decoding the rice genome. BioEssays 28(4):421–432
Wang D, Xia Y, Li X et al (2012a) The Rice Genome Knowledgebase (RGKbase): an annotation database for rice comparative genomics and evolutionary biology. Nucleic Acids Res 41(D1): D1199–D1205
Wang M, Yan J, Zhao J, Song W, Zhang X, Xiao Y, Zheng Y (2012b) Genome-wide association study (GWAS) of resistance to head smut in maize. Plant Sci 196:125–131
Wang X, Wang H, Liu S et al (2016) Genetic variation in ZmVPP1 contributes to drought tolerance in maize seedlings. Nat Genet 48(10):1233–1241

Wang P, Hendron RW, Kelly S (2017) Transcriptional control of photosynthetic capacity: conservation and divergence from Arabidopsis to rice. New Phytol 216(1):32–45

Ware D, Jaiswal P, Ni J et al (2002) Gramene: a resource for comparative grass genomics. Nucleic Acids Res 30(1):103–105

Wu Y, Messing J (2012) RNA interference can rebalance the nitrogen sink of maize seeds without losing hard endosperm. PLoS One 7(2):e32850

Yan H, Jiang J (2007) Rice as a model for centromere and heterochromatin research. Chromosome Res 15(1):77–84

Yanofsky MF, Ma H, Bowman JL et al (1990) The protein encoded by the Arabidopsis homeotic gene agamous resembles transcription factors. Nature 346(6279):35–39

Yu Q, Jalaludin A, Han H et al (2015) Evolution of a double amino acid substitution in the 5-enolpyruvylshikimate-3-phosphate synthase in Eleusine indica conferring high-level glyphosate resistance. Plant Physiol 167(4):1440–1447

Zhang X, Henriques R, Lin SS et al (2006) Agrobacterium-mediated transformation of Arabidopsis thaliana using the floral dip method. Nat Protoc 1(2):641

Zhang ZY, Fu FL, Gou L et al (2010) RNA interference-based transgenic maize resistant to maize dwarf mosaic virus. J Plant Biol 53(4):297–305

Zhang J, Zhang X, Chen R (2020) Generation of transgene-free Semidwarf maize plants by gene editing of gibberellin-oxidase20-3 using CRISPR/Cas9. Front Plant Sci 11:1048

Zheng J, Zhang H, Dong S et al (2014) High yield exfoliation of two-dimensional chalcogenides using sodium naphthalenide. Nat Commun 5(1):1–7

Zimin A, Stevens KA, Crepeau MW et al (2014) Sequencing and assembly of the 22-Gb loblolly pine genome. Genetics 196(3):875–890

RNA Interference Technology in Plants: Mechanisms and Applications in Crop Improvement

10

Priyanka Gautam, Rishabh Kumar, Zainab Feroz,
Sivakumar Vijayaraghavalu, and Munish Kumar

Abstract

Scientific breakthroughs in recent times have brought major advances in fundamental research, which eventually lead to their utilization for human welfare. One such discovery is of RNA interference (RNAi), in which double-stranded RNA (dsRNA) hinders gene expression, usually by binding to messenger RNA (mRNA) and triggering its degradation. Among the various biotechnological tools currently available, RNAi has been playing a significant role in crop improvement as it guarantees greater accuracy and fidelity to plant improvement. The invention of this phenomenon has changed it into a potent tool of genetic engineering and functional genomics. RNAi technology gives us an explicit methodology for downregulation of gene of interest without inhibiting the expression of any other gene in the plant. RNAi has been successfully applied in different plants to bring about modifications of numerous desirable traits. Nutritional improvements have led to the activation of defence mechanism against biotic and abiotic stresses. Alteration in morphology, reduced content of food allergens, crafting male sterility, enhanced secondary metabolite synthesis and production of seedless plant varieties are some of the other advantages of RNAi. In spite of these advantages, crop plants developed by RNAi strategy may generate biosafety risks. So, there is a need for risk assessment of genetically modified crops in order to make RNAi a better tool to develop crops probably with less biosafety issues.

P. Gautam · R. Kumar · Z. Feroz · M. Kumar (✉)
Department of Biochemistry, University of Allahabad, Prayagraj, Uttar Pradesh, India

S. Vijayaraghavalu
Advanced Research Centre, Narayana Medical College, Nellore, Andhra Pradesh, India

R. L. Singh et al. (eds.), *Plant Genomics for Sustainable Agriculture*,
https://doi.org/10.1007/978-981-16-6974-3_10

Keywords

RNA interference · Gene silencing · Crop improvement · siRNA · miRNA · Stress tolerance

10.1 Introduction

The global population is on the rise leading to an extensive increase in the demand to supply ratio of food. Over the past 10 years, the increasing harmful environmental conditions have been accountable for the inferior plant growth and observable crop loss all over the world (Mickelbart et al. 2015). The crop yield is declining excessively due to the scarcity of arable land and water resources, unpredictable weather patterns and destruction caused by various abiotic as well as biotic stresses. The growing population is raising the demand for global sustainable agricultural practices. Conventional plant breeding methods are being used for centuries as one of the best competent ways to improve the quality and quantity of crops, but these processes are very time- taking, labour some and have many other biological, physiological, and ecological limitations. To overcome these obstacles, modern breeding practices accompanied with molecular genetics, recombinant DNA technology, and biotechnological studies consisting of proteomics, genomics and transcriptomics are essential to develop high-yielding, disease- resistant, and environmentally stress-tolerant plant varieties (Mittler and Blumwald 2010; Tester and Langridge 2010). The fidelity of biotechnological procedures, principally genetic engineering, has contributed immensely for crop improvement by offering a vast range of novel genes and characteristics which can be effectively incorporated into plants through genetic manipulations, leading to enhanced nutritional value, crop yield, and also increased tolerance to biotic and abiotic stresses (Sharma et al. 2002).

But this technology has many apprehensions and public concerns pertinent to their use in present-day agriculture, biosafety standard procedures, and impact of the genetically engineered crops on the natural environment principally when the genes derived from organisms are used other than the plants (Wolfenbarger and Phifer 2000). Transgenic products always had ethical problems in terms of biosafety specifically in case of edible crops. Thus, transgenic plants are first exposed to intricate tests before releasing them for everyday use, to ensure complete public safety and figure out the risks. Development of transgenic crops therefore requires surplus time, expenditure, and competence. Hence, new and eco-friendly strategies have to be developed for crop advancement which could validate to be more acceptable by the common people.

RNA interference (RNAi) is a conserved, naturally occurring gene regulatory mechanism. RNAi silencing technology has proven to be a plausible substitute for crop improvement, apparently with lesser biosafety concerns as no transgene protein is expressed in transgenic lines (Rajam 2012). Moreover, it is also required for sustaining genomic stability, regulation of transposon movement, epigenetic alternations, and management of cellular processes at transcriptional and translational level (Ketting 2011; Castel and Martienssen 2013). RNAi pathway primarily includes small-interfering RNAs (siRNAs) and micro-RNAs (miRNAs). RNAi has helped the researchers to completely or partly silence the expression of a particular gene, allowing targeted gene knockout and gene knockdown. RNAi is a biological process in which double-stranded (dsRNA) prohibits gene expression, resulting in gene silencing through cleavage of mRNAs and inhibition of protein synthesis. In other words, RNAi is a process which leads to post-transcriptional gene silencing (PTGS) triggered by dsRNA molecules to obstruct the expression of certain genes (Bosher and Labouesse 2000). The discovery of RNAi gave a new tool in the hand of scientists to manipulate the plants through genetic engineering and to study the functional genomics. Detection of RNAi in mechanism of hormone signal transduction, forbearance to environmental stress, and obstruction of microbe invasion is a worth mentioning process. RNAi technology has been successfully exploited in plants for resistance against pathogens, pests, nematodes, and virus that cause heavy financial losses (Saha and Mishra 2018). RNAi has been effectively implemented to accomplish the modifications of innumerable preferred traits, including nutritional fortifications, allergen or reduction of toxic compounds, morphological changes, alternation of male sterility, enhancement of secondary metabolite, and for defense against various biotic and abiotic stresses in various crop plants (Saurabh et al. 2014). The present chapter therefore focuses on the discovery of RNAi and its mechanism of action and applications for crop improvement.

10.2 RNAi: Discovery and Basic Mechanism of Action

RNAi is one of the most interesting phenomenon in which short dsRNA inhibits the expression of certain specific genes by causing degradation of sequence of a particular target mRNA in the cytoplasm. The phenomenon of RNAi was first witnessed by Napoli et al. 1990 in plants, where amplified production of anthocyanin pigments was obtained in *Petunia hybrida* L. after introduction of the chalcone synthase gene (CHS A). Surprisingly, instead of dark purple flowers transgenic plants producing white or chimeric flowers were achieved. Scientists working on plant and fungal systems found that the introduction of transgene caused downregulation of transgene as well as the endogenous gene and this event came to be known as "co-suppression" (Napoli et al. 1990; Hannon 2002). A very similar phenomenon termed "quelling" was then identified in the fungus *Neurospora crassa* (Romano and Macino 1992). Later on, Fire et al. (1998) demonstrated the mechanism of RNAi in the nematode, *Caenorhabditis elegans*, and designed the term "RNAi" for the first time.

The RNAi is an RNA-dependent gene silencing process, which is initiated by 20- to 24-nucleotide-long (nt), small-interfering RNAs (siRNAs), which are developed intracellularly from long endogenous or exogenous dsRNA molecules through the

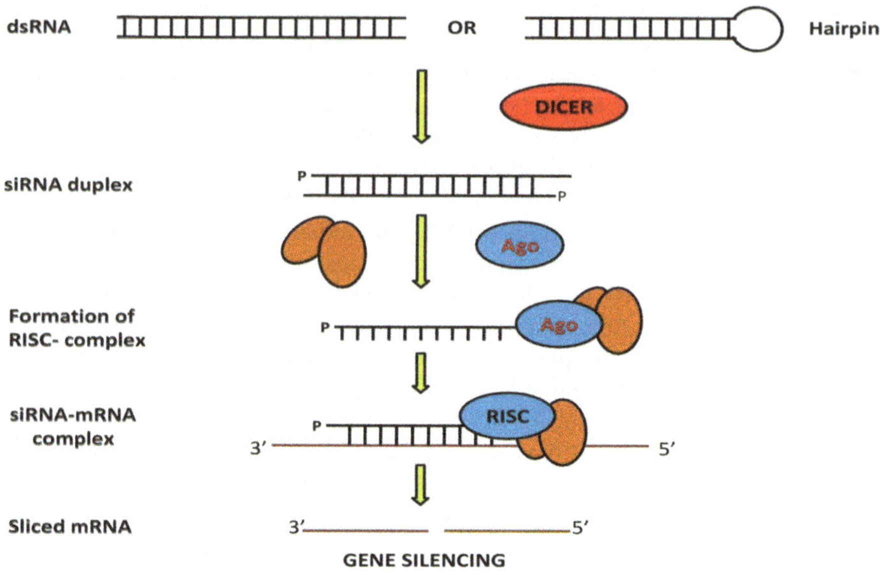

Fig. 10.1 The RNAi pathway for gene silencing- dsRNA or hairpin RNAs (hpRNAs) are first processed into 20–24 nt siRNA duplexes by the action of ribonuclease III-type enzyme termed as DICER. The siRNA are then incorporated into a multi-protein complex called RISC. The siRNA–RISC complex then targets a sequence, complementary to the siRNA, in a piece of mRNA. The protein synthesis is blocked either by inhibition of translation or degradation of mRNA

cleavage by a ribonuclease III-type enzyme termed as Dicer (Hamilton and Baulcombe 1999; Zamore et al. 2000). Each siRNA consists of a passenger (sense) strand and a guide (antisense) strand. These siRNAs (20–24 nt) are then integrated into a multi-protein complex called RNA-induced Silencing Complex (RISC) which comprised of Argonaute (AGO) proteins (Baumberger and Baulcombe 2005; Vaucheret 2008). The ATP-activated RISC separates the double-stranded siRNA. The passenger/sense strand of the siRNA duplex is degraded by the RNA helicase action and the antisense strand of siRNA molecule is contained in the RISC complex. The antisense strand of the siRNA–RISC complex then base pairs with the complementary mRNA target sequences and begins endonucleolytic cleavage through the activity of induced AGO protein (catalytic component of the RISC complex), thus hindering translation of the target transcript (Borges and Martienssen 2015) (Fig. 10.1).

RNAi technology is an excellent biotechnological approach having countless advantages as it is more specific, dominant, and sequence-based gene silencing. This huge capacity of RNAi has been successfully exploited for generating desirable characteristics.

10.3 Application of RNAi in Plant Improvement

10.3.1 Abiotic Stress Resistance

Abiotic stress conditions, for instance, drought, flooding, salinity, and variations in temperatures are the chief reasons for reduction in the productivity of numerous major crops all over the world (Shriram et al. 2016). The changing environment and rising food demands for growing population exert a lot of pressure on scientists for development of stress-tolerant crop varieties. Under stress conditions, plants manufacture various non-coding RNAs (ncRNAs) for gene regulation at transcriptional, post-transcriptional, and chromatin level. By the use of RNAi, the ncRNAs and their targets can be utilized for generation of abiotic stress-tolerant varieties. Response to abiotic stress is regulated by different miRNAs in economically important crops such as rice, wheat, legumes, sugarcane, etc. miRNAs generally regulate transcription, detoxification, and development processes. The expression levels of miRNAs and other related genes are seen to be transformed during various stress conditions. Sunkar and Zhu (2004) described the act of miRNAs in response to abiotic stresses like salinity, cold, drought, and oxidative stress in *Arabidopsis* seedlings due to manifestation of different abiotic stress conditions and revealed that miR393 was very heavily up-regulated by high salinity, cold, dehydration, and abscisic acid (ABA) treatments. Salt stress responsiveness of miRNAs is now known in several plant species, along with in many crops. The salinity responsive miRNAs are recognized in *Arabidopsis thaliana* (Barciszewska-Pacak et al. 2015), *Gossypium hirsutum* (Gao et al. 2016), *Populus tomentosa* (Ren et al. 2013), *Zea mays* (Fu et al., 2017), *Cicer arietinum* (Kohli et al. 2014), *Triticum aestivum* (Gupta et al. 2014), *Oryza sativa* (Mittal et al. 2016), and many more. Some major miRNA families presenting changed expressional actions under saline conditions consist of miR156, miR159, miR168, miR169, miR393, and miR398. In response to drought, various genes, along with their targets, have been examined by applying genome-wide expression studies (Kruszka et al. 2012). The miRNA expression profiles in response to drought are reported in Sorghum bicolor (Hamza et al. 2016), *Gossypium hirsutum* (Wang et al., 2013), *Solanum tuberosum* (Zhang et al. 2014), *Oryza rufipogon* (Zhang et al. 2016a, b), *Panicum virgatum* (Xie et al. 2014). Downregulation of receptor for activated C-kinase 1 (RACK1) through RNAi resulted in development of drought tolerance in rice (Da-Hong et al. 2009). RACK1 is a conserved scaffold protein that regulates expression of antioxidant-related enzymes such as superoxide dismutases (SODs) in plants. Inhibition of RACK1 enhances the accumulation of SODs and imparts tolerance against drought as well as reactive oxygen species (ROS). The transgenic rice plants were more tolerant to drought stress in comparison with the non-transgenic rice plants. Similarly, suppression of farnesyltransferase/squalene synthase (SQS) through siRNA generated from maize squalene synthase improved drought tolerance at both the reproductive and vegetative stages in rice (Manavalan et al. 2012). Increase in endogenous sterol level through silencing of SQS decreases the stomata density and impedes water loss through transpiration, thereby preventing the plant from wilting under drought condition. Silencing of OsTZF1 gene increases the tolerance of rice plants to high salt and low water conditions, illustrating its role in abiotic stress tolerance (Jan et al. 2013). Low expression of OsTZF1 gene maintains the inner homeostasis of plants through change in hormonal expression at cellular and molecular level under high salt condition. Plants change their gene expression pattern at post-transcriptional levels in response to inconsistent temperature variations. Various temperature-responsive miRNA species have been recognized in plants. *Panicum virgatum* (Hivrale et al. 2016), *Oryza sativa* (Li et al. 2015; Mangrauthia et al. 2017), and *Triticum aestivum* (Kumar et al. 2015) have deciphered the heat responsive alterations in different miRNA species; whereas chilling-responsive miRNAs have been exhibited in *Glycine max* (Xu et al. 2016), *Solanum habrochaites* (Cao et al. 2014), and *Zea mays* (Li et al. 2016). Thus, miRNAs maintain resistance to different abiotic stresses through up- and downregulation of the target gene transcripts.

10.3.2 Biotic Stress Tolerance

The overall crop production is badly affected by biotic stresses, which includes bacterial, viral, parasitic weeds, fungal pathogens, insects, and nematodes. The yearly crop losses due to various plant diseases are anticipated to be around $60 billion worldwide; therefore, there is a dire need to address this issue because this causes significant economic losses. Today, RNAi-induced gene silencing has come up as a very promising technique to engineer resistant genotypes (Saha and Mishra 2018).

10.3.2.1 Bacterial Disease Resistance

Bacteria are simple microscopic organisms and unlike viruses, bacteria do not require insects as vectors, instead they rely on rain, wind, soil, seed dispersal, or other means of transport to enter into plants. Bacterial pathogens cause various plant diseases which affects crop production enormously leading to major annual losses on a global scale. Bacterial diseases are one of the prime challenges faced in crop field such as banana, soybean, and tomato. RNAi has exhibited antibacterial defense against *Pseudomonas syringae* and *Agrobacterium tumefaciens*. Prokaryotes, for example, phytoplasmas and bacteria are deficient in conventional RNAi systems and thus RNAi cannot be induced inside the cells of these plant pathogens. Nevertheless, Escobar and his co-workers used RNAi to target particular mRNAs and grant resistance against crown gall disease (Escobar et al. 2001). The crown gall disease is caused by bacteria *Agrobacterium tumefaciens* which is widespread and affects dicot plants worldwide (Escobar and Dandekar 2003). Pathogenic strains of *A. tumefaciens* possess a large plasmid known as the tumor-inducing or Ti plasmid that carries genes mandatory for the development of the disease. The infection begins when *A. tumefaciens* penetrates through root of the plant via wounds. The wounds discharge signaling molecules such as acetosyringone which stimulate the *A. tumefaciens* to transport the T-DNA region from the Ti-plasmid into the plant cell and combine with the chromosomal DNA. Therefore, the wounded cells are transformed and T-DNA genes, along with the oncogenes, are transcribed. *iaaM* and *ipt* are two oncogenes that encode proteins involved in auxin and cytokinin synthesis. Overexpression of these products causes hyperplasia and hypertrophy which ultimately leads to tumorigenesis (gall formation). RNAi technique was utilized to target these two oncogene mRNAs in attempts to inhibit tumorigenesis. Transgenic tomato and *Arabidopsis thaliana* plants were created to transcribe hairpin RNAs that concurrently target iaaM and ipt oncogene mRNAs. When the transgenic plants were challenged with wild-type *A. tumefaciens*, many of the engineered lines were greatly immune to *A. tumefaciens* and did not develop galls, whereas non-transformed plants were 100% susceptible and developed normal galls. The transgene-derived hairpin RNA generated RNAi activity against the T-DNA-derived iaaM and ipt gene mRNAs and prevented gall development. The authors applied the same technique to develop genetically engineered walnuts, a commercial important crop for which crown gall disease is also a serious issue, and they exhibited excellent resistance in walnuts also (Escobar et al. 2002) (Table 10.1).

Infection caused by *Pseudomonas syringae* pv. tomato in *Arabidopsis* induced the production of natsiRNA (nat-siRNAATGB2) that downregulates a gene called PPRL which encodes a negative regulator of the RPS2 disease resistance pathway. The induction of nat-siRNAATGB2 enhances the RPS2-mediated race-specific resistance against *P. syringae* pv. tomato in *Arabidopsis* (Katiyar-Agarwal et al. 2007).

10.3.2.2 Viral Disease Resistance

Many plant viruses are extremely contagious and their effects on plants are often drastic. Some of them seriously decrease crop yield and quality as they have the ability to multiply and spread rapidly. They can transmit the disease either directly from parent to progeny or indirectly through insect vectors and thus their control becomes very burdensome. Amongst different strategies available in plants to fight against virus infections, pathogen-derived resistance (PDR) is one of the most powerful approaches. The successful application of the PDR concept has helped to engineer virus-resistant plants (Simon-Mateo and Garcia 2011). There is an additional line of attack that targets many regions of a viral gene showing a wide-ranging resistance against tospoviruses in tomato plants (Bucher et al. 2006). This strategy is very helpful and is dependent on the use of a miRNA construct that expresses multiple artificial miRNAs (amiRNAs) targeting multiple regions of a viral gene. It was first reported in potato, where simultaneous expression of both sense and antisense transcripts of the viral helper-component proteinase (HC-*Pro*) gene exhibited complete resistance to potato virus Y (PVY). RNAi targeting the coat protein (CP) gene of viruses has been found to be relatively effective in inducing resistance to the plant against viruses. There are several viral coat protein targeting RNAi-modified virus-resistant plants like Cucumber Green Mottle Mosaic Virus (CGMMV)-resistant N. benthamiana (Kamachi et al. 2007), Beet Necrotic Yellow Vein Virus (BNYVV)-resistant tobacco (Andika et al. 2005), Papaya Ring Spot Virus type W (PRSV-W)-resistant Cucumis melo L. var. cantalupensis cv. Sun Lady

Table 10.1 Application of RNAi technology in biotic-stress tolerance of plants

Type of biotic stress	Target organism	System used	Targeted Gene	Impact	Reference
Bacterial diseases	*Agrobacterium tumefaciens*	*Arabidopsis thaliana*	iaaM and ipt genes	Showed resistance to crown gall disease	Dunoyer et al. (2006)
	Xanthomonas citri subsp. citri (Xcc)	Lemon	PDS and CalS1		Enrique et al. (2011)
Fungal diseases	Blumeria graminis f. sp. Tritici	Wheat	Knockdown of MLO	Resistant	Riechen (2007)
	Phytophthora parasitica var. nicotianae	Tobacco	Glutathione S-transferase (GST) gene	Resistant	Hernandez et al. (2009)
Nematodes	*Meloidogyne javanica* (root-knot nematode)	Tobacco	Tis11	Resistant	Fairbairn et al. (2007)
Insects	Cotton bollworm (*Helicoverpa armigera*)	Cotton	P450 monooxygenase gene CYP6AE14	Sensitive to gossypol	Mao et al. (2007)
	Corn rootworm	Maize	V-type ATPase gene	Reduce feeding damage by corn rootworm	Baum et al. (2007)
Viral diseases	Mungbean yellow mosaic India virus (MYMIV)	*Vigna mungo* (black gram)	Bidirectional promoter	Recovery from infection	Pooggin et al. (2003)
	Potato spindle tuber viroid (PSTVd)	Tomato	PSTVd-specific siRNA	Recovery from infection	Sano and Matsuura (2004)
	Tobacco mosaic virus (TMV)	Tobacco	Replication-associated protein	Inhibition of TMV replication	Zhao et al. (2006)
	African cassava mosaic virus (ACMV)	Tobacco protoplast	Replication-associated protein gene	Reduced virus accumulation	Vanitharani et al. (2003)
	Alfalfa mosaic virus (AMV)	Tobacco	Arbitrary sequence	Recovery from infection	Tenllado et al. (2003)
	Potato virus Y (PVY)	Potato	HC-pro	Immunity	Waterhouse and Graham (1998)
	Rice dwarf virus (RDV)	Rice	PNS12		Shimizu et al. (2009)

(Krubphachaya et al. 2007), and Plum Pox virus (PPV)-resistant Nicotiana benthamiana and Prunus domestica (Hily et al. 2007). Commercial variety of potato (cv Spunta) was transformed with dsRNA derived from the 3′ terminal part of the coat protein gene of Potato Virus Y (PVY) (Missiou et al. 2004). Another example where RNAi technology may be abundantly applied is in the creation of banana varieties resistant to the Banana Bract Mosaic Virus (BBrMV), which affects the banana population in Southeast Asia and India (Rodoni et al. 1999). This virus infects banana plants damaging the fruit producing bract region, making them useless for the farmers. The virus is spread through small insects called aphids, as well as by infected plant materials. Nonetheless, by carefully designing an RNAi vector aimed at silencing the Coat Protein (CP) region of the virus, scientists may be able to develop a banana variety that is resistant to BBrMV and yet safe to consume. The CP region of the different strains of virus is extremely conserved and therefore silencing of this gene in other varieties of banana will not create a problem. In an another study, RNAi-mediated silencing of African cassava mosaic virus (ACMV) resulted in 66% reduction in viral DNA, the engineered transgenic cassava plants exhibited resistance to African cassava mosaic virus (ACMV), by expressing dsRNAs. Transgenic cassava lines with excessive levels of AC1-homologous small RNAs have ACMV replication linked with protein coding sequence imparting Rep/AC1-homologous hairpin double strain immunity (Vanderschuren et al. 2009). By using RNAi technique, several efforts have been made to control ssDNA viruses, particularly the geminiviruses, for instance, the non-coding intergenic region of the *Mungbean yellow mosaic India virus* (MYMIV) was expressed as an hp. construct under the control of the 35S promoter and used to biolistically inoculate MYMIV infected black gram (*Vigna mungo*) plants. Plants treated with the construct exhibited an absolute recovery from infection that lasted until senescence (Table 10.1).

10.3.2.3 Fungal Diseases

Fungal pathogens not only cause massive crop losses but they also generate harmful mycotoxins in crop plants. Intake of these mycotoxins even in smallest quantity can lead to serious health problems in living beings. Attempts have been made through research to attain RNAi-mediated resistance against fungal disease, caused by *Phytophthora parasitica* var. *nicotianae* by targeting glutathione S-transferase (GST) gene, which resulted in remarkable increment in resistance of tobacco (*Nicotiana tabacum*) to infection following gene silencing for GST-silenced plants in comparison with control plants. GSTs are a phase II metabolic enzymes that play a crucial role in cellular defense against various harmful chemicals produced both exogenously and endogenously. Silencing of GST was achieved by cloning a GST gene in sense and antisense orientation to an RNAi vector to stop the spread of black shank disease (Hernandez et al. 2009). It was also observed that few defense genes are up-regulated in GST-silenced plants during the interaction with the pathogen. Host-inducing gene silencing (HIGS) methods have shown successful defense against fungal pathogens. Nowara and his co-worker produced RNAi effects using HIGS, in the obligate bio-trophic plant pathogen *Blumeria graminis,* which causes powdery mildew of barley and wheat. *B. graminis* lives in a close association with the host cell because the haustorium *B. graminis*is very firmly intact with the plasma membrane of the host cell, and therefore the nutrients are transported into it. Thus, sRNAs can also be transferred from the plant into the *B. graminus* haustorium and once it enters inside the fungal cell, it is possible that the plant-derived sRNAs couple with the fungal RNAi components and generate RNAi effects against the target fungal mRNAs (Nowara et al. 2010) (Table 10.1).

Fusarium oxysporum affects a large variety of hosts. Tomato, cucurbits, sweet potatoes, tobacco, and banana are some of the most affected plants. *Fusarium oxysporum* f. sp. *cubense* (Foc) causes Fusarium wilt which is among the most harmful diseases of banana (Musa spp.) Ghag et al. 2014 reported that knockdown of important genes of fungus (*velvet* and *Fusarium transcription factor 1*) using RNAi showed successful resistance against Foc. After 6-week-long greenhouse bioassays, the transformed banana lines were found to be free from the external and internal symptoms of Foc. The 5 chosen transgenic lines for every construct exhibited resistance to Foc for 8 months post-inoculation.

10.3.2.4 Nematode Resistance

Annual crop loss because of phyto-parasitic nematodes is anticipated to be around US$125 billion worldwide. Nematodes are extremely harmful for foliage plants, vegetable crops, fruit and nut trees, and forest trees. Gheysen and Vanholme 2007 showed that expression of dsRNA in a host plant against parasitism genes in the root-knot nematode leads to nematode resistant in plants. Root-knot nematodes (*Meloidogyne* spp.) are plant parasites that live in the soil and cause severe injury to crops. Various parasitism proteins that are encoded by the parasitism genes are expressed in esophageal gland cells. Their function is to mediate infection and parasitism of plants by root-knot nematodes (RKN). Newly bioengineered crops expressing dsRNA obstruct the parasitic process by targeting the RKN parasitism genes. This represents a feasible and flexible means of developing new durable RKN-resistant crops as it will provide the crops broader resistance to RKN (Huang et al. 2006).

Sindhu et al. (2009) having host *A. thaliana* targeted all 4 nematode parasitism genes (3B05, 4G06, 8H07, and 10A06) of sugar beet cyst nematode (*Heterodera schachtii*), through host-induced RNAi. They revealed that no complete resistance was seen, but it led to 25–64% decline in the number of full-grown nematode females in different RNAi lines.

Parasites such as *Heterodera avenae*, *H. filipjevi*, and *H. latipons* cause excessive damage to wheat crops. The Virus-induced Gene Silencing (VIGS)-based RNAi approach targets the Ha18764 effector protein family genes of *H. avenae*. This decreased the parasitism and reproduction status of *H. avenae* in wheat (Yang et al. 2019). Also, downregulation of pat-10 and unc-87 genes by the application of RNAi, present on the Thorne's meadow nematode (*Pratylenchus thornei*), which infects wheat roots, lessened the reproduction rate of the worms significantly (Liu et al. 2021).

10.3.2.5 Parasitic Weed Resistance

Some researchers have also reported that through the application of RNAi technology weed-resistant plant varieties can be developed. Aly et al. (2009) created transgenic tomato plants bearing M6PR dsRNA-expression cassette. They observed that the underground shoots of *Orobanche aegyptiaca* grown on transgenic tomato plants and endogenous level of M6PR mRNA in the tubercles were decreased by 65–80% with a significant reduction in mannitol level and a significant increment in the percentage of dead *O. aegyptiaca* tubercles.

10.3.3 Increasing Nutritional Value

Plants offer maximum number of nutrients required in the human diet still the major staple crops often lack in vital mineral elements. Therefore, malnutrition, with respect to vitamins and micronutrient has affected more than 50% of the world's population. Recent advancement in modern breeding techniques, genetics, and biotechnology studies has provided the means and incentive to improve the nutritional value of plants. Another nutritional target involves the alteration in the fatty acid composition and the improvement in the antioxidant levels (Tucker 2003). RNAi technology offers new avenue for the bio-fortification of nutrients in plants for the development of bio-fortified foods.

RNAi was applied to intensify the content of β-carotene in potato by suppressing the gene β-carotene hydroxylase (BCH), that transforms β-carotene to zeaxanthin. *A. tumefaciens*-mediated transformation was applied to introduce 2 RNAi constructs having the tuber-specific granule bound starch synthase (GBSS) promoter, and the other contained the powerful constitutive cauliflower mosaic virus 35S (CaMV 35S) promoter into potato lines. The transformants obtained from the GBSS construct contained an increased β-carotene content than the CaMV 35S transformants. These result showed that in potato silencing of the BCH gene can upsurge the content of two carotenoids, β-carotene and lutein. In future this will provide a new tool for eliminating the prevalence of vitamin A deficiency in populations (Eck et al. 2007).

Recently, Kusaba along with his teammates have made outstanding contribution by applying RNAi to improve rice plants. They successfully reduced the level of glutenin and produced a rice variety called LGC-1 (low glutenin content 1). The low glutenin content was a great relief for the kidney patients who are unable to digest glutenin. The trait was stable and was transmitted for a number of generations. (Kusaba et al. 2003). Using RNAi the levels of sinapate esters in transgenic canola seeds (*Brassica napus*) were reduced by 76% of the T3 generation by obstructing UDP-Glc:sinapate glucosyltransferase gene activity (Husken et al. 2005).

The consumption of α-linolenic acid (18:3) is found to be unhealthy for humans as well as for animals. The reduction of α-linolenic acid (18:3) improves the flavor of soybean oil and stability with lesser need for its hydrogenation. The linoleic acid (18: 2) is converted into α-linolenic acid (18:3) in the presence of omega-3 fatty acid desaturase enzyme. Flores et al. (2008) constructed hairpin RNA for the downregulation of omega-3 fatty acid desaturase (GmFAD3A, GmFAD3B, and

GmFAD3C), using glycinin promoter for seed-specific silencing. Transgenic soybean seed has been reported to have 1–3% of α-linoleic acid in comparison with 7–10% in non-transgenic soybean seed.

RNAi technology can be used to raise the starch content in the leaves. It has been demonstrated that starch phosphorylation and dephosphorylation are important constituents of leaf starch degradation, where glucan, water dikinase (GWD) adds phosphate to starch and phosphoglucan phosphatase (SEX4) eliminates these phosphates. In maize, the route of leaf starch degradation is not very well characterized. In 2012, Weise et al. manipulated phosphate metabolism using RNAi constructs to enhance the starch content in *Zea mays* (maize) and *A. thaliana*. A new variety of apples known as the Arctic Apples were produced by suppression of PPO (polyphenol oxidase) gene using RNAi. Usually when an apple is cut it produces a browning effect because PPO reacts with the polyphenolics present in the fruit. But these PPO-silenced apples varieties will not undergo browning after being cut as they are not able to convert chlorogenic acid into quinone product (Kumar et al. 2017).

Overexpression of carotenoid or flavonoid synthetic genes or transcription factors escalates either carotenoid or flavonoid content. RNAi has helped to improve the level of both carotenoids and flavonoids in tomato fruit. DET1, which is a photomorphogenesis regulatory gene, represses various light-mediated signaling pathways. It has been reported that expression of dsRNA of DET1 under fruit-specific promoter in tomato suppressed endogenous expression of DET1 and resulted in high levels of flavonoids and carotenoids in tomato fruits (Davuluri et al. 2005). In the same way, downregulation of lycopene epsilon-cyclase (ε-CYC) gene expression through RNAi enhanced the carotenoid content in rapeseed (*Brassica napus*). Thus, RNAi has tremendous potential to eradicate the malnutrition across the world.

10.3.4 Development of Male Sterile Lines

The development of male sterility is one of the most essential traits chosen to enhance the crop productivity by the hybridization process. Hybridization leads to production of offsprings with superior characteristics in comparison to their parents, and the process is known as hybrid vigor or heterosis (Duvick 1999). When male sterility is not available naturally engineered male sterility can be an alternative option for developing hybrids. Genetic engineering today is used to generate male sterile plant varieties like tobacco and tomato using RNAi.

Downregulation of TA29 gene of tobacco by RNAi (which is an anther-specific gene, expressed in anthers at the time of microspore development) produced male sterile lines (Nawaz-ul-Rehman et al. 2007). Ten out of thirteen tobacco lines containing the TA29 sequences transformed with a hairpin RNAi construct were found to be male sterile. Transgenic plants were phenotypically very identical to the non-transgenic plants. Bcp1, is another anther-specific gene, active in diploid tapetum as well as haploid microspores. Silencing of Bcp1 (a male-specific gene) in the host *A. thaliana* resulted in male sterile lines. Transgenic plants were phenotypically similar to the non-transgenic plants, and after crossing with non-transgenic fertile pollens, successful seed set was observed (Tehseen et al. 2010).

Nucleases are a diverse class of enzymes playing crucial role in nucleic acid metabolism. In 2005, Moritoh et al. cloned OsGEN-L (OsGEN-like) gene, a new member of the RAD2/XPG nuclease family from rice (*O. sativa* L.). Transgenic rice plants expressing hairpin RNA for OsGEN-L, displayed low fertility and were male sterile.

S-adenosyl methionine decarboxylase (SAMDC) is a key enzyme in polyamine biosynthesis, essential during pollen maturation and germination (Sinha and Rajam 2013). Therefore, expression of chimeric (SAMDC) dsRNA under the control of tapetum-specific A9 promoter caused simultaneous silencing of three SAMDC isoforms in tapetum tissue, which resulted in formation of male sterile SAMDC-RNAi lines without affecting their female fertility. Zhu and Deng (2012) have highlighted the link between sRNA (osa-smR5864) with photoperiod- and temperature-regulated male sterility. MicroRNA-mediated regulation of male sterility has been reported in various plant species such as cotton (Yang et al. 2016) and radish. For instance, rsa-miR159a controls the expression of transcription factor required during anther and pollen development. Huge expression of ras-mir159a downturns the expression of MYB101 TF and, consequently, activates male sterility through prohibition of normal pollen development in radish plants (Zhang et al. 2016a, b).

10.3.5 Modification of Flower Color and Scent by RNAi-Mediated Gene Silencing

Alternation of flower color and scent is one of the most essential traits in floriculture as it has a great economic and aesthetic value. Flowers like rose, tulip, orchid, lotus, poppy, and petunia are always in demand for the purpose of decoration and making of cosmetics. By using RNAi researchers have suppressed several structural genes in anthocyanin biosynthesis, causing the inhibition of anthocyanin accumulation resulting in change of flower color in transgenic plants. Amongst anthocyanins precursors for all plant pigments, there are cyanidin, pelargonidin, and delphinidin. The cyanidin gene is in charge for a synthetic pathway that leads to formation of red pigment and a correspondent Delphinidin gene is responsible for the formation of blue color. Knockdown of the cyanidin genes in rose by RNAi and introduction of delphinidin genes produced flowers that accumulated delphinidin-based anthocyanins completely with a concomitant color change toward blue (Tanaka et al. 2009; Katsumoto et al. 2007). This result proved to be magical for the flower industry and the Horticultural Societies of Britain and Belgium has presented a prize of 500,000 francs to the first person to produce a blue rose (Van Uyen 2006).

Japanese gentians such as Gentiana triflora, Gentiana scabra, and their interspecific hybrid are one of the most popular ornamental flowers in Japan. Cool climate and huge mountain areas provide favorable conditions for the cultivation and breeding of gentians in Japan. More than 60% of gentian production occurs from the Iwate prefecture. Gentians flowers bloom in mid-summer to late autumn in Japan. Commercialization of several ornamental plants has been scaled up by the help of genetic engineering approaches. For instance, Florigene Ltd. and Suntory Ltd. have successfully developed blue flowered carnations, and they are commercialized in Japan, North America, and Australia. They have also manufactured white-flowered transgenic gentians using antisense technology by subduing the chalcone synthase (CHS) gene (Nishihara et al. 2006). In this particular case, only 3 out of 17 independent transgenic lines exhibited white-flowered phenotypes, the other remaining transformants did not lead to successful inhibition of CHS gene expression.

Suppression of biosynthetic genes engaged in flower color formation is a necessary approach for obtaining target flower colors. The flower color of commercially important garden plant *Torenia hybrida* was effectively modulated by RNAi-mediated gene silencing against a gene of chalcone synthase (CHS), which is the first committed enzyme for anthocyanin and flavonoid biosynthesis. Anthocyanins, are pigments that contributes color to the flowers. Artificial regulation of its biosynthesis could result in flowers of desired colors. By utilizing distinct mRNA sites (the coding region and the 3′ -untranslated region) as RNAi targets, gene-specific gene silencing was induced, and the original blue flower color was modulated to white and pale colors, respectively (Fukusaki et al. 2004).

The RNAi technology can act as excellent tool for the silencing of pigment synthesis genes, which are responsible for different flower color patterns. Nishihara et al. (2005) suppressed the genes coding for chalcone isomerase (CHI) in tobacco by using RNAi. After CHI suppression the transgenic tobacco plants exhibited change of flavonoid components and reduced pigmentation in flower petals. Due to accumulation of high levels of chalcone in pollens plants showed yellow coloration. This indicated that CHI plays an important role in the cyclization reaction from chalcone to flavanone.

In 2008, Nakatsuka et al. carried out RNAi-mediated suppression of 3 anthocyanin biosynthetic genes—chalcone synthase (CHS), anthocyanidin synthase (ANS), and flavonoid 3′5′-hydroxylase (F3′5′H)—in gentian plant. In transgenics with suppressed CHS, petals showed pure white to pale-blue color, whereas in ANS suppressed transgenics, petals were only pale blue. Suppression of the F3′5′H gene minimized delphinidin derivatives and amplified cyanidin derivatives which resulted into magenta color flowers. The same group demonstrated that downregulation of anthocyanin 5,3′-aromatic acyltransferase (5/3′AT) and flavonoid 3′,5′-hydroxylase (F3′5′H) activities in gentian plant produced modified flower color.

10.3.6 Enhanced Fruit Shelf Life

In comparison to cereals, fruits and vegetables are more susceptible to spoilage due to their nature and composition. According to the Agricultural Research Data Book 2004 India loses about 30–40% of total fruits and vegetables produced due to spoilage. Therefore, we need to apply techniques that could increase the shelf life of vegetables and fruits as another necessary agronomic trait which may decrease the spoilage of vegetables and fruits, consequently lowering the horticultural loss.

By using RNAi technology Xiong et al. (2005) increased the shelf life in tomato by targeting the genes coding for ethylene biosynthesis pathway. A unit of dsRNA was introduced in tomato and the expression of ACC oxidase gene was blocked. The ethylene production rate in ripened fruits of transgenic plants was found to be extensively inhibited, ensuring a prolonged shelf life of tomato. Meli et al. (2010) suppressed 2 ripening-specific N-glycoprotein modifying enzymes, namely α-mannosidase (α-Man) and β-d-N-acetylhexosaminidase (β-Hex) using RNAi technology, after their identification in tomato. This exhibited that suppression of these genes increases the shelf life of tomato, by reducing the rate of softening.

10.3.7 Manipulation of Secondary Metabolite

Plant secondary metabolites are important sources of drugs, food additives, fragrances, pesticides, and pigments. It is anticipated that about 80% people throughout world are dependent mainly on the traditional herbal medicines to meet their primary healthcare needs (Canter et al. 2005). Sometimes the production of secondary metabolite is blocked due to the synthesis of undesirable compounds which can be inhibited by RNAi. RNAi technology is an extremely versatile technique because it is capable of controlling multigenes which are responsible for metabolite production (Borgio 2009).

The content of artemisinin (an anti-malarial drug isolated from *Artemisia annua* L) was enhanced by utilizing a hairpin-RNA-mediated RNAi technique. The expression of squalene synthase (SQS) was suppressed which is the key enzyme of sterol pathway. As a result, artemisinin content of some transgenic plants was significantly increased by three-fold as compared to untransformed control plants (Zhang et al. 2009). California poppy (*Eschscholzia californica*) cells were modified with RNAi construct harboring berberine bridge enzyme (BBE) gene to repress the activity of the enzyme and in these transgenic cells, end-products of isoquinoline alkaloid biosynthesis, such as sanguinarine, were decreased considerably and reticuline accumulation at a maximum level (Fujii et al. 2007).

Recently, potato tubers have developed as bioreactors for making of human therapeutic glycoproteins, increasing the yield of recombinant proteins, targeting the produced proteins to specific cellular compartments, and diminishing expensive protein purification steps. To develop potato tubers as a more efficient protein expression system the potato patatins were knocked out almost completely via RNAi technology (Kim et al. 2008).

Cassava is the main staple food in many tropical countries but it consists of unnecessary glucosides. Jorgensen et al. 2005 applied RNAi to stop production of the cytochrome P450 enzyme that makes the first step in the biosynthesis of linamarin and lotaustralin and produced transgenic cassava (*Manihot esculenta*) plants with removal of cyanogenic glucosides in the leaves (<1% of non-transgenic amounts) and a 93% reduction of cyanogenic glucoside amount in tubers.

Caffeine is a natural stimulant found in beverages like tea and coffee. A normal cup of filter coffee usually contains between 65 and 140 mg of caffeine. A single espresso contains around 100 mg of caffeine, while decaffeinated coffee has about 1.5–3 mg per cup. Caffeine stimulates the central nervous system, the heart muscle and also has a diuretic effect. Consumption of excessive caffeine can cause health problems, such as insomnia, anxiety, restlessness, dehydration, and palpitations. RNAi technology has enabled the scientist to produce varieties of coffee with low to very low caffeine content by modulation of caffeine biosynthesis in plants and suppression of CaMXMT1 (7-N-methylxanthine methyltransferase or theobromine synthase) (Ogita et al. 2003, 2004).

10.3.8 Seedless Fruit Development

It is already known that the phytohormone plays a very important role in regulation of transition between fruiting, fertilization, and flowering. Parthenocarpy can certainly be very useful for producing vegetables and fruits when pollination or fertilization is affected due to extreme temperatures, like in winter (Tomes 1997) or, more generally, to ensure yield stability in case of adverse pollination conditions. Recent findings have shown that seedlessness can contribute to increase the quality, texture, and shelf life of the fruits. Addtionally, it has been shown that seed development in fruits restricts the yield in cucumber (Denna 1973) and tomato (Falavigna and Soressi 1987).

Two members of the AUXIN RESPONSE FAMILY (ARF8 of *Arabidopsis thaliana* and ARF7 of tomato) show high level of expression in non-pollinated flowers and are down-regulated after pollination. RNAi technology is used to subdue the function of ARF7 in tomato and ARF8 in *Arabidopsis thaliana*. The transgenic plants produced resulted in the production of seedless (parthenocarpic) fruits. De Jong et al. (2009) highlighted that SlARF7 acts as a modifier of both auxin and gibberellin responses during tomato fruit set and development. Normally, SlARF7 transcript levels turn down after pollination and fertilization (Vriezen et al. 2008). Reduction of SlARF7 transcript levels by an RNAi approach may release the repression of the auxin and GA signaling pathways that are imposed by SlARF7 independently of pollination and fertilization, resulting in the incomplete activation of these pathways and thus in parthenocarpic fruit growth in tomato. Therefore, the fertilization-dependent step of the auxin signaling transduction pathway may be bypassed, which might be compulsory to start cell division activity and excite GA biosynthesis.

10.3.9 Deletion of Allergens from Food Crops

Allergens are naturally occurring compounds found in different food crops, capable of producing allergic response even if consumed in minute quantities. Consumption of allergens containing food causes various health problems in humans, which even cannot be cured with the use of existing therapies. Although symptoms of a food allergy can range from mild to severe, but some food allergies can be very serious and could even be life threatening which can impair breathing and affect heart rate. When these harmful substances are ingested in significant quantities or when they are not processed properly they can be potentially dangerous to human health causing food poisoning. Milk, egg, fish, peanut, tree nuts (walnut, Brazil nut, cashew, etc.), soybean, and shellfish (Sicherer and Sampson 2010; Zuidmeer et al. 2008) are some of the food items that are majorly responsible for the food allergic reactions. The RNAi technology has already been effectively employed to delete allergens and toxic compounds from food items because it can alter the biosynthesis of allergens by changing its biochemical pathway to upgrade the quality of food by lowering the probability of food allergy and toxicity.

Peanut allergy is one of the largest life-threatening food allergies causing lethal food related anaphylaxis. RNAi technology was applied to silence Ara h 2 gene, which is the most immune dominant allergen causing over 86% allergic reactions. By infecting peanut hypocotyls explants with *A. tumefaciens* EHA 105 harboring the pDK28 construct transgenic peanuts were produced. The allergenicity of transgenic peanut seeds was expressed as IgE binding capacity which was assessed by ELISA, using sera of patients who were allergic to peanut. The result revealed a remarkable decline in the IgE binding capacity of selected transgenic seeds without disturbing the plant morphology, growth rate, and reproduction as compared to the wild type. The unpurified peanut extract from the transgenic plants exhibited about 25% decrease in Ara h 2 content, thus illustrating the possibility of reducing peanut allergy by the use RNAi (Dodo et al. 2008).

The RNAi strategy was also used to inhibit the expression of a very well-known apple allergen Mald1, which belongs to a group of pathogen-related protein PR10. Mald 1 induces IgE-mediated hypersensitive response in organisms. Expression of Mal d1 dsRNA sequence reduces the expression of endogenous gene in developed RNAi apple plants and lowered the allergic response upon consumption (Gilissen et al. 2005). Soybean is a leguminous edible plant. The soybean allergy is mostly dominated by the Gly m Bd 30 K protein, which is also known as P34. By using transgene-induced gene silencing method Herman et al. (2003)stopped the assemblage of immune dominant soybean allergen Gly m Bd 30 K protein in transgenic soybean seeds without showing any developmental, structural, or ultra-structural phenotypic differences when compared with control plants.

In countries like Bangladesh, Ethiopia, and India many people eat a leafy vegetable known as grass pea (*Lathyrus sativus*). Grass pea also known as chickling pea contains beta-N-oxalylaminoalanine-L-alanine (BOAA), which is a neurotoxin that can cause paralytic disease called lathyrism, yet people still consume this vegetable in times of famine (Spencer et al. 1986). BOAA also imparts immunity

Table 10.2 List of allergens that could be removed from the corresponding plants using RNAi technology

Crop Plants	Allergens to be removed
Glycine max (soybean)	Protease/amylase inhibitors
Brassica oleracea	Glucosinolates
Ricinus communis (Castor bean)	Ricin
Phaseolus lunatus (Lima bean)	Cyanogenic glycosides
Solanum tuberosum(potato)	Solanine
Solanum lycopersicum (tomato)	Tomatine

to this unique crop under excessive stressful conditions. RNAi technology helped in bringing down the level of BOAA to a suitable concentration, rather than completely silencing the concerned genes, which will make crop safe for its consumption (Angaji et al. 2010).

Nor-nicotine is the precursor of a carcinogenic, tobacco-specific nitrosamine (TSNA), N′-nitrosonornicotine (NNN) and the change of nicotine to nor-nicotine is facilitated by an enzyme called nicotine demethylase (Hecht 1998). By inhibiting the activity of nicotine demethylase the level of defined carcinogen can significantly be reduced. Gavilano et al. (2006) used RNAi for silencing the N-demethylase (CYP82E4) gene which prevented the conversion of nicotine to nor-nicotine in tobacco.

Le et al. (2006) applied RNAi technology to silence two allergens Lyc e 1 and Lyc e 3, a non-specific lipid transfer protein (ns-LTP) of tomato. The transgenic lines obtained exhibited around 10- fold reductions in Lyc e 1 accumulation in fruits as compared to wild plants. Conversely, the level of Lyc e 3 in transgenic fruit reduced to less than 0.6% to that of in wild-type fruits producing phenotypically normal plants. These results are very promising for producing hypoallergenic tomatoes. Consumption of heavy metals even in low concentration can be life threatening for human beings. Rice can accumulate cadmium (Cd) to a significant level in its seeds due to the presence of phytochelatin synthase (PCS) genes. RNAi-mediated suppression of phytochelatin synthase (OsPCS1) gene reduced the accumulation of Cd in rice (Li et al. 2007). Thus, accumulation of heavy metals in rice seeds can be regulated through RNAi even when plants are grown in heavy metal-polluted soil. Similarly, there are many other plants where RNAi technology could be exploited to make plants free from allergic substances and made suitable for human consumption as explained in Table 10.2.

10.3.10 Change in Plant Architecture

Plant architecture controls various important agronomic traits in plants. For instance, plant height, pattern of shoot branching, plant morphology, inflorescence, crop yield, and resistance to environmental stresses (Khush 2001; Wang and Li 2006). Plant architecture can also be altered in order to reduce the negative effects of climate change on crop productivity. Understanding of molecular basis of plant architecture has served as platforms for RNAi-mediated alternation in plant architecture. Shorter plants with erect leaf architecture were created through RNAi-mediated silencing of OsDWARF4 gene in rice helped (Feldmann 2006).

RNAi-mediated downregulation of ornithine decarboxylase (ODC) gene (involved in polyamine biosynthesis) resulted in significant physiological and morphological changes including reduced leaf size, decreased abiotic stress tolerance, delayed flowering, and early onset of senescence in tobacco (Choubey and Rajam 2017).

Plant architecture has been found to be regulated by miRNAs. The manipulation of miRNA expression directly or indirectly affected the plant architecture, biomass accumulation, and yield. Corngrass1 (Cg1) miRNA that belongs to the mir156 family regulates vegetative growth and flowering in plants. In maize, overexpression of Cg1 miRNA caused prolongation of vegetative phase and delay in flowering time (Chuck et al. 2011). Similarly, phenotype was also observed when Cg1 was overexpressed in other plant species, for instance, overexpressing Cg1miRNA in *Populus* plants showed significant shortening of internode length, increase in the growth of axillary meristem, and about 30% reduction in stem lignin content as compared to the untransformed control (Rubinelli et al. 2013). Biofuel production can be enhanced through low lignin content in plant material. Lignin makes the plant material recalcitrant for conversion to ethanol. Downregulation of lignin biosynthetic genes by RNAi can help in the production of low lignin-containing plants. For instance, RNAi-mediated downregulation of lignin-associated genes such as Cinnamate-4-hydroxylase, shikimate hydroxycinnamoyl transferase, and 4-coumarate-CoA ligase reduced the lignin content and increased its accessibility to cellulose to degradation (Hisano et al. 2009).

Thus, RNAi technology has a wide utility in manipulating the plant architecture for high yield, increase in biomass, flowering, and removal of undesirable phenotypes. Rose plant can be easily modified for its thorn characteristic or the plant architecture in mulberry, and tea plants can be manipulated for easy plucking of leaves.

10.4 Conclusion

The currently available agricultural technology requires additional molecular tools to decrease the crop loss and feed extra mouths, which will increase by 2 billion over the next 30 years, according to the latest estimation by the Food and Agriculture Organization (FAO). RNAi-mediated gene suppression is an area of new research nowadays leading to brand-new discoveries. In the last 15 years, it has emerged as a powerful innovation for gene silencing with a huge potential to improve the agricultural yield and engineer resistant genotypes, thereby improving the way of life without influencing other agronomic traits. The implementation of RNAi-mediated gene silencing has opened new avenues in the successful development of transgenic crops such as nicotine free tobacco, decaffeinated coffee, tea, nutrient bio-fortified cereals, low glutelin-containing wheat and hypoallergenic crops. It is being used in functional genomics studies to decipher the function of genes.

There are several additional benefits of RNAi technology compared to other biotechnological approaches. Sequence-specificity of the gene silencing process and multiple gene targeting at the same time are some of the advantages of RNAi. RNAi technology has been successfully applied to bring about the modifications of a variety of desired traits, including nutritional fortifications, reduction of allergen, morphological amendments, male sterility altering, secondary metabolite enrichment, and boosted defense against varying environmental conditions like extreme temperatures, flood and drought, oxidative stress, and changes in soil compositions (heavy metal accumulation, salinity, etc.).

However, every technology has certain limitations. The RNAi technology helps in sequence-specific targeting but there can be issues of off-target effects leading to unwanted traits. Despite the tremendous success, commercialization and the use of RNAi-based pathogen resistance in plants for practical disease control have not been as broadly adopted as might be expected. This may be likely due to both the costs related with regulatory approvals and public awareness.

Therefore it can be concluded that such sophisticated technology having revolutionary capabilities could be further exploited for functional analysis of target genes and regulation of gene expression for crop protection and improvement to conquer the problem of food security.

Acknowledgement Priyanka Gautam acknowledges financial support from University of Allahabad, Allahabad, India in the form of UGC- CRET research fellowship. All the authors also acknowledge DST-FIST and UGC-SAP facility of the Department of Biochemistry, University of Allahabad, Prayagraj, India.

Conflict of Interest The authors declare no conflicts of interest.

References

Aly R, Cholakh H, Joel DM, Leibman D, Steinitz B, Zelcer A, Naglis A, Yarden O, Gal-On A (2009) Gene silencing of mannose 6-phosphate reductase in the parasitic weed Orobanche aegyptiaca through the production of homologous dsRNA sequences in the host plant. Plant Biotechnol J 7:487–498

Andika IB, Kondo H, Tamada T (2005) Evidence that RNA silencing mediated resistance to beet necrotic yellow vein virus is less effective in roots than in leaves. Mol Plant-Microbe Interact 18: 194–204

Angaji SA, Hedayati SS, Hoseinpoor R, Samadpoor S, Shiravi S, Madani S (2010) Application of RNA interference in plants. Plant Omics J 3:77–84

Barciszewska-Pacak M, Milanowska K, Knop K, Bielewicz D, Nuc P, Plewka P, Pacak AM, Vazquez F, Karlowski W, Jarmolowski A, Szweykowska-Kulinska Z (2015) Arabidopsis microRNA expression regulation in a wide range of abiotic stress responses. Front Plant Sci 6:410

Baum JA, Bogaert T, Clinton W, Heck GR, Feldmann P, Ilagan O, Johnson S, Plaetinck G, Munyikwa T, Pleau M, Vaughn T, Roberts J (2007) Control of coleopteran insect pests through RNA interference. Nat Biotechnol 25:1322–1326

Baumberger N, Baulcombe DC (2005) Arabidopsis ARGONAUTE1 is an RNA slicer that selectively recruits microRNAs and short interfering RNAs. Proc Natl Acad Sci U S A 102: 11928–11933

Borges F, Martienssen RA (2015) The expanding world of small RNAs in plants. Nat Rev Mol Cell Biol 16:727–741

Borgio JF (2009) RNA interference (RNAi) technology: a promising tool for medicinal plant research. J Med Plant Res 3:1176–1183

Bosher JM, Labouesse M (2000) RNA interference: genetic wand and genetic watchdog. Nat Cell Biol 2(2):E31

Bucher E, Lohius D, van Poppel PM, Geerts-Dimitriadou C, Goldbach R, Prins M (2006) Multiple virus resistance at a high frequency using a single transgene construct. J Gen Virol 87:697–701

Canter PH, Thomas H, Ernst E (2005) Bringing medicinal plants into cultivation: opportunities and challenges for biotechnology. Trends Biotechnol 23:180–185

Cao X, Wu Z, Jiang F, Zhou R, Yang Z (2014) Identification of chilling stress-responsive tomato microRNAs and their target genes by high-throughput sequencing and degradome analysis. BMC Genomics 15:1130

Castel SE, Martienssen RA (2013) RNA interference in the nucleus: roles for small RNAs in transcription, epigenetics and beyond. Nat Rev Genet 14:100–112

Choubey A, Rajam MV (2017) Transcriptome response and developmental implications of RNAimediated ODC knockdown in tobacco. Funct Integr Genomics. https://doi.org/10.1007/s10142-016-0539-3

Chuck GS, Tobias C, Sun L, Kraemer F, Li C, Dibble D, Arora R, Bragg JN, Vogel JP, Singh S, Simmons BA, Pauly M, Hake S (2011) Overexpression of the maize corngrass1 microRNA prevents flowering, improves digestibility, and increases starch content of switchgrass. Proc Natl Acad Sci U S A 108:17,550–17,555

Da-Hong L, Hui L, Yan-li Y, Ping-ping Z, Jian-sheng L (2009) Down regulated expression of RACK1 gene by RNA interference enhances drought tolerance in rice. Rice Sci 16:14–20

Davuluri GR, Tuinen A, Fraser PD, Manfredonia A, Newman R, Burgess D, Brummell DA, King SR, Palys J, Uhlig J, Bramley PM, Pennings HMJ, Bowle C (2005) Fruit-specific RNAimediated suppression of DET1 enhances carotenoid and flavonoid content in tomatoes. Nat Biotechnol 23:890–895

De Jong M, Wolters-Arts M, Feron R, Mariani C, Vriezen WH (2009) The Solanum lycopersicum auxin response factor 7 (SlARF7) regulates auxin signaling during tomato fruit set and development. Plant J 5:160–170

Denna DW (1973) Effect of genetic parthenocarpy and gynoecious flowering habit on fruit production and growth of cucumber, Cucumis sativus L. J Am Soc Hortic Sci 98:602–604

Dodo HW, Konan KN, Chen FC, Egnin M, Viquez OM (2008) Alleviating peanut allergy using genetic engineering: the silencing of the immunodominant allergen Ara h 2 leads to its significant reduction and a decrease in peanut allergenicity. Plant Biotech J 6:135–145

Dunoyer P, Himber C, Voinnet O (2006) Induction, suppression and requirement of RNA silencing pathways in virulent Agrobacterium tumefaciens infections. Nat Genet 38(2):258–263

Duvick DN (1999) Heterosis: feeding people and protecting natural resources. In: Coors JG, Pandey S (eds) Genetics and exploitation of heterosis in crops. American Society of Agronomy/Crop Science Society of America, Madison, pp 19–29

Eck JV, Conlin B, Garvin DF, Mason H, Navarre DA, Brown CR (2007) Enhanced beta-carotene content in potato via RNAi silencing of the beta-carotene hydroxylase gene. Am J Potato Res 84:11,331–11,342

Enrique R, Siciliano F, Favaro MA, Gerhardt N, Roeschlin R, Rigano L, Sendin L, Castagnaro A, Vojnov A, Marano MR (2011) Novel demonstration of RNAi in citrus reveals importance of citrus callose synthase in defence against Xanthomonas citri subsp. Citri Plant Biotechnol J 9: 394–407

Escobar MA, Dandekar AM (2003) *Agrobacterium tumefaciens* as an agent of disease. Trends Plant Sci 8:380–386

Escobar MA, Civerolo EL, Summerfelt KR, Dandekar AM (2001) RNAi-mediated oncogene silencing confers resistance to crown gall tumorigenesis. Proc Natl Acad Sci USA 98(23): 13,437–13,442

Escobar MA, Leslie CA, McGranahan GH, Dandekar AM (2002) Silencing crown gall disease in walnut (*Juglans regia* L.). Plant Sci 163:591–597

Fairbairn DJ, Cavallaro AS, Bernard M, Mahalinga-Iyer J, Graham MW, Botella JR (2007) Host-delivered RNAi: an effective strategy to silence genes in plant parasitic nematodes. Planta 226: 1525–1533

Falavigna A, Soressi GP (1987) Influence of the pat-sha gene on plant and fruit traits in tomato (L. esculentum Mill.). In: Modern trends in tomato genetics and breeding, p 128. Proceedings of the 10th meeting of the EUCARPIA tomato working group FAO (2000). The state of food insecurity in the world 2000. FAO, Rome

Feldmann KA (2006) Steroid regulation improves crop yield. Nat Biotechnol 24:46–47

Fire A, Xu S, Montgomery MK, Kostas SA, Driver SE, Mello CC (1998) Potent and specific genetic interference by doublestranded RNA in Caenorhabditis elegans. Nature 391:806–881

Flores T, Karpova O, Su X, Zeng P, Bilyeu K, Sleper DA, Nguyen HT, Zhang ZJ (2008) Silencing of GmFAD3 gene by siRNA leads to low alpha-linolenic acids (18:3) of fad3-mutant phenotype in soybean [Glycine max (Merr.)]. Transgenic Res 17:839–850

Fujii N, Inui T, Iwasa K, Morishige T, Sato F (2007) Knockdown of berberine bridge enzyme by RNAi accumulates (S)-reticuline and activates a silent pathway in cultured California poppy cells. Transgenic Res 16:363–375

Fu R, Zhang M, Zhao Y, He X, Ding C, Wang S, Feng Y, Song X, Li P, Wang B (2017) Identification of salt tolerance-related microRNAs and their targets in maize (Zea mays L.) using high-throughput sequencing and degradome analysis. Front Plant Sci 8:1–13

Fukusaki E, Kawasaki K, Kajiyama S, An CI, Suzuki K, Tanaka Y, Kobayashi A (2004) Flower color modulations of Torenia hybrida by downregulation of chalcone synthase genes with RNA interference. J Biotechnol 111:229–240

Gao S, Yang L, Zeng HQ, Zhou ZS, Yang ZM, Li H, Sun D, Xie F, Zhang B (2016) A cotton miRNA is involved in regulation of plant response to salt stress. Sci Rep 6:19,736

Gavilano LB, Coleman NP, Burnley L, Bowman M, Kalengamaliro N, Hayes A, Bush LP, Siminszky B (2006) Genetic engineering of Nicotiana tabacum for reduced nornicotine content. J Agric Food Chem 54:9071–9078

Ghag SB, Shekhawat UKS, Ganapathi TR (2014) Host induced post-transcriptional hairpin RNA-mediated gene silencing of vital fungal genes confers efficient resistance against Fusarium wilt in banana. Plant Biotechnol J 12(5):541–553

Gheysen G, Vanholme B (2007) RNAi from plants to nematodes. Trends Biotechnol 25:89–92

Gilissen LJ, Bolhaar ST, Matos CI, Rouwendal GJ, Boone MJ, Krens FA, Zuidmeer L, Van Leeuwen A, Akkerdaas J, Hoffmann Sommergruber K, Knulst AC, Bosch D, Van de Weg WE, Van Ree R (2005) Silencing the major apple allergen Mal d 1 by using the RNA interference approach. J Allergy Clin Immunol 119:384–390

Gupta OP, Meena NL, Sharma I, Sharma P (2014) Differential regulation of microRNAs in response to osmotic, salt and cold stresses in wheat. Mol Biol Rep 41:4623–4629

Hamilton AJ, Baulcombe DC (1999) A species of small antisense RNA in posttranscriptional gene silencing in plants. Science 286:950–952

Hamza NB, Sharma N, Tripathi A, Sanan-Mishra N (2016) MicroRNA expression profiles in response to drought stress in Sorghum bicolor. Gene Expr Patterns 20:88–98

Hannon GJ (2002) RNA interference. Nature 418:244–251

Hecht SS (1998) Biochemistry, biology and carcinogenicity of tobacco specific N-nitrosamines. Chem Res Toxicol 11:559–603

Herman EM, Helm RM, Jung R, Kinney AJ (2003) Genetic modification removes an immunodominant allergen from soybean. Plant Physiol 132:36–43

Hernandez I, Chacon O, Rodriguez R, Portieles R, Pujol YLM, Borras-Hidalgo O (2009) Black shank resistant tobacco by silencing of glutathione S-transferase. Biochem Biophys Res Commun 387:300–304

Hily JM, Ravelonandro M, Damsteegt V, Basset C, Petri C, Liu Z, Scorza R (2007) Plum pox virus coat protein gene intron-hairpin-RNA (ihpRNA) constructs provide resistance to plum pox virus in Nicotiana benthamiana and Prunus domestica. J Am Soc Hortic Sci 132:850–858

Hisano H, Nandakumar R, Wang ZY (2009) Genetic modification of lignin biosynthesis for improved biofuel production. In Vitro Cell Dev Biol Plant 45:306–313

Hivrale V, Zheng Y, Puli COR, Jagadeeswaran G, Gowdu K, Kakani VG, Barakat A, Sunkar R (2016) Characterization of drought- and heat-responsive microRNAs in switchgrass. Plant Sci 242:214–223

Huang G, Allen R, Davis EL, Baum TJ, Hussey RS (2006) Engineering broad root-knot resistance in transgenic plants by RNAi silencing of a conserved and essential root-knot nematode parasitism gene. Proc Natl Acad Sci U S A 103:14,302–14,306

Husken A, Baumert A, Milkowski C, Becker HC, Strack D, Mollers C (2005) Resveratrol glucoside (Piceid) synthesis in seeds of transgenic oilseed rape (*Brassica napus* L.). Theor Appl Genet 111:1553–1562

Jan A, Maruyama K, Todaka D, Kidokoro S, Abo M, Yoshimura E, Shinozaki K, Nakashima K, Yamaguchi-Shinozaki K (2013) OsTZF1, a CCCH-tandem zinc finger protein, confers delayed senescence and stress tolerance in rice by regulating stress-related genes. Plant Physiol 161: 1202–1216

Jorgensen K, Bak S, Busk PK, Sorensen C, Olsen CE, Puonti-Kaerlas J, Moller BL (2005) Cassava plants with a depleted cyanogenic glucoside content in leaves and tubers. Plant Physiol 139: 363–374

Kamachi S, Mochizuki A, Nishiguchi M, Tabei Y (2007) Transgenic Nicotiana benthamiana plants resistant to cucumber green mottle mosaic virus based on RNA silencing. Plant Cell Rep 26: 1283–1288

Katiyar-Agarwal S, Morgan R, Dahlbeck D, Borsani O, Villegas JA, Zhu J, Staskawicz BJ, Jin H (2007) A pathogen-inducible endogenous siRNA in plant immunity. Proc Natl Acad Sci U S A 103:47–52

Katsumoto Y, Fukuchi-Mizutani M, Fukui Y, Brugliera FA, Holton T, Karan M, Nakamura N, Yonekura-Sakakibara K, Togami J, Pigeaire A, Guo-Qing Tao S, Nehra N, Chin-Yi Lu K, Dyson B, Tsuda S, Ashikari T, Kusumi TG, Mason J, Tanaka Y (2007) Engineering of the rose flavonoid biosynthetic pathway successfully generated blue-hued flowers accumulating Delphinidin. Plant Cell Physiol 48(11):1589–1600

Ketting RF (2011) The many faces of RNAi. Dev Cell 20:148–161

Khush GS (2001) Green revolution: the way forward. Nat Rev Genet 2:815–822

Kim YS, Lee YH, Kim HS, Kim MS, Hahn KW, Ko JH, Joung H, Jeon JH (2008) Development of patatin knockdown potato tubers using RNA interference (RNAi) technology, for the production of human-therapeutic glycoproteins. BMC Biotechnol 8:36

Kohli D, Joshi G, Deokar AA, Bhardwaj AR, Agarwal M, Katiyar-Agarwal S, Srinivasan R, Jain PK (2014) Identification and characterization of wilt and salt stress-responsive microRNAs in chickpea through high-throughput sequencing. PLoS One 9:e108851

Krubphachaya P, Jurˇícˇek M, Kertbundit S (2007) Induction of RNAmediated resistance to papaya ring spot virus type W. J Biochem Mol Biol 40:401–411

Kruszka K, Pieczynski M, Windels D, Bielewicz D, Jarmolowski A, Szweykowska-Kulinska Z, Vazquez F (2012) Role of microRNAs and other sRNAs of plants in their changing environments. J Plant Physiol 169:1664–1672

Kumar RR, Pathak H, Sharma SK, Kala YK, Nirjal MK, Singh GP, Goswami S, Rai RD (2015) Novel and conserved heat-responsive microRNAs in wheat (*Triticum aestivum* L.). Funct Integr Genomics 15:323–348

Kumar P, Mundiyara R, Jitarwal RC (2017) RNA interference: new approach of gene silencing in plants. Int J Pure Appl Biosci 5(3):421–425

Kusaba M, Miyahara K, Iida S, Fukuoka H, Takano T, Sassa H, Nishimura M, Nishio T (2003) Low glutelin content 1: a dominant mutation that suppresses the glutelin multigene family via RNA silencing in rice. Plant Cell 15:1455–1467

Le LQ, Mahler V, Lorenz Y, Scheurer S, Biemelt S, Vieths S, Sonnewald U (2006) Reduced allergenicity of tomato fruits harvested from Lyc e 1-silenced transgenic tomato plants. J Allergy Clin Immunol 118:1176–1183
Li JC, Guo JB, Xu WZ, Ma M (2007) RNA interference-mediated silencing of phytochelatin synthase gene reduces cadmium accumulation in rice seeds. J Integr Plant Biol 49:1032–1037
Li J, Wu LQ, Zheng WY, Wang RF, Yang LX (2015) Genome-wide identification of microRNAs responsive to high temperature in rice (*Oryza sativa*) by high-throughput deep sequencing. J Agron Crop Sci 201:379–388
Li C, Li Y, Bai L, He C, Yu X (2016) Dynamic expression of miRNAs and their targets in the response to drought stress of grafted cucumber seedlings. Hortic Plant J 2:41–49
Liu S, Geng S, Li A, Mao Y, Mao L (2021) RNAi technology for plant protection and its application in wheat. aBIOTECH. https://doi.org/10.1007/s42994-021-00036-3
Manavalan LP, Chen X, Clarke J, Salmeron J, Nguyen HT (2012) RNAi-mediated disruption of squalene synthase improves drought tolerance and yield in rice. J Exp Bot 63:163–175
Mangrauthia SK, Bhogireddy S, Agarwal S, Prasanth VV, Voleti SR, Neelamraju S, Subrahmanya D (2017) Genome-wide changes in microRNA expression during short and prolonged heat stress and recovery in contrasting rice cultivars. J Exp Bot 68:2399–2412
Mao YB, Cai WJ, Wang JW, Hong GJ, Tao XY, Wang LJ, Huang YP, Chen XY (2007) Silencing a cotton bollworm P450 monooxygenase gene by plant-mediated RNAi impairs larval tolerance of gossypol. Nat Biotechnol 25:1307–1313
Meli VS, Ghosh S, Prabha TN, Chakraborty N, Chakraborty S, Datta A (2010) Enhancement of fruit shelf life by suppressing N-glycan processing enzymes. Proc Natl Acad Sci U S A 107: 2413–2418
Mickelbart MV, Hasegawa PM, Bailey-Serres J (2015) Genetic mechanisms of abiotic stress tolerance that translate to crop yield stability. Nat Rev Genet 16:237–251
Missiou A, Kalantidis K, Boutla A, Tzortzakaki S, Tabler M, Tsagris M (2004) Generation of transgenic potato plants highly resistant to potato virus Y (PVY) through RNA silencing. Mol Breed 14:185–197
Mittal D, Sharma N, Sharma V, Sopory SK, Sanan-Mishra N (2016) Role of microRNAs in rice plant under salt stress. Ann Appl Biol 168:2–18
Mittler R, Blumwald E (2010) Genetic engineering for modern agriculture: challenges and perspectives. Ann Rev Plant Biol 61:443–462
Napoli C, Lemieux C, Jorgensen R (1990) Introduction of a chimeric chalcone synthase gene into petunia results in reversible cosuppression of homologous genes in trans. Plant Cell 2:279–289
Nawaz-ul-Rehman MS, Mansoor S, Khan AA, Zafar Y, Briddon RW (2007) RNAi-mediated male sterility of tobacco by silencing TA29. Mol Biotechnol 36:159–165
Nishihara M, Nakatsuka T, Yamamura S (2005) Flavonoid components and flower color change in transgenic tobacco plants by suppression of chalcone isomerase gene. FEBS Lett 579:6074–6078
Nishihara M, Nakatsuka T, Hosokawa K, Yokoi T, Abe Y, Mishiba K, Yamamura S (2006) Dominant inheritance of white-flowered and herbicide-resistant traits in transgenic gentian plants. Plant Biotechol 23:25–31
Nowara D, Gay A, Lacomme C, Shaw J, Ridout C, Douchkov D, Hensel G, Kumlehn J, Schweizer P (2010) HIGS: host-induced gene silencing in the obligate biotrophic fungal pathogen *Blumeria graminis*. Plant Cell 22:3130–3141
Ogita S, Uefuji H, Yamaguchi Y, Koizumi N, Sano H (2003) RNA interference: producing decaffeinated coffee plants. Nature 423:823
Ogita S, Uefuji H, Morimoto M, Sano H (2004) Application of RNAi to confirm theobromine as the major intermediate for caffeine biosynthesis in coffee plants with potential for construction of decaffeinated varieties. Plant Mol Biol 54:931–941
Pooggin M, Shivaprasad PV, Veluthambi K, Hohn T (2003) RNAi targeting of DNA virus in plants. Nat Biotechnol 21:131–132
Rajam MV (2012) Host induced silencing of fungal pathogen genes: an emerging strategy for disease control in crop plants. Cell Dev Biol 1:1

Ren Y, Chen L, Zhang Y, Kang X, Zhang Z, Wang Y (2013) Identification and characterization of salt-responsive microRNAs in Populus tomentosa by high-throughput sequencing. Biochimie 95:743–750

Riechen J (2007) Establishment of broad-spectrum resistance against Blumeria graminis f. sp. tritici in Triticum aestivum by RNAi mediated knock-down of MLO. J Verbr Leb 2:120

Rodoni BC, Dale JL, Harding RM (1999) Characterization and expression of the coat protein-coding region of the banana bract mosaic potyvirus, development of diagnostic assays and detection of the virus in banana plants from five countries in Southeast Asia. Arch Virol 144: 1725–1737

Romano N, Macino G (1992) Quelling: transient inactivation of gene expression in Neurospora crassa by transformation with homologous sequences. Mol Microbiol 22:3343–3353

Rubinelli PM, Chuck G, Li X, Meilan R (2013) Constitutive expression of the corn-grass1 microRNA in poplar affects plant architecture and stem lignin content and composition. Biomass Bioenergy 54:312–321

Saha K, Mishra S (2018) RNA interference technology: a novel approach for fruit and vegetable crop improvement. Res J Agri Sci 10(1):01–07

Sano T, Matsuura Y (2004) Accumulation of short interfering RNAs characteristic of RNA silencing precedes recovery of tomato plants from severe symptoms of potato spindle tuber viroid infection. J Gen Plant Pathol 70:50–53

Saurabh S, Vidyarthi AS, Prasad D (2014) RNA interference: concept to reality in crop improvement. Planta 239:543-5641-297

Sharma HC, Crouch JH, Sharma KK, Seetharama N, Hash CT (2002) Applications of biotechnology for crop improvement: prospects and constraints. Plant Sci 163:381–395

Shimizu T, Yoshii M, Wei T, Hirochika H, Omura T (2009) Silencing by RNAi of the gene for Pns12, a viroplasm matrix protein of Rice dwarf virus, resulted in strong resistance of transgenic rice plants to the virus. Plant Biotechnol J 7:24–32

Shriram V, Kumar V, Devarumath RM, Khare TS, Wani SH (2016) MicroRNAs as potential targets for abiotic stress tolerance in plants. Front Plant Sci 7:8173389–8173817

Sicherer SH, Sampson HA (2010) Food allergy. J Allergy Clin Immunol 125:116–125

Simon-Mateo C, Garcia JA (2011) Antiviral strategies in plants based on RNA silencing. Biochim Biophys Acta 1809:722–731

Sindhu AS, Maier TR, Mitchum MG, Hussey RS, Davis EL, Baum TJ (2009) Effective and specific in planta RNAi in cyst nematodes: expression interference of four parasitism genes reduces parasitic success. J Exp Bot 60:315–324

Sinha R, Rajam MV (2013) RNAi silencing of three homologues of S-adenosylmethionine decarboxylase gene in tapetal tissue of tomato results in male sterility. Plant Mol Biol 82:169–180

Spencer PS, Roy DN, Ludolph A, Hugon J, Dwivedi MP, Schaumburg HH (1986) Lathyrism: evidence for role of the neuroexcitatory aminoacid BOAA. Lancet 2:1066–1067

Sunkar R, Zhu JK (2004) Novel and stress-regulated micro RNAs and other small RNAs from Arabidopsis. Plant Cell 16:2001–2019

Tanaka Y, Brugliera F, Chandler S (2009) Recent progress of flower colour modification by biotechnology. Int J Mol Sci 10:5350–5369

Tehseen M, Imran M, Hussain M, Irum S, Ali S, Mansoor S, Zafar Y (2010) Development of male sterility by silencing Bcp1 gene of Arabidopsis through RNA interference. Afr J Biotechnol 9: 2736–2741

Tenllado F, Martínez-García B, Vargas M, Díaz-Ruíz JR (2003) Crude extracts of bacterially expressed dsRNA can be used to protect plants against virus infections. BMC Biotechnol 3: 3–14

Tester M, Langridge P (2010) Breeding technologies to increase crop production in a changing world. Science 327:818–822

Tomes DT (1997) Seedless hopes bode well for winter vegetables. Nat Biotechnol 15:1344–1345

Tucker G (2003) Nutritional enhancement of plants. Curr Opin Biotechnol 14:221–225

Van Uyen N (2006) Novel approaches in plants breeding RNAi technology. Proceedings of International Workshop on Biotechnology in Agriculture 12–16

Vanderschuren H, Alder A, Zhang P, Gruissem W (2009) Dose dependent RNAi-mediated geminivirus resistance in the tropical root crop cassava. Plant Mol Biol 64:549–557
Vanitharani R, Chellappan P, Fauquet CM (2003) Short interfering RNA-mediated interference of gene expression and viral DNA accumulation in cultured plant cells. Proc Natl Acad Sci U S A 100:9632–9636
Vaucheret H (2008) Plant ARGONAUTES. Trends Plant Sci 13:350–358
Vriezen WH, Feron R, Maretto F, Keijman J, Mariani C (2008) Changes in tomato ovary transcriptome demonstrate complex hormonal regulation of fruit set. New Phytol 177:60–76
Wang Y, Li J (2006) Genes controlling plant architecture. Curr Opin Biotechnol 17:123–129
Wang M, Wang Q, Zhang B (2013) Response of miRNAs and their targets to salt and drought stresses in cotton (Gossypium hirsutum L.). Gene 530:26–32
Waterhouse PM, Graham WMB (1998) Virus resistance and gene silencing in plants can be induced by simultaneous expression of sense and antisense RNA. Proc Natl Acad Sci U S A 95: 13,959–13,964
Wolfenbarger LL, Phifer PR (2000) Biotechnology and ecology-the ecological risks and benefits of genetically engineered plants. Science 290:2088–2093
Xie F, Stewart CN, Taki FA, He Q, Liu H, Zhang B (2014) High-throughput deep sequencing shows that microRNAs play important roles in switchgrass responses to drought and salinity stress. Plant Biotechnol J 12:354–366
Xiong A, Yao Q, Peng R, Li X, Han P, Fan H (2005) Different effects on ACC oxidase gene silencing triggered by RNA interference in transgenic tomato. Plant Cell Rep 23:639–646
Xu S, Liu N, Mao W, Hu Q, Wang G, Gong Y (2016) Identification of chilling-responsive microRNAs and their targets in vegetable soybean (*Glycine max* L.). Sci Rep 6:26,619
Yang X, Zhao Y, Xie D, Sun Y, Zhu X, Esmaeili N, Yang Z, Wang Y, Yin G, Lv S, Nie L, Tang Z, Zhao F, Li W, Mishra N, Sun L, Zhu W, Fang W (2016) Identification and functional analysis of microRNAs involved in the anther development in cotton genic male sterile line Yu98-8A. Int J Mol Sci 17:1677
Yang S, Dai Y, Chen Y, Yang J, Yang D, Liu Q, Jian H (2019) A novel G16B09-like effector from Heterodera avenae suppresses plant defenses and promotes parasitism. Front Plant Sci 10:66
Zamore PD, Tuschl T, Sharp PA, Bartel DP (2000) RNAi: double-stranded RNA directs the ATP-dependent cleavage of mRNA at 21 to 23 nucleotide intervals. Cell 101:25–33
Zhang L, Jing F, Li F, Li M, Wang Y, Wang G, Sun X, Tang K (2009) Development of transgenic Artemisia annua (Chinese wormwood) plants with an enhanced content of artemisinin, an effective anti-malarial drug, by hairpin-RNA-mediated gene silencing. Biotechnol Appl Biochem 52:199–207
Zhang N, Yang J, Wang Z, Wen Y, Wang J, He W, Liu B, Si H, Wang D (2014) Identification of novel and conserved microRNAs related to drought stress in potato by deep sequencing. PLoS One 9(4):e95489
Zhang F, Luo X, Zhou Y, Xie J (2016a) Genome-wide identification of conserved microRNA and their response to drought stress in Dongxiang wild rice (*Oryza rufipogon* Griff.). Biotechnol Lett 38:711–721
Zhang W, Xie Y, Xu L, Wang Y, Zhu X, Wang R, Zhang Y, Muleke EM, Liu L (2016b) Identification of microRNAs and their target genes explores miRNA-mediated regulatory network of cytoplasmic male sterility occurrence during anther development in radish (*Raphanus sativus* L.). Front Plant Sci 7:1054
Zhao MM, An DR, Zhao J, Huang G, He Z, Chen J (2006) Transiently expressed short hairpin RNA targeting 126 kDa protein of tobacco mosaic virus interferes with virus infection. Acta Biochim Biophys Sin 38:22–28
Zhu D, Deng XW (2012) A non-coding RNA locus mediates environment-conditioned male sterility in rice. Cell Res 22:791–792
Zuidmeer L, Goldhahn K, Rona RJ, Gislason D, Madsen C, Summers C, Sodergren E, Dahlstrom J, Lindner T, Sigurdardottir ST, McBride D, Keil T (2008) The prevalence of plant food allergies: a systematic review. J Allergy Clin Immunol 121:1210–1218

Use of Genomics to Improve Stress Tolerance

11

Ritu Mahajan, Nisha Kapoor, and Bijender K. Bajaj

Abstract

Plants being sessile are exposed to large number of biotic and abiotic stresses during their growth and development which are the major constraints in the agricultural productivity. Several stresses being complex in nature are controlled by networks of several genetic and environmental factors thus hindering the traditional breeding strategies. Unraveling the molecular mechanism behind the complex nature of stress is important to broaden the perspective of plant breeding for generating stress tolerant varieties under the current climate scenario and meeting the demands of forever increasing global population.

Recent advances in genomics have paved a significant role in understanding the response of a plant towards stress. Emergence of large high-throughput tools has deciphered the entire genome to gain insights about various genes and metabolic pathways/networks involved. Several genomic approaches like transcriptomics, metabolomics, and proteomics help in identification of novel metabolites in a particular cell. Development in these omics based approaches and their utilization can enhance the sensitivity of genomic tools and further bioinformatic studies will decode various regulatory mechanisms involved in stress biology. This will give a better understanding for utilizing the outcomes of valuable information in manipulating the mystery of four nucleotides and generating engineered based crops. Recent introduction of genome editing tools has revolutionized every aspect of crop plants in developing strategies for increasing crop productivity and quality.

Keywords

Genomics · Abiotic stress · Biotic stress · Sequencing transcription factors

R. Mahajan · N. Kapoor · B. K. Bajaj (✉)
School of Biotechnology, University of Jammu, Jammu, Jammu & Kashmir, India

R. L. Singh et al. (eds.), *Plant Genomics for Sustainable Agriculture*,
https://doi.org/10.1007/978-981-16-6974-3_11

11.1 Introduction

With the quick increase in the world's population, there is a corresponding increase in food demand, owing to concerns about the large impact on agriculture productivity. India is an agricultural country, where large number of people depends upon various crops both for sustenance and livelihood. Changes in environmental conditions have large effect on the crop productivity. Stresses either biotic or abiotic are the most important causes that results in yield losses in crop plants. Abiotic stresses like extreme temperature, drought, salinity, etc., are the major outcomes of global climate changes thereby reducing the yield of crops mostly by 50% (Raza et al. 2019). These stresses being complex in nature are controlled by several different interacting signaling pathways based on genetic and environmental factors. Plants show their response to stresses by altering molecular and physiological processes at molecular, cellular, and whole-plant levels. At the molecular level, several stress-responsive functional or regulatory genes are produced. Functional genes produce various metabolites, channel proteins, anti-oxidative enzymes which helps the plants to survive during stress and their subsequent post recovery after stress (Table 11.1). Determination of stress mediated signaling mechanisms at molecular level and interacting network of underlying regulatory proteins is the major focus of research. Explication of functional genes involved in various regulatory pathways is important to generate stress tolerant crops (Sharma and Pandey 2017).

However, regulatory genes encode certain regulatory proteins like transcription factors, stress receptors, and protein kinases that aids in controlling signal transduction pathways and modifying gene expression (Joshi et al. 2016). At the transcriptional level, several stress-responsive and/or stress-activated TFs along with transcriptional regulators (bZ IP, DREB, MYC, MYB, NAC, and WRKY families) are responsible for plant stress tolerance (Baillo et al. 2019). So it is necessary to adopt some recent and novel approaches to meet the demands of an ever-growing global population.

Research in plant genomics has increased tremendously in the past 30 years due to the availability of cost-effective and high-throughput tools. Emergence of DNA sequencing platforms for exploring the plant genomes has provided with broad implications for plant and applications in biology research. Also, omics based techniques have resulted in generation of huge amount of data and have considerably scale down the cost along with the time to perform large-scale activities including whole genome sequencing and gene expression analysis. These technological based approaches aim to gain insights into plant molecular responses to generate specific strategies for crop improvement (Fig. 11.1). In this chapter, genomics approaches for plant improvement against various biotic and abiotic stresses have been discussed.

Table 11.1 Gene(s) associated with abiotic stresses in plants

S. No.	Type of Stress	Plant	Gene(s)	Reference
1	Drought stress	*Solanum pimpinellifolium*	DREBA1	Rao et al. (2015)
		Alfalfa	GsZFP1	Tang et al. (2013)
		Soybean	GmCDPK3	Zhou et al. (2019)
			GmPPR4	Su et al. (2019)
		Arabidopsis and soybean	GmCAMTA12	Noman et al. (2019)
2	Abscisic acid stress	Wild tomatoes	Asr	Fischer et al. (2011)
		Arabidopsis	TaHsfA6f	Bi et al. (2020)
3	Salt stress	*Arabidopsis*	AtHKT1	Horie et al. (2009)
		Soybean	GmSALT3	Guan et al. (2014)
			GmVOZ1G	Li et al. (2020a, b, c)
		Wheat	TaWRKY13	Zhou et al. (2019)
		Rice	OsHKT1, OsHKT2, and OsVHA	Wei et al. (2021)
			LOC_Os02g49700, LOC_Os03g28300	Liu et al. (2019a)
4	Cold stress	*Solanum* sp.	CBF3 and CBF5	Pennycooke et al. (2008)
		Arabidopsis	TsnsLTP4	Sun et al. (2015)
			MbNAC25	Han et al. (2020)
		Radish	RsLOX	Wang et al. (2019b)
5	Heat stress	Chinese kale	BocMBF1c	Zou et al. (2019)
		Carnation	DcaHsfs	Li et al. (2019a, b)
6	Metal stress	Barley	HvPAA1	Wang et al. (2019a)
		Soybean	*GmIREG3*,	Cai et al. (2020)

11.2 Abiotic Stresses

11.2.1 Drought Stress

Drought stress has the most deleterious effect on crop productivity. Though many QTLs have been reported for drought tolerance in various food crops which has resulted in the improvement of stress tolerance using marker assisted selection, yet minor QTLs are not part of the selection process (Younis et al. 2020; Wang et al. 2019c). The effect of environmental and genetic factors on all the alleles across the genome is generally achieved by genomic selection, as it also addresses the effect caused by the small genes. Lorenzana and Bernardo (2009) reported that the genomic selection reduces the selection time by almost half per cycle compared to the phenotypic selection for almost all phenotypic traits in different crops. Even in wheat, the drought tolerance mechanism is complex and is influenced by variations controlled by QTLs. Hence, in such cases of the genetic and physiological basis of

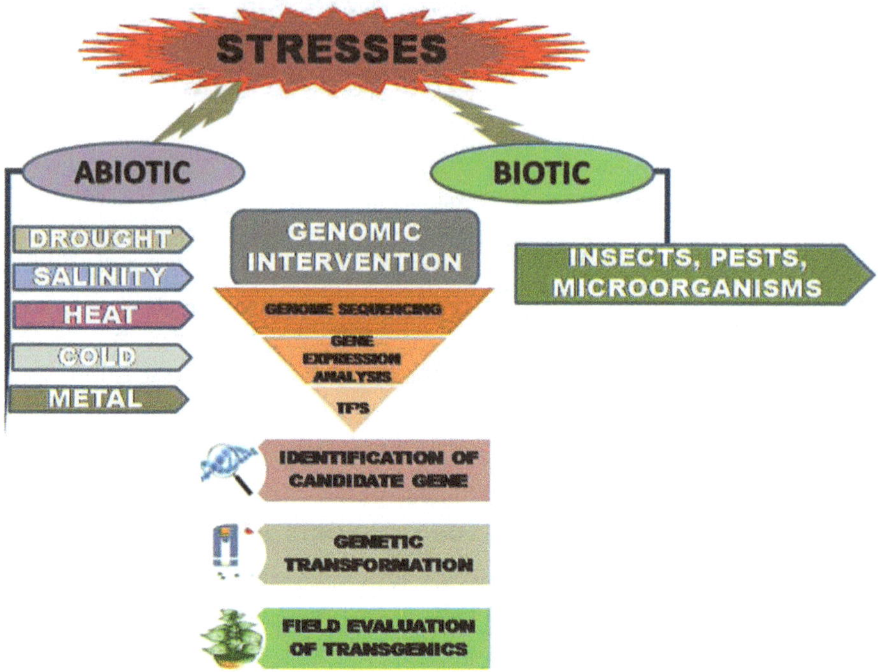

Fig. 11.1 Functional genomics approaches for plant improvement against stresses

tolerance to drought needs to be understood (Fleury et al. 2011). Crop breeders need to understand the molecular mechanisms and use genomic tools to understand and correlate the complex drought tolerance so as to enhance the crop productivity. Several nucleotide bases code for proteins that plays a significant role in detoxification, ion transporter, and osmolyte biosynthesis. Along with them regulatory genes such as TFs, signaling protein kinases, receptor protein kinases, ribosomal protein kinases also show their expression during stress (Joshi et al. 2016).

Shikha et al. (2017) tested the breeding values of maize subtropical lines for drought at different environments using SNPs. They observed that out of 1053 SNPs, 77 SNPs were associated with 10 drought-responsive transcription factors which have role in different physiological and molecular functions and are important for the selection of superior genotypes and candidate genes for selecting drought-tolerant maize hybrids.

Wang et al. (2018) studied and compared microsite evolution in drought-tolerant wild barley populations growing in Terra Rossa soil and basalt soil types where the genetic mechanisms behind tolerance were not known. Using genome re-sequencing and RNA-sequencing few genes were identified on chromosomes 6H and 7H that showed divergence in the wild barley genotypes that have significant role in plant drought tolerance. They also identified various adaptation mechanisms against drought in barley which are associated with the rhizosphere of the plant.

Xiang et al. (2007) reported putative *CIPK* genes for their transcriptional responses to various abiotic stresses. About 20 *OsCIPK* genes were differentially induced by at least one of the stresses. Also, few *CIPK* genes without stress-responsive cis-elements were also induced during stress. They overexpressed three stress-responsive genes in *japonica* rice and the transgenic plants overexpressing these genes showed significantly improved tolerance to stress.

Sahebi et al. (2018) revealed that certain WRKY transcription factors particularly zinc finger protein family play a role in plant development in rice. They reported a total of 89 WRKY genes and 97 WRKY genes in *japonic*a and *O. nivara,* respectively and mapped them onto individual chromosomes based on the physiological/morphological parameters. On the basis of various molecular mechanisms observed in the drought resistant parents, a strategy was developed where suitable germplasm adapted to a particular environment was selected.

11.2.2 Salinity Stress

Salinity stress is one of the main abiotic stresses which is caused due to the deposition of salt ions in the upper layer of the soil and thus affect the productivity of the land. Salt stress is mainly caused due to weathering of rocks, deforestation, and high rate of transpiration. Plant receptors after receiving the stress signals produce secondary messengers such as reactive oxidative stress, calcium, and inositol phosphates and hormones like ethylene, salicylic acid, and abscisic acid (Tuteja and Sopory 2008). As the intracellular Ca^{2+} level changes the stress-responsive genes or the transcription factors become activated and help the plant to survive the stress conditions. These stress-responsive genes are either induced early after the signals are received or are expressed slowly like membrane stabilizing proteins, osmolytes, or LEA proteins (Chen et al. 2019a, b).

Several sequence based approaches along with NGS based methods have been used for discovering and mapping markers and candidate genes in various crop plants. Also, genome-wide association studies have proved to be very useful tool for allele and marker discovery. Huang et al. (2010) obtained a genetic map for 14 different agronomic traits including grain quality and stress tolerance, using NGS based genotyping against 373 rice lines. Kumar et al. (2015) mapped three new QTLs against salinity tolerance using GWAS studies on 220 indica rice accessions and discovered few accessions with genes coding for novel salinity tolerance and superior alleles. Dabab Nahas et al. (2019) predicted drought and salt and drought tolerance genes in wheat using 6717 expressed sequence tags. After clustering and assembling them into contigs, the contigs were mapped on to International Wheat Genome Sequencing Consortium RefSeq v1.0 assembly and full-length gene sequences were designed which were further validated for molecular breeding programs against wheat drought and salinity research.

At the transcriptional level, Ding et al. (2013) divided stress response genes into memory genes and non-memory genes. The former genes showed significantly different levels of up- or down-regulation in subsequent stress while the latter had similar responses to each stress. Ding et al. (2014) on comparing dehydration stress memory genes between Arabidopsis and maize observed that these genes were conserved during the evolution process. However, Zhang et al. (2018a, b) in switchgrass suggested the existence of a complicated signaling network of plant hormones in response to repeated dehydration stresses which are conserved and show species-specific distribution.

Several genes responsive to salt stress have been reported both in model and non-model plant species. Wu et al. (1996) reported *salt overly sensitive 1* (SOS) mutants of Arabidopsis which were exceptionally hypersensitive to NaCl while Zhu et al. (1998) further identified mutants with limited shoot growth during salt stress were finally led to the identification of the fundamental components of the SOS pathway.

Sanchez et al. (2011) used comparative genomics approach to analyze response of salt stress in model and cultivated species of Lotus. They concluded a positive correlation between chloride ion concentration in shoots and stress tolerance. Also, they observed that a small set of salt-responsive genes were found to be conserved in all Lotus genotypes while response of other genes was confined to tolerant or sensitive cultivars. Similarly, Kumar et al. (2013) used functional genomics to understand the molecular basis against salinity tolerance in rice for its sustainable production. Their study involved the investigation of various osmoregulants, protein kinases, receptor kinases, and transcription factors. Several transcription factors like MYB, Zinc finger, WRKY, NAC-type have been reported from many plants that are involved in salinity stress (Baillo et al. 2019; Yoon et al. 2020). Dai et al. (2007) reported that the overexpression of MYB TFs (OsMYB3R-2) improved salt stress tolerance in *Arabidopsis*, while overexpression of TF, OsMYB2 in transgenic rice improved salt stress tolerance along with drought and cold stress tolerance (Yang et al. 2012). In cotton plant where salinity is the key limiting factor, Li et al. (2019a, b) observed that the down-regulation of *GhWRKY6* enhanced salt tolerance in cotton. As such these TFs can be manipulated to engineer crops with higher stress tolerance (Hoang et al. 2017).

11.2.3 Heat Stress

Heat stress is a major threat for global food security and being a polygenic trait it is regulated by several genes and transcriptional factors. The effects of increased temperature exhibit a larger impact on vegetative growth and metabolites present in the plants. Higher temperatures adversely affect the plant growth and different metabolic pathways involved in defense responses. Plants have to make adjustments regarding key metabolites such as sugars, fats, amino acids in response to heat stress (Wani et al. 2017). Plants respond to heat stress by activating complex molecular networks, metabolite production, and expressions of genes associated with heat stress (Singh et al. 2019a, b). Various functional genomic approaches like genome-wide association studies (GWAS) and gene expression profiling have led to the discovery of novel genes associated with heat stress (Duan et al. 2019). GWAS is a powerful tool which has been applied to major food crops, like wheat, rice, maize, *Brassica* to identify the genetic basis behind complex phenotypic traits associated with heat stress (Lafarge et al. 2017; Rahaman et al. 2018). Kumar et al. (2020) identified heat shock proteins (HSPs) in wheat to understand their role during plant development and under different stress conditions. They identified 753 HSPs using Position-Specific Scoring Matrix and then their confirmation by sequence homology. High number of HSPs were identified in wheat as compared to other grasses due to their high ploidy level. All identified HSPs showed random distribution on chromosomes with high number present in B and D sub-genomes as compared to the A sub-genome.

Even the use of transcriptomic studies to screen heat-responsive candidate genes is widely used these days. Singh et al. (2019a, b) detected Single Nucleotide Polymorphisms and Insertion-deletions (Indels) associated with heat stress in lentils. Further expression of genes under different combinations of heat stress led to the identification of candidate differentially expressed genes and pathways. Their studies revealed that heat stress had a severe effect on cell wall and secondary metabolite pathways, including pollen phenotyping related genes.

Fu et al. (2020) reported that the high temperatures affected growth and production of lilies. Using transcriptomics, they observed that several genes which are involved in metabolic pathways, phenylpropanoid biosynthesis, plant–pathogen interactions, and kinase signaling pathways were down-regulated in *L. distichum* during heat stress. Their studies on discovery of heat-resistance genes and key components that are responsive to heat stress can facilitate heat-resistance breeding in lily in near future.

Several NAC proteins which are plant-specific transcription factors have been recognized as an important regulator in stress responses in rice and *Arabidopsis*, but their molecular mechanisms are still unknown. In rice, stress-responsive *NAC* genes are induced by drought, high temperature. Fang et al. (2015a, b) observed that the transgenic rice plants overexpressed *SNAC3* gene that conferred tolerance to heat stress by modulating ROS homeostasis. Since high temperature results in failure of grain filling during caryopsis development in rice so Ren et al. (2020) reported that the two caryopsis-specific NAC gene regulates the caryopsis filling through sugar transportation and abiotic stress responses in rice. They observed that the transcriptional regulatory networks involving ONAC127 and ONAC129 coordinate multiple pathways to modulate caryopsis development and heat stress response at rice filling stage.

El-Esawi and Alayafi (2019) investigated that rice overexpression improved rice tolerance to drought and heat stress. They cloned the gene and obtained transgenic rice plants and observed that the expression of four genes encoding for reactive oxygen species scavenging enzymes and eight genes against stress tolerance were up-regulated in the transformed rice lines as compared to their expression in wild-type. Also the overexpression of *Rab7* enhanced grain yield in rice.

Mitogen-activated protein kinases (MAPKs) are highly conserved serine and threonine protein kinases that participate in signal transduction in response to diverse environmental stresses in plants. Yu et al. (2019) identified *SlMAPK3* genes in tomato. They observed knock out *slmapk3* mutants exhibited more tolerance to heat stress than wild types. Also, *slmapk3* mutants exhibited less membrane damage with lower content of reactive oxygen species. The transcript showed enhances levels of antioxidant enzymes, as well as heat shock proteins.

Heat shock proteins (HSPs) are important molecular chaperones in plants that contribute to restore the protein homeostasis by their folding, assembly, and degradation, which is critical for a plant to survive under heat stress (Young 2010). In plants five major families of HSPs (HSP100, HSP90, HSP70, HSP60, and small HSP) are mainly located in the cytoplasm that responds to abiotic and biotic stresses (Park and Seo 2015). In addition HSPs are also located in ER, chloroplasts, mitochondria, and nucleus, where they play dynamic roles in protein homeostasis. Transcription for heat shock protein genes is controlled by heat stress transcription factors which are primary metabolism enzymes and signal transduction components with specific role in regulation (Jacob et al. 2017). Guo et al. (2020) isolated a heat-responsive gene *TaHsfA2–10* which expressed itself in different organs at different stages of development in wheat. The gene enhanced the thermotolerance in transgenic *Arabidopsis thaliana*. The transactivation activity of *TaHsfA2–10* was also revealed by yeast one-hybrid assay.

11.2.4 Cold Stress

Decrease in temperature causes cold stress that limits the productivity of crop species. Plant requires optimum temperature for its growth and development. Plants exhibit changes in gene expression and proteins in response to low temperature. Many plant species show chilling injury like chlorosis, necrosis, or are killed. Use of genomic tools has revealed that the alteration in expression of genes is followed by enhancement in several metabolites that provide defense to the plants against cold stress. Several C-repeat binding factors, dehydration-responsive element binding (CBF/DREB1) transcription factors which are part of a large number of stress-responsive regulatory genes provide protection (Agarwal et al. 2017).

Several inducible genes like osmolyte biosynthesis, LEA proteins, and detoxification enzymes have been isolated from plants which are overexpressed in response to cold stress (Sanghera et al. 2011). Wang et al. (2009) studied ten transcripts in tea leaves which were classified into three functional groups. Out of three transcripts two encoded for early light proteins while one transcript encoded for beta-amylase which has a role in cold acclimatization. Barah et al. (2013a, b) reported response of ten *A. thaliana* ecotypes from different geographical locations using genome-scale transcript. On comparing the transcriptomes of ten ecotypes, a total of 6061 transcripts which were cold regulated were observed along with transcription factors and transposable elements. Ecotype specific expression pattern was observed with about 75% transcripts. Also, the cold stress regulating genes showed non-synonymous amino acid changes in the coding region. In silico constructed transcriptional regulatory network model contained 178 transcription factors and 1331 target genes with 1275 nodes and 7720 connections.

Also, many lipid molecules play important role in signal transduction during cold stress (Hou et al. 2016). Lipid molecules such as phosphatidic acid and diacylglycerol kinase which constitutes a minor portion of membrane lipids are membranous messenger molecules (Arisz et al. 2009). Apart, phosphatidic acid is generated rapidly in response to cold stress in where it regulates the enzymatic activity of target proteins (Arisz et al. 2013). Chen and Thelen (2013) reported that acyl-lipid desaturase-2 plays an important role by adjusting the compositions of organelle membrane lipids in *Arabidopsis* during low temperature. A galactolipid remodeling enzyme located on the outer chloroplast membrane causes lipid remodeling and membrane stabilization during freezing and hence imparts cold tolerance to the plants (Moellering et al. 2010).

Phytohormones being small chemical molecules are also involved in the abiotic stress responses (Peleg and Blumwald 2011). Plant hormones such as auxin, cytokinins, gibberellins, abscisic acid, jasmonic acid, and ethylene have significant role in regulating plant tolerance to chilling stress by either CBF-dependent or -independent pathways. To induce changes in cold-responsive gene expression several complex signaling cascades directed by hormones are utilized by the plants that help them to withstand chilling or freezing temperatures. Thus, crosstalk among plant hormones is important for downstream gene activation which is essential for protection against cold stress.

Barah et al. (2013a, b) studied diversity at genome scale in *A. thaliana* ecotypes collected from different geographical locations. They observed that a high number of transcripts including transcription factors and transposable elements were significantly cold regulated. They constructed an in silico transcriptional regulatory network model containing nodes and connections with target genes in response to cold stress.

11.2.5 Metal Stress

Heavy metals are an essential component of earth's crust. During the last few decades, their levels have increased significantly due to activities like mining, smelting, industrial applications, use of pesticides and as natural geogenic activities (Shukla et al. 2017) Metals being non-biodegradable, the deposition of heavy metals is increasing continuously where industrialization is at boom. As a result, a number of heavy metals are entering into plants and human body along with certain elements like Arsenic and Cadmium which are toxic and are transported along with essential divalent cations like Zinc (Zn^{2+}), Iron (Fe^{2+}), Calcium (Ca^{2+}), and Manganese (Mn^{2+}) (Mendoza-Cozat et al. 2011). They cause hindrance in certain physiological processes in plants like mineral nutrient uptake, seed germination, and photosynthesis herby effecting overall growth and biomass accumulation (Finnegan and Chen 2012).

However, plants have evolved several strategies to overcome heavy metal detoxification due to the presence of heavy metal-associated proteins (HMPs) which are metalloproteins or metallochaperone-like proteins (Zhang et al. 2018a, b). These proteins contain heavy metal-associated (HMA) domains which are conserved, consisting of ~30 amino acid residues and two cysteine residues for binding and transferring heavy metal ions (Gitschier et al. 1998). Plant proteins containing HMA domains fall into one of the following groups: HPPs (heavy metal-associated plant proteins), HIPPs (heavy metal-associated isoprenylated plant proteins) ATX1-like and P1B-ATPase (Li et al. 2020a, b, c; Pedersen et al. 2012).

Li et al. (2020a, b, c) identified HMPs in rice and *Arabidopsis* and based on the characteristics of their heavy metal-associated domains, divided them into six clades on the basis of their different gene structures and motifs. Miyadate et al. (2011) identified gene OsHMA3 in rice that controlled root-to-shoot Cd translocation, while OsHMA4 was the gene responsible for controlling Cu accumulation in rice grain (Huang et al. 2016).

ATX1-like metallochaperones already identified from *Arabidopsis* and rice (Zhang et al. 2018a, b) have been transferred to yeast Ccc2 P1B-type ATPase to enhance its antioxidant mechanism (Puig et al. 2007). Functionally investigation has revealed that the HMPs are identified as an upstream controller of stress-related regulatory networks. They also have a positive role in salicylate-dependent pathogen response pathway and in flower development and seed setting (Banday and Nandi 2015).

Several Reactive Nitrogen Species (RNS) plays a crucial role in regulating plant responses during stress. Mahmood et al. (2009) observed that during Cd stress, there was increase in Nitric oxide (NO) production in *Triticum aestivum* roots while Cd stress causes the over-production of peroxisomes in *Arabidopsis* (Corpas and Barroso 2014). Similarly, Feigl et al. (2015) observed a strong relationship among two species of *Brassica* subjected to Zn stress for ROS metabolism while NO production was strongly depressed by Cd toxicity in pea but Cd treatment on plants had positive effect for pathogen-related proteins which were regulated by jasmonic acid and ethylene (Rodriguez Serrano et al. 2009). Hence, RNS metabolism plays an important role to understand the mechanisms involved in the defense of plant cells against metal stress.

11.3 Biotic Stress

Apart from abiotic factors several biotic constraints also poses threat to plants. Fungi, bacteria, viruses, and nematodes cause several diseases and reduction in yield. Being sessile, plants lacks strong immune system but have strong defense mechanism that is stored in its genetic code. After infection pathogen spread is decreased when ROS is generated and also increase in plant cell wall lignification and suberization reduces the pathogen entry and spread (He et al. 2011).

Other defense mechanisms to biotic stress also include morphological changes and production of certain proteins and enzymes. As a result, certain volatile chemical compounds give strength and rigidity to the plant. Morphological changes are the first line of plant defense causing barrier against insects, pests like the formation of a waxy cuticle, development of spines, and trichomes and incorporation of granular minerals into plant tissues that reduces the palatability and digestibility (Chamarthi et al. 2010).

Apart, several plant hormones such as salicylic acid, jasmonic acid (JA), and ethylene play a significant role during biotic stress. JA does not work independently but acts in a complex signaling network combined along with other plant hormone signaling pathways (Hu et al. 2017). JA signaling plays an important role against necrotrophic pathogens or the necrotrophic stage of hemi-biotrophic pathogens (Pandey et al. 2016). Several reports on the mechanism of crosstalks between jasmonic acid and other plant hormones during stress responses regulating the balance between plant growth and defense response. Thus, phytohormone crosstalk can open new avenues for genetic improvement of plants needed during biotic stress to meet the future global food production. Suppressor proteins like JASMONATE ZIM DOMAIN PROTEIN (JAZ) and MYC2 as the key components in the crosstalks (Verma et al. 2016; Yang et al. 2019).

Plants have microbial-associated molecular-pattern-triggered immunity (MTI or PTI) and effector-triggered immunity (ETI). Both have similar defense responses, but ETI is much faster and is associated with a localized cell death termed the hypersensitive response that further prevents the spread of microbial attack. However, PTI is important for non-host resistance and for basal immunity as it recognizes conserved microbial elicitors. There are several receptors for microbe-associated molecules that result in the production of antimicrobial peptides/compounds (War et al. 2011). Felix et al. (1999) reported a 22 amino acid peptide-spanning region in the N-terminal part of flagellin of *Pseudomonas syringae* which is sufficient to elicit a typical immune responses in a variety of plants. However, in *Arabidopsis* leucine-rich repeat receptor-like kinase is responsible for flagellin perception (*FLS2*). Functional *FLS2* homologs have been identified and reported from several plants suggesting that the receptors for the epitope of bacterial flagellin are ancient and conserved (Trda et al. 2014). Another flagellin *flgII-28* was identified from *Solanaceae* which was physically linked by a stretch of 33 amino acid residues to *flg22* from rice, suggesting that both these molecules are detected by the same receptor, *FLS2* (Clarke et al. 2013).

Another, bacterial protein which is isolated from *Escherichia coli* is the Elongation factor (EF). In rice elongation factor resistance is mediated by heteromeric complex formation where a complex is formed between SOMATIC EMBRYOGENESIS RECEPTOR KINASEs and XA21 binding protein 24, both of them being important component of XA21-mediated defense response (Chen et al. 2014). Similarly, *Arabidopsis* PRR EF-Tu receptor recognizes the bacterial PAMP elongation factor Tu triggers immunity in wheat when inoculated with bacterial pathogen *Pseudomonas syringae pv. oryzae*, the transgenic wheat lines showed reduced lesion size and low rate of bacterial multiplication (Schoonbeek et al. 2015) .

Chitin is a homopolymer and a major constituent of fungal cell walls. On pathogen attack the plant chitinases break down microbial chitin polymers. *CEBiP* was the first chitin-elicitor binding protein that localizes in the plasma membrane and is identified in rice plant (Kouzai et al. 2014). Upon chitin binding, *CEBiP* forms a hetero-oligomeric complex with the *Chitin-Elicitor Receptor Kinase 1*. This binding results in the formation of sandwich-type receptor system for chitin and varies from plant to plant.

β-Amino-*n*-butyric acid (BABA) is a non-protein amino acid, which has a significant role in accelerating defense to pathogen infection. Recently, induction of resistance by BABA has been reported against several crop plants against diverse bacterial and fungal species in recent decades (Ben Rejeb et al. 2018; Ye et al. 2020). Mode-of-action of BABA includes crosstalk with other defense signaling pathways like in *Arabidopsis* against P*ectobacterium carotovorum* ssp. *carotovorum*. However, salicylic acid is required for BABA resistance against *Phytophthora infestans* in potato (Eschen-Lippold et al. 2010). Similarly, in *Arabidopsis* resistance by BABA to *Alternaria brassicicola* and *Plectospharella cucumerina* infection is dependent on abscisic acid mediated signaling pathways (Ton and Mauch-Mani 2004) while in lettuce against *Bremia lactucae*, BABA resistance is independent of ABA pathway (Cohen et al. 2010).

11.4 Role of Transcription Factors against Biotic Stress

Plants have evolved several responses to unfavorable conditions including interconnected networks at the molecular level which are controlled by signal cascades. These signals activate several transduction pathways that involve phosphatases and protein kinases. Specific transcription factors (TFs) which are DNA-binding proteins that bind to specific DNA sequences called *cis*-elements in the gene promoters are either up-regulated or down-regulated by protein kinases or phosphatases. There are generally five TF families (WRKY, ethylene responsive factor (ERF), basic-domain leucine-zipper (bZIP), basic helix-loop-helix (bHLH), and NAC) that play roles in plant defense mechanisms. TFs belonging to NAC family were first reported from Petunia (Souer et al. 1996).

NAC TFs in the N-terminal regions have highly conserved DNA-binding domain and diverse C-terminal transcription regulatory domains. NAC proteins act both via an ABA-dependent and ABA-independent pathway. Jin et al. (2020) using genome-wide studies have identified 93 *NAC* genes in tomato and divided them into five groups on basis of phylogenetic studies. Gene expression analysis revealed different expression levels in various tissues and at different fruit development stages. Similarly, Liu et al. (2019b) reported 80 *NAC* genes (FtNAC) from buckwheat and named them as per their distribution on chromosomes. Phylogenetic analysis revealed FtNAC proteins are widely distributed in 15 subgroups with one unclassified subgroup in both Tartary buckwheat and *Arabidopsis* though the structural diversity was low in *NAC* genes. Li et al. (2019a, b) observed 90 *NACs* in quinoa and divided them into 14 distinct subfamilies phylogenetically. Diversities in gene proportions, exon–intron structures, and motif compositions were observed in subfamilies with non-random duplication events due to selection pressure with limited functional divergence.

Ethylene response factor (ERF) is another family of TFs that plays a key role when plant encounters the biotic stress. It results in defense responses in the form of accumulation of antimicrobial phytoalexin and synthesis of pathogenesis-related (PR) proteins. Cao et al. (2006) and Eulgem et al. (2000) have studied the presence of cis-acting elements (GCC box and W box) and involvement of WRKY transcription factors, respectively, in the regulation of PR gene expression. ERFs can either activated or repress the plant defense response during the biotic stress. Barah et al. (2013a, b) studied the role of ERF for enhancement of resistance against *Pseudomonas syringae* attack while Zhu et al. (2013) recorded overexpression of ERFs in transgenic tobacco against both bacterial pathogen *Ralstonia solanacearum* and fungal pathogen *Phytophthora parasitica*.

Similarly, WRKY TF family has large number of members that are present in plant genomes. *Cis*-element bound by WRKY TFs is called the W-box, with a consensus sequence of TTGACT/C. Recently studies have been performed on several plants involving genome-wide identification and characterization of WRKY TFs (Chen et al. 2020; Li et al. 2020a, b, c). WRKY TFs are involved in MAPK cascades that play significant signaling roles in multiple defense responses, especially against pathogen effectors (Chen et al. 2019a, b) and their role in defense have been extensively studied, mainly in *Arabidopsis*. In *Arabidopsis*, 20 MAPKs, 10 MAPKKs, and 60 MAPKKKs are reported (MAPK-Group 2002), and their role has been identified via targeted experiments and systematic approaches like phosphoproteomics and protein array screening (Rayapuram et al. 2017).

11.5 Effect of Gene Targeting on Abiotic and Biotic Stress

Studying of major candidate genes is necessary to overcome various stresses that a plant comes across. The two types of genes, viz. structural and regulatory play significant role in producing stress tolerant crops. While proteins are encoded by the structural genes, regulatory genes control the expression of these genes. Targeting of these regulatory genes by a robust and efficient method such as CRISPR (clustered regularly interspaced short palindromic repeat) associated protein 9 (Cas9) system can help in developing crop varieties which will be resilient to climate change. The CRISPR/CAS 9 technique is different from ZFN and TALEN in terms of the DNA-binding system. It is a type of adaptive immune system, which degrades exogenous DNA. It was discovered in 1987 in *Escherichia coli* (Ishino et al. 1987), officially named by the scientist who identified CRISPR-associated genes (Jansen et al. 2002), but became popular after 2012 when its potential for genome editing was considered (Jinek et al. 2012).

A CRISPR-Cas locus consists of a CRISPR array that comprised of short repetitive elements intercalated with invader DNA-targeting spacers along with an AT-rich leader sequence, and an operon of Cas genes encoding the Cas proteins. Based on the different types of Cas proteins that are participating, CRISPR-Cas systems are divided into three main types: type I and type III systems that use a large multi-Cas protein complex for binding and targeting, while the type II system requires only a single CRISPR-associated protein 9 for the recognition of RNA-guided double-stranded DNA and cleavage using its two distinct domains,

RuvC and HNH31 (Wang et al. 2019d). Thus the CRISPR-Cas9 system has made remarkable progress in crop improvement (Osakabe et al. 2016).

Some genes contribute to stress tolerance by disrupting the pathways that are involved. In rice, a stress-related RING finger protein 1 (OsSRFP1) acts as a negative regulator by enhancing the level of H_2O_2. Knockdown of OsSRFP1 enhanced stress tolerance by disrupting H_2O_2 biosynthesis (Fang et al. 2015a, b). Osakabe et al. (2016) produced site-directed modifications in *Arabidopsis* using truncated-gRNAs in the CRISPR/Cas9 system. They observed modifications in transgenic plants for OPEN STOMATA 2 (OST2) gene that encodes a major plasma membrane H+-ATPase in stomata response. Using tissue-specific promoters, high mutation rate was observed in the germ lines. Thus, the new mutant alleles obtained for OST2 displayed altered stomatal closing in response to environmental conditions.

He et al. (2017) observed that *ANAC069* gene in *Arabidopsis* functions as a negative regulator by decreasing the ROS-scavenging capability, thereby enhancing the level of proline biosynthesis that results in increased sensitivity to salt and osmotic stress. Thus the knockdown mutants of *ANAC069* obtained via T-DNA insertion showed improved tolerance to salt and osmotic stress. Shkryl et al. (2021) observed that HIGH EXPRESSION OF OSMOTICALLY RESPONSIVE GENES 1 (HOS1) functions as a main regulator against cold stress in *Arabidopsis*. Targeted mutagenesis of the HOS1 gene study was performed by CRISPR/Cas9 that resulted in frameshift indels which disrupted the open reading frame and resulted in the appearance of premature stop codons. The mutant plants on exposure to cold stress enhanced tolerance and expression of cold-responsive genes. Further the content of glucosinolates was down-regulated by 1.5-times, while flavonol glycosides were up-regulated by 1.2–4.2 times in transgenic plants.

Osakabe et al. (2016) produced site-directed modifications in *Arabidopsis* using truncated-gRNAs in the CRISPR/Cas9 system. They observed modifications in transgenic plants for OPEN STOMATA 2 (OST2) gene that encodes a major plasma membrane H+-ATPase in stomata response. Using tissue-specific promoters, high mutation rate was observed in the germ lines. Thus the new mutant alleles obtained for OST2 displayed altered stomatal closing in response to environmental conditions.

Tashkandi et al. (2018) engineered tomato plants which were resistant to the tomato yellow leaf curl virus used the CRISPR-Cas9 system. They targeted the coat protein and replicase loci of the genome and the transgenic tomato showed less accumulation of viral genomic DNA than the wild-type plants and the immunity was passed over to multiple generations. Dinkins et al. (2021) observed that isoflavones accumulation and biosynthesis have a significant importance in the rhizosphere of red clover, a forage legume. Isoflavone plays a potential role in medicinal, antimicrobial, and environmental insinuations. Using CRISPR/Cas9, a key enzyme isoflavone synthase (IFS) that functions in the biosynthesis of isoflavones was knocked out. Further they inter-crossed a hemizygous plant with 9-bp deletion in the IFS gene to obtain homozygous mutant plants. After inoculating wild-type and mutant plants with rhizobia no significant differences were observed on nodulation, suggesting that the isoflavones have no important roles in nodulation. However, enhanced gene expression was observed for upstream genes, viz. phenylalanine ammonium lyase and chalcone synthase which are precursors for IFS and also for ethylene response, associated with biotic stress.

In rice, genome editing has enhanced the crop yield by knock out those genes that affected the crop productivity. Li et al. (2016) knocked out four negative yield genes, viz. *Gn1a*, *DEP1*, *GS3*, and *IPA1* by CRISPR/Cas9 in the cultivar Zhonghua 11. Knocking out of the three genes (*Gn1a*, *DEP1*, *GS3*) enhanced the grain number and grain size. Similarly, Xu et al. (2016) observed increase in grain weight by knocking out three major negative regulators of grain weight (*GW2*, *GW5*, and *TGW6*) using a CRISPR/Cas9-mediated system.

Genome editing techniques have been widely used to engineer plants resistant to various pathogens (Mushtaq et al. 2019). Peng et al. (2017) engineered canker-resistant plants through CRISPR/Cas9-targeted editing of the susceptibility gene *CSLOB1* promoter in citrus. Wang et al. (2014) improved resistance to fungal pathogens by targeting homologs of *MILDEW-RESISTANCE LOCUS* in bread wheat. Similarly, powdery mildew free tomatoes were developed by targeting the *SlMlo1* gene using CRISPR/Cas9 (Nekrasov et al. 2017) PMR_4 gene (Santillán Martínez et al. 2020) while CRISPR/Cas9-mediated mutagenesis of VvMLO3 results in enhanced resistance to powdery mildew in grapevine (Wan et al. 2020). Thus the technique of genome editing is highly applicable to plants as target sites of either plants or biotic agent can be edited. It can result as a powerful tool to develop non-transgenic crops highly resistant to biotic stresses.

11.6 Conclusion and Future Prospects

Plants in natural ecosystem are exposed to several abiotic and biotic stresses that occur simultaneously. So, recent developments in genomics have facilitated the identification of various stress-responsive factors in plants. Use of high-throughput approaches genomic approaches has contributed to the identification of candidate gene and their role in abiotic and biotic stress tolerance, thus helping the breeders, physiologists, and molecular biologists to design stress-resistant crops. Functional genomic tools have greatly facilitated the study of gene functions and engineering abiotic and biotic stress tolerance in many plants. Further understanding of the complexity of cell signaling networks and role of various gene regulatory networks will help in developing stress-tolerant plants.

References

Agarwal PK, Gupta K, Lopato S et al (2017) Dehydration responsive element binding transcription factors and their applications for the engineering of stress tolerance. J Exp Bot 68:2135–2148

Arisz SA, Testerink C, Munnik T (2009) Plant PA signaling via diacylglycerol kinase. Biochem Biophys Acta 1791:869–875

Arisz SA, Van Wijk R, Roels W et al (2013) Rapid phosphatidic acid accumulation in response to low temperature stress in *Arabidopsis* is generated through diacylglycerol kinase. Front Plant Sci 4:1

Baillo EH, Kimotho RN, Zhang Z et al (2019) Transcription factors associated with abiotic and biotic stress tolerance and their potential for crops improvement. Genes (Basel) 10:771

Banday ZZ, Nandi AK (2015) Interconnection between flowering time control and activation of systemic acquired resistance. Front Plant Sci 6:174

Barah P, Jayavelu ND, Rasmussen S et al (2013a) Genome-scale cold stress response regulatory networks in ten *Arabidopsis thaliana* ecotypes. BMC Genomics 14:722

Barah P, Winge P, Kusnierczyk A, Tran DH, Bones AM (2013b) Molecular signatures in *Arabidopsis thaliana* in response to insect attack and bacterial infection. PLoS One 8:e58987

Ben Rejeb I, Pastor V, Mauch-Mani B et al (2018) Impact of β-aminobutyric acid on induced resistance in tomato plants exposed to a combination of abiotic and biotic stress. J Agric Sci Bot 21:2–23

Bi H, Zhao Y, Li H, Liu W (2020) Wheat heat shock factor TaHsfA6f increases ABA levels and enhances tolerance to multiple abiotic stresses in transgenic plants. Int J Mol Sci 21:3121

Cai Z, Xian P, Lin R et al (2020) Characterization of the soybean GmIREG family genes and the function of GmIREG3 in conferring tolerance to aluminum stress. Int J Mol Sci 21:497

Cao Y, Song F, Goodman RM, Zheng Z (2006) Molecular characterization of four rice genes encoding ethylene-responsive transcriptional factors and their expressions in response to biotic and abiotic stress. J Plant Physiol 163:1167–1178

Chamarthi SK, Sharma HC, Sahrawat KL et al (2010) Physico-chemical mechanisms of resistance to shoot fly, *Atherigona soccata* in sorghum, *Sorghum bicolor*. J Appl Entomol 135:446–455

Chen M, Thelen JJ (2013) ACYL-LIPID DESATURASE2 is required for chilling and freezing tolerance in *Arabidopsis*. Plant Cell 25:1430–1444

Chen X, Zuo S, Schwessinger B et al (2014) An *XA21*-associated kinase (OsSERK2) regulates immunity mediated by the *XA21* and *XA3* immune receptors. Mol Plant 7:874–892

Chen X, Li C, Wang H et al (2019a) WRKY transcription factors: evolution, binding, and action. Phytopathol Res 1:13

Chen Y, Li C, Zhang B et al (2019b) The role of the late Embryogenesis-abundant (LEA) protein family in development and the abiotic stress response: a comprehensive expression analysis of potato (*Solanum Tuberosum*). Genes (Basel) 10:148

Chen C, Chen X, Han J et al (2020) Genome-wide analysis of the *WRKY* gene family in the cucumber genome and transcriptome-wide identification of WRKY transcription factors that respond to biotic and abiotic stresses. BMC Plant Biol 20:443

Clarke CR, Chinchilla D, Hind SR et al (2013) Allelic variation in two distinct *Pseudomonas* syringae flagellin epitopes modulates the strength of plant immune responses but not bacterial motility. New Phytol 200:847–860

Cohen Y, Rubin AE, Kilfin G (2010) Mechanisms of induced resistance in lettuce against *Bremia lactucae* by DL-β-amino-butyric acid (BABA). Eur J Plant Pathol 126:553–573

Corpas FJ, Barroso JB (2014) Peroxynitrite (ONOO-) is endogenously produced in *Arabidopsis* peroxisomes and is overproduced under cadmium stress. Ann Bot 113:87–96

Dabab Nahas L, Al-Husein N, Lababidi G, Hamwieh A (2019) *In-silico* prediction of novel genes responsive to drought and salinity stress tolerance in bread wheat (*Triticum aestivum*). PLoS One 14:e0223962

Dai X, Xu Y, Ma Q et al (2007) Overexpression of an R1R2R3 MYB gene, *OsMYB3R-2*, increases tolerance to freezing, drought, and salt stress in transgenic *Arabidopsis*. Plant Physiol 143: 1739–1751

Ding Y, Liu N, Virlouvet L et al (2013) Four distinct types of dehydration stress memory genes in *Arabidopsis thaliana*. BMC Plant Biol 13:229

Ding Y, Virlouvet L, Liu N et al (2014) Dehydration stress memory genes of *Zea mays*; comparison with *Arabidopsis thaliana*. BMC Plant Biol 14:141

Dinkins RD, Hancock J, Coe BL et al (2021) Isoflavone levels, nodulation and gene expression profiles of a CRISPR/Cas9 deletion mutant in the isoflavone synthase gene of red clover. Plant Cell Rep 40:517–528

Duan S, Liu B, Zhang Y et al (2019) Genome-wide identification and abiotic stress-responsive pattern of heat shock transcription factor family in *Triticum aestivum* L. BMC Genomics 20:257

El-Esawi MA, Alayafi AA (2019) Overexpression of rice Rab7 gene improves drought and heat tolerance and increases grain yield in rice (Oryza sativa L.). Genes (Basel) 10:56

Eschen-Lippold L, Altmann S, Rosahl S (2010) DL-beta-aminobutyric acid-induced resistance of potato against *Phytophthora infestans* requires salicylic acid but not oxylipins. Mol Plant-Microbe Interact 23:585–592

Eulgem T, Rushton PJ, Robatzek S, Somssich IE (2000) The WRKY superfamily of plant transcription factors. Trends Plant Sci 5:199–206

Fang Y, Liao K, Du H et al (2015a) A stress-responsive NAC transcription factor SNAC3 confers heat and drought tolerance through modulation of reactive oxygen species in rice. J Exp Bot 66: 6803–6817

Fang H, Meng Q, Xu J et al (2015b) Knock-down of stress inducible OsSRFP1 encoding an E3 ubiquitin ligase with transcriptional activation activity confers abiotic stress tolerance through enhancing antioxidant protection in rice. Plant Mol Biol 87:441–458

Feigl G, Lehotai N, Molnar A et al (2015) Zinc induces distinct changes in the metabolism of reactive oxygen and nitrogen species (ROS and RNS) in the roots of two *Brassica* species with different sensitivity to zinc stress. Ann Bot 116:613–625

Felix G, Duran JD, Volko S et al (1999) Plants have a sensitive perception system for the most conserved domain of bacterial flagellin. Plant J 18:265–276

Finnegan PM, Chen W (2012) Arsenic toxicity: the effects on plant metabolism. Front Physiol 3: 182

Fischer I, Camus-Kulandaivelu L, Allal F, Stephan W (2011) Adaptation to drought in two wild tomato species: the evolution of the Asr gene family. New Phytol 190:1032–1044

Fleury D, Jefferies S, Kuchel H et al (2011) Genetic and genomic tools to improve drought tolerance in wheat. J Exp Bot 61:3211–3222

Fu Y, Yang L, Gao H et al (2020) Comparative transcriptome analysis reveals heat stress-responsive genes and their signalling pathways in lilies (Lilium longiflorum vs. Lilium distichum). PLOS. https://doi.org/10.1371/journal.pone.0239605

Gitschier J, Moffat B, Reilly D et al (1998) Solution structure of the fourth metal-binding domain from the Menkes copper-transporting ATPase. Nat Struct Biol 5:47–54

Guan R, Qu Y, Guo Y et al (2014) Salinity tolerance in soybean is modulated by natural variation in GmSALT3. Plant J 80:937–950

Guo X, Yuan SN, Zhang HN et al (2020) Heat-response patterns of the heat shock transcription factor family in advanced development stages of wheat (*Triticum aestivum* L.) and thermotolerance-regulation by TaHsfA2–10. BMC Plant Biol 20:364

Han D, Du M, Zhou Z et al (2020) Overexpression of a *Malus baccata* NAC transcription factor gene *MbNAC25* increases cold and salinity tolerance in *Arabidopsis*. Int J Mol Sci 21:1198

He J, Chen F, Chen S et al (2011) Chrysanthemum leaf epidermal surface morphology and antioxidant and defense enzyme activity in response to aphid infestation. J Plant Physiol 168: 687–693

He L, Shi X, Wang Y et al (2017) *Arabidopsis* ANAC069 binds to C[A/G]CG[T/G] sequences to negatively regulate salt and osmotic stress tolerance. Plant Mol Biol 93:369–387

Hoang XLT, Nhi DNH, Thu NBA et al (2017) Transcription factors and their roles in signal transduction in plants under abiotic stresses. Curr Genomics 18:483–497

Horie T, Hauser F, Schroeder JI (2009) HKT transporter-mediated salinity resistance mechanisms in *Arabidopsis* and monocot crop plants. Trends Plant Sci 14:660–668

Hou Q, Ufer G, Bartels D (2016) Lipid signalling in plant responses to abiotic stress. Plant Cell Environ 39:1029–1048

Hu Y, Jiang Y, Han X et al (2017) Jasmonate regulates leaf senescence and tolerance to cold stress: crosstalk with other phytohormones. J Exp Bot 68:1361–1369

Huang X, Wei X, Sang T et al (2010) Genome-wide association studies of 14 agronomic traits in rice landraces. Nat Genet 42:961–967

Huang XY, Deng F, Yamaji N et al (2016) A heavy metal P-type ATPase OsHMA4 prevents copper accumulation in rice grain. Nat Commun 7:12138

Ishino Y, Shinagawa H, Makino K, Amemura M, Nakata A (1987) Nucleotide sequence of the IAP gene, responsible for alkaline phosphatase isozyme conversion in *Escherichia coli*, and identification of the gene product. J Bacteriol 169:5429–5433

Jacob P, Hirt H, Bendahmane A (2017) The heat-shock protein/chaperone network and multiple stress resistance. Plant Biotechnol J 15:405–414

Jansen R, Embden JDAV, Gaastra W, Schouls LM (2002) Identification of genes that are associated with DNA repeats in prokaryotes. Mol Microbiol 43:1565–1575

Jin JF, Wang ZQ, He QY et al (2020) Genome-wide identification and expression analysis of the NAC transcription factor family in tomato (*Solanum lycopersicum*) during aluminum stress. BMC Genomics 21:288

Jinek M, Chylinski MK, Fonfara I, Hauer M, Doudna JA, Charpentier E (2012) A programmable dual-RNA: guided DNA endonuclease in adaptive bacterial immunity. Science 337:816–821

Joshi R, Wani SH, Singh B et al (2016) Transcription factors and plants response to drought stress: current understanding and future directions. Front Plant Sci 7:1029

Kouzai Y, Nakajima K, Hayafune M et al (2014) CEBiP is the major chitin oligomer-binding protein in rice and plays a main role in the perception of chitin oligomers. Plant Mol Biol 84: 519–528

Kumar K, Kumar M, Seong-Ryong Kim SR et al (2013) Insights into genomics of salt stress response in rice. Rice (NY) 6:27

Kumar V, Singh A, Mithra SV et al (2015) Genome-wide association mapping of salinity tolerance in rice (*Oryza sativa*). DNA Res 22:133–145

Kumar A, Sharma S, Chunduri V et al (2020) Genome-wide identification and characterization of heat shock protein family reveals role in development and stress conditions in *Triticum aestivum* L. Sci Rep 10:7858

Lafarge T, Bueno C, Frouin J et al (2017) Genome-wide association analysis for heat tolerance at flowering detected a large set of genes involved in adaptation to thermal and other stresses. PLoS One 12:e0171254

Li M, Li X, Zhou Z et al (2016) Reassessment of the four yield-related genes Gn1a, DEP1, GS3, and IPA1 in rice using a CRISPR/Cas9 system. Front Plant Sci 7:377

Li Z, Li L, Zhang Y et al (2019a) GhWRKY6 acts as a negative regulator in both transgenic Arabidopsis and cotton during drought and salt stress. Front Genet. https://doi.org/10.3389/fgene.2019.00392

Li W, Wan XL, Yu JY (2019b) Genome-wide identification, classification, and expression analysis of the *Hsf* gene family in carnation (*Dianthus caryophyllus*). Int J Mol Sci 20:5233

Li J, Islam F, Huang Q, Wang J, Zhou W, Xu L et al (2020a) Genome-wide characterization of WRKY gene family in Helianthus annuus L. and their expression profiles under biotic and abiotic stresses. PLoS One 15:e0241965

Li J, Zhang M, Sun J et al (2020b) Heavy metal stress-associated proteins in rice and *Arabidopsis*: genome-wide identification, phylogenetics, duplication, and expression profiles analysis. Front Genet 11:477

Li B, Zheng JC, Wang TT et al (2020c) Expression analyses of soybean VOZ transcription factors and the role of *GmVOZ1G* in drought and salt stress tolerance. Int J Mol Sci 21:2177

Liu C, Chen K, Zhao X et al (2019a) Identification of genes for salt tolerance and yield-related traits in rice plants grown hydroponically and under saline field conditions by genome-wide association study. Rice 12:88

Liu M, Ma Z, Sun W et al (2019b) Genome-wide analysis of the NAC transcription factor family in Tartary buckwheat (*Fagopyrum tataricum*). BMC Genomics 20:113

Lorenzana RE, Bernardo R (2009) Accuracy of genotypic value predictions for marker-based selection in biparental plant populations. Theor Appl Genet 120:151–161
Mahmood T, Gupta KJ, Kaiser WM (2009) Cadmium stress stimulates nitric oxide production by wheat roots. J Bot 41:1285–1290
MAPK-Group (2002) Mitogen-activated protein kinase cascades in plants: a new nomenclature. Trends Plant Sci 7:301–308
Mendoza Cozat DG, Jobe TO, Hauser F et al (2011) Long-distance transport, vacuolar sequestration, tolerance, and transcriptional responses induced by cadmium and arsenic. Curr Opin Plant Biol 14:554–562
Miyadate H, Adachi S, Hiraizumi A et al (2011) OsHMA3, a P1B-type of ATPase affects root-to-shoot cadmium translocation in rice by mediating efflux into vacuoles. New Phytol 189:190–199
Moellering ER, Muthan B, Benning C (2010) Freezing tolerance in plants requires lipid remodeling at the outer chloroplast membrane. Science 330:226–228
Mushtaq M, Sakina A, Wani SH et al (2019) Harnessing genome editing techniques to engineer disease resistance in plants. Front Plant Sci 10:550
Nekrasov V, Wang C, Win J, Lanz C et al (2017) Rapid generation of a transgene-free powdery mildew resistant tomato by genome deletion. Sci Rep 7:482
Noman M, Jameel A, Qiang WD et al (2019) Overexpression of GmCAMTA12 enhanced drought tolerance in *Arabidopsis* and soybean. Int J Mol Sci 20:4849
Osakabe Y, Watanabe T, Sugano S et al (2016) Optimization of CRISPR/Cas9 genome editing to modify abiotic stress responses in plants. Sci Rep 6:26,685
Pandey D, Rajendran SR, Gaur CK et al (2016) Plant defense signaling and responses against necrotrophic fungal pathogens. J Plant Growth Regul 35:1159–1174
Park CJ, Seo YS (2015) Heat shock proteins: a review of the molecular chaperones for plant immunity. Plant Pathol J 31:323–333
Pedersen CNS, Axelsen KB, Harper JF et al (2012) Evolution of plant P-type ATPases. Front Plant Sci 3:31
Peleg Z, Blumwald E (2011) Hormone balance and abiotic stress tolerance in crop plants. Curr Opin Plant Biol 14:290–295
Peng A, Shanchun C, Tiangang L et al (2017) Engineering canker-resistant plants through CRISPR/Cas9-targeted editing of the susceptibility gene CSLOB1 promoter in citrus. Plant Biotechnol J 15:1509–1519
Pennycooke JC, Cheng H, Roberts SM et al (2008) The low temperature-responsive, *Solanum* CBF1 genes maintain high identity in their upstream regions in a genomic environment undergoing gene duplications, deletions, and rearrangements. Plant Mol Biol 67:483–497
Puig S, Mira H, Dorcey E et al (2007) Higher plants possess two different types of ATX1-like copper chaperones. Biochem Biophys Res Commun 354:385–390
Rahaman M, Mamidi S, Rahman M (2018) Genome-wide association study of heat stress-tolerance traits in spring-type *Brassica napus* L. under controlled conditions. Crop J 6:115–125
Rao ES, Kadirvel P, Symonds RC et al (2015) Variations in DREB1A and VP1. 1 genes show association with salt tolerance traits in wild tomato (*Solanum pimpinellifolium*). PLoS One 10:e0132535
Rayapuram N, Bigeard J, Alhoraibi H, Bonhomme L, Hesse AM, Vinh J et al (2017) Quantitative phosphoproteomic analysis reveals shared and specific targets of *Arabidopsis* MPK3, MPK4 and MPK6. Mol Cell Proteomics 17:mcp.RA117.000135
Raza A, Razzaq A, Mehmood SS et al (2019) Impact of climate change on crops adaptation and strategies to tackle its outcome: a review. Plants (Basel) 8:34
Ren Y, Huang Z, Jiang H et al (2020) A heat stress responsive NAC transcription factor heterodimer plays key roles in rice caryopsis filling. J Exp Bot. https://doi.org/10.1101/2020.02.08.939728
Rodriguez Serrano M, Romero Puertas MC, Pazmino DM et al (2009) Cellular response of pea plants to cadmium toxicity: cross talk between reactive oxygen species, nitric oxide, and calcium. Plant Physiol 150:229–243

Sahebi M, Hanafi MM, Rafii MY et al (2018) Improvement of drought tolerance in rice (*Oryza sativa* l.): genetics, genomic tools, and the WRKY gene family. Biomed Res Int:3158474

Sanchez DH, Pieckenstain FL, Szymanski J et al (2011) Comparative functional genomics of salt stress in related model and cultivated plants identifies and overcomes limitations to translational genomics. PLoS One 6:e17094

Sanghera GS, Wani SH, Hussain W et al (2011) Engineering cold stress tolerance in crop plants. Curr Genomics 12:30–43

Santillán Martínez MI, Bracuto V, Koseoglou E et al (2020) CRISPR/Cas9-targeted mutagenesis of the tomato susceptibility gene *PMR4* for resistance against powdery mildew. BMC Plant Biol 20:284

Schoonbeek HJ, Wang HH, Stefanato FL et al (2015) *Arabidopsis* EF-Tu receptor enhances bacterial disease resistance in transgenic wheat. New Phytol 206:606–613

Sharma M, Pandey GK (2017) Editorial: genomics and functional genomics of stress-mediated signaling in plants: volume I. Curr Genomics 18:467–468

Shikha M, Kanika A, Rao AR et al (2017) Genomic selection for drought tolerance using genome-wide SNPs in maize. Front Plant Sci 8:550

Shkryl Y, Yugay Y, Avramenko T, Grigorchuk V, Gorpenchenko T, Grischenko O, Bulgakov V (2021) CRISPR/Cas9-mediated knockout of *HOS1* reveals its role in the regulation of secondary metabolism in *Arabidopsis thaliana*. Plants (Basel) 10:104

Shukla A, Srivastava S, Suprasanna P (2017) Genomics of metal stress-mediated signalling and plant adaptive responses in reference to phytohormones. Curr Genomics 18:512–522

Singh B, Salaria N, Thakur K et al (2019a) Functional genomic approaches to improve crop plant heat stress tolerance. F1000Res 8:1721

Singh D, Singh CK, Taunk J et al (2019b) Genome wide transcriptome analysis reveals vital role of heat responsive genes in regulatory mechanisms of lentil (*Lens culinaris* Medikus). Sci Rep 9: 12,976

Souer E, van Houwelingen A, Kloos D et al (1996) The no apical meristem gene of Petunia is required for pattern formation in embryos and flowers and is expressed at meristem and primordia boundaries. Cell 85:159–170

Su HG, Li B, Song XY (2019) Genome-wide analysis of the DYW subgroup PPR gene family and identification of *GmPPR4* responses to drought stress. Int J Mol Sci 20:5667

Sun W, Li Y, Zhao YX, Zhang H (2015) The TsnsLTP4, a nonspecific lipid transfer protein involved in wax deposition and stress tolerance. Plant Mol Biol Rep 33:62–974

Tang L, Cai H, Ji W et al (2013) Overexpression of GsZFP1 enhances salt and drought tolerance in transgenic alfalfa (*Medicago sativa* L.). Plant Physiol Biochem 71:22–30

Tashkandi M, Ali Z, Aljedaani F, Shami A, Mahfouz MM (2018) Engineering resistance against Tomato yellow leaf curl virus via the CRISPR/Cas9 system in tomato. Plant Signal Behav 13(10):e1525996

Ton J, Mauch-Mani B (2004) Beta-amino-butyric acid-induced resistance against necrotrophic pathogens is based on ABA-dependent priming for callose. Plant J 38:119–130

Trda L, Fernandez O, Boutrot F et al (2014) The grapevine flagellin receptor VvFLS2 differentially recognizes flagellin-derived epitopes from the endophytic growth-promoting bacterium *Burkholderia phytofirmans* and plant pathogenic bacteria. New Phytol 201:1371–1384

Tuteja N, Sopory SK (2008) Chemical signaling under abiotic stress environment in plants. Plant Signal Behav 3:525–536

Verma V, Ravindran P, Kumar PP (2016) Plant hormone-mediated regulation of stress responses. BMC Plant Biol 16:86

Wan DY, Guo Y, Cheng Y, Hu Y, Xiao S, Wang Y, Wen YQ (2020) CRISPR/Cas9-mediated mutagenesis of *VvMLO3* results in enhanced resistance to powdery mildew in grapevine (*Vitis vinifera*). Hortic Res 7:116

Wang L, Li X, Zhao Q et al (2009) Identification of genes induced in response to low-temperature treatment in tea leaves. Plant Mol Biol Rep 27:257–265

Wang Y, Cheng X, Shan Q, Zhang Y, Liu J, Gao C et al (2014) Simultaneous editing of three homoeoalleles in hexaploid bread wheat confers heritable resistance to powdery mildew. Nat Biotechnol 32:947

Wang X, Chen ZH, Yang C et al (2018) Genomic adaptation to drought in wild barley is driven by edaphic natural selection at the Tabigha Evolution Slope. PNAS 115:5223–5228

Wang X-K, Gong X, Cao F et al (2019a) *HvPAA1* encodes a P-type ATPase, a novel gene for cadmium accumulation and tolerance in barley (*Hordeum vulgare* L.). Int J Mol Sci 20:1732

Wang J, Hu T, Wang W, Hu H, Wei Q, Wei X, Bao C (2019b) Bioinformatics analysis of the lipoxygenase gene family in radish (*Raphanus sativus*) and functional characterization in response to abiotic and biotic stresses. Int J Mol Sci 20:6095

Wang D, Liu YX, Yu Q (2019c) Functional analysis of the soybean *GmCDPK3* gene responding to drought and salt stresses. Int J Mol Sci 20:5909

Wang T, Zhang H, Zhu H (2019d) CRISPR technology is revolutionizing the improvement of tomato and other fruit crops. Hortic Res 6:77

Wani SH, Kapoor N, Mahajan R (2017) Metabolic responses of medicinal plants to global warming, temperature and heat stress. In: Ghorbanpour M, Varma A (eds) Medicinal plants and environmental challenges. https://doi.org/10.1007/978-3-319-68717-9_4

War AR, Paulraj MG, War MY et al (2011) Herbivore and elicitor-induced resistance in groundnut to Asian armyworm, *Spodoptera litura* (Fab.) (Lepidoptera: Noctuidae). Plant Signal Behav 6: 1769–1777

Wei H, Wang X, Yuqing He Y et al (2021) Clock component OsPRR73 positively regulates rice salt tolerance by modulating OsHKT2;1-mediated sodium homeostasis. EMBO J 40:e105086

Wu SJ, Ding L, Zhu JK (1996) A genetic locus essential for salt tolerance and potassium acquisition. Plant Cell 8:617–627

Xiang Y, Huang Y, Xiong L (2007) Characterization of stress-responsive CIPK genes in rice for stress tolerance improvement. Plant Physiol 144:1416–1428

Xu R, Yang Y, Qin R et al (2016) Rapid improvement of grain weight via highly efficient CRISPR/Cas9-mediated multiplex genome editing in rice. J Genet Genomics 43:529–532

Yang A, Dai X, Zhang WH (2012) A R2R3-type MYB gene, *OsMYB2*, is involved in salt, cold, and dehydration tolerance in rice. J Exp Bot 63:2541–2556

Yang J, Duan G, Li C et al (2019) The crosstalks between Jasmonic acid and other plant hormone signaling highlight the involvement of Jasmonic acid as a core component in plant response to biotic and abiotic stresses. Front Plant Sci 10:1349

Ye B, Wu Y, Zhai X et al (2020) Beneficial effects of endophytic fungi from the *Anoectochilus* and *Ludisia* species on the growth and secondary metabolism of *Anoectochilus roxburghii*. ACS Omega 5:3487–3497

Yoon Y, Seo DH, Shin H et al (2020) The role of stress-responsive transcription factors in modulating abiotic stress tolerance in plants. Agronomy 10:788

Young JC (2010) Mechanisms of the Hsp70 chaperone system. Biochem Cell Biol 88:291–300

Younis A, Ramzan F, Ramzan Y et al (2020) Molecular markers improve abiotic stress tolerance in crops: a review. Plan Theory 9:1374

Yu W, Wang L, Zhao R et al (2019) Knockout of *SlMAPK3* enhances tolerance to heat stress involving ROS homeostasis in tomato plants. BMC Plant Biol 19:354

Zhang Y, Chen K, Zhao FJ et al (2018a) *OsATX1* interacts with heavy metal P1B-type ATPases and affects copper transport and distribution. Plant Physiol 178:329–344

Zhang C, Peng X, Guo X et al (2018b) Transcriptional and physiological data reveal the dehydration memory behavior in switchgrass (*Panicum virgatum* L.). Biotechnol Biofuels 11:91

Zhou S, Zheng WJ, Liu BH et al (2019) Characterizing the role of *TaWRKY13* in salt tolerance. Int J Mol Sci 20:5712
Zhu JK, Liu J, Xiong L (1998) Genetic analysis of salt tolerance in *Arabidopsis*. Evidence for a critical role of potassium nutrition. Plant Cell 10:1181–1191
Zhu Z, Shi J, Xu W, Li H et al (2013) Three ERF transcription factors from Chinese wild grapevine *Vitis pseudoreticulata* participate in different biotic and abiotic stress-responsive pathways. J Plant Physiol 170:923–933
Zou L, Yu B, Ma XL et al (2019) Cloning and expression analysis of the *BocMBF1c* gene involved in heat tolerance in Chinese Kale. Int J Mol Sci 20:5637

Genetics of Plant Organelles: Plastid and Mitochondrial Genomes

12

Samar Singh, Jogindra Naik, and Ashutosh Pandey

Abstract

Plant organelles like chloroplasts and mitochondria are essential organelles serving critical functions like photosynthesis and respiration, respectively, in plants. While most of the processes and the components required by the functioning of these organelles are contributed by nuclear DNA, they have few of their own components encoded by their respective genome. Mitochondrial and chloroplast genomes give a real insight into the evolution of land plants, as evident by several studies. Few studies have successfully conducted gene transfer technology into these organelles' genomes. Although extensive research on plant organelle genome is yet to be done, recent research has shown the probability of these organelles as a target of genome engineering. From targeting individual genes of their genome to incorporating new genes from other species, they hold promises to produce improved traits. Packaging of their genome, which varies significantly in various hierarchies of land and primitive plants, has also been studied in few plant species. This chapter summarizes the current studies and findings in the study of the organellar genome concerning their structure, organization, distribution, regulatory mechanism, and gene transfer technologies. This chapter provides an updated account of the evolution of these organelle genomes.

Keywords

Chloroplast genome · Evolution · Genome diversity · Mitochondrial genome · Organelle genome engineering

S. Singh · J. Naik · A. Pandey (✉)
National Institute of Plant Genome Research, Aruna Asaf Ali Marg, New Delhi, India
e-mail: ashutosh@nipgr.ac.in

R. L. Singh et al. (eds.), *Plant Genomics for Sustainable Agriculture*,
https://doi.org/10.1007/978-981-16-6974-3_12

12.1 Introduction

The genome is defined as an organism's complete set of genetic materials. Tom Roderick coined the term genomics in 1920, which is the study of the genome and is a more recent idea than genetics and genes. Genetics entails studying genetic information, the transmission of genetic information from generation to generation in the form of DNA, and the study of a restricted number of genes. In a pre-genomic era, the mechanism of information transfer to new cells at the functional level was not precise (Goldman and Landweber 2016; Hjort et al. 2010), which could be explored in the last few decades due to the advancement in molecular techniques. The human genome consists of the nuclear genome inside the nucleus, having 3.2 billion base pairs and 35000 genes (Schneider and Grosschedl 2007). The DNA packaging into the chromosome occurs in a highly systematic manner and is regulated at various levels. The genome inside the nucleus in base-pair information is a highly compact nucleosome structure made up of histone and other proteins and nuclear DNA. The highly repetitive and large number of noncoding sequences are involved in the regulatory mechanism. Apart from the nuclear genome, the extra-nuclear genome, which includes the mitochondrial and chloroplastic genome, plays an essential role like the initiation of apoptosis, aging, a large amount of ATP production, amino acid biosynthesis, steroid biosynthesis, β-oxidation of lipid, in mitochondria and photosynthesis, and important biochemical reactions in the chloroplast

According to endosymbiont theory, these organelles evolved separately via endosymbiosis, holding and merging free-living bacteria and cyanobacteria with the host eukaryotic cell around 1.4 billion years ago (Smith and Keeling 2015). This mechanism led to structural diversity as well as functional diversity of both the organelles. When an α-proteobacterial endosymbiont integrates into the eukaryotic host organism existing at the moment, mitochondria were formed, which evolved with due course of time. Because all eukaryotic cells include mitochondria, we may certainly propose that α-proteobacteria are the ancestors of all extant eukaryotic organisms. Photosynthetic cyanobacteria merged with the eukaryotic progenitor of archiplastida via chloroplast-derived letter via initial endosymbiosis to form chloroplast

12.2 Organelles Genome

The first mitochondrial genome was sequenced entirely in 1981 (Bibb et al. 1981; Anderson et al. 1981), whereas the chloroplastic genome of *Marchantia polymorpha* and tobacco was entirely sequenced five years later in 1986 (Shinozaki et al. 1986; Ohyama et al. 1986). Because of their evolutionary histories, mitochondrial and chloroplastic genomes share high sequence similarities. The organelle genome size (Mt genome, Pt genome) gradually decreased over billions of years in comparison to free-living α-proteobacteria and cyanobacteria through the transfer of several genes involved in DNA repair mechanisms, replication, from the nucleus to organelle

(Timmis et al. 2004). It is also important for functional and structural stability, as well as a variety of metabolic processes. The mitochondrial genome's regulatory element is not confined to noncoding regions within genes (Lee and Han 2017)

12.3 Plant Mitochondrial Genome

Mitochondria are an important cellular organelle in the plant responsible for growth, fitness, reproduction, energy generation, metabolism, and cell homeostasis due to its semi-autonomous genetic system. Plant mitochondrial genome encodes several critical polypeptides involved in oxidative phosphorylation chain complex formation. Plant mitochondrial DNA (mtDNA) varies from that of animals and fungi in several ways. Higher plants have a larger mitochondrial genome than lower ones. In addition, mitochondria have nucleoids that replicate independently of the nuclear chromosome in many plants although many components of the replication machinery are derived from nuclear-encoded proteins. Plant mtDNA sequences evolved deliberately as compared to the mtDNA of animals, where point mutations were infrequent.

The mitochondrial DNA (mtDNA) originated from the symbiotic ancestral genomes of α-proteobacteria through the endosymbiosis process. The genomes of the mitochondria have an array of distinct characteristics in higher plants. Plant mitochondrial genomes are significantly bigger and vary greatly in size, even among relatively similar. (Allen et al. 2007; Kubo and Newton 2008). While the mitochondria of most mammals have circular DNAs having around 15–17 kb, plant mitochondrial genomes are quite larger, which vary significantly in terms of their size. In angiosperms, they typically range from 200 to 750 kb (Anderson et al. 1981) although some lineages have larger genome size, e.g., the organization of the mitochondrial genome into three independent chromosomal structures having 1556, 84, and 45 kb, respectively, is attributed to the expansion of dispersed repeats and existing introns, and accretion of nuclear, plastidial, viral, and bacterial sequences in cucumber (*Cucumis sativus*) (Alverson et al. 2011). The mitochondrial genome of the plants is enormous, yet their ploidy seems to be less, which is surprising. Plants have significantly lower mtDNA levels than human cells, which can have thousands of copies of mtDNA (Preuten et al. 2010). The *Arabidopsis thaliana* mtDNA is 367 kb in size, encodes 32 protein-coding genes, 22 transfer RNAs, and three ribosomal RNAs (5S, 18S, and 26S). On the other hand, the 16.5 kb human mtDNA code for 13 proteins, two rRNAs (12S and 16S), and 22 tRNAs (Anderson et al. 1981; Unseld et al., 1997; Stupar et al. 2001). Simpler organisms can have more mitochondrial genes, e.g., *Reclinomonas americana* having a 69 kb mtDNA encodes around 100 genes (Lang et al. 1997). The structure of the mitochondrial genome of the *Arabidopsis thaliana* has been shown in Fig. 12.1a. Mitochondrial genome sequences of plants have become more widely available, but the origin and role of noncoding DNA remain unknown, and thus, it is difficult to compare various species. The analysis of whole mitochondrial genomes of two *Arabidopsis thaliana* ecotypes, C24 and Columbia-0 (Col-0) (Davila et al. 2011), allows for noncoding

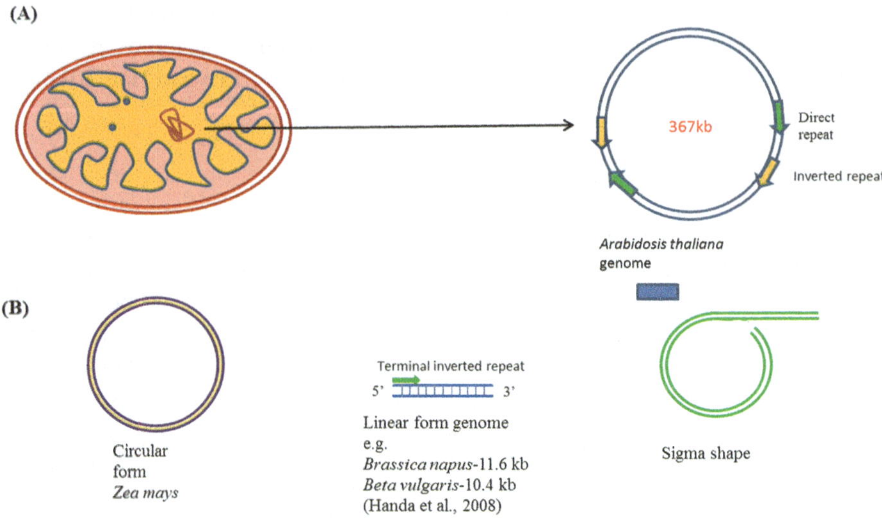

Fig. 12.1 The basic structure and types of plant mitochondrial genome. Arabidopsis mitochondrion with its circular DNA and mitoribosomes inside the matrix. A circular 367 kb Arabidopsis thaliana mitochondrial DNA having inverted repeats and direct repeats (**a**). Different conformations of mitochondrial DNA found in angiosperms, e.g., circular, linear, and sigmoid type (**b**)

sequence comparison and molecular evolution organelle genome in plants. Plant mtDNA contains numerous introns and repetitive sequences (accounting for 90 percent of the total sequences). It is susceptible to various gene gain, gene loss, gene transfer, and duplication, as well as genomic rearrangements (Li et al. 2020). Although mitochondrial genomes' structure and gene composition vary from species to species in plants, gene-coding sequences of most seed plants species develop gradually, with synonymous substitution rates 100 times lesser than that of mammalian mitochondria (Kitazaki and Kubo 2010).

Despite the fact that the nucleoid of the plant mitochondria is less well-known, it most likely possesses the PolI-like DNA polymerases1A and PolI-like DNA polymerases1B, the replicative DNA primase-helicase (TWINKLE), the type II topoisomerase Gyrase, the RecA-like recombinases (RECA2, RECA3), the SSB-like ssDNA-binding proteins (SSB1, SSB2), the RecA-like recombinases (RECA2, RECA), RPOTm and RPOTmp phage-type RNA polymerases, the MutS-like homolog MSH1, and some of the additional proteins (Palmer and Herbon 1988; Xu et al. 2011). DNA repair and homologous recombination are also facilitated by several of these proteins.

12.4 Structure of the Chloroplast Genome

A plant biologist discovered the chloroplast genome for the first time in 1950. Due to the lack of advanced techniques at that time, comparative restriction site mapping and cloning were being used to study genome organization, structure, and gene order. The genome size is more variable than the mitochondrial genome size and highly conserved, ranging from 35 to 217 kb in most plants species (Chumley et al., 2006). The chloroplast genome consists of 79 protein-coding genes out of 113 total genes, where 30 genes code for transfer-RNA and 4 genes encode rRNA genes in tobacco. It is a circular and quadripartite structure having a unique sequence like a small single-copy region(SSC), a large single-copy region(LSC), and one pair of inverted repeats (IRs) (Yang et al., 2010). The structure of the chloroplast genome of *Nicotiana tabacum* with its different regions has been provided in Fig. 12.2a. In the inverted repeat (IR), 5 protein-coding, 4 rRNA, and 7 tRNA genes are present, which evolved during the duplication event. The LSC region consists of 62 protein-coding, 22 tRNA genes in contrast to the SSC region, which contains 12 protein-coding genes and a single tRNA gene. In the chloroplast genome, only 12 genes possess introns, out of which *ycf3* and *clpP* genes contain two introns, and the other genes possess a single intron. In the LSC region of 5′end, trans-splicing occurs in *rps12*

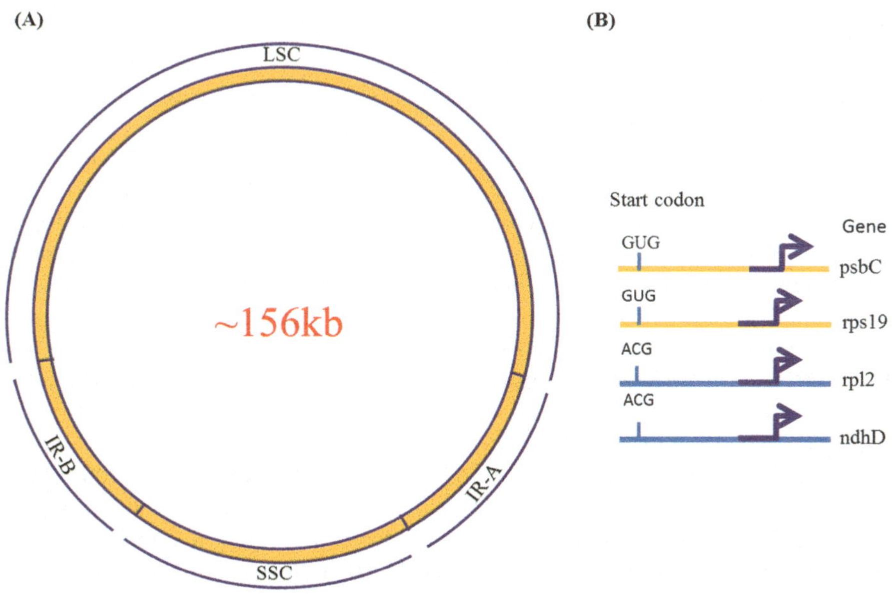

Fig. 12.2 Structure and strat codon diversity in tobacco chloroplast genome A circular 156 kb tobacco chloroplast genome with different regions, i.e., LSC (large single copy), IR-A (inverted repeat A), IR-B (inverted repeat B), and SSC (Small single copy) regions (**a**). Variation in the start codon in various genes of the chloroplast genome, e.g., GUG as start codon for psbC (photosystem II 44 kDa protein) and rps19 (ribosomal protein subunit 19) gene; ACG as start codon for rpl2 (ribosomal protein large subunit 2) and ndh2 (NADH dehydrogenase subunit 4) gene (**b**)

genes. Duplication event has also been reported in the IR regions of the chloroplast genome (Liu et al. 2016). The chloroplast genome possess some protein-coding region having several initiator codons. The *psbC* and *rps19* use the GUG initiator codon, while the *rpl2* and *ndhD* use ACG initiator codon. The schematic representation of the start codon diversity has been given in Fig. 12.2b. The rest 75 genes code universal tRNA. Translational efficiency is higher in the non-canonical initiator GUG codon than the canonical initiation codon AUG. In *Oryza minuta*, the initiator codon of the *rpl2* and *rps19* genes are ACG and GUG, respectively. AUG was used as the initiator codon in the *rps19* and *psbC* genes in plants belonging to the Brassicaceae family (Kuroda et al. 2007; Hu et al. 2015). The chloroplast genome has a large molecular marker to study large-scale evolutionary concepts in different taxonomic groups. One of the advanced techniques is DNA barcoding, in which a small segment of DNA from a given gene(s) is used for species identification. Some chloroplast genes like *matK, rbcL,* and *ycf1* act as barcodes to study evolutionary conservation in the plant kingdom via DNA barcoding (Asaf et al. 2017; Wu et al. 2010).

12.5 Mitochondrial Genome Diversity in Angiosperm Plants

Rearrangements among large repetitive sequences within the genome often occur in mitochondria of most angiosperms, and the genome exists in different conformations inside the mitochondria among different species (Mower et al. 2012). Consequently, mitochondrial genomes in angiosperms have multipartite genome maps that can be depicted as a single "master circular" or a collection of subgenomes (Palmer and Shields 1984; Sugiyama et al. 2005; Arrieta-Montiel and Mackenzie 2011). In flowering plants, mitochondrial genomes highly vary in structure and size (Mower et al. 2012). While the genomes are typically depicted as single circular rings, plant mitochondrial chromosomes have also been found in a variety of shapes and sizes, which include linear or circular forms, highly branched or σ-like structures, and multi-chromosomal structures capable of co-existing in a single plant (Kitazaki and Kubo 2010). The different conformations of the genome inside the mitochondria have been depicted in Fig. 12.1b.

Some CMS (Cytoplasmic Male sterility) lines in maize and rice, as examples, exhibit linear mitochondrial genome (Allen et al. 2007; Notsu et al. 2002). The mitochondrial genome in the case of *Eruca sativa* is multipartite, with six bigger circular and four smaller subgenomic circular DNA, indicating that repeat-induced genomic rearrangement is possible (Wang et al. 2014). *Brassica oleracea*, which has a tripartite structure having a 220 kb circular genome divided into two smaller circular genomes of 170 and 50 kb by homologous recombination of repetitive sequences, has a similar arrangement (Grewe et al. 2014).

12.6 Mitochondrial Genome Stability

The mitochondrial genome has a high risk of DNA damage due to the high ROS produced during electron transport in the respiratory mechanism (Møller 2001). Plants use different mitigating strategies for their genome stability and proper functioning. Repeated sequences play a significant part in higher plants for the stability of mtDNA (Kmiec et al. 2006). It may also function in homologous recombination and, as a result, have a significant influence on the structure of mtDNA. Massive repeated sequences (>1 kb), average-size repeats (50–500 bp), and micro-homologies (50 bp) are the three types of mtDNA repeats. Large size repeats are frequently engaged in changeable reciprocal recombination processes that regulate mtDNA flexibility in plants, which is generally made up of a combination of interconvertible subgenomes (Negruk et al. 1986; Oda et al. 1992). Homologous recombination (HR) appears to be the main DNA repair mechanism in plant mitochondria, with the end-joining mechanism being uncommon. HR is also required for the replication and segregation of mtDNA and is accountable for the genome's fast evolution (Cappadocia et al. 2010). As an outcome, the large number of HR must precisely be controlled to circumvent intragenomic rearrangements that might be harmful to the mitochondria because of repetitive sequences in mtDNA (Wallet et al. 2015).

Nuclear genes that play a crucial role in the stability of mtDNA have been discovered through studies in mutants with variegated leaves or by altering genes previously thought to be involved in organelle DNA metabolism (Maréchal and Brisson 2010). RECA1 was found in the mitochondria of the bryophyte *Physcomitrella patens* which are the homologs of bacterial RecA protein (Terasawa et al. 2007; Inouye et al. 2008). A RECA1 knock-out strain has abnormalities in development and mitochondrial morphology and a decreased rate of mtDNA repair (Terasawa et al. 2007; Odahara et al. 2009). Furthermore, the RECA1 knock-out mutant exhibits large rearrangements due to abnormal recombination between short repeats ranging from 62 to 84 bp dispersed across mtDNA, indicating that RECA1 keeps mt genome stable by inhibiting gross rearrangements (Odahara et al. 2009). Mutations in plant-specific single-strand DNA-binding proteins, such as WHY2 from the whirly protein family (Cappadocia et al. 2010) and organellar single-stranded DNA-binding protein 1 (OSB1), also cause abnormal recombination between repeats (Dong et al., 2013). Recombination occurs between short repetitions (30 bp) in the OSB1 mutant and is gyrase inhibitor-dependent in the WHY2 mutant (Cappadocia et al. 2010).

12.7 Chloroplast Genetic Engineering

There are currently new technologies for sophisticated chloroplast genome engineering. Through enabling genome engineering of chloroplast genomes in major crops and increased expression of foreign genes using modular vectors, RNA interference (RNAi), and crop-specific vectors, multigene engineering can be utilized to create

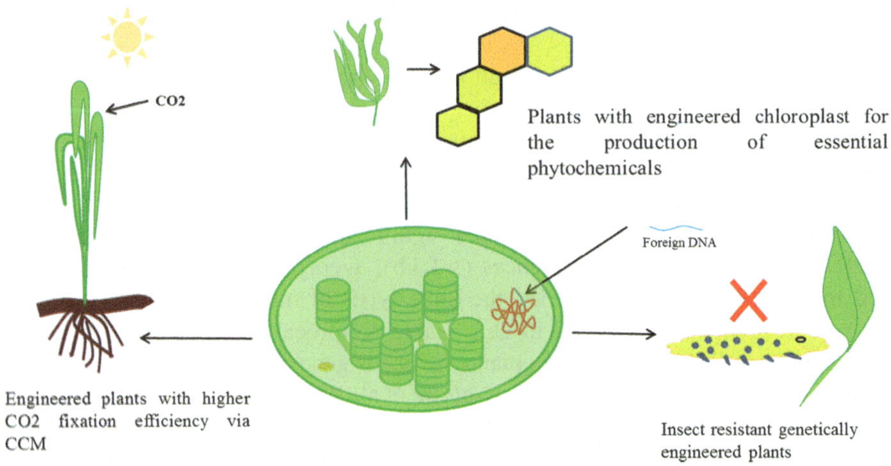

Fig. 12.3 Application of chloroplast genome engineering for various purposes, i.e., for increased CO_2 fixation efficiency via CCM mechanism, production of essential volatile compounds necessary for unique traits, and modulation of insect resistance in target host plants

high-value desired bioproducts. Introgression of a single gene into the chloroplast genome can be employed to give biotic/abiotic stress tolerance or increase biomass (Jin and Daniell et al. 2015). Retrograde signaling allows genes expressed in chloroplasts to regulate the nuclear genome. Photosynthesis is the process of producing vital energy sources for the whole world as a form of ATP, oxygen, fuel, food, etc., done in the chloroplast. The chloroplast genome can be modulated by the genetic engineering and biotechnology tool for enhanced agronomic traits, medicines, and industrial enzymes for a beneficial approach. Chloroplast genome engineering is used for stable introgression and expression of the foreign gene from the heterologous system like fungus, virus, animal for producing biopharmaceutical proteins, industrial enzymes, antibiotics, and benefitted agronomic traits (Chen et al. 2014). Plastid transformation has various advantages like the very less unwanted effect of transgene due to compartmentalization, transgene confinement via maternal or uniparental inheritance. Double homologous recombination is required for chloroplast transformation (Verma and Daniell 2007; Verma et al. 2008). The diagrammatic representation of chloroplast genome engineering and its applications for plant trait enhancement has been provided in Fig. 12.3.

Hundreds of genes from other organisms have been expressed in chloroplasts, which exhibit substantially elevated expression than expression systems inside the nucleus in many situations. However, the requisite expression level has still not been obtained in a few cases. Protein N-terminal degradation is a well-known phenomenon in heterologous systems. Insulin, the most renowned chimeric protein in the blood, has not been expressed without N-terminal fusion proteins in any expression vector (Lee et al. 2011). As a result, numerous human therapeutic proteins have been discovered. To confer stability, green fluorescent protein (GFP) has been expressed in chloroplasts by tagging with cholera toxin B (CTB) to aid oral delivery and stability (Lee et al. 2011; Kohli et al. 2014; Shenoy et al. 2014).

12.7.1 Agronomic Trait Enhancement Via Chloroplast Modulation

Synthetic antimicrobial peptide production by chloroplast genome engineering for protection against various fungal infections, bacterial, pathogens, viral particles, and abiotic stress has also been established in few studies (Lee et al. 2011; Kwon et al. 2013). The expression of β-glucosidase in chloroplast activates the ester conjugates to produce hormones (indolyl-3-acetic acid, zeatin, gibberellic acid). It has been used for high biomass, height, internode length, increase leaf area in the targeted plants. It also provides protection against whitefly and aphid via sugar esters which are produced in high amounts in the dense globular trichomes on the leaf surface (Jin et al. 2011).

12.7.2 Engineering of the Metabolic Pathway in Chloroplast for the Beneficial Product

The heterologous expression of chloroplastic enzymes involved in the isoprenoid pathway (mevalonate pathway) without any regulatory sequences like (promoter, UTR) in the host chloroplast shows an increase in the amount of mevalonate, carotenoids, sterols, squalene, and triglycerols, in the transformed plants (Kumar et al. 2012). Carotenoids are photoprotective compounds synthesized by the terpenoids pathway. Astaxanthin, a carotenoid, acts as an antioxidant and is responsible for the pigmentation of salmon and few other organisms. Transfer of gene *isopentenyl diphosphate isomerase*, *β-carotene hydroxylase*, and *β-carotene ketolase* from marine bacteria to the chloroplast genome of lettuce produced key carotenoids (ketocarotenoids) astaxanthin fatty acid esters (Harada et al. 2014). Lipid soluble tocopherols act as antioxidants in plants and are known for scavenging of the reactive oxygen species (ROS). The α-tocopherol, an isoform of the tocopherol, was synthesized by the expression of *tocopherol cyclase (TC)* and *γ-tocopherol methyl-transferase (γ-TMT)* in the chloroplast. Some rate-limiting enzymes like *homogentisate phytyltransferase (HPT)*, TC, and γ-TMT elevate the total tocopherol content up ten-fold. Enhanced synthesis of α-TOC in transplastomic plants enhances the nutritional, biotic, and abiotic stress tolerance by reducing the ROS level, lipid peroxidation, and ion leakage in plants (Lu et al. 2013; Jin and Daniell 2014).

12.7.3 Enhancement in Photosynthesis Efficiency Via Plastid Engineering

Ribulose-1,5-bisphosphate carboxylase/oxygenase, abbreviated as RuBisCO, is a major enzyme operating in the Calvin cycle that has sparked interest to improve the catalytic activity, carbon fixation efficiency, and decrease oxygenase activity. The assembly of small subunits of Rubisco from the nucleus to chloroplast forms functional Rubisco. Early efforts entailed relocating the small subunit gene to the chloroplast genome (Dhingra et al. 2004). The heterologous expression of Rubisco subunits in chloroplasts has been the subject of more recent investigations. The CO_2-concentrating mechanism (CCM) from cyanobacteria has recently been introduced into transplastomic plants, which has resulted in a breakthrough in crop improvement (Lin et al. 2014). Within chloroplasts, hybrid assembly of Rubisco from *Synechococcus elongates* Se7942 and CcmM35, a β-carboxysomal protein, led to enhanced carbon fixation efficiency but decreased growth. This is a significant footstep toward chloroplast genetic engineering to increase photosynthesis. Other discoveries increased recombinant Rubisco biogenesis by simultaneous expression of the RuBisCO accessory chaperone RAF1 and RuBisCO (Whitney et al. 2015).

12.7.4 Chloroplast Genome Engineering for Insect Resistance

Although significant progress has been witnessed in expressing biopesticide genes from a gram-positive bacteria *Bacillus thuringiensis* in the chloroplast for the production of Bt toxin crystals, plastidial expression of these genes in many crops has yet to reach commercial development because of the saturation of the market value of Bt crop products. However, a new revelation of successful Bt resistance has prompted the US Environmental Protection Agency (EPA) regulations to plant Bt maize (De Cosa et al. 2001; Dufourmantel et al. 2005). As a result, recent attention has shifted to identify novel features or methodologies to aid commercial development.

The chloroplast genome has recently been engineered using RNA interference technology (RNAi). *Chitin synthase (CHS)*, *V-ATPase*, and *cytochrome P450 monooxygenase (CYP450)* found in lepidopterans were chosen as the target for RNAi in this work. Cleaved and processed dsRNA had a higher quantity than the native psbA transcript, which was robustly expressed (Redick et al. 2015). The expression level of the targeted genes was lowered significantly in the mid-gut of the insects feeding on leaves, having silenced *CYP450*, *CHI*, and *V-ATPase* gene, most likely due to additional processing of engulfed siRNA in the insect gut. The larvae's net weight and their development and pupation rates were drastically decreased. In separate research, Bock and colleagues used the stably expressed dsRNA in the chloroplast genome against the insect β-actin gene and evoked resistance to the potato beetle; this groundbreaking work was confirmed as effective in the field trials.

12.8 Evolution of Organelle Genome

Both organelle genomes (Mt genome and Pt genome) originated from prokaryotes (alfa proteobacteria and cyanobacteria, respectively) through the endosymbiosis process. Both are independent events in eukaryotic evolution over a billion years ago. RNA editing refers to mechanisms that add or remove the specific nucleotides in RNA molecules and mechanisms that add or remove nucleotides. The information in the mature mRNA is different from the gene that encodes this mRNA (Aphasizhev and Aphasizheva 2011). The pentatricopeptide repeat (PPR) proteins are nuclear-encoded factors necessary for editing in sites of different mitochondria and plastids. RNA editing occurs in viruses, early eukaryotes, mammals, fungi, and plants in various ways. RNA editing mechanisms are employed as checkpoints, and they can maintain the function of the encoded protein and produce new proteins (Koito and Ikeda 2012). Comprehensive studies of numerous processes in diverse species that detail their mechanistic and functional features and their origins have recently been studied (Nishikura 2010).

12.8.1 Mitochondria Genome for Phylogeny Analysis

Phylogenetic and phylogenomic reconstructions suggest mitochondrial origin from a single ancestor, which is also known as monophyly. Two further pieces of genetic study back up the theory that all mitochondrial genomes originated from a single ancestor. The eubacterial origins of this genome have been adequately confirmed by studies on mitochondrial DNA (mtDNA). The mtDNA sequences have facilitated tracing the evolution of mitochondria from a single ancestor connected to the Proteobacteria (Timmis et al. 2004). *Anaplasma, Rickettsia,* and *Ehrlichia* are the closest eubacterial relatives of mitochondria in the rickettsial subgroup of the α-Proteobacteria, which are a group of intracellular obligate parasites that also includes the above 3 genera (Gillham 1994).

In several areas of biology, next-generation sequencing (NGS) has evolved as a valuable tool. One of the uses in evolutionary and phylogenetic investigations is the ability to swiftly and cheaply assemble organelle genomes. In evolutionary investigations, metazoan mitochondrial genome possessing 13 protein-coding genes, 22 tRNA, and 2 rRNA have been proved to be particularly helpful biomarkers (Yang et al. 1985). It may also be helpful in evolutionary studies by comparing the rates of base substitution among species, in addition to the other gene sequences. Moreover, the mitochondrial gene ordering can also help study phylogenetic inferences (Gissi et al. 2008).

RNA editing was discovered in the mitochondria of the flowering plants (angiosperms) because of the sequence variations between DNA and RNA in 1989 (Covello and Gray 1989). This variation in the locations of U nucleotides in RNA and C nucleotides in DNA was discovered to be caused C-to-U substitution in the RNA. After editing, the codons are more comparable to those found in orthologous proteins from other species at the same locations (Gualberto et al. 1989; Covello and Gray 1989).

In understanding species-level phylogenies, mitochondrial DNA data can be extremely useful. The arrangement of genes in the mitochondrion varies, and the extensive stretches of noncoding DNA divide them. The mitochondrial genome rearrangement occurs more often, resulting in several altered forms in a single cell.

Because of the advancements in isolation methods of mtDNA, the use of restriction endonucleases to recognize specific nucleotide differences, PCR methodologies, the applicability of universal primers for DNA amplification in phylogenetics, and population genetic studies, mitochondrial genes are becoming progressively demanded in phylogenetics and population genetic studies (Borsch and Quandt 2009).

Cytochrome oxidase I/II, the electron transport-chain enzyme cytochrome c oxidase, is present in both bacteria and mitochondria. This gene, commonly employed to estimate molecular phylogenies, is gradually changing compared to other mitochondrial genes that encode protein (Lavrov and Lang 2005). The mitochondrial 12S rRNA sequence analysis is widely used in phylogenetic studies and molecular taxonomy. Its sequences have previously been proposed as a tool for determining intermediate to lengthy divergence dates (Lavrov and Meyer 1996). The cytochrome-b gene is the most helpful marker for resolving phylogenetic relationships between closely related species although it lacks clarity at deeper nodes. Although it has been reported to be beneficial in retrieving phylogenetically significant information about taxonomic levels, its use is lineage-dependent and diminishes as evolutionary depth increases.

12.8.2 Evolution of Chloroplast Genome and Its Use in Phylogeny Analysis

The gene content, structure, and organization of the chloroplast genome are largely conserved in comparison to the mitochondrial and nuclear genome. The rate of substitution in nucleotide sequences is higher in the chloroplast genome than mitochondrial genes but lower than the nuclear genome (Burger et al. 2003). However, a number of studies have shown evolutionary processes like gene duplications, mutations, deletions, and rearrangements. This organellar genome has long been regarded as a suitable model for comparative and evolutionary genomic investigations owing to its small size and preserved gene content. Comparative studies of chloroplast genomes have been conducted on several focused species, genera, or plant groups in recent years (Drouin et al. 2008; Dong et al., 2013).

RNA editing, including C-to-U alterations, has also been observed in the chloroplast, apart from (Hoch et al. 1991). All the plant lineages from simpler bryophytes to advanced angiosperms show this type of editing in plastids (Sugita et al. 2006). Several species from the order Marchantiales, for example, have mRNAs that stay as dictated by the chloroplast and mitochondrial genome (Rüdinger et al. 2008). RNA editing has not yet been discovered in plant cytoplasmic RNAs. It appears that the process is limited to these two organelles.

Comparative examinations of chloroplast genomes at higher taxonomic levels are valuable for phylogenetic research and in comprehending the evolution of the genomes in the context of genome size changes, gene deletions, and nucleotide changes. However, choosing a gene with the right length and substitution rate is critical. The *atpB, matK, ndhF, rbcL, rpl16*, and many more genes were currently utilized CpDNA genes for this purpose.

12.8.2.1 rbcL Gene

RuBisCO, the first enzyme in the C3 cycle, is the world's most abundant and significant protein and a key component of the global carbon cycle (Raven 2013). The *rbcL* gene is a single-copy gene found on the cp-genome that has a lot of phylogenetic usefulness. The *rbcL* gene has a length of 1428 bp and is located in all plants, excluding some parasites (Dong et al. 2018). It is straightforward to examine, align. Its secondary confirmation is well-studied and exists as multiple copies with few insertions and deletions. The *rbcL* gene codes for a large subunit of RuBisCO, whereas the rbcS gene in the nucleus encodes a small subunit. The first and one of the most commonly sequenced regions of the plant is the *rbcL* gene. It has frequently been employed in systematic research of terrestrial plants, particularly angiosperms. Phylogenetic connections among angiosperms and extant seed plants were investigated using around 500 *rbcL* sequences. Even though *rbcL* is conserved and easily alignable among the taxa, it has a greater replacement rate than 18S rDNA (Chase et al. 1993).

12.8.2.2 *MatK* Gene

MatK (maturase), a maturase enzyme, is involved in the splicing mechanism of type II introns in the RNA transcripts. It was situated within the intronic region of the chloroplast gene *trnK,* which encodes lysine tRNA. This gene's use in rectifying the intergeneric or interspecific connections among angiosperms has recently been demonstrated in research. The gene has high rates of substitution in comparison to other grass systematics genes and has a high proportion of transversion mutations. The three sections of its coding region help construct phylogenies in the Poaceae at the subfamily level (Patwardhan et al., 2014).

12.8.2.3 *ndhF* Gene

The ndhF genes, which encode for NADP dehydrogenase subunit F, are approximately 1100 bp long and are found in a single-copy region. The sequence variation among these genes was employed to rebuild a phylogenetic tree between 282 species representing 78 monocot groups. Furthermore, based upon rbcL alone or in combination with *atpB* and 18S rDNA, they demonstrated that relationships within orders are consistent. However, this gene provides more informative characters than rbcL and other genes (Givnish et al. 2006).

12.9 Conclusion

Besides serving as a tool for phylogenetic and evolutionary studies, the organellar genome can be used as a target for the improvement of essential agronomic traits. Plant chloroplast genome contributes mainly to the synthesis of carbohydrates, in addition to few other molecules. Our review gives a comprehensive and updated account of the basic structural organization of chloroplast and mitochondrial genome as studied in several plant species. A regulatory mechanism like RNA editing, DNA damage repairs inside the two organelles has also been discussed as studied in few plant species. Many researches have focussed on improving the photosynthetic efficacy and growth performance through genome engineering via transgene technology. However, more detailed studies on the various regulatory mechanism of gene expression need to be studied in major crop species.

Acknowledgments This work was supported by the core grant of the National Institute of Plant Genome Research and the Department of Science and Technology-SERB for a Startup research grant to AP. SS and JN acknowledge the University Grant Commission and Council of Scientific and Industrial Research, Government of India, for Junior and Senior Research Fellowships, respectively. The authors are thankful to DBT-eLibrary Consortium (DeLCON) for providing access to e-resources.

Conflict of Interest Statement The authors declare no conflict of interest.

References

Allen JO, Fauron CM, Minx P, Roark L, Oddiraju S, Lin GN, Meyer L, Sun H, Kim K, Wang C, Du F (2007) Comparisons among two fertile and three male-sterile mitochondrial genomes of maize. Genetics 177(2):1173–1192

Alverson AJ, Rice DW, Dickinson S, Barry K, Palmer JD (2011) Origins and recombination of the bacterial-sized multichromosomal mitochondrial genome of cucumber. Plant Cell 23(7): 2499–2513

Anderson S, Bankier AT, Barrell BG, de Bruijn MH, Coulson AR, Drouin J, Eperon IC, Nierlich DP, Roe BA, Sanger F, Schreier PH (1981) Sequence and organization of the human mitochondrial genome. Nature 290(5806):457–465

Aphasizhev R, Aphasizheva I (2011) Uridine insertion/deletion editing in trypanosomes: a playground for RNA-guided information transfer. Wiley Interdiscip Rev RNA 2(5):669–685

Arrieta-Montiel MP, Mackenzie SA (2011) Plant mitochondrial genomes and recombination. InPlant mitochondria. Springer, New York NY, pp 65–82

Asaf S, Waqas M, Khan AL, Khan MA, Kang SM, Imran QM, Shahzad R, Bilal S, Yun BW, Lee IJ (2017) The complete chloroplast genome of wild rice (Oryza minuta) and its comparison to related species. Front Plant Sci 8:304

Bibb MJ, Van Etten RA, Wright CT, Walberg MW, Clayton DA (1981) Sequence and gene organization of mouse mitochondrial DNA. Cell 26(2):167–180

Borsch T, Quandt D (2009) Mutational dynamics and phylogenetic utility of noncoding chloroplast DNA. Plant Syst Evol 282(3):169–199

Burger G, Gray MW, Lang BF (2003) Mitochondrial genomes: anything goes. Trends Genet 19(12):709–716

Cappadocia L, Maréchal A, Parent JS, Lepage É, Sygusch J, Brisson N (2010) Crystal structures of DNA-Whirly complexes and their role in Arabidopsis organelle genome repair. Plant Cell 22(6): 1849–1867

Chase MW, Soltis DE, Olmstead RG, Morgan D, Les DH, Mishler BD, Duvall MR, Price RA, Hills HG, Qiu YL, Kron KA (1993) Phylogenetics of seed plants: an analysis of nucleotide sequences from the plastid gene rbcL. Nat Biotechnol 1:528–580

Chen PJ, Senthilkumar R, Jane WN, He Y, Tian Z, Yeh KW (2014) Transplastomic Nicotiana benthamiana plants expressing multiple defence genes encoding protease inhibitors and chitinase display broad-spectrum resistance against insects, pathogens, and abiotic stresses. Plant Biotechnol J 12(4):503–515

Chumley T, Palmer JD, Mower JP, Fourcade HM, Calie PJ, Boore L, Jansen RK (2006) The complete chloroplast genome sequence of Pelargonium hortorum organization and evolution of the largest and most highly rearranged chloroplast genome of land plants. Mol Biol Evol 23 (11):2175–2190

Covello PS, Gray MW (1989) RNA editing in plant mitochondria. Nature 341(6243):662–666

Davila JI, Arrieta-Montiel MP, Wamboldt Y, Cao J, Hagmann J, Shedge V, Xu YZ, Weigel D, Mackenzie SA (2011) Double-strand break repair processes drive evolution of the mitochondrial genome in Arabidopsis. BMC Biol 9(1):1–4

De Cosa B, Moar W, Lee SB, Miller M, Daniell H (2001) Overexpression of the Bt cry 2Aa2 operon in chloroplasts leads to formation of insecticidal crystals. Nat Biotechnol 19(1):71–74

Dhingra A, Portis AR, Daniell H (2004) Enhanced translation of a chloroplast-expressed RbcS gene restores small subunit levels and photosynthesis in nuclear RbcS antisense plants. Proc Natl Acad Sci U S A 101(16):6315–6320

Dong W, Xu C, Cheng T, Zhou S (2013) Complete chloroplast genome of Sedum sarmentosum and chloroplast genome evolution in Saxifragales. PLoS One 8:e77965

Dong WL, Wang RN, Zhang NY, Fan WB, Fang MF, Li ZH (2018) Molecular evolution of chloroplast genomes of orchid species: insights into phylogenetic relationship and adaptive evolution. Int J Mol Sci 19(3):716

Drouin G, Daoud H, Xia J (2008) Relative rates of synonymous substitutions in the mitochondrial, chloroplast and nuclear genomes of seed plants. Mol Phylogenet Evol 49(3):827–831

Dufourmantel N, Tissot G, Goutorbe F, Garcon F, Muhr C, Jansens S, Pelissier B, Peltier G, Dubald M (2005) Generation and analysis of soybean plastid transformants expressing Bacillus thuringiensis Cry1Ab protoxin. Plant Mol Biol 58(5):659–668

Gillham NW (1994) Organelle genes and genomes. Oxford University Press

Gissi C, Iannelli F, Pesole G (2008) Evolution of the mitochondrial genome of Metazoa as exemplified by comparison of congeneric species. Heredity 101(4):301–320

Givnish TJ, Pires JC, Graham SW, McPherson MA, Prince LM, Patterson TB, Rai HS, Roalson EH, Evans TM, Hahn WJ, Millam KC (2006) Phylogenetic relationships of monocots based on the highly informative plastid gene ndhF Aliso. J Syst Evol B 22(1):28–51

Goldman AD, Landweber LF (2016) What is a genome. PLoS Genet 12(7):e1006181

Grewe F, Edger PP, Keren I, Sultan L, Pires JC, Ostersetzer-Biran O, Mower JP (2014) Comparative analysis of 11 Brassicales mitochondrial genomes and the mitochondrial transcriptome of Brassica oleracea. Mitochondrion 19:135–143

Gualberto JM, Lamattina L, Bonnard G, Weil JH, Grienenberger JM (1989) RNA editing in wheat mitochondria results in the conservation of protein sequences. Nature 341(6243):660–662

Harada H, Maoka T, Osawa A, Hattan JI, Kanamoto H, Shindo K, Otomatsu T, Misawa N (2014) Construction of transplastomic lettuce (Lactuca sativa) dominantly producing astaxanthin fatty acid esters and detailed chemical analysis of generated carotenoids. Transgenic Res 23(2): 303–315

Hjort K, Goldberg AV, Tsaousis AD, Hirt RP, Embley TM (2010) Diversity and reductive evolution of mitochondria among microbial eukaryotes. Philos Trans R Soc Lond Ser B Biol Sci 365(1541):713–727

Hoch B, Maier RM, Appel K, Igloi GL, Kössel H (1991) Editing of a chloroplast mRNA by creation of an initiation codon. Nature 353(6340):178–180

Hu S, Sablok G, Wang B, Qu D, Barbaro E, Viola R, Li M, Varotto C (2015) Plastome organization and evolution of chloroplast genes in Cardamine species adapted to contrasting habitats. BMC Genomics 16(1):1–4

Inouye T, Odahara M, Fujita T, Hasebe M, Sekine Y (2008) Expression and complementation analyses of a chloroplast-localized homolog of bacterial RecA in the moss Physcomitrella patens. Biosci Biotechnol Biochem 72(5):1340–1347

Jin S, Daniell H (2014) Expression of γ-tocopherol methyltransferase in chloroplasts results in massive proliferation of the inner envelope membrane and decreases susceptibility to salt and metal-induced oxidative stresses by reducing reactive oxygen species. Plant Biotechnol J 12(9): 1274–1285

Jin S, Daniell H (2015) The engineered chloroplast genome just got smarter. Trends Plant Sci 20(10):622–640

Jin S, Kanagaraj A, Verma D, Lange T, Daniell H (2011) Release of hormones from conjugates: chloroplast expression of β-glucosidase results in elevated phytohormone levels associated with significant increase in biomass and protection from aphids or whiteflies conferred by sucrose esters. Plant Physiol 155(1):222–235

Jin S, Singh ND, Li L, Zhang X, Daniell H (2015) Engineered chloroplast dsRNA silences cytochrome p450 monooxygenase, V-ATPase and chitin synthase genes in the insect gut and disrupts Helicoverpa armigera larval development and pupation. Plant Biotechnol J 13(3): 435–446

Kitazaki K, Kubo T (2010) Cost of having the largest mitochondrial genome: evolutionary mechanism of plant mitochondrial genome. J Bot. https://doi.org/10.1155/2010/620137

Kmiec B, Woloszynska M, Janska H (2006) Heteroplasmy as a common state of mitochondrial genetic information in plants and animals. Curr Genet 50(3):149–159

Kohli N, Westerveld DR, Ayache AC, Verma A, Shil P, Prasad T, Zhu P, Chan SL, Li Q, Daniell H (2014) Oral delivery of bioencapsulated proteins across blood–brain and blood–retinal barriers. Mol Ther 22(3):535–546

Koito A, Ikeda T (2012) Apolipoprotein B mRNA-editing, catalytic polypeptide cytidine deaminases and retroviral restriction. Wiley Interdiscip Rev RNA 3(4):529–541

Kubo T, Newton KJ (2008) Angiosperm mitochondrial genomes and mutations. Mitochondrion 8(1):5–14

Kumar S, Hahn FM, Baidoo E, Kahlon TS, Wood DF, McMahan CM, Cornish K, Keasling JD, Daniell H, Whalen MC (2012) Remodeling the isoprenoid pathway in tobacco by expressing the cytoplasmic mevalonate pathway in chloroplasts. Metab Eng 14(1):19–28

Kuroda H, Suzuki H, Kusumegi T, Hirose T, Yukawa Y, Sugiura M (2007) Translation of psbC mRNAs starts from the downstream GUG, not the upstream AUG, and requires the extended Shine–Dalgarno sequence in tobacco chloroplasts. Plant Cell Physiol 48(9):1374–1378

Kwon KC, Verma D, Jin S, Singh ND, Daniell H (2013) Release of proteins from intact chloroplasts induced by reactive oxygen species during biotic and abiotic stress. PLoS One 8(6):e67106

Lang BF, Burger G, O'Kelly CJ, Cedergren R, Golding GB, Lemieux C, Sankoff D, Turmel M, Gray MW (1997) An ancestral mitochondrial DNA resembling a eubacterial genome in miniature. Nature 387(6632):493–497

Lavrov R, Meyer (1996) Phylogenetic performance of mitochondrial protein-coding genes in resolving relationships among vertebrates. Mol Biol Evol 3(7):933–942

Lavrov DV, Lang BF (2005) Poriferan mtDNA and animal phylogeny based on mitochondrial gene arrangements. Syst Biol 54(4):651–659

Lee SR, Han J (2017) Mitochondrial nucleoid: shield and switch of the mitochondrial genome. Oxidative Med Cell Longev. https://doi.org/10.1155/2017/8060949

Lee SB, Li B, Jin S, Daniell H (2011) Expression and characterization of antimicrobial peptides Retrocyclin-101 and Protegrin-1 in chloroplasts to control viral and bacterial infections. Plant Biotechnol J 9(1):100–115

Li X, Li L, Bao Z, Tu W, He X, Zhang B, Ye L, Wang X, Li Q (2020) The 287,403 bp mitochondrial genome of ectomycorrhizal fungus Tuber calosporum reveals intron expansion, tRNA loss, and gene rearrangement. Front Microbiol 11:3048

Lin MT, Occhialini A, Andralojc PJ, Parry MA, Hanson MR (2014) A faster Rubisco with potential to increase photosynthesis in crops. Nature 513(7519):547–550

Liu TJ, Zhang CY, Yan HF, Zhang L, Ge XJ, Hao G (2016) Complete plastid genome sequence of Primula sinensis (Primulaceae): structure comparison, sequence variation and evidence for accD transfer to nucleus. Peer J 4:e2101

Lu Y, Rijzaani H, Karcher D, Ruf S, Bock R (2013) Efficient metabolic pathway engineering in transgenic tobacco and tomato plastids with synthetic multigene operons. Proc Natl Acad Sci U S A 110(8):E623–E632

Maréchal A, Brisson N (2010) Recombination and the maintenance of plant organelle genome stability. New Phytol 186(2):299–317

Møller IM (2001) Plant mitochondria and oxidative stress: electron transport, NADPH turnover, and metabolism of reactive oxygen species. Annu Rev Plant Biol 52(1):561–591

Mower JP, Sloan DB, Alverson AJ (2012) Plant mitochondrial genome diversity: the genomics revolution. Plant Genom Divers 1:123–144

Negruk VI, Eisner GI, Redichkina TD, Dumanskaya NN, Cherny DI, Alexandrov AA, Shemyakin MF, Butenko RG (1986) Diversity of Vicia faba circular mtDNA in whole plants and suspension cultures. Theor Appl Genet 72(4):541–547

Nishikura K (2010) Functions and regulation of RNA editing by ADAR deaminases. Annu Rev Biochem 79:321–349

Notsu Y, Masood S, Nishikawa T, Kubo N, Akiduki G, Nakazono M, Hirai A, Kadowaki K (2002) The complete sequence of the rice (Oryza sativa L.) mitochondrial genome: frequent DNA sequence acquisition and loss during the evolution of flowering plants. Mol Genet Genom 268(4):434–445

Oda K, Kohchi T, Ohyama K (1992) Mitochondrial DNA of Marchantia polymorpha as a single circular form with no incorporation of foreign DNA. Biosci Biotechnol Biochem 56(1):132–135

Odahara M, Kuroiwa H, Kuroiwa T, Sekine Y (2009) Suppression of repeat-mediated gross mitochondrial genome rearrangements by RecA in the moss Physcomitrella patens. Plant Cell 21(4):1182–1194

Ohyama K, Fukuzawa H, Kohchi T, Shirai H, Sano T, Sano S, Umesono K, Shiki Y, Takeuchi M, Chang Z, Aota SI (1986) Chloroplast gene organization deduced from complete sequence of liverwort Marchantia polymorpha chloroplast DNA. Nature 322(6079):572–574

Palmer JD, Herbon LA (1988) Plant mitochondrial DNA evolved rapidly in structure, but slowly in sequence. J Mol Evol 28(1):87–97

Palmer JD, Shields CR (1984) Tripartite structure of the Brassica campestris mitochondrial genome. Nature 307(5950):437–440

Patwardhan A, Ray S, Roy A (2014) Molecular markers in phylogenetic studies-a review. J Evol Biol 2:2

Preuten T, Cincu E, Fuchs J, Zoschke R, Liere K, Börner T (2010) Fewer genes than organelles: extremely low and variable gene copy numbers in mitochondria of somatic plant cells. Plant J 64(6):948–959

Raven JA (2013) Rubisco: still the most abundant protein of Earth. New Phytol 198(1):1–3

Redick TP, Galey MR, Feitshans TA (2015) Litigation and regulatory challenges to innovation in biotech crops. Drake J Agric L 20:71

Rüdinger M, Polsakiewicz M, Knoop V (2008) Organellar RNA editing and plant-specific extensions of pentatricopeptide repeat proteins in jungermanniid but not in marchantiid liverworts. Mol Biol Evol 25(7):1405–1414

Schneider R, Grosschedl R (2007) Dynamics and interplay of nuclear architecture, genome organization, and gene expression. Genes Dev 21(23):3027–3043

Shenoy V, Kwon KC, Rathinasabapathy A, Lin S, Jin G, Song C, Shil P, Nair A, Qi Y, Li Q, Francis J (2014) Oral delivery of Angiotensin-converting enzyme 2 and Angiotensin-(1-7) bio encapsulated in plant cells attenuates pulmonary hypertension. Hypertension 64(6):1248–1259

Shinozaki K, Ohme M, Tanaka M, Wakasugi T, Hayashida N, Matsubayashi T, Zaita N, Chunwongse J, Obokata J, Yamaguchi-Shinozaki K, Ohto C (1986) The complete nucleotide sequence of the tobacco chloroplast genome: its gene organization and expression. EMBO J 5(9):2043–2049

Smith DR, Keeling PJ (2015) Mitochondrial and plastid genome architecture: reoccurring themes, but significant differences at the extremes. Proc Natl Acad Sci U S A 112(33):10,177–10,184

Stupar RM, Lilly JW, Town CD, Cheng Z, Kaul S, Buell CR, Jiang J (2001) Complex mtDNA constitutes an approximate 620-kb insertion on Arabidopsis thaliana chromosome 2: implication of potential sequencing errors caused by large-unit repeats. Proc Nat Acad Sci U S A 98(9): 5099–5103

Sugita M, Miyata Y, Maruyama K, Sugiura C, Arikawa T, Higuchi M (2006) Extensive RNA editing in transcripts from the PsbB operon and RpoA gene of plastids from the enigmatic moss Takakia lepidozioides. Biosci Biotechnol Biochem 70(9):2268–2274

Sugiyama Y, Watase Y, Nagase M, Makita N, Yagura S, Hirai A, Sugiura M (2005) The complete nucleotide sequence and multipartite organization of the tobacco mitochondrial genome: comparative analysis of mitochondrial genomes in higher plants. Mol Genet Genomics 272(6): 603–615

Terasawa K, Odahara M, Kabeya Y, Kikugawa T, Sekine Y, Fujiwara M, Sato N (2007) The mitochondrial genome of the moss Physcomitrella patens sheds new light on mitochondrial evolution in land plants. Mol Biol Evol 24(3):699–709

Timmis JN, Ayliffe MA, Huang CY, Martin W (2004) Endosymbiotic gene transfer: organelle genomes forge eukaryotic chromosomes. Nat Rev Genet 5(2):123–135

Verma D, Daniell H (2007) Chloroplast vector systems for biotechnology applications. Plant Physiol 145(4):1129–1143

Verma D, Samson NP, Koya V, Daniell H (2008) A protocol for expression of foreign genes in chloroplasts. Nat Protoc 3(4):739–758

Wallet C, Le Ret M, Bergdoll M, Bichara M, Dietrich A, Gualberto JM (2015) The RECG1 DNA translocase is a key factor in recombination surveillance, repair, and segregation of the mitochondrial DNA in Arabidopsis. Plant Cell 27(10):2907–2925

Wang Y, Chu P, Yang Q, Chang S, Chen J, Hu M, Guan R (2014) Complete mitochondrial genome of Eruca sativa mill. (Garden rocket). PloS one 9(8):e105748

Whitney SM, Birch R, Kelso C, Beck JL, Kapralov MV (2015) Improving recombinant Rubisco biogenesis, plant photosynthesis and growth by coexpressing its ancillary RAF1 chaperone. Proc Natl Acad Sci U S A 112(11):3564–3569

Wu FH, Chan MT, Liao DC, Hsu CT, Lee YW, Daniell H, Duvall MR, Lin CS (2010) Complete chloroplast genome of Oncidium Gower Ramsey and evaluation of molecular markers for identification and breeding in Oncidiinae. BMC Plant Biol 10(1):1–2

Xu YZ, Arrieta-Montiel MP, Virdi KS, de Paula WB, Widhalm JR, Basset GJ, Davila JI, Elthon TE, Elowsky CG, Sato SJ, Clemente TE (2011) MutS HOMOLOG1 is a nucleoid protein that alters mitochondrial and plastid properties and plant response to high light. Plant Cell 23(9): 3428–3441

Yang D, Oyaizu Y, Oyaizu H, Olsen GJ, Woese CR (1985) Mitochondrial origins. Proc Natl Acad Sci U S A 82(13):4443–4447

Yang M, Zhang X, Liu G, Yin Y, Chen K, Yun Q, Yu J (2010) The complete chloroplast genome sequence of date palm (Phoenix dactylifera L.). PLoS One 5(9):e12762

DNA Barcoding in Plants: Past, Present, and Future

13

Pradosh Mahadani, Majusha Dasgupta, Joshitha Vijayan, Chandan Sourav Kar, and Soham Ray

Abstract

DNA barcoding is a method of identifying biological specimens and assigning them to their respective species. It involves sequencing of single/multiple short stretch/stretches of previously agreed-upon genomic region which evolves fast enough to allow species-level discrimination. Thus, obtained sequence(s) of unknown samples serve as a molecular identifier which is compared to a reference database of museum samples using specialized algorithms to reveal the identity of the specimen under study. In effect it complements classical taxonomy to quickly identify any newly obtained sample and aid in describing, naming, and classifying it to species. Unlike in animals where DNA barcoding is well standardized utilizing mitochondrial gene *CO1*, DNA barcoding in plants has perpetually been a matter of concern due to low substitution rates of plant mitochondrial genome. Alternatively, plastid genome has been targeted in case of plants for DNA barcoding purpose with some amount of success but ambiguities remain regarding selection of barcode region that can provide best possible resolution. A large number of studies tested the efficiency of seven leading candidate plastid DNA regions (*matK, rbcL, rpoB, rpoC*1 genes and *atpF–atpH*, *psbK–psbI*, *trnH–psbA spacers*) as the standard DNA barcode for plants. Based on universality, sequence quality, and species discrimination rate, a

P. Mahadani
Bioinformatics Centre, National Tea Research Institute, Kolkata, India

M. Dasgupta
Department of Biotechnology, Gauhati University, Jalukbari, Guwahati, Assam, India

J. Vijayan
ICAR-Central Island Agricultural Research Institute, Port Blair, Andaman & Nicobar Islands, India

C. S. Kar · S. Ray (✉)
ICAR-Central Research Institute for Jute & Allied Fibres, Kolkata, West Bengal, India

R. L. Singh et al. (eds.), *Plant Genomics for Sustainable Agriculture*,
https://doi.org/10.1007/978-981-16-6974-3_13

double-locus barcode (*rbcL+ matK*) system is suggested to perform best in plants. However, internal transcribed spacer (ITS) region of plastid gene *ycf1* has recently been suggested as the most promising single-locus plant DNA barcode. On contrary, a recent study argues that with an ever-growing sequence database even double-locus barcode (*rbcL+ matK*) system might become unfit for precise discrimination purpose. Hence, with the availability of next-generation sequencing technologies, partial genome representation-based barcoding, genome skimming based barcoding, full-length multi-barcoding (FLMB), etc. might be the preferred approaches to improve diagnostic power. DNA barcoding in plants not only speeds up writing the encyclopedia of life, but also opens up the possibility of establishing Digital Plant Identification System (DPIS) which works independent of type, age, or developmental stage of the sample under study. Hence, if used properly, DNA barcoding can be an effective and efficient tool for exploring and protecting biodiversity, expedite bioprospecting, and defending against bio-piracy.

Keywords

DNA Barcoding · ITS · *matK* · Species identification

13.1 Introduction

Plant biodiversity is a product of natural evolution and selection pressure that has sustained humankind's security for thousands of years. It is recognized as a valuable gene pool for various traits, which may stand as a potential solution in the face of rising environmental and anthropogenic challenges. Still, a vast amount of biodiversity remains undiscovered to the world. In this aspect, traditional taxonomy has been serving to identify and classify species over many years. As many valuable species and gene pools face the risk of extinction, the first step in preserving biodiversity is assessment. Recent estimates suggest that around 70,000 flowering plant species await to be discovered (Gross 2011). Unfortunately, there are not enough taxonomists to catalog species throughout the world. DNA taxonomy and DNA barcoding are accessory technologies that have helped speed up the process and emerged as a conservation practice tool. By harnessing advances in molecular genetics, sequencing technology, and bioinformatics, DNA barcoding was initially proposed by (Hebert et al. 2003)and has emerged as a vital new tool for taxonomists who take care of inventory and management of our planet's immense and changing biodiversity (Kress and Erickson 2008). DNA barcoding equips the taxonomist with the ability to quickly and cheaply (relatively) provide diagnostic identifications of species present in specific locations with immediate conservation and environment-related implications. Therefore, this diagnostic tool was developed as an aid to the taxonomic identification of species. It uses a standardized DNA region from the genome, which ideally has sufficient sequence variation to discriminate among species (CBOL et al. 2009). It has been advocated as a more efficient approach than traditional taxonomic practices (Blaxter 2004; Tautz et al. 2003). The classical techniques of plant identification involving the conventional keys are tricky and time-consuming. As it involves micro-and macroscopic characters as well as chemical profiling, which did not evolve successfully. In this aspect, DNA barcoding has rapidly achieved recognition as an essential tool with the power to aid much basic research and applied endeavors in taxonomy and species identification (Hajibabaei et al. 2007; Hebert et al. 2003; Savolainen et al. 2005).

13.2 The Genesis of Concept

In 2003, Paul D. N. Hebert, a professor at the University of Guelph, Ontario, Canada, for the first time proposed the concept of DNA barcoding with an announcement that it would serve as a basic tool for species identification of global biological samples. His announcement is based on his observation and analysis with the class Hexapoda, representing the greatest biodiversity on the planet. The technique involved selective amplification of only 648 bp of mitochondrial *Cytochrome Oxidase SubunitI* (COI) gene near its 5′end. He coined this segment DNA barcode for species-level identification. He justified its universality based on rapid evolution properties of the COI gene and variability properties of A, T, G, and C nucleotides of DNA. He argued that integrating DNA barcodes into traditional taxonomic tools could efficiently reveal unexplored biodiversity more swiftly and more securely in an authenticated way than traditional methods alone. Since the genesis, it has been successfully used for rapid biodiversity assessment studies, bio-monitoring, investigation of the illegal trade of endangered species, feeding ecology studies and for conservation of medicinal and poisonous plants, etc. (Muellner et al. 2011; Hollingsworth et al. 2011). The use of nucleotide sequence variations to investigate biodiversity, however, is not a new concept. It has been long realized that the changes in the four nucleotides A, T, G, and C set the backbone of molecular evolution, leading to discrete variation patterns among organisms. Thus, during the evolution, initial changes accumulate at the molecular level, which in the long-term lead to visible morphological variations. However, even if two organisms are morphologically alike, they may bear substantial variation at the molecular level, and the phenotypic similarity or dissimilarity between organisms is not a true reflection of actual variation. This dilemma often leads to misinterpretation and is a major drawback in biodiversity research where morphological keys are the basis. Several enthusiasts who were inclined to explore variation at the molecular level proposed using nucleotide segments, genes, rDNA, allozymes, etc., as markers to characterize organisms. However, the propositions were mainly suitable for a group of organisms while lacking broad range utility or universal application. DNA barcoding is a comparatively easy, quick, reprove able, universal approach for species identification. The principal requirement for barcoding is judicious locus/loci for DNA barcoding and should be prioritized and standardized so that large databases of sequences for that locus can be generated. Sequences are able to generate without species-specific PCR primers from the taxa of interest. Three essential principles of DNA barcoding are standardization, minimalism, and extensibility.

The leading DNA barcoding bodies and resources are (1) Consortium for the Barcode of Life (CBOL) http://www.barcodeoflife.org established in 2004. Worldwide DNA barcoding efforts have resulted in the formation of CBOL which promotes DNA barcoding through more than 200 member organizations from 50 countries, (2) International Barcode of Life (iBOL) http://www.ibol.org launched in October 2010, iBOL represents a not -for-profit effort to involve both developing and developed countries in the global barcoding effort, establishing commitments and working groups in 25 countries. It is the largest biodiversity genomics initiative ever undertaken, which maintains a barcode reference library, (3) The Barcode of Life Data systems (BOLD) consists of different institutional nodes from several nations clustered into separate working groups which works coordinately for the development of a specialized repository for DNA barcode sequences and has emerged as a global bio-identification system for species. BOLD is a web-based system for DNA barcodes, combines a barcode repository, analytical tools, an interface for submission of sequences to GenBank, a species identification tool and connectivity for external web developers and bioinformaticians (Ratnasingham and Hebert 2007). As of 2017, BOLD included over 5.9 million DNA barcode sequences from over 542,000 species.

13.3 Technical Know-How of DNA Barcoding

The development of reference data sources for each taxa of the world and thus creation of a reference database is an important step in the realm of barcoding research. It involves either mass participation of renowned taxonomists across the globe for the construction of a sound reference database. Another way by which this is achieved is by focusing on the museum specimens identified by various experts and using their barcode sequences as the standards or references for those taxa. However, all resources are not cataloged in museums and hence, new collections and explorations are also considered vital. As museum specimens maintain some standard data, new collections of specimens were undertaken maintaining some standards records such as collector name, collection date, geographical location, elevation/depth, collection gear, notes on habitat and microhabitat, sex of specimen, life stage, specimen imaging, identifier, etc. Practically cataloguing the total biodiversity of Earth in a museum is not feasible and even if it can be done gradually with time, the specimens get deformed as no fixatives can guarantee total preservation of the samples. Under such circumstances, the specimen's information is maintained in a database. The second part of DNA barcoding involves access of the reference data by enthusiasts for subsequent analysis and interpretation. Additional favorable factors are short length of barcode loci facilitated routine sequencing, even with sub-optimal material, lack of heterozygosity enabling direct polymerase chain reaction followed by sequencing without cloning, ease of alignment that enables the use of character-based data analysis methods, lack of problematic sequence composition, such as regions with several microsatellites, that reduces sequence quality, universal capability to get amplified/sequenced with standardized primers, easy align ability and capability to get recovered easily from herbarium samples and other degraded DNA samples (Hollingsworth et al. 2009). From the preparation of the data to the final analysis, DNA barcoding technology comprises several practical steps, which will be discussed below briefly.

13.3.1 Sampling

The DNA barcoding is a molecular concept, where focus is on the DNA molecules that remain embedded within each cell. Hence, sampling for DNA barcoding involves both specimen sampling and DNA sampling. Specimen sampling is done from a taxonomist point of view, where a complete coverage of morphological and geographical information is gathered. DNA sampling can be done from any tissue of the organism; however, the areas which bear the key morphological characters for the specimens are always kept intact. The specimen sampling is immediately followed by sampling of a small part of tissue for DNA sampling.

13.3.2 DNA Extraction, PCR Amplification, and Sequencing

The DNA molecule which is the principal component of DNA barcoding concept has to be extracted and preserved. Nowadays there are several technologies involved for isolation of DNA, however, the best technique is adopted which will keep the DNA intact. The isolation is followed by PCR amplification of the targeted DNA barcode segment using available published primers and then sequencing.

13.3.3 Analysis and Interpretation

PCR amplification and sequencing of the barcode segment are followed by analysis of the sequence using bioinformatics tools. The major part of the analysis involves checking the quality of the sequence and its maximum similarity with the reference database.

The prerequisite criteria for any DNA barcode loci are a large amount of sequence variation between species; however, variation should be low enough within species so that a gap between intra- and inter-specific genetic variations can be defined and also known as barcode gap. Besides that, conserved flanking regions for universal primer are required to enable routine amplification across highly divergent taxa. In practice, an unidentified organism's specific standardized portion DNA sequence acts as a repository signal which is compared to the reference sequences databases of known species. The similarity of sequence, i.e., unknown organism to one of the reference sequences leads to a rapid and reproducible identification. Some large group of linkage or association should exist as support for species monophyly and

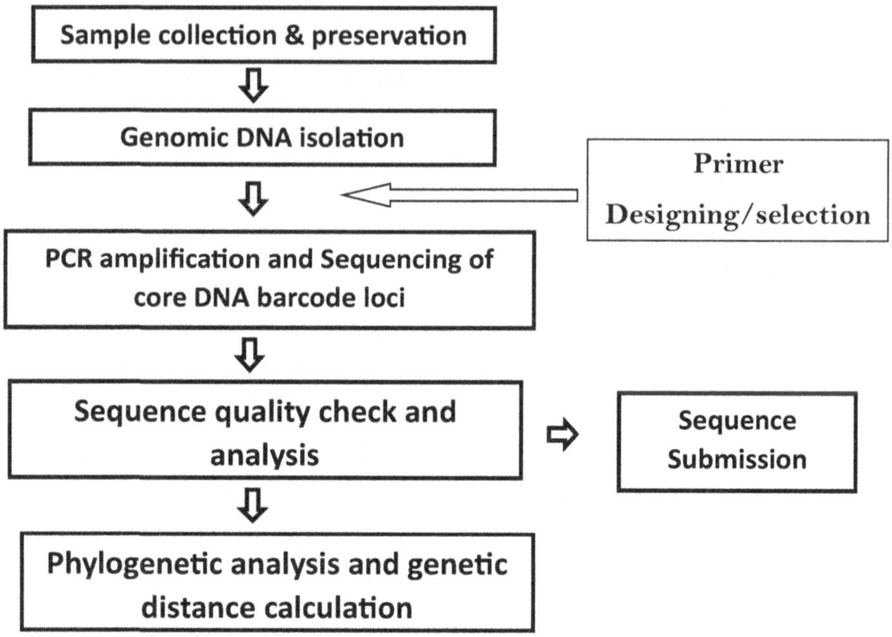

Fig. 13.1 Basic workflow of DNA barcoding approach

the ability of DNA barcoding marker systems to differentiate or distinguish species (Fig. 13.1).

13.4 Promising Plant DNA Barcoding Loci

Chloroplasts are organelles of prokaryotic origin and house of photosynthetic apparatus and also play a crucial role in sulfur and nitrogen metabolism. Plant DNA barcoding involves sequencing a standard region of the chloroplast genome as a tool for species identification. The chloroplast is a nearly autonomous organelle because it contains the biochemical machinery necessary to replicate and transcribe its own genome and carry out protein synthesis. The DNA of chloroplast is a circular that ranges in size from 120 to 190 kb depending on the species. The chloroplast genome is symbiotic in its origin from both algal and protistan lineages; its gene expression machinery is an assembly of prokaryotic, eukaryotic, and phage-like components, resulting in the acquisition of a significant number of regulatory proteins during evolution. Comparative evaluation indicates gene order and gene content of land plants chloroplast genomes are highly conserved. Traditionally, the plastid genome has been a more readily choice for phylogenetic studies in plants than the nuclear genome. As the chloroplast genome is maternally inherited, no recombination occurs, and, in general, structurally they are more stable with high copy number. Several candidate regions have been proposed as barcoding loci, including some coding genes (*matK*, *rbcL*, *rpoB*, and *rpoC1*) non-coding region (*psbA-trnH*, *atpF-atpH*, *ycf*) or a combination of several regions.

Maturase K of the chloroplast genome is the most conserved gene in the plant kingdom and is involved in Group II intron splicing. *matK* gene sequence is about 1500 bp long and encodes maturase like protein. Due to the high substitution rates, *matK* is emerging as a potential candidate for DNA barcoding (Hilu and Liang 1997). The *matK*gene has ideal size, large proportion of variation at open reading frame level at first and second codon position. The *matK* gene is rapidly evolving and considered as a good DNA barcode region (Mahadani et al. 2013; Sun et al. 2012). Thus, *matK* sequence plays a vital role in phylogenetic and evolutionary studies. Lahaye et al. (2008) collected more than 1600 plant samples from Mesoamerica and southern Africa, biodiversity hotspot. This was the first large scale study to compare eight potential barcodes in all the samples. As a universal plant DNA barcode, Plastid *matK* gene showed easy amplification, alignment, discrimination power. In addition, analyzing >1000 species of Mesoamerican orchids, DNA barcoding with *matK* alone revealed cryptic species and proved useful in identifying species listed in Convention on International Trade of Endangered Species (CITES) appendixes.

Several researchers proposed *rbcL* as a potential plant barcode region, as large amounts of information are already available in the sequence databases. About 1300 bp long, *rbcL* sequences showed a fair degree of success in discriminating species (Newmaster et al.; 2006). But the *rbcL* marker, which is easy to amplify sequence and align, has a limited discrimination power, especially when among closely related species. The Consortium for the Barcode of Life (Plant Working Group) recognized a combination of *matK* and *rbcL* as the universal plant barcode (CBOL et al. 2009) although the levels of variation are sometimes low and insufficient to recognize species with these two specific markers. In large scale studies, these loci provide a discriminatory efficiency at the species level of 72% and 49.7%, respectively. In some instances, they have failed to differentiate closely related species (Hollingsworth et al. 2009). As a result, other chloroplast regions, e.g., *trnH-psbA, trnL, trnL-F* and the nuclear ribosomal Internal Transcribed Spacer (ITS) are routinely used in combination with *matK* and *rbcL.*

In higher plants, two plastidal RNA polymerases referred as plastid encoding polymerases or PEP (α-, β-, β'-, and β''-subunits) encoded by *rpoA*, *rpoB*, *rpoC1,* and *rpoC2* genes are promising candidates (Serino and Maliga 1998). In the chloroplast genomes, *ndhF* is located at one end of the small single-copy region and encoding the ND5 protein of chloroplast NADH dehydrogenase complex. *ndhF* contains more phylogenetic information than *rbcL* (Kim and Jansen 1995).

Among the non-coding region, *trnH-psbA* has highly conserved PCR priming sites, high numbers of substitutions and is often used as an additional marker, especially when DNA barcoding is applied to closely related plant taxa (Kress and Erickson 2007). However, mononucleotide repeats in the *trnH-psbA* region cause constraints in PCR and sequencing. Although *trnH-psbA* shows high levels of inter-specific variation, it has found only limited use in species-level phylogenetic reconstruction due to its short length as well as difficulty of alignments resulting from a high number of insertion and deletion (indels). The intergenic spacer *trnL-F* has a long history of use in studies on plant phylogenetic and species identification studies (Wallander and Albert, 2000). In some groups this region often contains ploy A and T structures and affects sequence quality (Shaw et al. 2005). In the chloroplast genome, the pseudogene, *ycf1* is located in the boundary regions between IRb/S and IRa/SSC, respectively. *ycf1* is the first open reading frame coded by Tic214 (part of TIC core complex). The lack of their protein-coding ability is due to partial gene duplication. This gene is related to ATP synthase, and much more closely related to the *rbcL* gene with respect to its genetic structure. Recently, two regions of the plastid gene ycf1, ycf1a, and ycf1b were recognized as most variable loci in plastid. Dong et al. (2015) designed primers for amplification of these regions and analyzed the potential of these regions as DNA barcode in 420 tree species. The study showed ycf1a or ycf1b perform better than any of the *matK*, *rbcL*, and *trnH-psbA* for a large group of plant taxa (Dong et al. 2015) (Fig. 13.2).

In case of plants, mitochondrial genes are poor candidates for species-level discrimination due to low rate of sequence change. As the plastid genome evolves very slowly relative to other genomes and shows intra-molecular recombination, more than one barcode is necessary to provide enough to work. Although it is widely accepted that a single ("universal") set of barcode regions should be adopted to establish a reference barcoding database for all plants. The seven plastid regions *rpo*C1, *rpo*B, *rbc*L, *mat*K, *trn*H-*psb*A, *atp*F-*atp*H, and *psb*K-*psb*I were evaluated in three divergent groups of land plants (Newmasteret al., 2006. Hollingsworth et al. 2009). Study reports that 92% to 96% of plant specimens can be distinguished by combining the two core barcode markers *rbcL* and *matK*. In general, the genes used in angiosperms are *matK*, *rpo*C1, *rpo*B, *acc*D, *YCF5*, and *ndh*J, whereas in non-angiosperms *mat*K, *rpo*C1, *rpo*B, *acc*D, and *ndh*J. The Plant researcher from Consortium for the Barcode of Life (CBOL) proposed additional combinations of non-coding and coding plastid. In plant systematics for phylogenetic, *rbcL* sequenced most commonly, followed by the *trnL-F* intergenic spacer, *matK*, *ndhF*, and *atpB-rbcL* has been suggested as a candidate for plant barcoding.

Internal transcribed spacer regions (*ITS*) of nuclear ribosomal DNA are often highly variable in angiosperms at the generic and species level and divergent copies are often present within single individuals. About 400–800 bp long ITS regions are the most commonly sequenced region among the nuclear ribosomal cistron regions (18S-5.8S-26S), across the plants defined barcode gap between inter- and intra-specific variations (Group C P B et al. 2011). Chen et al. 2010 reported that The ITS1 and ITS2 (each <300 bp) adjoining the 5.8 S locus have a higher degree of variation than the rRNA genes (Chen et al. 2010). These genes contain enough phylogenetic signal for discrimination of both plants and animals. The ITS2 in comparison with ITS1 is more suitable for amplification and sequencing due to its shorter length of the target region which is referred to as a mini-barcode. The ITS of nuclear DNA has been used as a target for analyzing fungal diversity in environmental samples, and has been selected as the standard marker for fungal DNA barcoding (Schoch et al., 2012).

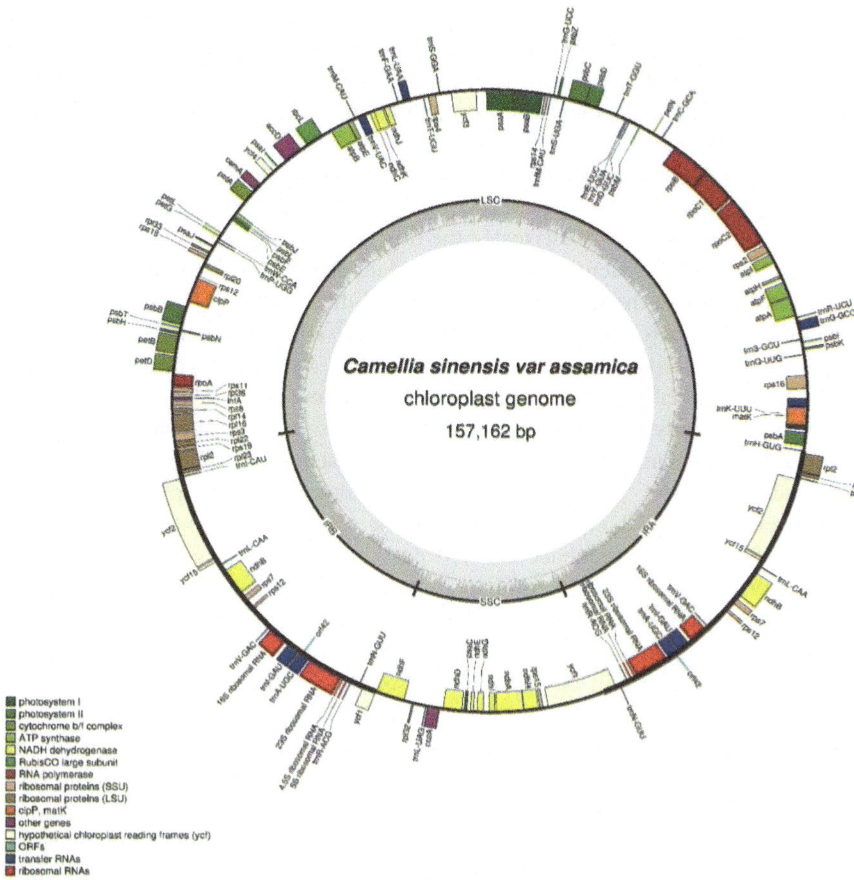

Fig. 13.2 Schematic diagram of Chloroplast genome. The graphical map *Camellia sinensis* var. *assamica* (Accession No: JQ975030) was drawn using OGDRAW (https://chlorobox.mpimp-golm.mpg.de/OGDraw.html)

13.5 Utility of Plant DNA Barcoding

The additional power of the DNA barcode speeds up writing the encyclopedia of life. It opens up the way to develop an electronic field guide which workes at all stages of life. It can deal with fragments, unmask the look-alikes, reduce ambiguity, democratize access, and thus sprouts a new leaves on the tree of life. Forensic investigators have also applied these plant DNA barcodes in the regulatory areas of traffic in endangered species and monitoring commercial products, such as foods and herbal supplements. Categories of use include species-level taxonomy(Mahadani et al. 2013), biodiversity inventories (Lahaye et al. 2008; Hollingsworth 2008), phylogenetic evaluation (Hajibabaei et al. 2007), conservation assessment and environmental preservation(Hollingsworth et al. 2011), species interactions and ecological networks (Erickson et al. 2008), cryptic diversity information (Fazekas et al. 2008), ecological forensics(Mishra et al. 2016), community assembly, traffic in endangered species, and monitoring of commercial products (Stoeckle et al. 2011; Mahadani and Ghosh 2013).

DNA barcoding assists in identification by expanding its power to detect species by including all life history stages of life, like pollen, seed, seedling and unstructured plant parts, etc. Kool et al. (2012) tested the functionality, efficiency, and accuracy of the use of barcoding for identifying 110 medicinal plant roots combining *rpc1*, *trnH-psbA*, and *ITS*. These three candidates identified the majority of samples up to the genus level. DNA barcoding helps to find out undiscovered species that are potentially new to encyclopedia of life (Kool et al. 2012).Over the last decade, four plant DNA barcode markers, viz. *rbcL*, *matK*, *trnH-psbA*, and ITS2, have been tested, and used to address many questions in systematics, ecology, evolutionary biology, and conservation. Mahadani et al. (2013) examined the sequences of core DNA barcode *matK* and *trnH-psbA* for differentiation of ethnomedicinal plants of family Apocynaceae from North east India. Among the selected medicinal Rauvolfioideae species, ~758 bp *matK* sequence showed easy amplification, alignment, and high level of discrimination value in comparison to the *trnH-psbA* spacer sequences. The partial *matK* sequences exhibited 3 indels in multiple of 3 at 5′ ends, but clustered cohesively, with their conspecific Genbank sequences. The possessing indels in multiple of 3 could be utilized as molecular markers in further studies both at the intra-specific and inter-specific levels (Mahadani et al. 2013).

Reliable identification of plant material by regulatory authorities is often of vital essence. Their domain includes identification of pests, pathogens and invasive species, illegal trades, identifying food or herbal medicine labeling errors/fraud. In this aspect, DNA barcoding approaches to assess the plant components of herbal medicines and dietary supplements, and evidence of market adulteration have been reported from many findings. A large array of commercial tea products were first time authenticated through *rbcL* and *matK* barcode sequences. Matching DNA identification to listed ingredients was limited by incomplete databases for the two markers, shared or nearly identical barcode among some species and lack of standard names for the plant species. About 1/3 of herbal tea generated DNA identification were not found on levels. This study demonstrated the importance of plant barcoding (Stoeckle et al. 2011). Six *Sabia* species and their seven adulterants were investigated DNA barcodes (*trnH-psbA*, *rbcL-a*, *matK*). Based on sequence alignments, they concluded that not only *trnH-psbA* spacer sequence distinguished *S. parviflora* from other *Sabia* species, but the *matK+rbcL-a* sequence also differentiated it from the substitute and adulterants. The three candidate barcodes identified *S. parviflora* and distinguished it from common substitutes or adulterants (Sui et al. 2011).

In traditional taxonomy-based identification, as seedlings, roots, seeds, and pollen and other gametophytes of many species appear similar, it is difficult to identify species from individual tissue types/juvenile life stages. Thus, with paleo barcoding, even barcode datasets with imperfect species resolution can still provide knowledge gains. Moreover, the field of pollen barcoding is growing rapidly, and even modest increases in discriminatory power beyond morphological identification holds great promise to enhance understanding of the dynamics and consequences of pollination and pollen movement (Bell et al. 2016).

13.6 Challenges of Plant Barcodes

Several factors can potentially contribute towards a lack of unique species identification with DNA barcodes. To successfully implement DNA barcoding, sufficient time since speciation is required for point mutations or genetic drift. It leads to developing of a set of genetic characters that "group" or outgroup conspecific individuals are together unique from other species. In phylogenetic evaluation, barcode sequences are shared among related taxa or species in clades where speciation has been very recent. These problematic scenarios arise mainly in groups like woody species with long generation times and/or slow mutation rates and groups with evidence of recent radiation. Composition of monophyletic species is more in animal (>90%) than plants (~70%) using barcode markers. Both animals and plant systems have barcode gaps based on intra- and inter-specific genetic distances. However, animal species showed larger barcode gaps than plants. However, overall fine scale species discrimination in plants is relatively more difficult than animals because species boundaries are less well defined (Fazekas et al. 2008). Polyploid speciation may cause divergence between barcode sequences and taxon concepts where multiple allopolyploid species share a common parent species. In such cases, they may show similarity in plastid sequences, whereas independent origins of allopolyploid species can lead to taxa treated as conspecifics possessing divergent haplotypes. At least initially, plastid haplotype(s) with a diploid progenitor will be shared by the species that have originated by allo or autoploidy (Wang et al. 2018). The complexity of taxonomically complex groups (TCGs) cannot be solved using one or few markers, as these groups result from recurrent ecotypic origin of taxa, or recurrent ploidy transitions, apomixes, or recent hybrid speciation. Species discrimination success can be predicted by its dispersal ability and in that case an inverse correlation between intra- and inter-specific gene flow may rise. In case of *Solanum* sect. *Petota* (wild potatoes), *ITS*, *trnH-psbA*, *matK* regions showed too much intra-specific variation and lacked sufficient polymorphism in plastid markers (Spooner 2009). The universal barcode concept in plants is not working in Indian *Berberis* and two other genera, *Ficus* and *Gossypium*. Even the most promising plant DNA barcode loci (one from nuclear genome—*ITS*, and three from plastid genome—*trnH-psbA, rbcL,* and *matK*) failed to resolve species identification in these plant groups (Roy et al. 2010). Mahadani and Ghosh (2014) provide an alternative approach to identify the species using indel polymorphism as a species-level marker in Citrus.

13.7 Prospect of DNA Barcoding

The major challenge for DNA barcoding in plants is to achieve the proportion of unique species identifications. The selection of markers often depends on the nature of the application or research queries. For instance, single specimen-based studies tend to use a blend of the traditional DNA markers, while to recover a higher number of taxa from degraded or mixed DNA samples, metabarcoding approaches are taken which aims for shorter, easy to amplify fragments. The criteria of using multiple loci or multi-tiered system increase sample handling, preparation time, and costs in plant barcoding. Various limitations of traditional plant DNA barcoding has also been overcome by the advancement of high-throughput sequencing technologies which expedited the progress of plant genomics, particularly chloroplast genomics. Recently complete chloroplast genomes have also been shown to discriminate closely related species successfully. Until now, most DNA barcoding methods follow a traditional PCR-based approach followed by dideoxy chain termination (Sanger)-based sequencing. Alternatively, next-generation sequencing (NGS) technologies which decrease the cost of sequencing solve the problem partially by sequencing large portions of genomes (genome skimming) or whole genomes (organellar or otherwise "genome skimming" approaches (Coissac et al. 2016; Li et al. 2015). Short universally primed amplicon is ideal for sequence characterization through new parallelized high-throughput sequencing technologies, allowing inexpensive but comprehensive studies of biodiversity to be a realistic goal. These methods generate millions of sequence reads in a single run, they are still expensive for many research groups with regard to consumables, informatics, computational power, and data storage. In this aspect, though traditional Sanger-based sequencing technology is more expensive than next-generation sequencing (NGS) and is generally hampered by the need for relatively high target amplicon yield, coamplification of nuclear mitochondrial pseudogenes, confusions with sequences from intracellular endosymbiotic bacteria and instances of intracellular variability (i.e., heteroplasmy). Due to all these limitations, the high-throughput technology of next-generation sequence-based DNA barcoding has recently showed promising outcome for the elucidation of plant genetic diversity and its conservation.

13.8 Next-Generation Sequencing and DNA Barcoding

NGS technology allows for the sequencing of millions of DNA fragments from thousands of DNA templates in parallel and produces millions of short reads. NGS is a term loosely applied to the set of technologies used for genome-scale sequencing, viz. Roche 454, Illumina, Ion Torrent, SMRT, etc. It finds vast implementation, because of its protocol simplicity, reduced cost per read, high throughout and added information, sequencing sensitivity and accuracy by enabling the simultaneous detection of co-amplified products such as homologues, prologues, and contaminants. In this aspect, 454 pyro sequencing was the first NGS platform that came into the market. It permits the analysis of mixtures of DNA fragments that are co-amplified during PCR or obtained by pooling different PCR products. Parallel sequencing of PCR amplicons is most effective when limited sequence data are targeted per specimen. NGS is also a powerful tool to detect numerous DNA sequence polymorphism based markers within a short timeframe and triggered numerous ground-breaking discoveries from many organisms (Van et al. 2013). The information that has emerged serves as a strong molecular tool for species exploration, progression, transformation studies, and the conservation of biodiversity.

13.9 DNA Barcode-Based High-Resolution Melting Curve Analysis (Bar-HRM)

In combination, high-resolution melting (HRM) analysis and DNA barcoding has emerged as a potential molecular method for plant species authentication, commonly known as Bar-HRM approach. The Bar-HRM has proven to be a reliable method for detection of contamination of different plant mixtures, particularly at the early stages of production like industrial quality control procedures for herbal medicines, etc. (Fernandes et al. 2020; Lee et al. 2019; Madesis et al. 2012). It is a novel DNA-based, cost-effective, and reliable quick identification assay that detects single base changes between samples. DNA dissociation ("melting") kinetics is monitored to detect the point mutations, indels, and methylated DNA.

The denaturation thermodynamics of individual double-stranded DNA to single strands is based on individual nucleotide pairs' binding affinities. Moreover, melting patterns will vary due to variations in product sizes, GC contents, and nucleotide composition, which vary due to indels, mutations, and methylations, inferred in terms of varying melting temperatures (T_m). In HRM in addition to standard PCR equipment and reagents, it requires a generic DNA intercalation fluorescent dye which is added to the previously amplified PCR products. As the double-stranded DNA samples dissociate with increasing temperature, the dye is progressively released and fluorescence diminishes. These differences of fluorescent measurements collected at standard temperature increments, which are plotted as a melting curve. So, variations in length, GC content, and base sequence will alter the melting profile defined by a plot. This plotted curve is generated between melting temperature and fluorescent level due to the release of intercalating SYBR Green I dye in a real-time PCR system. This melting curve's shape and peak are characteristic for individual specimen sample, allowing for comparison and discrimination among samples.

The HRM analysis has many advantages (1) As the sequencing is not required for Bar-HRM, which is generated by combining DNA barcode with HRM (called Bar-HRM), limitations of DNA barcoding technique could be minimized, (2) HRM analysis method is quite sensitive detecting 0.1%–1% presence of adulterated sample, (3) It is a high-throughput technology that is capable of analyzing multiple samples at the same time, (4) Post-PCR processes are not needed thus cross-contamination could be avoided, (5) The sample genotype can be traced by evaluating HRM curve analysis. Bar-HRM has thus been proven to be a powerful tool for species identification capable of identifying species and quantitatively detecting adulterants from mixtures of samples of different specimens.

To optimize HRM conditions, care should be taken in terms of primer designing, PCR reagents, and cycling conditions since small differences in melting curves can arise from sources other than the nucleotide sequence. Factors like genomic DNA (gDNA) quality, amplicon length, primer design, dye selection, and PCR conditions are all predicted to affect the melting behavior (Montgomery et al., 2007; van der Stoep et al., 2009).

13.10 High-Throughput Plant DNA Barcoding Using Microfluidic Enrichment Barcoding *(ME Barcoding)*

ME Barcoding is a precious tool for DNA metabarcoding. It is a cost-effective method for high-throughput DNA barcoding that uses microfluidic PCR-based target enrichment (Gostel et al. 2020), for species-level phylogenetic reconstruction. Microfluidic PCR-based barcoding might be preferable to molecular phylogenetics because of its efficiency, with minimal starting template size. There was a very low amount of template and reagents needed for PCR reactions (0.033 mL in the Access Array™ System). Nowadays, Fluidigm Access Array and Illumina MiSeq are used in M.E barcoding. The barcode can be generated from 96 or even more samples for each of the four primary DNA barcode loci in plants: *rbcL*, *matK*, *trnH-psbA*, and ITS. Fluidigm Access Array simultaneously amplifies targeted regions for 48 DNA samples and thus hundreds of PCR primer pairs (producing up to 23,040 PCR products) during a single thermal cycling protocol. This technique is emerging as an alternative to traditional PCR and Sanger sequencing to generate large amounts of plant DNA barcodes and build more comprehensive barcode databases. Microfluidic PCR amplification followed by high-throughput sequencing can produce by locus sequence with minimal resource investment. However, there are two limitations of this approach, viz. (1) A high initial equipment cost, (2) lower sequencing success compared to Sanger methods (Uribe-Convers et al. 2016).

Alternative HTS platforms (e.g., Pacbio SMRT) could be better suited to build plant DNA barcode libraries due to the *matK* region's length. The single molecule, real-time (SMRT) sequencing implemented on the SEQUEL platform to sequence barcode sequence libraries for COI. The instrument had capacity to sequences from more than 5 million DNA extracts a year (Hebert et al. 2018). Combining Pacbio with ME Barcoding could help determine whether the longer sequence read length provided by this single molecule, real-time (SMRT) sequencing approach (up to 60 kb) can improve the recovery success of all four plant DNA barcode loci (Fig. 13.3).

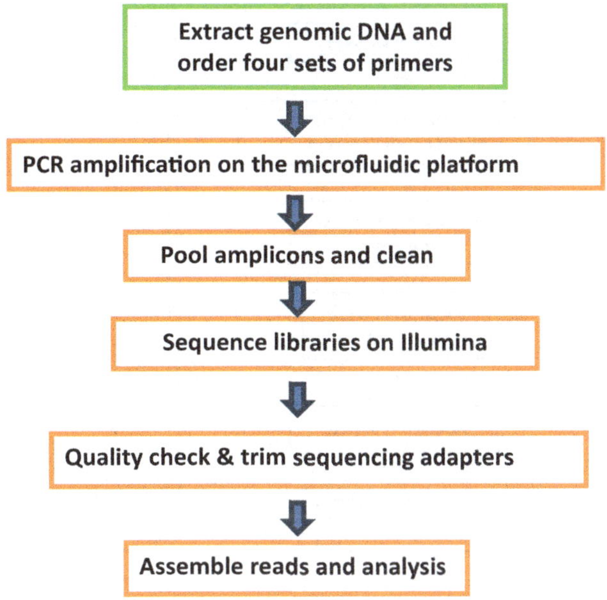

Fig. 13.3 Steps of microfluidic enrichment barcoding

13.11 Full-Length Multi-Barcoding (FLMB)

FLMB is a superior but feasible approach for identifying complex biological mixtures, which shows perfect interpretation for DNA barcoding that could lead to its application in multi-species mixtures. This full-length multi-barcoding (FLMB) via long-read sequencing is employed to identify biological compositions inadequate and well-controlled studies. For instance, in recent years, using various science, engineering, and biotechnology tools, the market foods are modified to improve their taste, color, and flavor, making them commercially more attractive. In this aspect, FLMB can detect most commercially processed foods and herbal mixtures for quantitative analysis of unknown fruit mixtures. It can also determine the composition of mixed spices, flavored teas, vegetable stock cubes, curry, deep-frozen vegetables, food supplements, and health drinks, as well as comprehensive identification of biological origin for herbs (Zhang et al. 2019). Overall, this tool has the potential to provide novel insights into biodiversity analysis in many research areas.

To test the efficacy of FLMB based DNA metabarcoding, DNA is extracted from individual and mixed biological samples and then individually quantified using qPCR, followed by library preparation and SMRT based Sequencing. Bioinformatics analysis is done for proper authentication of individual samples from the contaminated or mixture of samples. The working principle of FLMB is depicted in Fig. 13.4.

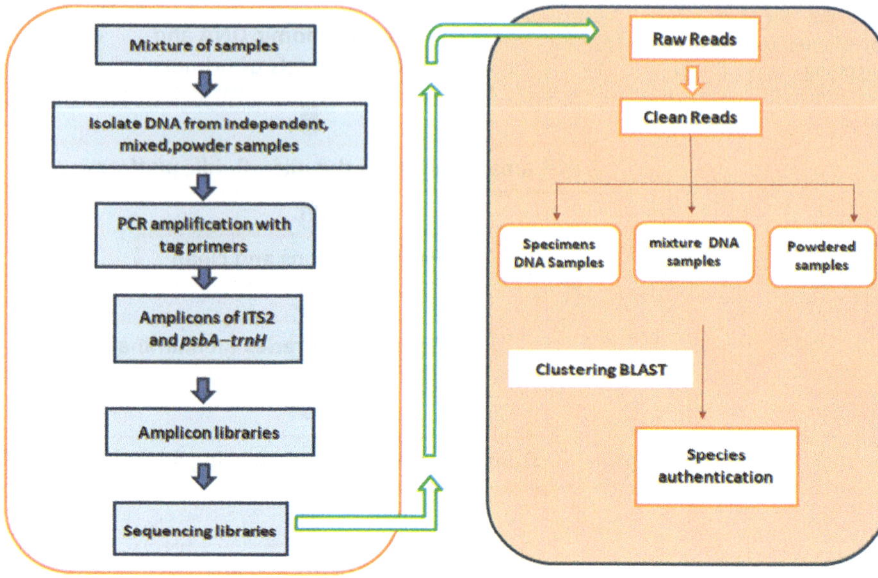

Fig. 13.4 Steps of full-length multi-barcoding

13.12 Genome Skimming Based Barcoding

One approach which offers a relatively straightforward mechanism to improve and extend DNA barcodes is genome skimming (Dodsworth 2015). As a genomic DNA extract typically contains a mix of both nuclear and organellar DNA (plastid and mitochondria), NGS generates data across the three genomes. Therefore, genome skimming deals with the ultimate goal of assembling organellar reference genomes. Through genome skimming, there is also potential to make a highly fragmented nuclear genome assembly. Overall, genome skimming is scalable, cost-effective, and can be used effectively with degraded DNAs from herbarium specimens or highly fragmented nuclear genome assembly (Nevill et al. 2020). This approach recovers simultaneously all of the different "standard" plant barcoding regions and provides a direct link with all other phylogenetically informative genomic regions. The second benefit of genome skimming is that it is compatible with the standard plant barcodes and genome sequencing. Genome skimming can recover plastid barcode loci and ITS, thus adding to the standard barcoding loci's growing reference database. Many genome skimming studies only assemble the organellar and ribosomal DNA, excluding the nuclear reads.

The key challenges to widespread adoption of genome skimming as an extended barcode will be dependent on the efficacy of its implementation at a vast scale, cost implications for library preparation (which is also time-consuming), consumables, computational power, and data storage. Another major challenge of genome skimming DNA barcoding is how to use nuclear data effectively (Coissac et al. 2016) because 99% genome sequence data are discarded. Coissac et al. (2016) proposed DNA mark pipeline which will enable future DNA-based identification.

13.13 Restriction Site-Associated DNA Sequencing (RAD)

Restriction site-**A**ssociated **D**NA sequencing (RADseq) and its derivative methods have been applied mostly for assessing population structure, hybridization, demographic history, phylogeography of organisms (Baird et al. 2008). The reduced representation of genome scale has the potential to be implemented as an alternative species identification tool. This method accesses large numbers of sequence variations adjacent to restriction-enzyme cut sites and sequence the homologous regions across hundreds of individuals, without genome sequence information. RAD sequencing is one promising approach which has already been used to authenticate complex species. RAD showed huge phylogenetic resolution among temperate bamboo species which has less molecular variation due to their recent origin (Wang et al. 2017).

13.14 Conclusion

The primary aim of DNA barcoding is to identify known specimens and to help flag possible new species, thereby making taxonomy more useful for science and society. Thus, it is based on conventional and inexpensive protocols for DNA extraction, amplification, and sequencing. DNA barcoding is an approach to developing a global, open-access library of standardized DNA barcode sequences, which would help non-expert identify specimens up to species level. Certain limitations (low PCR efficiency, inadequate variation in single-locus barcode) restrict achievement of a universal DNA barcode system for land plants. However, multi-locus DNA barcoding approach is still one of the most effective strategies for barcoding some complex plants groups. With the advancement of next-generation sequencing technologies, genome skimming RAD seq, etc. were evolved to sampling variation throughout the genome and help identify the complex plant species with better species resolution. Integrations of genome skimming RAD seq, HRM, ME Barcoding, FLMB approaches have further paved the way in overcoming the present limitations of plant DNA barcoding which would play a vital role towards the development of Digital Plant Identification System.

References

Baird NA, Etter PD, Atwood TS, Currey MC, Shiver AL, Lewis ZA, Selker EU, Cresko WA, Johnson EA (2008) Rapid SNP discovery and genetic mapping using sequenced RAD markers. PLoS One 3(10):e3376

Bell KL, De Vere N, Keller A, Richardson RT, Gous A, Burgess KS, Brosi BJ (2016) Pollen DNA barcoding: current applications and future prospects. Genome 59(9):629–640

Blaxter ML (2004) The promise of a DNA taxonomy. Philos Trans R Soc Lond B Biol Sci 359(1444):669–679

CBOL, Hollingsworth PM, Forrest LL, Spouge JL, Hajibabaei M, Ratnasingham S, van der Bank M, Chase MW, Cowan RS, Erickson DL (2009) A DNA barcode for land plants. Proc Natl Acad Sci 106(31):12,794–12,797

Chen S, Yao H, Han J, Liu C, Song J, Shi L, Zhu Y, Ma X, Gao T, Pang X (2010) Validation of the ITS2 region as a novel DNA barcode for identifying medicinal plant species. PLoS One 5(1): e8613

Coissac E, Hollingsworth PM, Lavergne S, Taberlet P (2016) From barcodes to genomes: extending the concept of DNA barcoding. Wiley Online Library

Dodsworth S (2015) Genome skimming for next-generation biodiversity analysis. Trends Plant Sci 20(9):525–527

Dong W, Xu C, Li C, Sun J, Zuo Y, Shi S, Cheng T, Guo J, Zhou S (2015) ycf1, the most promising plastid DNA barcode of land plants. Sci Rep 5(1):1–5

Erickson DL, Spouge J, Resch A, Weigt LA, Kress JW (2008) DNA barcoding in land plants: developing standards to quantify and maximize success. Taxon 57(4):1304–1316

Fazekas AJ, Burgess KS, Kesanakurti PR, Graham SW, Newmaster SG, Husband BC, Percy DM, Hajibabaei M, Barrett SC (2008) Multiple multilocus DNA barcodes from the plastid genome discriminate plant species equally well. PLoS One 3(7):e2802

Fernandes TJ, Amaral JS, Mafra I (2020) DNA barcode markers applied to seafood authentication: an updated review. Crit Rev Food Sci Nutr:1–32

Gostel MR, Zúñiga JD, Kress WJ, Funk VA, Puente-Lelievre C (2020) Microfluidic Enrichment Barcoding (MEBarcoding): a new method for high throughput plant DNA barcoding. Sci Rep 10(1):1–13

Gross M (2011) Herbaria source of new plant species. Curr Biol 21:R6–R7

Group C P B, Li D-Z, Gao L-M, Li H-T, Wang H, Ge X-J, Liu J-Q, Chen Z-D, Zhou S-L, Chen S-L (2011) Comparative analysis of a large dataset indicates that internal transcribed spacer (ITS) should be incorporated into the core barcode for seed plants. Proc Natl Acad Sci 108(49):19,641–19,646

Hajibabaei M, Singer GA, Hebert PD, Hickey DA (2007) DNA barcoding: how it complements taxonomy, molecular phylogenetics and population genetics. Trends Genet 23(4):167–172

Hebert PD, Cywinska A, Ball SL, Dewaard JR (2003) Biological identifications through DNA barcodes. Proc R Soc Lond Ser B Biol Sci 270(1512):313–321

Hebert PD, Braukmann TW, Prosser SW, Ratnasingham S, DeWaard JR, Ivanova NV, Janzen DH, Hallwachs W, Naik S, Sones JE (2018) A Sequel to Sanger: amplicon sequencing that scales. BMC Genomics 19(1):1–14

Hilu KW, Liang G (1997) The matK gene: sequence variation and application in plant systematics. Am J Bot 84(6):830–839

Hollingsworth P (2008) DNA barcoding plants in biodiversity hot spots: progress and outstanding questions. Nature Publishing Group

Hollingsworth ML, Andra Clark A, Forrest LL, Richardson J, Pennington RT, Long DG, Cowan R, Chase MW, Gaudeul M, Hollingsworth PM (2009) Selecting barcoding loci for plants: evaluation of seven candidate loci with species-level sampling in three divergent groups of land plants. Mol Ecol Resour 9(2):439–457

Hollingsworth PM, Graham SW, Little DP (2011) Choosing and using a plant DNA barcode. PLoS One 6(5):e19254

Kim K-J, Jansen RK (1995) ndhF sequence evolution and the major clades in the sunflower family. Proc Natl Acad Sci 92(22):10,379–10,383

Kool A, de Boer HJ, Krüger Å, Rydberg A, Abbad A, Björk L, Martin G (2012) Molecular identification of commercialized medicinal plants in southern Morocco. PLoS One 7(6):e39459

Kress WJ, Erickson DL (2007) A two-locus global DNA barcode for land plants: the coding rbcL gene complements the non-coding trnH-psbA spacer region. PLoS One 2(6):e508

Kress WJ, Erickson DL (2008) DNA barcodes: genes, genomics, and bioinformatics. Proc Natl Acad Sci 105(8):2761–2762

Lahaye R, Van der Bank M, Bogarin D, Warner J, Pupulin F, Gigot G, Maurin O, Duthoit S, Barraclough TG, Savolainen V (2008) DNA barcoding the floras of biodiversity hotspots. Proc Natl Acad Sci 105(8):2923–2928

Lee SY, Lamasudin DU, Mohamed R (2019) Rapid detection of several endangered agarwood-producing Aquilaria species and their potential adulterants using plant DNA barcodes coupled with high-resolution melting (Bar-HRM) analysis. Holzforschung 73(5):435–444

Li X, Yang Y, Henry RJ, Rossetto M, Wang Y, Chen S (2015) Plant DNA barcoding: from gene to genome. Biol Rev 90(1):157–166

Madesis P, Ganopoulos I, Anagnostis A, Tsaftaris A (2012) The application of Bar-HRM (Barcode DNA-High Resolution Melting) analysis for authenticity testing and quantitative detection of bean crops (Leguminosae) without prior DNA purification. Food Control 25(2):576–582

Mahadani P, Ghosh SK (2013) DNA barcoding: a tool for species identification from herbal juices. DNA Barcodes 1:35–38

Mahadani P, Sharma GD, Ghosh SK (2013) Identification of ethnomedicinal plants (Rauvolfioideae: Apocynaceae) through DNA barcoding from Northeast India. Pharmacogn Mag 9(35):255

Mishra P, Kumar A, Nagireddy A, Mani DN, Shukla AK, Tiwari R, Sundaresan V (2016) DNA barcoding: an efficient tool to overcome authentication challenges in the herbal market. Plant Biotechnol J 14(1):8–21

Muellner A, Schaefer H, Lahaye R (2011) Evaluation of candidate DNA barcoding loci for economically important timber species of the mahogany family (Meliaceae). Mol Ecol Resour 11(3):450–460

Nevill PG, Zhong X, Tonti-Filippini J, Byrne M, Hislop M, Thiele K, Van Leeuwen S, Boykin LM, Small I (2020) Large scale genome skimming from herbarium material for accurate plant identification and phylogenomics. Plant Methods 16(1):1–8

Ratnasingham S, Hebert PD (2007) BOLD: the barcode of life data system. Mol Ecol Notes 7(3):355–364. (http://www.barcodinglife.org)

Roy S, Tyagi A, Shukla V, Kumar A, Singh UM, Chaudhary LB, Datt B, Bag SK, Singh PK, Nair NK (2010) Universal plant DNA barcode loci may not work in complex groups: a case study with Indian Berberis species. PLoS One 5(10):e13674

Savolainen V, Cowan RS, Vogler AP, Roderick GK, Lane R (2005) Towards writing the encyclopaedia of life: an introduction to DNA barcoding. Philos Trans R Soc B Biol Sci 360(1462):1805–1811

Serino G, Maliga P (1998) RNA polymerase subunits encoded by the plastid rpoGenes are not shared with the nucleus-encoded plastid enzyme. Plant Physiol 117(4):1165–1170

Shaw J, Lickey EB, Beck JT, Farmer SB, Liu W, Miller J, Siripun KC, Winder CT, Schilling EE, Small RL (2005) The tortoise and the hare II: relative utility of 21 noncoding chloroplast DNA sequences for phylogenetic analysis. Am J Bot 92(1):142–166

Spooner DM (2009) DNA barcoding will frequently fail in complicated groups: an example in wild potatoes. Am J Bot 96(6):1177–1189

Stoeckle MY, Gamble CC, Kirpekar R, Young G, Ahmed S, Little DP (2011) Commercial teas highlight plant DNA barcode identification successes and obstacles. Sci Rep 1(1):1–7

Sui X-Y, Huang Y, Tan Y, Guo Y, Long C-l (2011) Molecular authentication of the ethnomedicinal plant Sabia parviflora and its adulterants by DNA barcoding technique. Planta Med 77(05):492–496

Sun X-Q, Zhu Y-J, Guo J-L, Peng B, Bai M-M, Hang Y-Y (2012) DNA barcoding the Dioscorea in China, a vital group in the evolution of monocotyledon: use of mat K gene for species discrimination. PLoS One 7(2):e32057

Tautz D, Arctander P, Minelli A, Thomas RH, Vogler AP (2003) A plea for DNA taxonomy. Trends Ecol Evol 18(2):70–74

Uribe-Convers S, Settles ML, Tank DC (2016) A phylogenomic approach based on PCR target enrichment and high throughput sequencing: resolving the diversity within the south American species of Bartsia L. (Orobanchaceae). PLoS One 11(2):e0148203

Van K, Rastogi K, Kim K, Lee S (2013) Next-generation sequencing technology for crop improvement. SABRAO J Breeding Genet 45(1):84–99

Wang X, Ye X, Zhao L, Li D, Guo Z, Zhuang H (2017) Genome-wide RAD sequencing data provide unprecedented resolution of the phylogeny of temperate bamboos (Poaceae: Bambusoideae). Sci Rep 7(1):1–11

Wang X, Gussarova G, Ruhsam M, de Vere N, Metherell C, Hollingsworth PM, Twyford AD (2018) DNA barcoding a taxonomically complex hemiparasitic genus reveals deep divergence between ploidy levels but lack of species-level resolution. AoB Plants 10(3). https://doi.org/10.1093/aobpla/ply026

Zhang P, Liu C, Zheng X, Wu L, Liu Z, Liao B, Shi Y, Li X, Xu J, Chen S (2019) Full-length multi-barcoding: DNA barcoding from single ingredient to complex mixtures. Genes (Basel) 10(5):343. https://doi.org/10.3390/genes10050343

Advances in Epigenetics for Crop Improvement and Sustainable Agriculture 14

Venkateshwaran Krishnaswami, Munish Kumar, and Sivakumar Vijayaraghavalu

Abstract

To feed the ever-increasing human population globally, continuous attempts have been made to enhance the agricultural productivity significantly through the application of modern breeding techniques. Agronomic approaches such as increased abiotic/biotic stress tolerance, reduced toxicity, superior nutritional quality, delayed ripening, better post-harvest quality, etc. were attempted. However, loss of agricultural lands due to urbanization in both developing and under-developed countries, drastic climate changes, plant diseases, and farmer's non-compliance to modern agricultural methods are some of the major hindrances to these strategies. Plant-epigenetic modifications play vital roles in acclimatization, stress tolerance, adaptation, and evolution processes. The agronomically important traits of crop life such as flowering time, fruit development, risk avoidance from environmental factors, and crop immunity are attained by the epigenetic modifications such as DNA methylation, histone post-translational modifications, and small non-coding RNAs modification. This chapter emphasizes on the epigenetic changes and its effect on crop growth, yield, and herbicide resistance. As well as develop plants that are resistant to harsh weather conditions such as cold, heat, and rain and also to increase the crop yield in poorly-arable places like deserts, marshy lands, backwaters, high

V. Krishnaswami
Department of Allied Health Sciences, Noorul Islam Center for Higher Education, Kumaracoil, Kanyakumari, Tamil Nadu, India

M. Kumar
Department of Biochemistry, University of Allahabad, Prayagraj, Uttar Pradesh, India

S. Vijayaraghavalu (✉)
Narayana Translational Research and Incubation Center, Narayana Medical College, Nellore, Andhra Pradesh, India

R. L. Singh et al. (eds.), *Plant Genomics for Sustainable Agriculture*,
https://doi.org/10.1007/978-981-16-6974-3_14

altitudes and areas with high/low soil pH. Hence, the potential of epigenetics in crop improvement and sustained agriculture is highly recognized by researchers both in academia and industry and by policy makers and government.

Keywords

Plant Epigenetics · Sustained agriculture · Crop improvement · Epigenetic regulation · Plant immunity · Plant breeding

14.1 Introduction

14.1.1 Epigenetics

The term epigenetics was coined by Conrad Hal Waddington in 1942, it is defined as phenotypic changes due to heritable pattern of alteration in chromatin without any changes in the underlying DNA sequences. The mechanism by which the gene expression turned on or off was poorly understood at that time. However in the last few decades numerous scientific studies revealed the mechanism of epigenetic modifications and its influence on gene expression profile in normal development and disease conditions in an organism. Unlike the genetic mutations the epigenetic modifications can be reversed, as well can be manipulated to obtain the desired phenotype, this phenomenon allows epigenesist to understand the differentiation and development process of a seed to plant and has high potential to generate phenotypic variations that could well adapt to abiotic and biotic stress and favorable agronomical traits such as disease resistant and high yielding quality crops. Plants being sessile developed several sophisticated mechanisms of gene regulation including epigenetic regulation to respond, adapt, and thrive in a constantly changing ecosystem. The three major epigenetic marks are methylation of fifth carbon of cytosine in Guanine–Cytosine (GC) rich sequence in a gene, acetylation / deacetylation of histone proteins and small RNAs. To alter the gene expression all these can act alone or in combination with each other. DNA-methylation in plants is done by three different enzymes, namely (1) MET1 DNA methyltransferase at GC rich sequence; (2) CHROMOMETHYLASE3—CMT3 at *CHG* (H = A, C, or T); (3) DOMAINS REARRANGED METHYLTRANSFERASE—DRM1/DRM2 or CMT2 methyltransferase are responsible for methylating CHH (Weinhold 2006; Simmons 2008; Gibney and Nolan 2010). DNA methylation frequently happens in repetitive regions such as transposable elements (TEs), centromeric repeats and in rDNA—genes that encode ribosomal RNA. The TE transposition can disrupt genomic integrity and could alter the gene expression if it integrates within or near the gene; hence plants epigenetically silence them by methylation. TE silencing is commonly found in epigenome of almost all the investigated plants that ranges from moss -Physcomitrella to gymnosperms and angiosperms.

Manipulation of DNA methylation in plants can alter its phenotype, for example, rice treated with DNA demethylating agents results in dwarf progeny. Reduction in global DNA methylation was observed in vernalization treatment and it might be one of the factors for initiating flowering in plants. Similarly, DNA methylation pattern changes were observed in cold stressed-maize and Arabidopsis plants. Constant response to the environmental stress induces methylation changes in plants; such epigenetic modifications can be either reversed and the gene expression returns to its near normal levels or it is carried as epigenetic stress memory to its progeny. Thus the DNA methylation plays significant role in various stages of plant development which will be discussed in several following sub-headings.

The regulation of gene activity has been achieved by reversible histone acetylation and deacetylation at the N-terminus of histone tails catalyzed by histone acetyltransferases (HATs) and histone deacetylases (HDACs), respectively. Regulation of gene expression in plant development and its responses to environmental stresses can be also achieved by HATs and HDACs due to its interaction with various chromatin-remodeling factors and transcription factors. Around 80% of proteins in plants undergo acetylation.

14.1.2 RNAs—miRNA, ShRNA, si RNA, Non-coding RNA

Precursors of miRNAs or siRNAs are generated in several ways. DNA-dependent RNA polymerase II (RNA Pol II) transcripts with extensive self-complementarity fold back on themselves to form stem-loop structures in the case of miRNAs. The double-stranded precursors can be generated via convergent, bidirectional transcription by a DNA-dependent RNA polymerase such as RNA Pol II, thereby generating transcripts that overlap and base-pairing the case of siRNAs. RNA transcripts can be used as templates for an RNA-dependent RNA polymerase (RdRP), to generate a complementary strand. The criticality for plant development requires miRNAs which supports the functions of complementary mRNAs to a specific subset of cells by bringing about their posttranscriptional degradation or translational repression in adjacent cells. In the context of a multi-protein RISC complex miRNAs silence gene expression by base-pairing to target mRNAs, which facilitates mRNA cleavage, "slicing," or the inhibition of mRNA translation. Transcriptional gene silencing by inducing DNA methylation and histone H3K9 demethylation are regulated by heterochromatic siRNAs.

14.1.3 Small RNAs Can Trigger DNA Methylation and Chromatin Modification

Large number of long non-coding RNAs (lncRNAs) are produced by many eukaryotes, including plants which holds regulatory roles in various developmental processes. lncRNAs are versatile regulatory molecules.

14.1.4 Chromatin Remodeling/Condensation

The basal state of chromatin remodeling can be altered by promoting either an "open" (activation of transcription) or a "closed" (repression of transcription) chromatin configuration. According to the size of transcripts NcRNAs can be classified into two categories they are Nc RNAs with more than 200 nucleotides are considered as long ncRNAs (lncRNAs), whereas short ncRNAs are less than 200 nucleotides. The micro RNA (miRNAs), small interfering RNAs (siRNAs), and Piwi-interacting RNAs (piRNAs) are included in the short ncRNAs. The stable inheritance of chromatin structures can be invoked by transcriptional states to daughter cells through mitosis or even meiosis. The existence of post-translational modifications (mainly acetylation, methylation, and phosphorylation) of the amino-terminal tails of the histones is demonstrated in various plants based studies. In plants, histone H4 can be acetylated at Lys 5, 8, 12, 16, and 20, whereas histone H3 is acetylated at positions 9, 14, 18, and 23. To a lower extent histone H2A and H2B are also acetylated.

14.1.5 Polycomb Proteins

The formation of conserved regulatory structures that can suppress genes through a variety of physiological roles and types of epigenetic patterning is the major role of polycomb proteins (PcG). PcG proteins bind nucleosomes thereby alters the intrinsic structure of chromatin to initiate epigenetic modifications and maintain these modifications during development. These multi-protein complexes modify chromatin structure to form flexible, repressive chromatin configurations that include numerous targeted genes and maintain silencing. The paradigmatic model for epigenetic regulation of gene silencing has been controlled by the PcG of proteins. PcG proteins are a collection of transcriptional regulatory factors that can control gene expression, whose transcriptional imposed silencing can be transmitted from embryos to adulthood.

14.1.6 Fungal Prions and Epigenetics

Due to self-perpetuating abilities which enable to record and reproduce the memory of acquired alteration that initially caused prion formation prions are built as perfect machines of molecular memory. Prion-like elements trigger environmental signals to cellular and organismal processes. Two-level regulatory mechanism, with a prion-like domain in the regulatory protein for initial sensing and recording a signal can be achieved by epigenetic pathways with regulatory change in the chromatin organization or in the mode of gene transcription.

14.1.7 Global Hunger and Crop Production

Chronic hunger and malnutrition are undue burden on humanity; global statistics are alarming with 821 million people who are in hunger. WHO's Global Hunger Index (GHI) shows that the many of the African countries are in the serious (GHI; 20–34.9) to extremely alarming (GHI $\geq$ 50) category. Developing economies like India is also in serious condition with the GHI of 30.3. Hence urgent measures are needed to address this global crisis. One of the major factors for the hunger is crop failure due to natural disasters, unpredictable rainfall, drought, drastic climate changes. It also hinders livestock management and adds up to food shortage. As well the on-farm losses due to pests, rodents, birds, diseases, weather, harvesting and storage methods (Chen 1990; Wu et al. 2014). Additionally, human activities such as urbanization, deforestation lead to reduction in crop land. The environmental factors such as soil erosion, fall of water tables, and aquifer pose threat to sustained agriculture. Hence, governments all over the world make policies and programs to protect the agricultural lands and prevent on-farm loss to maximize crop yield and utilization of agricultural inputs to combat famine (Acevedo 2011). Genetic and epigenetic approaches resulted in crops with high nutritional values, drug and drought varieties and increased the crop yield; which in-turn could reduce the global hunger.

14.2 Plant Epigenetics

The study of genes, genetic variation, and heredity specifically in plants is the plant genetics. Plant epigenetics based technologies dramatically increase our capacity to understand the molecular basis of traits and utilize the available resources for accelerated development of stable high-yielding, nutritious, input-use efficient and climate-smart crop varieties. The highly diversified group of sessile organisms, plants gets evolved with extreme ultraviolet light, salinity, and hypoxia interactors. Some biotic interactors are beneficial for the plant (pollinators, rhizobia, mycorrhiza), whereas some interactors (herbivores, pathogens, or strong competitors) are detrimental to plants. Molecular level changes of plants constantly are required to adapt themselves to the changing environments via epigenetic regulations which may improve the survivability of plants in terms of tolerance toward external stress. Gene expression changes in plants may respond to environmental stimuli/chromatin-based gene regulation. The chromatin modification in plants through DNA methylation is the major part of plant epigenetics. Direct modification of DNA genes in plants occurs through mitosis and amitosis (Chen et al. 2010). DNA methylation and histone modification cause the direct phenotypic characterization of epigenetic mutants which possess the ability to characterize whole-plant traits and organ traits (Alonso et al. 2019). Specific genetic and epigenetic variants dynamic interplay in case of plant stress responses (Eriksson et al. 2020). The epigenetic modified plant is shown in Fig. 14.1.

Epigenetic regulation in plants reflects their mode of development, lifestyle, and evolutionary history. Plants grow by continuously producing new organs from

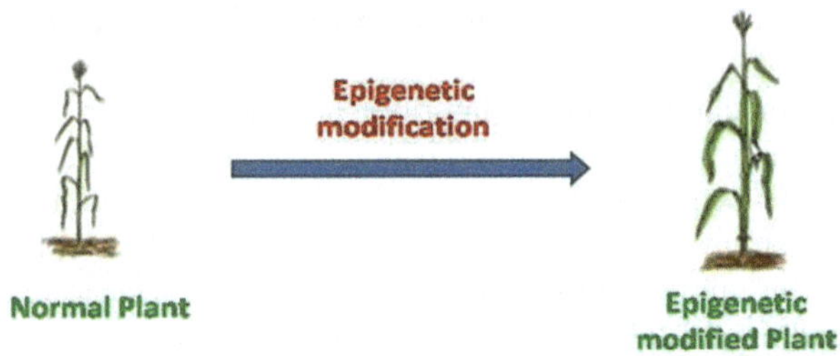

Fig. 14.1 Epigenetic modification in plant growing in drought condition

self-sustaining stem cell populations known as meristems. Unlike growth in mammals, in which organ and tissue formation is largely specified during embryonic development. The high degree of phenotypic plasticity can be achieved by the postembryonic development in plants in a continuous process shaped by environmental influences. Because plants are unable to escape their surroundings, they are forced to cope with changeable and often unfavorable growth conditions. Epigenetic regulatory mechanisms can facilitate metastable changes in gene activity and fine-tune gene expression patterns, thus enabling plants to survive and reproduce successfully in unpredictable environments. Genome size, genome complexity, and the ratio of heterochromatin to euchromatin in seed plants are generally comparable to mammals. In case of gene regulation plants and mammals make similar use of DNA methylation and histone post-translational modifications (PTMs).

DNA methylation consists mostly in adding a methyl group at the fifth carbon position of a cytosine ring, and, different to what happens in animals, plants have three sites that frequently can suffer methylation: CG, CHG (where H is A, C, or T), and CHH. Modification of DNA methylation profiles in plant can cause phenotypic variation. For instance, demethylation of rice genomic DNA causes an altered pattern of gene expression, inducing dwarf plants. Thus, the knowledge of epigenetic contributions in phenotypic plasticity and hereditable variation is important to understand how natural population can adapt in different environmental condition, especially in a world context of climate change.

14.3 Epigenetics to Increase the Crop Yield and Sustained Agriculture

Improved crop yield with a goal for sustained agriculture may be achieved through epigenetic regulation (Fig. 14.1) through epigenetic mechanisms leading to changes in chromatin architecture, such as DNA methylation, post-translational histone modifications, and the action of non-coding RNA molecules which are either small RNAs (small interfering RNAs, siRNAs and microRNAs, miRNAs) or long non-coding RNAs (long snRNAs) (Kapazoglou et al. 2018). The understanding of epigenetics supports towards the crop improvement and ultimately this will get replicated in the nutritional management and plant breeding. The molecular and mechanistic basis of genotype along with the environmental interactions facilitates the epigenetic mechanisms. Epigenetics thus supports plant cellular/physiological processes during its development with increased fitness/adaptability to the plant in the changing environmental conditions (Kumar 2019). Recently the newer technologies of gene editing accomplish things in shorter periods of time than a natural breeding process. The genetically modified crops may open path for convenient and more economical production practices for farmers.

14.4 Epigenetics for Crop Growth

Prevailing conditions for epigenetic changes for sustained agriculture was shown in Fig. 14.2. The two major types of small RNAs, small interfering RNAs (siRNAs) and microRNAs (miRNAs) in plants are likely to be associated with the silencing of gene expression. miRNAs regulate the plant growth, development, organogenesis, and responses to various biotic and abiotic stresses. The critical step for successful grain production in rice is the flowering time control with criticality in the day length as a key factor for rice flowering. Shorter day length will greatly induce rice flowering with criticality in the photoperiod. Liu et al. 2014 reported that the accurate control of flowering based on photoperiod is regulated by key mechanisms that involve the regulation of flowering genes such as Heading date1 (Hd1), Early hd1 (Ehd1), Hd3a, and RFT1. Rice enhancer of zeste [E(z)] genes SDG711 and SDG718, which encode the polycomb repressive complex2 (PRC2) key subunit is required for trimethylation of histone H3 lysine 27 (H3K27me3) which supports the long day (LD) and short day (SD) regulation of key flowering genes (Liu et al. 2014).

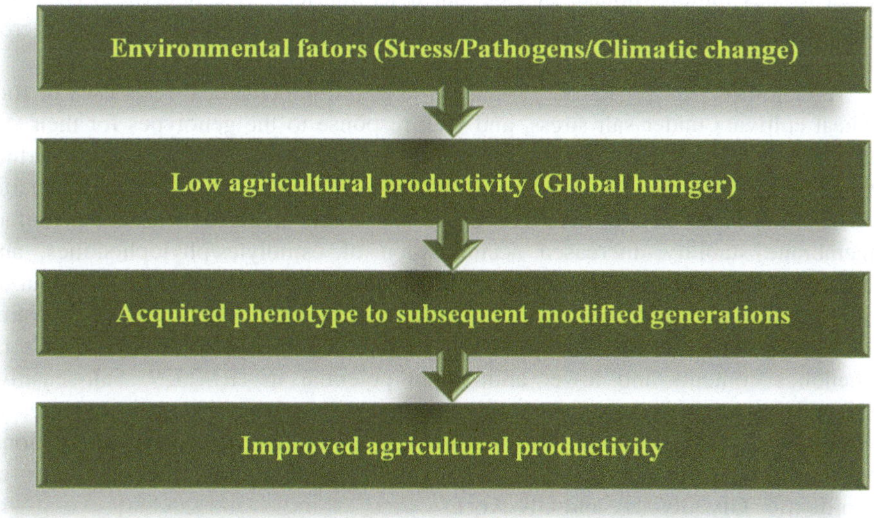

Fig. 14.2 Epigenetics for improved agricultural productivity

Kapazoglou et al. 2013 emphasized that epigenetic factors such as DNA methylation and histone modifications may regulate plant development. Gene silencing or activation occurs by cytosine methylation and demethylation. They reported that in seed development of Arabidopsis, cytosine demethylation is achieved by specific DNA glycosylases, including AtDME (DEMETER) and AtROS1 (REPRESSOR OF SILENCING1). They studied the DME homologue from barley (HvDME), during seed development and drought conditions. Their phylogenetic analysis revealed high degree of homology to other monocot DME glycosylases, and sequence divergence from the ROS1, DML2, and DML3 orthologs. They concluded that expression analysis during seed development and under dehydration conditions provides role for HvDME in endosperm development, seed maturation, and in response to drought.

The daily protein requirements have been increased due to wheat consumption and wheat becomes one of the most important cereal crops having global production of $>$700 million tones (Kumar et al. 2017). Limited success in growing wheat on salt-affected soils has been achieved because only a few salt-tolerant bread wheat genotypes have been identified. The accumulation of excessive salt contents in the soil may lead to the salt stress which eventually results in the crop growth inhibition and ultimately leads to crop death which creates very dangerous for agricultural productivity. The impairment of water potential cells, ion toxicity, membrane integrity and function, and delayed uptake of essential mineral nutrients occurs in crop growth due to salt stress. This ultimately affects the metabolic processes in plants. Understandings on biochemical, physiological, genetic, and epigenetic aspects of salt tolerance along with cloning of the genes involved in salt tolerance and development of transgenic may support toward the better breeding strategy for the crops growing under saline conditions. Accumulation of sodium ion (Na^+) in plant tissues inhibits uptake of essential macronutrients like potassium (K^+) and calcium (Ca^{2+}) from soil which creates the one of the detrimental effects in crop growth (Very and Sentenac 2003; Shi et al. 2003).

Hamamoto et al. (2015) investigated the physio-biochemical responses of four wheat cultivars under salt stress, and with response to the genotypes for their salt tolerance level to identify the most contrasting salt-responsive genotypes. Further they investigated the mechanisms responsible for genotype and tissue specific differential expression of *TaHKTs* genes and found better antioxidant potential, membrane stability, increased accumulation of osmolytes/phytophenolics, and higher K^+/Na^+ ratio under 200 mMNaCl stress-induced condition of Kharchia-65 and found the same to be the most salt-tolerant cultivar. The sensitivity towards stress may be ensured by reduced soluble sugar, proline, total chlorophyll, total phenolics contents and lower antioxidant potential in HD-2329. Salt-sensitive (HD-2329) genotype may be revealed by genetic and bioinformatic analyses of *HKT1;4* of contrasting genotypes (Kharchia-65 and HD-2329) revealed deletions, transitions, and transversions resulting into altered structure and loss of conserved motifs (Ser-Gly-Gly-Gly and Gly-Arg).

In order to gain insights into the genomic basis of apple (Malus domestica) evolution and domestication a newer high quality apple WGS, GDDH13 v1.1, was released which serves as the reference genome for apple. WGSs apple shown enormous impact on apple biological functioning, trait physiology, and inheritance, with the valued outcome of high quality crop. Here in WGS apple shown improvement associated with innovative approach to obtain durable, environmentally sound, productive, and consumer desirable apples (Peace et al. 2019).GDDH13 based apple reference genome supports in the understanding of epigenetic mechanisms involved in fruit size regulation which helps in the fruit size regulation. Kumar et al. (2017) investigated the physio-biochemical, molecular indices, and defense responses of wheat cultivars to identify the most contrasting salt-responsive genotypes and the mechanisms associated with their differential responses. For multivariate analysis in order to identify the most contrasting genotypes the physio-biochemical traits specifically membrane stability index, antioxidant potential, osmoprotectants, and chlorophyll contents are measured at vegetative stage. They observed that Kharchia-65 to be the most salt-tolerant cultivar based on the genetic and epigenetic analyses with mechanisms associated with differential response of the wheat genotypes under salt stress such as improved antioxidant potential, membrane stability, increased accumulation of osmolytes/phytophenolics, and higher KC/NaC ratio under 200 mM NaCl stress. Sensitivity to the stress associated with HD-2329 strain was elicited by increased MDA level, reduced soluble sugar, proline, total chlorophyll, total phenolics contents, and lower antioxidant potential. The contrasting genotypes can be revealed by deletions, transitions, and transversions resulting into altered structure, loss of conserved motifs (Ser-Gly-Gly-Gly and Gly-Arg) and function in salt-sensitive (HD-2329) (Kharchia-65 and HD-2329). Further the tissue- and genotype-specific changes were explained by their epigenetic variations in cytosine methylation. Overall they indicated that abiotic stress exerts significant impact on plant's growth, development, and productivity, which can be overcomed by epigenetic and molecular mechanisms of salt tolerance in plants.

Recognization and respond of plants to pathogens have been evolved by its innate immune systems which includes pathogen-associated molecular pattern (PAMP)-triggered immunity and effector-triggered immunity. The key signaling molecules involved in defense and growth, including jasmonic acid, salicylic acid, and reactive oxygen species which includes the plant–pathogen interactions and neighboring plants perception via light-quality receptors influence affects the profile of plant secondary metabolites and emitted volatile organic compounds shown greater impact on plant herbivore and plant pollinator interactions which in turn affects the fitness of the plant (Alonso et al. 2019). Cai et al. (2018) reported that host *Arabidopsis* cells secrete exosome-like extracellular vesicles to deliver sRNAs into fungal pathogen *Botrytis cinerea* which accumulate at the infection sites and are taken up by the fungal cells. These sRNA-containing vesicles accumulate at the infection sites and are taken up by the fungal cells. The silencing of fungal genes critical for pathogenicity has been induced by transferred host sRNAs. They reported that *Arabidopsis* has adapted immune response based exosome-mediated cross-kingdom RNA interference during the evolutionary arms race with the pathogen. They emphasized that functional studies of host transferred sRNAs support the identification of important virulence genes in interacting pathogens and pests.

Crespo-Salvador et al. (2018) determined the histone marks' profile in two differentially expressed genes in response to *B. cinerea*, as well as to oxidative stress, given its relevance in this infection. Here in both the induced CYP71A13 essential against this necrotroph and the repressed EXL7 (Exordium-like 1), which encodes a cytochrome P450 involved in camalexin synthesis. On both the promoter and the body of the highly induced PR1 in Arabidopsis plants infected with *B. cinerea* at 24 and 33 h after inoculation they performed chromatin immunoprecipitation analysis associated with activating marks H3K4me3, H3K9ac, and the repressor one H3K27me3. They observed that at 24 hpi, H3K4me3 level increased on the promoter associated with different locations of the body of the genes induced upon *B. cinerea*, including DES (divinyl ethyl synthase), LoxD (lipoxygenase D), DOX1 (α-dioxygenase 1), PR2 (pathogenesis-related protein2), WRKY53 and WRKY33. They further analyzed the genes as *B. cinerea* potential biomarkers for infection in crops. Pathogen infection induces the expression of histone deacetylase in Arabidopsis which indicates that histone acetylation/deacetylation has an important role in the plant pathogenic response. In this regimen De-La-Pena et al. (2012) observed that the loss of function of ASHH2 and ASHR1 resulted in faster hypersensitive responses to both mutant (hrpA) and pathogenic (DC3000) strains of *P. syringae*, whereas control (Col-0) and ashr3 mutants appeared to be more resistant to the infection after 2 days with highest expression level of PR1 gene on infection with DC3000 with increased resistance against this pathogen. The histone H3 lysine 4 dimethylation (H3K4me2) levels decreased at the promoter region of PR1 on infection with the DC3000 strain in both the ashr1 and ashh2 backgrounds, suggesting that an epigenetically regulated PR1 expression is involved in the plant defense. They concluded that histone methylation is essential in the signaling and defense processes of microbes against plants (De-La-Pena et al. 2012). They hypothesized that microorganism upon contact with a plant switch on different signaling pathways. If the microorganism is nonpathogenic, JA-related genes, such as JAR1, are induced.

Plants are continuously exposed to different biotic and abiotic pressures. Newer tools based plants to build resistance or tolerance to environmental stress to plants can be achieved by DNA mutations. The epigenetic mark DNA methylation is based on the covalent and reversible modifications to DNA/histone proteins, may alter the chromatin structure and, in specific cases, can be inherited to the next generation. By controlling the expression of several resistance genes plant response to biotic stressors can be activated by epigenetic modulations. Here in upon the first stress exposure, plants can be primed and subsequently activate defense genes more efficiently against the next stress encounter based on the DNA methylation and histone modifications of immune memory state of plants. Non-coding RNAs (ncRNAs) such aslnc RNAs, siRNA, and miRNAs can act as direct and indirect modulators in epigenetics research. The restoration of the stress-induced epigenetic modifications to initial levels when the stress is removed will occurs. But in certain conditions inheritance across mitotic or even meiotic cell divisions will occur and makes the stress to be stable. Epigenetic modifications may support the plants to relief the stable stress (Fig. 14.3). Even though plants have a peculiar characteristics

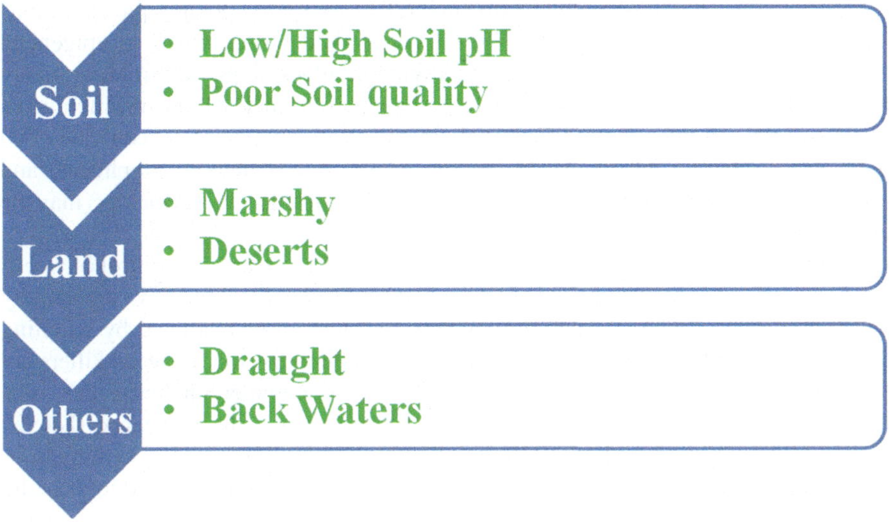

Fig. 14.3 Epigenetic modifications improves crop growth in adverse conditions

of priming system which may offer enhanced resistance in which the previous localized attack can be memorized by the plants at distal sites (Ramirez-Prado et al. 2018).

In order to detect the effectors plants use transmembrane/intracellular receptors known as "resistance proteins" for the effector recognition which may support the effector-triggered immunity. The concept of transcriptional gene silencing also supports towards the more stable layer of defense against DNA viruses. Development of herbicide-resistant weed populations is the major problem affecting the agricultural productivity. Herbicide resistance occurs due to mutation which reduces the herbicide binding at the protein target of the herbicide. A strong abiotic stress associated with the herbicides will be achieved in which the weeds respond by activating stress-signaling networks that reprogram gene expression.

Kim et al. (2017) performed methyl C sequencing of Arabidopsis thaliana leaves that developed after either mock treatment or two different sub-lethal doses of the herbicide glyphosate treatments which causes herbicide injury of 9205 differentially methylated regions across the genome among which 5914 of these differentially methylated regions were induced in a dose-dependent manner with positive correlation of methylation levels with that of the herbicide injury. They observed that out of 3680 genes associated with glyphosate induced differentially methylated regions only 7% were implicated in methylation changes following biotic or salinity stress. They concluded that plants responding to herbicide stress through changes in methylation patterns are associated with dose-sensitivity and stress specificity. Globally horseweed (Conyza canadensis) is one of the most commonly encountered weed species that developed resistance to herbicide glyphosate. Wherein glyphosate application, to horseweed shown phenotypic plasticity in response to selection pressure. In this strategy Gunjune Kim et al. (2017) observed phenotypic alterations and differential expression of the EPSPS gene could be attributed to epigenetic changes. They observed that DNA methylation plays a pivotal role in many biological procedures (gene expression, differentiation, and cellular proliferation). Their results shown that differential methylation pattern between the two populations wherein the increased resistance of C. canadensis to glyphosate and future development strategies that restrict weed resistance to herbicides may be elucidated (Margaritopoulou et al. 2018).

Synthetic herbicides manage the weeds affecting the plants. The control of weeds has been achieved by approaches of laborious manual weeding and environmentally damaging tillage. Improved agricultural productivity can be achieved by adopting synthetic herbicides with increased efficacy of weeding. Synthetic weeds killers are affected by wide spread evolution of resistance to commercial herbicides. The mechanisms of target-site resistance (TSR) and non-target-site resistance (NTSR) have been associated with most herbicide classes. Mutations (non-synonymous single-nucleotide polymorphisms, polymorphisms in more than one codon or entire codon deletions in genes) encoding the protein targets of herbicides are encountered with TSR which affects the binding of the herbicide (near catalytic domains or in regions affecting access), whereas NTSR target-site mechanisms involve absorption or translocation and increased sequestration or metabolic degradation (Gaines et al. 2011). The black-grass (*Alopecurus myosuroides*) and annual rye-grass (*Lolium perenne*) hold a global problem of loss of chemical weed control due to multiple-herbicide resistance. In this strategy Cummins et al. 2013 found that in both annual rye-grass and black-grass, multiple-herbicide resistance was observed with the increased expression of an evolutionarily distinct plant phi (F) GSTF1 that had a restricted ability to detoxify herbicides. They emphasized that upon black-grass *A. myosuroides* (Am) AmGSTF1 expression in Arabidopsis thaliana, the transgenic plants acquired resistance to multiple herbicides and showed similar changes in their secondary, xenobiotic, and antioxidant metabolism to those determined in MHR weeds. Their results of transcriptome array experiments showed that these changes in biochemistry were not due to changes in gene expression. Here in the AmGSTF1 showed a direct regulatory control on metabolism that led to an accumulation of protective flavonoids. They conclude that specific GSTFs in multiple-herbicide resistance in weeds with similar roles for its potentiality as targets for chemical intervention in resistant weed management (Cummins et al. 2013).The response of plant to environmental stresses has been achieved by variation in the expression of numerous genes. The major threat to grass weed is the non-target-site-based resistance to herbicides (NTSR). Duhoux and Delye (2013) identified a set of reference genes with a stable expression to be used as an internal standard for the normalization of quantitative PCR data in studies investigating NTSR to herbicides inhibiting aceto-lactate synthase (ALS) in the major grass weed *Lolium* sp., in these references gene has been used to check the herbicide response. In resistant plants their results indicate that herbicide application enhanced *CYP* gene expression with up-regulation of all *CYP* genes (Duhoux and Delye 2013)

14.5 Epigenetic Modifications to Sustain Agriculture for Crops Growing in Deserts

Plants get activated when they are exposed to the stressful conditions allowing to react under epigenetic regulation for improving crop productivity associated with stress conditions. Plant adaptation ability can be checked by analysis of candidate genes and studying their regulation in response to abiotic stresses. With response to abiotic stresses the analysis of stress related genes and their regulation of expression are commonly employed for enhanced understanding of the plants ability to adapt under abiotic stress environments. The different abiotic stress associated with plants is shown in Fig. 14.4. The exposure to abiotic stresses may cause demethylation of certain functionally inactive genes. Transgenerational inheritance can be achieved by the expression of certain genes modified by epigenetic mechanism.

Mousavi et al. (2019) performed a multidisciplinary approach, including physiological, epigenetic, and genetic studies to clarify the salt tolerance mechanisms in olive varieties (Koroneiki, Royal de Cazorla, Arbequina, and Picual) and a related form (*O. europaea* subsp. *cuspidata)* by growing in a hydroponic system under different salt concentrations from zero to 200 mM. They checked the photosynthesis, gas exchange, and relative water content at different time points, chlorophyll and leaf concentration of Na+, K+, and Ca2+ ions, quantified at 43 and 60 days after treatment. Their results indicate that several fragments differentially get methylated among genotypes, treatments, and time points. Significant expression changes related to plant response to salinity were observed by the real time quantitative PCR (RT-qPCR) analysis. Four genes (OePIP1.1, OePetD, OePI4Kg4, and OeXyla) were identified, as well as multiple retrotransposon elements usually targeted by

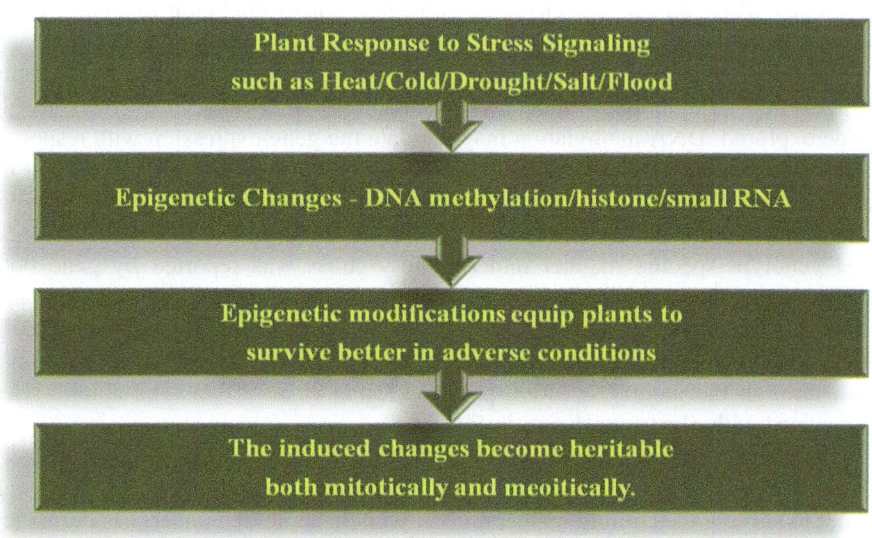

Fig. 14.4 Plant stress-signaling strategies

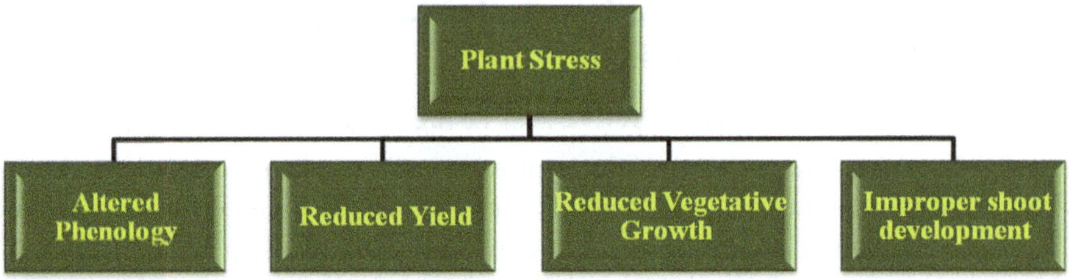

Fig. 14.5 Abiotic stress associated with plants

methylation under stress conditions. Overall their studies concluded that olive cultivation may successfully face the new scenarios driven by the climate change through the selection of tolerant varieties to salt and drought stresses.

Guarino et al. (2015) investigated the epigenetic diversity of the poplar populations of 83 white poplar trees at different sample locations on the island of Sardinia by determining their DNA methylation status; to assess if and how methylation status influence population clustering; to shed light on the changes that occur in the epigenome of ramets of the same poplar clone. They checked the methylation sensitive amplified polymorphism on the genomic DNA extracted from leaves at the same juvenile stage. Their results showed that the genetic biodiversity of poplars is quite limited but it is counterbalanced by epigenetic inter-population molecular variability. Their results of variable epigenetic status of Sardinian white poplars shown a decreased number of population clusters. Ramets of the same clone were differentially methylated in relation to their geographic position as observed during the landscape genetics analyses. They concluded that genetic biodiversity of the Sardinian white poplar is limited, by epigenetic inter-population diversity, which supports white poplars to grow in very large areas of the island of Sardinia.

Salinity is the major environmental factor which limits agricultural productivity. Epigenetic modulations under environmental stresses cause rice to be underexplored. DNA methylation may regulate the gene expression with responses to environmental stresses. In this regimen Ratna Karan et al. (2012) checked the effect of salt stress on DNA methylation in four genotypes of rice differing in the degree of salinity tolerance. Important role in regulating gene expression in organ and genotype-specific manner under salinity stress was reported by gene body methylation. They concluded that natural genetic variation for salt tolerance observed in rice germplasm may be independent of the extent and pattern of DNA methylation which may have been induced by abiotic stress followed by accumulation through the natural selection process (Karan et al. 2012). The abiotic stress associated with plants are shown in Fig. 14.5.

The important fruit tree and cash crop is the Apple (*Malus domestica*) globally. The crucial role of low temperature of winter season is the great risk for apple tree. In order to survive the chilling temperature of winters apple tree undergoes dormancy. Kumar et al. (2016) studied the cytosine methylation based epigenetic regulation of chilling mediated dormancy release in apple by employing methylation sensitive amplified polymorphism (MSAP) approach to analyze the changes in cytosine methylation pattern during dormancy break and subsequent fruit set. They found that under high chill conditions, total methylation gets decreased from 27% in dormant bud to 21% in fruiting stage, with no significant reduction under low chill conditions. Higher expression of DNA methyl transferases and histone methyl transferases during dormancy and fruit set, and lower expression of DNA glycosylases during active growth under low chill conditions were elicited during RNA-Seq analysis. Their results shown significant association between chilling and methylation changes which suggesting chilling acquisition during dormancy in apple is likely to affect the epigenetic regulation through DNA methylation.

The drought tolerant, low input, and high yielding sweet potato (*Ipomoea batatas*), produces more nutrients and has higher edible energy than most staples such as rice, cassava, wheat, and sorghum. The devastating loss of sweet potato productivity has been reported to get improved by micro-propagation techniques (meristem or nodal tip culture, coupled with thermotherapy or cryotherapy). Akomeah et al. (2019) checked the extent of in vitro culture induced soma-clonal variation, at a phenotypic, compositional, and genetic/epigenetic level, by comparing field-maintained and micropropagated lines of three elite Ghanaian sweet potato genotypes grown in a common garden. Their results indicates that micropropagated plants shown no observable morphological abnormalities compared to field-maintained plants which shown significantly lower levels of iron, total protein, zinc, and glucose. Their results of methylation sensitive amplification polymorphism analysis shown higher level of in vitro culture induced molecular variation in micropropagated plants. Overall they concluded that clonal fidelity of the micropropagated bio-fortified lines may reduce potential losses in the nutritional value. Benoit et al. (2019) developed a bioinformatics approach for the functional annotation of retrotransposons containing long terminal repeat and defined all full-length Rider elements in the tomato genome. They reported that accumulation of Rider transcripts and transposition intermediates in the form of extrachromosomal DNA is triggered by drought stress and relies on abscisic acid signaling. They concluded that Rider as an environment-responsive element and a potential source of genetic and epigenetic variation in plants. Herrera et al. (2013) investigated the transgenerational constancy of epigenetic structure in three populations of the perennial herb stinking hellebore (*Helleborus foetidus*). Their extensive epigenetic differentiation between sporophyte populations was revealed by single-locus and multilocus analyses. They observed 75% of epigenetic markers persisted unchanged through gametogenesis during locus-by-locus comparisons of methylation status in individual sporophytes and descendant gametophytes. Their findings indicate that individuals and populations of *H. foetidus* indicate that epigenetic marks acquired during the sporophyte life stage in response to biotic or abiotic stress would enhance parental fitness if passed unchanged to the germline.

Du et al. (2020) studied 91 bud mutations of "Fuji" apple using the genetic variation within "Fuji" as the control and examined the characteristics of epigenetic variation at different levels in both varieties and mutant groups. They observed a global genomic DNA methylation level of the 91 bud mutants of "Fuji" ranged from

29 to 45%, with an average of 36%. The main DNA methylation pattern is the internal cytosine methylation. They observed methylation level variation in the color mutant group; however, variation in methylation pattern was more obvious in both the early maturation and spur mutant groups. Their study concluded that abundant changes in methylation levels and patterns between bud mutants and their mother "Fuji" indicate the possibility of epigenetics mediated DNA methylation in "Fuji" bud mutation line.

14.6 Epigenome Engineering Novel Techniques for Crop Improvement

The novel techniques based on epigenome engineering for crop improvement include the genetic modification with Agrobacterium tumefaciens (Agrobacterium), which introduces a piece of its own DNA into the plant genome, in terms of tumor-inducing (Ti) plasmid into the plant cell causing genome integration. Hence the possibility of incorporation in to plant genome from distantly related/related organisms in terms of transgenesis/cisgenesis has been achieved (Sedeek et al. 2019). Epigenome editing using mobile RNA has the potential to allow breeding of artificial sport cultivars in vegetative crop propagation (Kasai et al. 2016). The clustered regularly interspaced palindromic repeats (CRISPR)/Cas9 system offers Cas9 nuclease complex direction to specific sites in the genome as determined by complementary base-pairing between the DNA and a short single guide RNA (sgRNA) (Lee et al. 2019). Genome editing technologies improve crops and ensure global food security. Genome-edited crops of natural/artificial mutagenesis based may support biotechnology companies to adopt genome editing (Sedeek et al. 2019). By adopting several approaches the plant genome can be altered which includes targeted introduction of nucleotide changes, deleting DNA segments, introducing exogenous DNA fragments, and epigenetic modifications. Targeted changes are mediated by sequence specific nucleases (SSNs), such as zinc-finger nucleases (ZFNs), transcription activator-like effector nucleases (TALENs), and CRISPR (clustered regularly interspersed short palindromic repeats)-Cas (CRISPR associated protein) systems. Improvement in the delivery of exogenous DNA into plant cells also increases successful gene editing events rate. Crop breeding based newer approaches for enhancing the resistance to abiotic stress, as knowledge of stability and heritability features of epigenetic marks and epigenetic regulatory mechanisms are crucial for breeding applications.

14.7 Epigenetics in Agricultural Sector Patents

Oliver et al. (1994) patented a technology of creating transgenic plant containing gene whose expression can be controlled by application of an external stimulus. Gene expression with positive control may be achieved by an external stimulus to maintain gene expression. Here in technology of gene expression in case of plant phenotype which is possible to grow favorable or unfavorable conditions may be selected, based on the selection plant may be grown accordingly. Bucher and Thieme (2016) patented a method for the mobilization of a transposable element by providing an inhibitor of DNA methylation, and/or an inhibitor of transcription, and by contacting the inhibitor(s) with a cell comprising inactivated transposable elements, yielding a cell with mobilized transposable elements using inhibitor of DNA methylation and/or an inhibitor of transcription. During 1990s, the first epigenetic patents has been granted. These patents claimed laboratory methods that lay the foundation for future epigenetic advances by describing how to detect and manipulate DNA methylation. Nari, a company announced that it has secured exclusive patent licenses for epigenetics from the University of California, with Los Angeles (UCLA) the revolutionizing plant breeding by tapping natural genetic diversity. The agreement, through UCLA's Technology Development Group, gives Inari access to tools that will positively influence crop performance without altering a plant's genetic code. U.S. Patent No. 5,871,917 claims methods of detecting hypomethylation (decreased DNA methylation) or hypermethylation in a CpG sequence. Prior to 2000 only nine epigenetic patents were granted. Epigenetic patenting expanded rapidly in recent years. During the last 5-year period from 2000 to 2004 a sharp increase in the number of successful epigenetic patent applications filed has been recorded.

14.8 Challenges and Opportunities in Phyto-Epigenetics

The genetic diversity is associated with wild-type plants collected from different geographical origins. The environmental conditions such as photoperiod/temperature changes provoke changes causing RNA- or chromatin-based transcriptional regulation. Altered chromatin and gene expression states also produce challenges even if the plant returns to the original environmental condition, as in the case of vernalization. Degree of epigenetic variation between ecotypes is also substantial, which supports to explore whether epigenetic adaptations contribute to plant form, survival, and performance under different conditions. Switches in epigenetic states such as mutations allow read-through transcripts to bring about the silencing of adjacent genes, including tumor-suppressor genes in humans. There is also the possibility that environmentally or pathogen-induced epigenetic states might be transmitted to progeny if the changes occur in meristems and can be maintained through meiosis. Numerous changes at the level of the chromatin that lead to activating or repressing specific gene expression are caused by environmental factors. The epigenetic changes may be inherited over the generation that often results in phenotypic variations. It is becoming evident that epigenetic changes play important roles in acclimatization, stress tolerance, adaptation, and evolution processes. Epigenetic variations suggest their effect on gene expression, with epigenetic machinery of gene regulation in plants, and its possible use in epigenome engineering/editing for crop improvement is found to be the crucial steps. Defining the molecular basis of transgenerational epigenetic inheritance could ultimately lead to development of epialleles designed for specific environmental conditions through targeted epigenetic modifications in genes of interest.

14.9 Conclusion

The innate capability to survive plants under drastic climatic conditions has been hold by plants. Improvement of plant capacity to produce more nutritious food and capacity to survive under drastic climatic conditions are helpful. Here in the gene expression profile of plants offers major impact towards the transformation of epigenotype to phenotypes. In the forthcoming years monitoring and manipulation of crop epigenome may offer wider scope which may support the development of superior crops with improved agricultural productivity. Even though research has paid way for better understanding for stress tolerance novel genetic engineering tools application may still provide wider scope in future.

References

Acevedo MF (2011) Interdisciplinary progress in food production, food security and environment research. Environ Conserv 38(2):151–171

Akomeah B, Quain MD, Ramesh SA, Anand L, Rodrıguez-Lopez CM (2019) Common garden experiment reveals altered nutritional values and DNA methylation profiles in micropropagated three elite Ghanaian sweet potato genotypes. PLoS One 14(4):e0208214

Alonso C, Cruz DR, Becker C (2019) The role of plant epigenetics in biotic interactions. New Phytol 221:731–737

Benoit M, Drost HG, Catoni M, Gouil Q, Lopez-Gomollon S, Baulcombe D et al (2019) Environmental and epigenetic regulation of *Rider* retrotransposons in tomato. PLoS Genet 15(9): e1008370

Etienne Bucher, Michael Thieme (2016) Mobilisation of transposable elements to enhance genetic and epigenetic variability in a population. WO2017093317A1

Cai Q, Qiao L, Wang M, He B, Lin FM, Palmquist J, Huang SD, Jin H (2018) Plants send small RNAs in extracellular vesicles to fungal pathogen to silence virulence genes. Science 360(6393):1126–1129

Chen RS (1990) Global agriculture, environment, and hunger: past, present, and future links. Environ Impact Assess Rev 10(4):335–358

Chen M, Lv S, Meng Y (2010) Epigenetic performers in plants. Develop Growth Differ 52:555–566

Crespo-Salvador Ó, Escamilla-Aguilar M, López-Cruz J, López-Rodas G, González-Bosch C (2018) Determination of histone epigenetic marks in *Arabidopsis* and tomato genes in the early response to Botrytis cinerea. Plant Cell Rep 37(1):153–166

Cummins I, Wortley DJ, Sabbadin F, He Z, Coxon CR, Straker HE, Sellars JD, Knight K, Edwards L, Hughes D et al (2013) Key role for a glutathione transferase in multiple-herbicide resistance in grass weeds. Proc Natl Acad Sci USA 110:5812–5817

De-La-Pena C, Rangel-Cano A, Alvarez-Venegas R (2012) Regulation of disease responsive genes mediated by epigenetic factors: interaction of *Arabidopsis–Pseudomonas*. Mol Plant Pathol 13: 388–398

Du X, Wang Y, Liu M, Liu X, Jiang Z, Zhao L et al (2020) The assessment of epigenetic diversity, differentiation, and structure in the 'Fuji' mutation line implicates roles of epigenetic modification in the occurrence of different mutant groups as well as spontaneous mutants. PLoS One 15(6):e0235073

Duhoux A, Delye C (2013) Reference genes to study herbicide stress response in *Lolium* sp.: up-regulation of P450 genes in plants resistant to acetolactate-synthase inhibitors. PLoS One 8(5):e63576

Eriksson MC, Szukala A, Tian B, Paun O (2020) Current research frontiers in plant epigenetics: an introduction to a virtual issue. New Phytol 226:285–288

Gaines TA, Duke SO, Morran S, Rigon CAG, Tranel AK, Dayan FE (2011) Mechanisms of evolved herbicide resistance. J Agric Food Chem 59(11):5886–5889

Gibney E, Nolan C (2010) Epigenetics and gene expression. Heredity 105:4–13

Guarino F, Cicatelli A, Brundu G, Heinze B, Castiglione S (2015) Epigenetic diversity of clonal white poplar (*Populus alba* L.) populations: could methylation support the success of vegetative reproduction strategy. PLoS One 10(7):e0131480

Hamamoto S, Horie T, Hauser F, Deinlein U, Schroeder J, Uozumi N (2015) HKT transporters mediate salt stress resistance in plants: from structure and function to the field. Curr Opin Biotechnol 32:113–120

Herrera CM, Medrano M, Bazaga P (2013) Epigenetic differentiation persists after male gametogenesis in natural populations of the perennial herb *Helleborus foetidus* (Ranunculaceae). PLoS One 8(7):e70730

Kapazoglou A, Drosou V, Argiriou A, Tsaftaris AS (2013) The study of a barley epigenetic regulator, HvDME, in seed development and under drought. BMC Plant Biol 13:172. https://doi.org/10.1186/1471-2229-13-172

Kapazoglou A, Ganopoulos I, Tani E, Tsaftaris A (2018) Epigenetics, epigenomics and crop improvement. Adv Bot Res 86:287–324

Karan R, DeLeon T, Biradar H, Subudhi PK (2012) Salt stress induced variation in DNA methylation pattern and its influence on gene expression in contrasting Rice genotypes. PLoS One 7(6): e40203

Kasai A, Bai S, Hojo H, Harada T (2016) Epigenome editing of potato by grafting using transgenic tobacco as siRNA donor. PLoS One 11(8):e0161729

Kim G, Clarke CR, Larose H, Tran TH, Haak DC, Zhang L, Askew S, Barneyand J, Westwood JH (2017) Herbicide injury induces DNA methylome alterations in Arabidopsis. PeerJ 5:e3560

Kumar S (2019) Epigenomics for crop improvement: current status and future perspectives. J Genet Cell Biol 3(1):128–134

Kumar G, Rattan UK, Singh AK (2016) Chilling-mediated DNA methylation changes during dormancy and its release reveal the importance of epigenetic regulation during winter dormancy in apple (malus x domestica Borkh.). PLoS One 11(2):e0149934

Kumar S, Beena AS, Awana M, Singh A (2017) Physiological, biochemical, epigenetic and molecular analyses of wheat (*Triticum aestivum*) genotypes with contrasting salt tolerance. Front Plant Sci 8:1151

Lee JE, Neumann M, Duro DI, Schmid M (2019) CRISPR-based tools for targeted transcriptional and epigenetic regulation in plants. PLoS One 14(9):e0222778

Liu X, Zhou C, Zhao Y, Zhou S, Wang W, Zhou DX (2014) The rice enhancer of zeste [E(z)] genes SDG711 and SDG718 are respectively involved in long day and short day signaling to mediate the accurate photoperiod control of flowering time. Front Plant Sci 5:1–9

Margaritopoulou T, Tani E, Chachalis D, Travlos I (2018) Involvement of epigenetic mechanisms in herbicide resistance: the case of Conyza Canadensis. Agriculture 8:17

Mousavi S, Regni L, Bocchini M, Mariotti R, Cultrera NM, Mancuso S, Googlani J, Chakerolhosseini MR, Guerrero C, Albertin E, Baldoni L, Proietti P (2019) Physiological, epigenetic and genetic regulation in some olive cultivars under salt stress. Sci Rep 9:1093, 1–17

Oliver MJ, Quisenberry JE, Trolinder NLG, Keim DL (1994) Control of plant gene expression. US5723765A

Peace CP, Bianco L, Troggio M, Weg EV, Howard NP, Cornille A et al (2019) Apple whole genome sequences: recent advances and new prospects. Hort Res 6(59):1–24

Ramirez-Prado JS, Abulfaraj AA, Rayapuram N, Benhamed M, Hirt H (2018) Plant immunity: from signaling to epigenetic control of defense. Trends Plant Sci 23(9):833–844

Sedeek KEM, Mahas A, Mahfouz M (2019) Plant genome engineering for targeted improvement of crop traits. Front Plant Sci 10:114

Shi H, Lee BH, Wu SJ, Zhu JK (2003) Overexpression of a plasma membrane Na+/H+ antiporter gene improves salt tolerance in *Arabidopsis thaliana*. Nat Biotechnol 21:81–85

Simmons D (2008) Epigenetic influence and disease. Nature. Education 1(1):6

Very AA, Sentenac H (2003) Molecular mechanisms and regulation of Na+ transport in higher plants. Annu Rev Plant Biol 54:575–603

Weinhold B (2006) Epigenetics: the science of change. Environ Health Perspect 114(3):A160–A167

Wu SH, Ho C, Nah SL, Chau CF, Global Hunger (2014) A challenge to agricultural, food, and nutritional sciences. Crit Rev Food Sci Nutr 54(2):151–162

Ethical Aspects and Public Perception on Plant Genomics

15

S. C. Srushti, Shivaleela Biradar, S. J. Shreedevi, M. Naveen Kumar, S. Mondal, and R. L. Babu

Abstract

Plant genomics is the study of a whole plant genome's architecture, functions, organization, genetic compositions, networks, and interactions. Internally, bioinformatics, system biology, metagenomics, proteomics, transgenomics, metabolomics, genomic selection, phenomics, contemporary instrumentation, epigenomics, and robotic sciences are all intertwined. In the country of cheap and completely sequenced over a hundred plant genomes, it has achieved tremendous progress in high-throughput sequencing over the last three decades. Because of the utilization of modification methods and unique genomic selection, these advances have a wide range of consequences in plant biology and breeding, while also introducing a host of new challenges and responsibilities. This article informs readers on plant genomics review, laws, transfer, ethics, public views, and a wide range of databases.

Keywords

Genomics · Proteomics · Metabolomics · Bioinformatics · System biology · Genomic selection · Epigenomics

S. C. Srushti · S. Biradar · S. J. Shreedevi · R. L. Babu (✉)
Department of Bioinformatics and Biotechnology, Karnataka State Akkamahadevi Women's University, Vijayapura, Karnataka, India

M. Naveen Kumar
Department of Biotechnology and Genetics, M.S. Ramaiah College of Arts, Science and Commerce, Bengaluru, India

S. Mondal
ICAR-National Institute of Animal Nutrition and Physiology, Adugodi, Bengaluru, Karnataka, India

R. L. Singh et al. (eds.), *Plant Genomics for Sustainable Agriculture*,
https://doi.org/10.1007/978-981-16-6974-3_15

15.1 Introduction

Ethics are generally used in all living things, philosophical justification of plant ethics consider for study of genetics. The biochemical composition of DNA understands the significant physiological sequence to read (Greely 2015). Development of plant DNA growth improved by advanced techniques that is testing, sequencing, and manipulating to get which genes code for a specific protein, enzyme, or a transporter in a specific plant and to search any mutation, SNP, or copy number. The sequencing starts from individual gene to whole or partial genomes. This leads to moving the scientific language from genetics to genomics. Genomics disciplines in plants takes many ways, genome editing, MAS (marker assisted selection), plant breeding, genetic engineering, transgenic, etc. The flowchart of process in ethics as follows (Fig. 15.1) (Mathaiyan et al. 2013).

Biotechnology began in the early 1970s. At the time, researchers anticipated that genetic code technology would be helpful not just for study, but also for chemical production, medicine, and agricultural and animal breeding. Scientists Paul Berg and Maxine Singer are concerned about a new biotechnology innovation and if it is hazardous to native plant species, including humans. After 20 years, Nobel Laureate

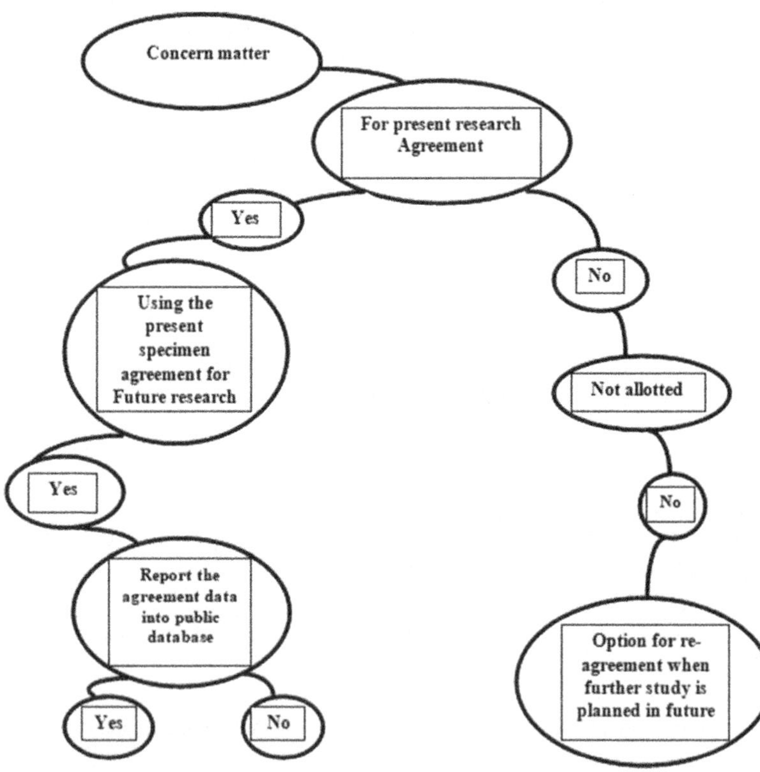

Fig. 15.1 Flowchart of ethical process

Paul Berg and Maxine Singer (another prominent biologist) published a retrospective study claiming that the revolutionary new technology in biological research is not detrimental to wild plants. Natural changes prompted the development of new biotechnology techniques. Mutation is caused by genetic alterations, and similar processes may be found in biotechnology including gene editing, mutagenesis induction, and gene transfer from other genomes. These things are caused in certain plants that produce various yields, such as *Agrobacterium* to sweet potatoes and rice to millet (Hansson 2019).

15.2 Plant Genomics

Plant genomes represent gene size, gene content, the amount of repetitive sequences, and polyploidy events in terms of gene size, gene content, and polyploidy events. In plants, molecular analysis is constructed utilizing a single level gene concentrate. Later, sophisticated technology was used to examine the plant's genome structure, expression, and interaction. The study of the movement of genes and genetic information throughout the genome, as well as how it affects the structure and function of organisms, is known as genomics. The study of genomics provides knowledge about DNA codes and expands all areas of plant life. New methods have been developed in plant genomics to study the biological function of genes and to bridge the genotype–phenotype gap.

Fossil plants, archeological fabrics, maize cobs, sediment cores, and herbarium sheets all these made extensive use of PCR in the 1990s. Furthermore, biomolecular development research is being conducted at the forefront of ancient plant development. Maize was the first DNA sequence that was successfully captured. Ancient RNA was initially discovered in cress and then sequenced in maize. The RNA virus was the first to be resurrected in barley. All of these studies were focused on DNA selection in a home setting (Kistler et al. 2020). Ancient plant study was inspired by the use of the technique called PCR in plants, and subsequently plant DNA was studied using a population genetic approach using an ancient DNA dataset to increase agricultural yield.

Over the course of 15 years, genomics has been evaluated. Whole genome doubling was significant in the evolution of polyploidy in plants. According to estimates based on conventional cytogenetic research and stomatal guard cell counts, chromosome doubling is common in many plant species. To better comprehend polyploidy, expressed sequence tags (EST) were used for the first time in various plant species (Wendel et al. 2016). These analyses showed that all of the expressed sequences have a common ancestor. Plant genomic evolution is divided into three waves. The single gene sequencing method, Restriction Fragment Length Polymorphism (RFLP) markers, dot blot types, one-gene, and one-phenotype method are all considered in the first evolution. The second evolution ended with whole genome sequencing and single nucleotide polymorphism (SNP) markers, which entailed identifying genes associated with particular traits. Comparative whole genome sequencing from several related species is included in the third or present evolution

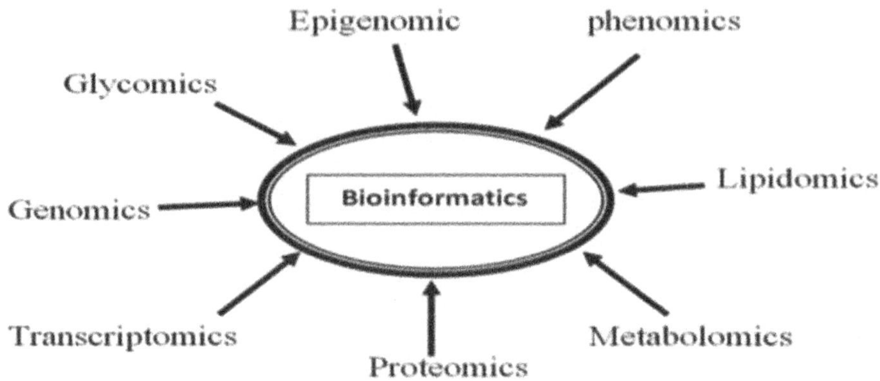

Fig. 15.2 Database study undertaken in Bioinformatics

(Borevitz and Ecker 2004). Genome analysis covers the response of DNA levels and gene expression levels in polyploidy. DNA-level responses include reciprocal or non-reciprocal homologous exchange, mutational loss of duplicated genes, inter sub-genomic dispersion of TEs (caused by genome merger and polyploidization), and divergence in molecular evolutionary rate.

Through comparative genome mapping of related plant species, genome colinearity was found between well-sequenced model crops and their related species (for example, *Arabidopsis* for dicots and rice for monocots) (Campos-de Quiroz 2002). In the inter- or intrastages of species, genomic information from genetic and physical maps is shown. Later, whole genome sequencing based on techniques such as fluorescence-based DNA sequencing is considered, which yields at least 500 bases per read. After that, the automated process will begin with sample selection and purification from people. Finally, the accessible data is handled in software and hardware for processing.

Improvement in reading genome undertakes plant genome assembly, repeat regions, De novo assembly, genome annotation, genome re-sequencing, phenotype study, data analysis. All these software are considered for analyzing the genome, the genome assembly consider tool the SOAP denovo, genome annotation consider tool to annotate the plant genome with AUGUSTUS, MAKER-P. Re-sequencing is used to study heterozygosity and ploidy levels assemblies (Bolger et al. 2017). To crop improvement through the biological tools so many data bases created to handle genome of plant species. The available of plant comprehensive information is provided to public utilization through databases. These database study undertaken in Bioinformatics (Fig. 15.2) (Bolger et al. 2017). The flow chart of bioinformatics database will be as follows for public usage of database in DNA, RNA, and proteins (Dhanapal and Govindaraj 2015).

15.3 Ethical Aspects of Plant Genomics

The public's knowledge of the science and technology that underpins the development of transgenic crops is critical to the breeding technique's future success. Without accurate information, the public will be unable to make fair choices on the benefits and drawbacks of GM food production and consumption. This is precisely the problem: scientific literacy is almost non-existent, and a lot of different groups are taking firm positions on both the safety and the advantages of this technology.

The word "intrinsic" was used to describe the basic ethical concerns, which included theology, respect for nature, and naturalness (Straughan 1995a). When it comes to GE, words and phrases are often at the core of issues. Though they are almost synonyms, "transgenic" sounds similar to "eugenic," and therefore may incriminate by association, while "engineered" sounds more evil than "domesticated" (Jones and Balasubramanian 1995). More likely, fundamental resistance to transgenic stems from the belief that crossing species borders is immoral, but it is important to remember that categorization is a man-made notion. The forces of natural selection result in the formation of new species, according to evolutionary theory, which is founded on dynamic ideas and gradualism. Creationist theory depicts life forms as fixed and immutable, determined by god, with small changes such as mutations occurring over long periods of time. Evolutionary theory is based on dynamic concepts and gradualism, with small changes such as mutations occurring over long periods of time, whereas creationist theory depicts life forms as fixed and immutable, determined by god, with small changes such as mutations occurring over long periods of time. Hybridization will be more prevalent in nature than is generally recognized, with about 20% of plant species hybridizing spontaneously (Brookes 1996). Another area of basic ethical concern is environmental respect (Straughan 1995a; Reiss and Straughan 2001). Some people object to reductionism because it dehumanizes humans by reducing their life to a sequence of gene products that may be transplanted into other animals.

The underlying resistance to biotechnology (Weil 1996), nature commercialization, and human evolution control—"evolution engineered"(Jones and Balasubramanian 1995) as a challenge to traditional beliefs and world views—posed a slippery downward scenario which necessarily leads to the acceptance of more advantages, one becomes increasingly unconvinced by the acceptability of controversial technologies. It is also suggested that ethical boundaries be established and that acceptable and unacceptable behaviors be explained. Fundamental issues, on the other hand, do not have to be theological, and they do not have to be anti-transgenics. Transgenics could be viewed as naturally acceptable, as they aid evolution and provide new information about the natural world; science is regarded as intrinsically great. While the consequence of scientific endeavors may be acceptable, this does not imply that the actions of scientists are necessarily acceptable, just as scientists' activities are not essentially awful regardless of whether scientific activities may result in horrific results (Kealey and Nelson 1996).

Assumptions regarding transgenic crops are clearly dependent on value judgments rather than empirically proven facts, characteristics and views are prone to shift with circumstance, time, and theoretical system changes. Beneficiaries of research findings may have ethical systems and beliefs that vary significantly from those of the scientific community performing the study (Hawtin 1997) from a philosophical standpoint, the difficulties of making ethical choices and creating moral innovations were addressed (Robinson 1999), including an explanation of how law, religion, and custom define right and wrong, as well as how deontological, utilitarian, and naturalistic moral theories can be used to aid rule and judgment making. It emphasized that ethical theories not to be confused with reality and that makes the ethical choices difficult (Carr and Levidow 1997). Opposed (Straughan 1995a) for separating intrinsic and extrinsic ethical issues, theoretically divorcing crop engineering from its effects, and so removing ethics from risk. Despite the fact that it is the result of a debate about the values that drive biotechnology research and development, "official policy downplays ethical choices by addressing risk as if it were an objective thing," according to the article, "official policy downplays ethical choices by addressing risk as if it were an objective thing." As a consequence, "state-sponsored ethics" will be labeled. From a scientific perspective, knowledge is always important in evaluating risk; ethical implications arise only when information must be assessed in order to make a decision, and only when responsibility, justification, and obligation must be given (Straughan 1995b).

15.4 Transgenic Plants Regulation in India

India's regulatory system for assessing the biosafety of genetically modified organisms (GMOs) and their derivatives is well-defined. In 1989, India became one of the first nations to establish a biosafety regulatory framework for genetically modified organisms (GMOs). The top regulations for regulating all GMO-related activities are published under the Environment (Protection) Act of 1986.

In 1986, India's Union Government passed the Environment Act (EPA 1986). This legislation was passed to preserve the environment and minimize the danger of pollutants and toxins disrupting the ecosystem's flora and fauna. The government subsequently established a set of regulations to handle a number of problems, including hazardous chemicals, hazardous wastes, solid wastes, biomedical wastes, and so on. Furthermore, the danger of harming animal and human health must be taken into account. Genetically modified organisms were intended to play a significant role in the country's economic development in a variety of ways including health care system of human and animal, agriculture, environmental management and industrial products. Simultaneously, it was recognized that the use of GMOs and their products could pose unintended hazards and risks if the new technology was not properly assessed and implemented. With this in mind, the Indian government issued Rules and Procedures for Handling GMOs and Hazardous Organisms on December 5, 1989, via the Union Ministry of Environment and Forests' Gazette Notification No. GSR 1037(E)(MoEFCC 1989). The rules apply to all genetically modified organisms (GMOs) and their products that are regulated commodities in the USA and must be handled and used by the Environmental Protection Agency (EPA). In addition to the Rules of 1989, several Acts and Rules relate to particular activities/products using GMOs (MoEFCC 2015). One of them is the Plant Quarantine Order of 2003. It is managed by the Ministry of Agriculture and Farmers Welfare and controls the import of germplasm, GMOs, and transgenic plant material for research purposes. The Biological Diversity Act of 2002, which regulates the use of biological resources, such as genes, to enhance crops and animals via genetic intervention, was implemented by the National Biodiversity Authority. Similarly, India's Food Safety and Standards Authority adopted the Food Safety and Standards Act of 2006, which regulates food manufacturing, storage, distribution, sale, and import, including genetically modified foods.

15.4.1 Framework for Implementation of Regulations for Handling with GMOs

The guidelines clearly identify the authorities in charge of handling all aspects of genetically modified organisms (GMOs) and their products, as well as their organized compositions. There are currently six competent authorities, who are listed below, along with their general roles and authorities:

(i) The Recombinant DNA Advisory Committee (RDAC)
(ii) The Review Committee on Genetic Manipulation (RCGM)
(iii) Institutional Biosafety Committee (IBSC)
(iv) Genetic Engineering Approval Committee (GEAC)
(v) State Biotechnology Coordination Committee (SBCC)
(vi) District Level Committee (DLC)

The DBT established the Recombinant DNA Advisory Committee (RDAC) to keep track of biotechnology breakthroughs both at home and abroad. The RDAC is a consultative body tasked with making recommendations on GMO safety rules and applications on a regular basis. In 1990, this group published the Recombinant DNA Biosafety Guidelines, which the Indian government has accepted as guidelines for GMO research and administration.

The DBT's Review Committee on Genetic Manipulation (RCGM) is in charge of assuring the safety of current research programs and activities using GE organisms/ hazardous microorganisms. The RCGM is also obliged to produce instruction manuals detailing the regulatory processes for operations using GE organisms in research in order to guarantee environmental safety. The RCGM is made up of delegates from the Indian Council of Medical Research (ICMR), the Indian Council of Agricultural Research (ICAR), the Council of Scientific and Industrial Research (CSIR), and other professionals acting individually.

Each university that wishes to conduct research involving genetic alteration of microbes, animals, or plants must join the Institutional Biosafety Committee (IBSC).

The IBSCs consist of the leaders of the organization, DNA scientists, a medical specialist, and a DBT nominee. The IBSC serves as a hub for internal communication about how the guidelines are being implemented.

The MoEFCC's top committee, the Genetic Engineering Approval Committee (GEAC), is made up of representatives from relevant ministries/agencies as well as experts. A senior MoEFCC officer chairs the GEAC, which is co-chaired by a DBT authority. From an environmental perspective, GEAC is in charge of authorizing activities involving the widespread use of hazardous microbes and recombinant products in research and commercial manufacturing.

SBCCs (State Biotechnology Coordination Committees) are formed in every state that conducts GMO research and applications. The Chief Secretary of State leads the SBCC, which is in charge of overseeing.

District Level Committee (DLC): When necessary, DLCs are formed in districts to monitor safety rules in facilities that use GMOs/hazardous microorganisms and their environmental uses. A District Collector (an officer in charge of a district's administration) leads each DLC, which includes officers in charge of public health, the environment, pollution management, and other district level issues. Interactive methods between committees are also included in the 1989 Rules. All IBSCs must evaluate the applications and provide the RCGM with suggestions and reports. The RCGM reviews the situation and makes recommendations to the GEAC for a variety of activities, field testing, and environmental release. On a regular basis, DLCs must also submit a report to the SBCC/GEAC (Ahuja 2018).

15.5 International Regulations of Transgenic Plants

26 nations have planted 190 million hectares of transgenic crops in the past 2 years, with five industrialized and 21 developing countries almost equally divided. The 9 industrialized nations that grow 46 percent of all biotech crops are the USA, Australia, Canada, Portugal, and Spain. India, Argentina, and Brazil are three of the top five countries with the most biotech crop production land, accounting for 54% of the total (ISAAA 2020b, 2018).

There are variations in authorization for breeding GM crops, import and export, and usage of GM food and feed products when it comes to biotech crops. Due to the numerous risks connected with growing, trading, and eating, several regulatory processes are needed. A variety of governmental organizations are often tasked with evaluating the authorization request. The end yield in the USA may come within the authority of the United States Department of Agriculture (USDA), the Environmental Protection Agency (EPA), or the Food and Drug Administration (FDA), or several agencies, depending on the nature of the finished item.

Government legislators all over the world work to pass laws that protect environment, society, and citizen. Similarly, laws governing flora and crops intended for food, feed, and industry are based on similar goals. It depends on the country or location how this is accomplished. GM laws are usually divided into two categories: process and product (Callebaut 2015; Medvedieva and Blume 2018; Eckerstorfer et al. 2019). Process-oriented regulations see GM technology as a new methodology in contrast to traditional ways, necessitating the application of particular legislation. The focus is on the manufacturing method for the unique product. The product-oriented laws, on the other hand, stress the product's unique qualities in compared to those produced by traditional breeding (McHughen 2016). Too far, Canada is the only country whose whole GM legislation is centered on the commodity rather than the procedure.

The discussion has focused on determining which regulatory framework is best for products made using gene editing methods (Kuzma 2016). Eckerstorfer et al. (2019) discovered that each approach has its own set of advantages and disadvantages, with neither technique being better to the other. Biotechnology professionals from across the globe, on the other hand, largely endorse the product-based evaluation procedure as the most scientific method (Scheben and Edwards 2018). Effective risk management is built on science and scientific assessments, as McHughen (2016) reminds out, and laws depend on risk management to safeguard society and the environment. As a result, research must have an impact on regulatory regulations, which cannot and will not happen in a vacuum.

15.5.1 Regulation of European Union

Regulation (EC) No 1829/2003 on genetically modified food and feed binds all 27 EU member states immediately, with a focus on GM food and feed produced "from" a GMO in the EU (EU). The goal of the Regulation is to guarantee that the processes for licensing GM foods and feeds safeguard human, animal, and environmental health. In conjunction with Regulation 1830/2003 on GM product traceability and labeling, this Regulation primarily applies to food and feed goods, as well as their imports. Cultivation of GMO crops, on the other hand, is a decision made by Member States in accordance with Directive 2001/18/EC on the deliberate release of genetically modified organisms into the environment (also known as the "Cultivation Directive"). This tool allows the production of genetically modified foods and plants after a thorough evaluation of potential harmful impacts on the environment as well as human health.

Member states may "provisionally restrict or ban the use and/or sale of such GMO as or in a product on their territory" (Hundleby and Harwood 2019) under "the Cultivation Directive." Article 23 allows Member States to prohibit or limit the cultivation of a certain GM crop in all or part of their territory if the EU authorizes it. Since the release of the safety note in 2015, a significant number of EU countries and regions have boycotted, if not outright banned, the growing of GM crops (Lombardo and Grando 2020). Only insect-resistant maize (MON810), one of the two instances approved for cultivation in the EU in the last 25 years, is extensively grown in Spain and Portugal (ISAAA 2018).

The concept of "genetically modified organism" (Eckerstorfer et al. 2019; Marchant and Stevens 2015; Sprink et al. 2016) used by the European Union is often cited as an example of a process-driven regulatory framework. An organism is considered genetically modified under Article 2(2) of the Cultivation Directive if the process of altering genetic material is not natural mating and/or recombination. According to the European Court of Justice (ECJ), organisms changed through site-directed mutagenesis, such as CRISPR/Cas9, are included in the explanation of a GMO (ECJ 2018). According to Wasmer (2019) (Wasmer 2019), the decision implies that the amount or kind of genetic change is immaterial; if there is mutagenesis, whether random or intentional, large or little, the organism is legally classed as a GMO. Despite the fact that this criterion is the starting point, the ECJ accepted the Cultivation Directive's relevant exemptions because of their long track record of safety (a result of time and experience) (ECJ 2018).

The EU legislation catches most changed plant products, with the exception of plants produced using exempted methods, which include mutation breeding based on techniques in use when the Directive entered into force in 2001, but not newer forms of mutagenesis. Examine the European Court of Justice's judgment on the interpretation of exempted methods in depth (2019). When a legislative instrument only allows minor deviations from the language of the law, the interpretive outcome can only be characterized as arbitrary, which happens often. In other words, arbitrary decisions indicate that the rules in question are no longer suitable for their original purpose (Smyth and Lassoued 2019; Wasmer 2019; Jorasch 2020). The EU's position has a significant impact on countries that export to Europe like former European colonies (Paarlberg 2010; Paarlberg 2014).

In light of the ECJ's ruling, the European Union's Council has ordered a study and proposal on the status of "novel genomic techniques" by April 2021. This is a good start since the legislation will be developed based on solid data and regulatory criteria. The European Network of GMO Laboratories (ENGL) earlier released a paper (ENGL 2019) outlining the different options and difficulties for recognizing NBT-created food and feed products. The EU presently relies on GM regulations to regulate commodities entering the country; as a result, the creator of the gene-altered product is responsible for developing a working detection system although no such items have been submitted for market approval (Commission 2019). The research seems to be mainly focused on gene-modified food and feed items, rather than cultivation, which is stated as one of the project's goals (Commission 2019), since the EU imports the bulk of its GM goods.

15.5.2 Regulation of Non-European Union

Both Norway and Switzerland have laws prohibiting the manufacturing of genetically modified crops. Their opposing approaches to limiting GM crops are diametrically opposed. Switzerland has had a temporary ban on the planting and processing of genetically modified crops since 2006, which has been extended until 2021, although GM crops for animal feed continue to be received by the Federal Office for the Environment (FOEF 2018). The Swiss Cabinet included a plan to establish distinct GM agricultural zones starting in 2021 when the prohibition was extended for the third time in 2016, based on farmer desire. The government hopes to improve public acceptance of GMOs while still keeping the door open for future employment possibilities by recommending that GM crops coexist with traditional agriculture (Chandrasekhar 2016).

Despite the fact that GM crops are allowed under the Gene Technology Act, no GM food or feed crops have been grown or imported in Norway. The Norwegian Food Safety Authority has yet to authorize any goods or their intentional distribution, with the exception of a single kind of decorative purple carnation (ISAAA 2020a). In addition to the EU's health and environmental safety standards, Norwegian legislation mandates the evaluation of three non-safety categories: social benefit, sustainability, and morally correct goods. These three major groupings' content and interpretative problems are addressed by the three categories (2009) (Rosendal and Myhr 2009; Rosendal 2008). They are mostly focused on farmers and GM crop producers in underdeveloped nations, with a little emphasis on the Norwegian consumer.

Despite Norway's adamant rejection to genetically modified foods, the Norwegian Biotechnology Advisory Board has developed and presented to the government a proposal requesting a change to the law governing the deliberate release of GMOs (Bratlie et al. 2019). One of the main reasons for the publishing was to respond to accusations that EU standards were no longer appropriate. The Advisory Board's goal is to bridge the gap between science and law by acknowledging the complexities of biotechnology use and recommending a multi-tiered regulatory framework.

Changes to Federal Law No. 358-FZ in July 2016 and the subsequent adoption of the new Food Security Doctrine in January 2020 in the Russian Federation restricted the farming of GM plants and the breeding of GM animals. Even though media claims to the adverse, the modifications are similar to those in the EU, which prohibits cultivation but allows authorized GM food and feed impor (FAS 2016; The Moscow Times 2016). The new restrictive attitude, altered by the anti-GMO movement and backed by the Minister of Agriculture (Galata Bickell 2019), puts a stop to cultivation plans for 2023 and 2024 (FAS 2016).

15.5.2.1 North America

The USA leads the world in the marketing and development of genetically modified crops, accounting for almost 30% of the global market (Reportlinker 2020). Unlike most other countries, the USA lacks comprehensive federal laws regulating genetically modified organisms (GMOs). Newly manufactured GM products are instead submitted to specialist regulatory bodies under the Coordinated Framework for Biotechnology Regulation. This means that genetically modified foods are subject to the same health, safety, and environmental standards as conventional foods, enabling authorized agencies 13 to treat them similarly. New GM agricultural plant products may be evaluated by the FDA, EPA, and USDA, among others. The Animal and Plant Health Inspection Service (APHIS) of the USDA is in charge of ensuring that the introduction of genetically modified plants does not harm them. The plant product's status is either regulated or uncontrolled, with the latter allowing manufacture, import, and transit without APHIS oversight. The non-regulated status of APHIS solely applies to the introduction of the GM plant for cultivation and transportation reasons. The FDA is in charge of assessing the safety of GM plants that are meant for human consumption. At the time of writing, 128 GM plant varieties were classified as non-regulated since they lacked foreign DNA from "plant pests" such as bacteria, fungi, viruses, insects, and other microbes (APHIS 2020a). In 2016, a common button mushroom (*Agaricus bisporus*) was granted non-regulated status after being engineered to resist browning and therefore reduce spoilage. This was also true of CRISPR/Cas9-edited food crops (Waltz 2016). Since then, CalynoTM, a high-oleic soybean oil, SU (sulfonylurea) CanolaTM, a herbicide-resistant canola, and waxy corn have all been released (Lassoued et al. 2019; APHIS 2020b). Wolt and Wolf (2018) examine the legislation regulating genome editing in the USA in detail.

Canada is also one of the world's top five biotech crop producers, accounting for about 6.6 percent of global biotech crop area in 2018 (ISAAA 2018). It is worth mentioning that the legislation in Canada has a product-oriented approach, which some think promotes agricultural biotechnology research (Atanassova and Keiper 2018; Whelan et al. 2020). The mere presence of a new feature, rather than how it was implemented, distinguishes Canadian law from other product-based regulatory regimes. Whether the novel feature was produced via conventional breeding techniques, traditional mutagenesis, or targeted mutagenesis, the new plant product is subject to the same risk assessment criteria (CFIA 2020).

According to (Smyth 2017), When it comes to new plants, Canada has maintained a comprehensive science-based risk assessment, concentrating on allergenicity, toxicity, and off-target effects of the product. When a certain characteristic in the plant expresses at least 20–30% lower or higher than typical types, the controls are activated. As a result, rather of being categorized as a genetically modified organism (GMO), the plant is classified as a plant with novel traits (also known as PNT) (CFIA 2020). All commercialization applications for unconfined environmental discharge must be submitted with the Canadian Food Inspection Agency (CFIA). Plant products intended for human consumption must also undergo a Health Canada inspection as well as an Animal Feed Division examination by the CFIA (Canada 2020). An example is the simplest method to illustrate Canada's distinct approach. Cibus Canada Inc. developed Falco™ Canola (Cibus Canola Event 5715), a herbicide-tolerant canola (Asmatulu 2020). It was created by causing a single nucleotide mutation in two genes using an NBT, oligonucleotide-directed mutagenesis (ODM). In that it is a gene editing technique, the ODM approach is comparable to CRISPR/Cas9. In 2013, the Canadian Food Inspection Agency determined that the new canola variety was identical to non-modified (conventional) canola varieties, certifying it as a non-GM crop (CFIA 2013; Canada 2013).

15.5.2.2 Latin America

Brazil and Argentina are two of the top five countries in the world for the production of genetically modified organisms. Bolivia, Chile, Colombia, Costa Rica, Honduras, Mexico, Paraguay, and Uruguay accounted for 42.7 percent of the global GM crop area (ISAAA 2018). There has also been a significant push in South America to harmonize GM product regulations. The agricultural ministers of Argentina, Brazil, Chile, Paraguay, and Uruguay signed a declaration14 on innovative breeding techniques in 2017, acknowledging and attempting to minimize unequal approvals throughout the area (Norero 2018; Benítez Candia et al. 2020). Eight of the 12 Latin American nations have written documents for this purpose in the past 5 years. The overall approach is one of case-by-case evaluation, allowing for the exclusion of some gene-edited products from tight control (Whelan and Lema 2015; Gatica-Arias 2020) Ecuador, Venezuela, and Peru, which do not allow commercial cultivation of GM crops, remain adamant in their opposition. Ecuador's Constitution, which was enacted in 2008, declares the nation "transgenic crop and seed free." The President may authorize GM seeds to be introduced into the nation if he or she thinks it is in the country's best interests. The Ecuadorian government took advantage of this exception to enact legislation allowing the import and cultivation of genetically modified seeds only for scientific purposes (Gatica-Arias 2020; Norero 2017). Despite the transgenic-free declaration, Ecuador joined a group of countries seeking to harmonize regulations to embrace new breeding techniques in May of this year, adopting Executive Decree No. 752. Article 230(a) exempts species that do not include foreign or recombinant DNA from the risk assessment that would otherwise be required for GM organisms (Gatica-Arias 2020).

In 2011, Peru enacted a 10-year ban on GM crops, banning the importation and manufacturing of genetically modified seeds (Branford 2013). The Peruvian Congress approved a 15-year extension of the ban as the deadline approaches in 2021. Without the signature of the President, whose position is now under political upheaval, the extension is not yet official (Montaguth 2020). Although (Dondanville and Dougherty 2020) say that the prohibition was enacted to give the government time to develop regulations that would enable agricultural biotechnology to be implemented, it is clear that the Peruvian government has no intentions to regulate genetically modified foods (Gatica-Arias 2020). In 2015, Venezuela passed the Seed Law, which made all GM plants and seeds illegal, even those used in research (APBREBES 2016; Agriculture 2016). According to the strategic plan for conservation efforts (Gómez et al. 2010), the introduction of GMOs is one of four main drivers of biodiversity loss in Venezuela. One of the fundamental reasons driving the restriction on environmental discharge is Venezuela's "agroecology" (Herrera et al. 2017). For food and feed, Venezuela continues to rely significantly on GM soybean and maize imports from Brazil, Argentina, and the USA (FAS 2018). Chile's regulatory framework for GM and gene-edited plants has evolved in a unique way. On the one hand, authorities moved swiftly to develop a case-by-case approach for plants produced via novel breeding methods, becoming just the second nation in the world to do so after Argentina. The Agricultural and Livestock Service (SAG) decides whether a variety or product is a GMO in part by looking for foreign DNA (Sánchez 2020). So far, eight products have been identified as non-GMO and may be sold as traditionally produced plants (Sánchez 2020; Eriksson et al. 2019). There is no comprehensive biotechnology framework in place for conventional GM plants (i.e., those that were not produced through innovative breeding techniques and fall under the definition of a "GMO") unless the plant is labeled as a

GMO (FAS 2020a) SAG is essential to the tight control of GM seed reproduction for the export market, despite the absence of a complete regulatory framework for GM plants. According to (Salazar et al. 2019; ISF 2020), Chile is the "southern seed nursery for the GM industry." Chile ranks tenth in the world for seed exports, with a significant portion of them being GM seeds. To monitor and control the import, manufacture, field testing, and export of genetically modified seeds, SAG relies on Resolution 1523 from 2001. Because there is presently no legislation regulating the domestic use of these GM seeds for food and feed, GM seeds cannot be produced in the country as a domestic product (Salazar et al. 2019). Surprisingly, there are no limitations on importing GM food and feed produced in other nations, with Brazil providing the majority of their soybean and maize imports (Sánchez and León 2016).

15.5.2.3 Africa

GM crops are grown in Ethiopia, Kenya, Malawi, Nigeria, South Africa, Sudan, and, most recently, eSwatini (formerly Swaziland), despite substantial challenges to food security presented by population growth and climate change. South Africa is Africa's major GM crop producer, with the tenth largest biotech crop area in the world, and was the first African country to establish a regulatory framework enabling GM crop import, export, and production (ISAAA 2018). Despite the fact that commercial cultivation is not yet practiced in Burkina Faso, the Biosafety Law of 2012 permits it. Schnurr (2019) examines the historical, political, and scientific developments, as well as the regulation, of conventional GM crops in Africa. Regulatory responses in Africa are divided into three groups by the author: early, medium, and late.

Emerging adopters (Uganda, Ghana, Nigeria, Cameroon, Ethiopia, Malawi, Mozambique, and eSwatini), resisters (Zambia, Zimbabwe, and Tanzania), and renegades (Uganda, Ghana, Nigeria, Cameroon, Ethiopia, Malawi, Mozambique, and eSwatini), and renegades (Zambia, Zimbabwe, and Tanzania) (Kenya and Sudan). When it comes to new breeding methods and the rules that govern them, African countries are collaborating and discussing harmonization measures, according to the African Biosafety Network of Expertise (ABNE 2019; Isaac 2019). South Africa is the first African country to approve white maize as a direct-consumption GM staple food crop. The first nations to authorize Bt cotton and Bt maize cultivation were Egypt and Burkina Faso. Egypt, on the other hand, outlawed the production of GM crops in 2012 (Gakpo 2019), while Burkina Faso outlawed them in 2016. Dowd-Uribe and Schnurr (Dowd-Uribe and Schnurr 2016; Schnurr 2019). Several researches (Adenle et al. 2013; Mabaya et al. 2015; Jawo et al. 2020; Luna 2020) have looked at the reasons for the poor adoption of GM crops, and there are obviously convincing arguments. (Glover and Paarlberg 2009; Paarlberg 2010; Paarlberg 2014) have frequently blamed the affluent global North and its loud anti-GMO organizations for the problem. Others believe that a mix of social, political, legal, and economic reasons have slowed the adoption of GM crops in Africa (Scoones and Glover 2009; Komen et al. 2020; Rock and Schurman 2020). Ethiopia, Kenya, Malawi, and Nigeria have just received environmental release licenses for GM cotton. Farmers in Ethiopia started planting in 2019, with seed supply expected in 2020 in both countries (Komen et al. 2020). Ghana and Uganda are also trying to move their field experiments closer to commercialization, while Burkina Faso intends to do the same with Bt cowpea (Gakpo 2020; Komen et al. 2020). These movements and discussions are promising indications of a rising acceptance of conventional GM crops, and even more so, plants produced via innovative breeding techniques, but there must be a delicate line established to prevent overregulation, which may impede innovation (Qaim 2020; Smyth 2020).

15.5.2.4 Asia and the Pacific

In Asia and the Pacific, commercial production of genetically modified crops is allowed in the following countries, in order of size: India, China, Pakistan, Australia, the Philippines, Myanmar, Vietnam, Bangladesh, and Indonesia are just a few of the countries involved (ISAAA 2018). With a BT cotton adoption rate of 95%, India is the world's largest cotton grower as well as the world's largest BT cotton producer (ISAAA 2018; Shahbandeh 2020). In a typical bottom-up16 creation of the law, thousands of small-scale Indian farmers were arrested illegally cultivating BT cotton in 2001, before the government authorized it in 2002 (Ramaswami et al. 2012). Although non-food GM cotton has been authorized for cultivation, Bt brinjal, a GM food crop, remains under a de facto ban (Kumar et al. 2011). The Minister of Environment and Forestry rejected the permission proposal of the Genetic Engineering Approval Committee (GEAC) in 2010, resulting in a "temporary" ban that is still in place today (Cao 2018). Nonetheless, there have been reports of Bt brinjal, stacked IR and HR cotton, and virus-resistant papaya being planted illegally (Todhunter 2019; Blakeney 2020; Rao 2013). Another bottom-up change to the prohibition is conceivable if farmers in India continue to cultivate Bt brinjal illegally. Ahuja (2018) believes that authorities may use existing laws to address gene-edited crops on a case-by-case basis, as long as they are not constrained by the Cartagena Protocol's definition of "modern biotechnology." In January 2020, the Indian government's Department of Biotechnology published proposed gene editing regulations for public comment. The suggested suggestions include a tiering scheme, with a higher number of DNA changes requiring more assessments (Fernandes 2020). China, like India, is the world's second-largest cotton producer, with a Bt cotton adoption rate of about 95% (Shahbandeh 2020; ISAAA 2017). Since the birth of GM crops, China has sponsored biotech research with substantial financing in a two-pronged effort to provide food security and world-leading agricultural biotechnology (Cao 2018). China started commercialization of a virus-resistant tobacco in 1990 (Raman 2017). Since their introduction in 1997, Bt cotton seeds have been enthusiastically received, and the majority of them are now grown domestically. According to (Cao 2018), Bt cotton was authorized swiftly (in just 2 years) for a variety of reasons, the most significant of which is that, unlike Bt rice, there were no global GMO concerns at the time. Only Bt cotton and virus-resistant papaya are being cultivated on a big basis in China, out of the seven crops authorized for production. Before starting to manufacture new genetically modified crops, the applicant must complete a three-phase trial procedure that includes field, environmental release, and preproduction studies (Jin et al. 2019). Following that, the applicant may submit to the Ministry of Agriculture and Rural Affairs for an Agricultural GMOs Safety Certificate (a Biosafety Certificate). The development of two domestically developed Bt rice cultivars, GM Shanyou 63 and Huahui-1/TT51-1, may be stopped even with a Biosafety Certificate (ISAAA 2020a). Despite the fact that both kinds of rice received short-term Biosafety Certificates in 2009, which were extended once and would expire in 2019, the Bt rice was never officially cultivated. At the end of 2019, a list of 192 GM crops slated for biosafety clearance was published for public comment, including GM soybean and maize (Cremer 2020; Xiaodong 2020). China has made significant investments in CRISPR/Cas research and development, including the use of additional Cas proteins, matching advances in transgenic crops (Cohen 2019). China accounted for 42 percent of CRISPR/Cas-related publications in agriculture (more than double the USA) and 69 percent of CRISPR/Cas patent applications in agriculture (the USA came in second with 19 percent) between 2014 and 2017 (Cohen 2019; Martin-Laffon et al. 2019). Regardless, China presently lacks a regulatory framework for evaluating gene-altered crops for commercial distribution, with some expecting that China would adopt the US model (Cohen 2019), while others think that the Japanese approach would be more suitable (Zhang et al. 2020). One of the most well-known GM crops that have yet to be authorized for distribution is Biofortified Golden Rice (event name: GR2E). Golden Rice has a gain-of-function characteristic in Africa and Southeast Asia that generates vitamin A precursor molecules to cure critical vitamin A deficiency in young babies and pregnant women (World Health Organization 2020b). From the early beginnings in 2000, it took 17 years for a few countries to gain authorization (Ye et al. 2000). In Australia, Canada, New Zealand, the USA, and the Philippines, golden rice is presently permitted for direct human consumption, but not for production (ISAAA 2020a). The Philippines, ironically, is the only country in the target group to provide such authorization (World Health Organization 2020a).

15.5.2.5 Regulatory Updates for Gene Editing in Asia-Pacific

When it comes to agricultural regulations involving genetically modified organisms, Japan takes a distinctive approach. Despite the approval of 141 GM events for cultivation by 2020 (save for the decorative blue rose flower), Japan ranked second behind the USA in terms of the number of GM events authorized for food, feed, and culture in 2018 (ISAAA 2018), and no GM crop planting will take place (save for the decorative blue rose flower) (ISAAA 2020a; FAS 2020b). According to the legislation, cultivation authorization is only needed for imported products intended for food, feed, or processing in Japan. As a result, environmental hazards connected with that GM crop, such as spilt GM grain or inadvertent mixing with non-GM seeds, have been assessed by authorities (Matsushita et al. 2020). Japan, like Europe, is a significant importer of GMO crops, importing almost all of its maize and 94% of its soybeans (FAS 2020c). In recent years, both Japan and Australia have worked to clarify their regulatory systems governing gene-edited crops and goods, with comparable regulatory results. The explanation was provided in Japanese through an interpretation document. According to the Japanese Ministry of Environment, goods that do not include inserted DNA or RNA are not deemed "living modified organisms" under the Cartagena Law. This means that SDN-1 organisms are no longer considered LMOs since they are generated by unguided repair of site-directed nuclease activity (Tsuda et al. 2019). In Australia, amendments to the Gene Technology Regulations 2001, which included a new exemption, provided clarity in 2019. As a consequence, the Office of the Gene Technology Regulator does not consider SDN-1 organisms to be GMOs, as defined under the Gene Technology Act of 2000 (OGTR 2020). In practice, this implies that the crop is no longer subject to the Gene Technology Act's regulatory oversight. Rather, it must adhere to the Department of Agriculture, Water and the Environment's rules, as well as the Australia New Zealand Food Standards Code if it manufactures food. New Zealand, unlike its neighbor Australia, does not manufacture genetically modified foods and opposes animals created via gene editing methods. The Hazardous Substances and New Organisms Act (HSNO), one of the most thorough in the world, is administered by the Environmental Protection Agency (EPA), which has strict minimum requirements for approval assessment (Fritsche et al. 2018). The EPA must decide if the benefits of the GMO outweigh the risks, which include the new plant's impact on Maori culture and traditions, especially in connection to their valued fauna and flora, ancestral lands, water, holy places, and valuable goods (Hudson et al. 2019). New Zealand was one of the first countries to change its legislation to distinguish plants generated by conventional mutagenesis from plants grown by conventional mutagenesis in terms of organism regulation as a result of innovative breeding techniques. This implies that even if new plants created via novel breeding techniques lack foreign DNA; they are nonetheless subject to GMO regulations (Ishii and Araki 2017).

15.6 Plant Genome Databases

Plant genome databases are accessible in a variety of formats. Plant genomic databases (Table 15.1) provide molecular sequence data for all plant species that have undergone extensive sequencing. The database divides EST sequences into contigs, each of which represents a potential distinct gene. Contigs are annotated and connected to the genomic DNA that they belong to.

Brassica.info compiles data from a number of researches to estimate genome sizes for different *Brassica* species. Ensembl Plants is a genome-centric portal that provides data such as genome size and base pairs for plants of scientific interest. It contains details on assembly, regulation, annotation, sequencing, and variants. BLAST or a gene identification sequence search may be used to search this data. The data that is stored in the database genome size in Asteraceae is a comprehensive collection of genome size statistics for the Asteraceae family. Genome sizes are presently available for 1219 species, covering roughly 5% of species, 40% of tribes, 50% of subfamilies, and 10% of genera, based on 2768 data from 133 publications. NCBI makes genetic data available to scientists in order to improve research. This database contains 16,326 species that may be organized into groups, subdivisions,

Table 15.1 List of Plant databases

S. No.	Name of the database	Website	Reference
01	*Brassica*	http://brassicadb.org	Wang et al. (2015)
02	*Ensembl Plants*	http://plants.ensembl.org/species.html	Bolser et al. (2017)
03	*The Genome Size in Asteraceae Database*	http://www.etnobiofic.cat/gsad_v2/	Garnatje et al. (2011)
04	*The National Center for Biotechnology Information* (NCBI)	https://www.ncbi.nlm.nih.gov	Omnibus (n.d.)
05	*The PGDJ DNA Marker and Linkage Database*	http://pgdbj.jp/plantdb/plantdb.html	Asamizu et al. (2014)
06	*Phytozome*	https://phytozome.jgi.doe.gov/pz/portal.html	Goodstein et al. (2012)
07	*Plant DNA C-values database*	https://cvalues.science.kew.org	Pellicer and Leitch (2020)
08	*The Plant rDNA Database*	https://www.plantrdnadatabase.com/	Garcia et al. (2014)
09	*Plant GDB Genome Browser*	http://www.plantgdb.org	Duvick et al. (2007)
10	*PTGBase*	http://ocri-genomics.org/PTGBase/	Yu et al. (2015)
11	Gramene	https://www.gramene.org/	Ware et al. (2002)
12	PLAZA	http://bioinformatics.psb.ugent.be/plaza/	Vandepoele (2017)
13	EMBL Nucleotide Sequence database	http://www.ebi.ac.uk/embl/	Kanz et al. (2005)
14	UK CropNet	http://www.ukcrop.net	Dicks et al. (2000)
15	National BioResource Project (NBRP)	http://www.nbrp.jp	Yamazaki and Sugawara (2009)
16	The IPK Crop EST Database (CR-EST)	http://www.ipk-gatersleben.de	Künne et al. (2005)
17	AFRICANCROPS.NET	http:www.africancrops.net	New (2007)
18	J. Craig Venter	http://www.jcvi.org J. Craig	Bhatia et al. (1997)
19	Harvest	http:www.harvest.ucr.edu	Close et al. (2005)
20	TAIR	http:www.arabidopsis.org	Poole (2005)
21	OryzaBase	http:www.shigen.nig.ac.jp/rice/oryzabase	Kurata and Yamazaki (2006)
22	Graingenes	http:www.wheat.pw.usda.gov	O'Sullivan (2007)
23	Panzea	http:www.Panzea.org	Zhao et al. (2006)
24	Soybase	http:www.Soybase.org	Grant et al. (1996)
25	CottonFGD	http:www.Cottondb.org	Zhu et al. (2017)

and kingdoms. The PGDJ DNA Marker and Linkage Database compile data from a variety of sources, including publications and smaller databases. This database contains information about the genome, markers, organisms, and genome-specific databases. At important phylogenetic nodes, the Phytozome database is a plant comparative genomics web portal in which families of related genes reflecting contemporary offspring of ancient genes are created. This resource contains 58 green plant genomes that have been sequenced and annotated. The Plant DNA C-values database contains C-values for roughly 8510 bryophyte, pteridophyte, angiosperm, gymnosperm, and algal species. Users may examine C-value data across various groups of plants or explore just a portion of the database by choosing a particular plant group of interest. The plant rDNA database includes information from over 3000 plant species and over 600 publications. It also includes data on the location and quantity of ribosomal DNA signals, as well as their structures, as well as chromosomal number, ploidy level, life cycle, and genome size. The gene structure models and transcript evidence from spliced alignment of cDNA sequences and EST are shown in the plant GDB genome browser. It also shows GSS contigs that match, microarray probes that are comparable, and community annotations. PTGBase is a database that aids in the research of tandem repeating genes in plants. This database now includes 39 plant species organized into 54,130 tandem repeat gene clusters with 129,652 genes. From this, a complete list of repeat genes, their coding and protein sequences, as well as entire genome sets for sequenced plant species, may be extracted. A search tool may be used to discover the name of a tandem array, the target gene, and the tandem array gene number in a particular species. Gramene is a web database that is accessible through the internet. Based on Ensembl technology, it provides a resource for plant comparative genomics and pathway research. PlantsDB is a database that stores and analyzes genetic and genomic data from a variety of plants, as well as providing tools for querying such data. It conducts comparative analysis with the assistance of in-house tools. PLAZA is a comparative genomics online resource that combines plant sequencing data with comparative genomics techniques. It conducts evolutionary study within the green plant lineage (Viridiplantae). TAIR provides an online database of *Arabidopsis thaliana*, a model higher plant.

EMBL Nucleotide Sequence database is data as well as tools used to analyze the genes, proteins, gene expression, small molecules, and pathways. UK CropNet is a database containing the information about barley, *Brassica*, *Arabidopsis*, millets, forage grasses and it also used as comparative analysis tool. National BioResource Project (NBRP) is a Resource center and repository for genomics projects. The IPK Crop EST Database (CR-EST) is an EST database for potato, barley, pea, tobacco, petunia, and wheat. AFRICANCROPS.NET is the database which contains the details of African crop breeding networks, seed sources, news, databases, seed source, training programs, and free access portals. J. Craig Venter is the database containing the information about Castor bean, Brassicaceae, rice genomics and MedicagoHarvest is the database having the EST datasets and tools for rice, barley, cowpea, citrus, wheat, soybean, coffee, and *Brachypodium*. TAIR is the database having the access of complete genome browser, genetic stocks, and extensive suite of tools. OryzaBase is the database having the information about integrated rice science, mutant database, and genetic map. Graingenes is the database containing details about genomic, phenotypic, and genomic information about the Triticeae and Avena. Panzea database gives the information about maize and teocinte genome data and it is also works as software for genome and association analysis. Soybase database will provide us the details of genetics data and also works as a tool for legumes and soybean. It is also called as Breeders Toolbox. CottonDB is a Cotton Genome Browser of variety of genes, genetics, and taxonomy.

15.7 Public Perception on Plant Genomics

Gene editing in agricultural crops is becoming more common. Nations like the USA have already taken an active stance on using these methods; nevertheless, many countries still need time to position the goods and technologies transparently beyond technical, ethical, political, and normative issues (Kato-Nitta et al. 2019). Gene editing and genetic modification are two distinct technologies, and if there are variations in public views between the two, investing in such disparities in a nation where genetically modified foods are less accepted is a risky proposition. Several studies have been conducted on both expert and popular views of the danger of biotechnology applications on food, with experts concluding that there is less risk than the general population (Savadori et al. 2004). As a consequence of these results, new technology has high expectations. Researchers with domain-specific scientific expertise in biotechnology would support the implementation of product-based policies. The public, according to a recent interview study by (Kato-Nitta et al. 2019), is in favor of process-based policy. However, quantifiable data on people's risk perceptions of genome editing applications in agricultural crops in connection to various degrees of scientific understanding is conspicuously lacking. Observing how these many levels influence technology, even the most recent, can help researchers better comprehend public views of new science and technology, as well as advance related studies.

Public perception is very much necessary when it comes to genetic engineered foods and crops. The advantages of agricultural biotechnology must be communicated within the constraints of reality. It is critical not to oversell the technology by concentrating only on potential advantages, such as consumer benefits, while the vast bulk of the cumulative benefits have gone to the farmers and international corporations that patented the gene. There is the need of evidence based open debate for every party to reach others and decision taken. Information related to the results of the researches on genetically modified crops should be more discussed and widespread. It is suggested that farmers should involve in the early research on genetically modified crops and the related standard settings and structures. The mass media is the main source of information for all the consumers and provides complete information about food safety and nutrition issues but public information source like government agencies and scientist are not the popular sources of information. Thus, it will be admirable if an inter-agency collaboration is settled to enable the media get properly informed in these kind of matters. In consideration of the benefit of the technology, benefit cost analysis should be considered the cost of non-adaptation. GMOs trade should be monitored properly. It was discovered that a lack of knowledge about GMOs causes misunderstanding regarding the advantages and dangers of GM foods, resulting in a negative assessment of the technology as a whole. Non-experts' capacity to understand scientific uncertainty connected with technological risk assessments was previously underestimated by elite scientific organizations (Frewer 2004). It was once thought that disclosing this information to the general public would harm public perceptions and attitudes, but now that consumers can rationalize, weigh risks against benefits, develop a positive attitude, and act on this in an informed purchase decision, disclosing objective information to the general public will be a solid rock on which to build trust. Consumer acceptability of genetically modified foods is likely to be influenced by societal values, and this must be included in the discussion over product regulation as well as a combined communication approach. These methods are based on non-tangible human emotions and will be utilized to communicate biotechnology to the general public. Control and information circulation models driven by science and technology rather than public requirements seem to have limited use in the near to medium term. It does not function merely by disseminating knowledge to the general population in any manner. It is critical that science's structure, techniques, and disciplinary variety, as well as the various institutional contexts in which it is performed, all combine to make it a particularly powerful trigger for political conflict. There is now worry about transnational corporations dominating the global food system under the pretext of combating global food insecurity. Furthermore, there is skepticism among the public about the significance of adopting GE crops, which mostly consist of non-native crops capable of reducing food diversity and local food kinds. These kinds of concerns should be addressed in public forums with a lot of media coverage and in people's vernacular. Finally, the policy options for resolving the GM crop debate are sufficiently wide to attract political agreement, allowing scientists' participation to be restricted to scientific concerns to prevent over-scientification, which may weaken science's social value in the long term. An interdisciplinary approach to the conceptualization of GMOs is suggested by the confluence of major value chain players, policymakers, scientists, and consumers. The ability of these many parties to come together to resolve the conflict will allow mankind to enjoy all of the benefits of technology.

Despite the fact that there have been no recorded instances of damage arising from the use of this technology (people have been eating biotech food for years), public opinion of GMOs and their products has not always been a success. For example, when consumers are given information on consumer advantages, their perception of danger associated with eating GM soybeans is lower than when they are not given such information (Brown and Ping 2003). As a result, the direct related advantages disclosed to the customer influence the consumer's perception of risk connected with the use of GM-derived goods. The reaction in Western Europe to the new technology cannot simply be mitigated by scientists claiming that there is no danger. GM food safety cannot be determined purely on the basis of scientific risk assessments unless there is a scientific agreement that the environmental effect is minimal. As a result, approaches like the one described at http://www.ars.usda.gov/sites/monarch/index.html allowing science to guide choices in the case of butterflies and BT corn may not be particularly successful in influencing public opinion.

To make the adoption of new technology acceptable, there must be some real advantage to the customer, not simply the manufacturer or provider. In the developing world, the advantages to small farmers from increased economic activity and improved living standards for a significant part of the population may be enough of a perceived benefit for consumers to encourage adoption of new technology. However, this situation creates new issues in terms of the structure and resources in place in many poor nations to allow for the cultivation and dissemination of GMOs.

15.8 Conclusion

Plant genomics emerged in the last three decades as a result of advances and approaches in traditional genetics and breeding, molecular breeding, molecular biology, molecular genetics, and molecular biotechnology in the field of high-throughput DNA sequencing technologies, energizing the plant research community to sequence and comprehend the genetic structures, compositions and functions in plant. Plant genomics development has been driven by improvements in equipment and technology, as well as the desire and necessity to feed a rising human population, maintain agricultural output in the face of global climate change, social globalization, and biosecurity problems. All of this resulted in the sequencing and collection of entire plant genomes, as well as gene function annotation, very complex polyploid plants, linking sequence variations to phenotypes, and exploiting sequence variations in crop or plant advancement on a genome-wide scale/through highly sequence specific native modification of plant genomes. Over a hundred plant genomes have been sequenced to far, including flowering and crop plants, as well as non-flowering, model and non-model, crop wild relatives. Over 1100 Arabidopsis accessions from diverse eco-geographic origins, as well as experimental populations, have been fully sequenced, providing plant scientists with more analytical tools and aiding in the exploitation and identification of physiologically significant variants. Using next generation genome editing, innovative genomic selections, and manipulation technologies, all of these accomplishments have automatically boosted agricultural improvement. More intriguing than the current avalanche of functional and sequence data in plant genome research are the unanswered questions that will provide fodder for thought experiments in the coming years. The new molecular tools that are becoming available should enable us to answer some of the questions about plant genomics in the future. Governments of industrialized nations should invest in public genomics research and implement urgent data storage policies.

References

ABNE ABNoE (2019) African biosafety regulators improve their knowledge on genome editing regulations and establish the african union biosafety regulators forum. http://nepad-abne.net/african-biosafety-regulators-improve-theirknowledge-on-genome-editing-regulations-and-establish-the-african-unionbiosafety-regulators-forum/

Adenle AA, Morris EJ, Parayil GJFP (2013) Status of development, regulation and adoption of GM agriculture in Africa: views and positions of stakeholder groups. Food Policy 43:159–166

Agriculture G (2016) Venezuela passes new seed law banning genetically modified crops. https://www.globalagriculture.org/whatsnew/news/en/31519.html

Ahuja V (2018) Regulation of emerging gene technologies in India. BMC Proc 8:5–11

APBREBES (2016) The new seed law of Venezuela. https://www.apbrebes.org/news/new-seed-law-venezuela?pk_campaign=NL22&utm_content=buffer34dc3&utm_medium=social&utm_source=facebook.com&utm_campaign=buffer

APHIS. U (2020a) Petitions for determination of nonregulated status

APHIS. U (2020b) Regulated article letters of inquiry. https://www.aphis.usda.gov/aphis/ourfocus/biotechnology/am-i-regulated/Regulated_Article_Letters_of_Inquiry

Asamizu E, Ichihara H, Nakaya A, Nakamura Y, Hirakawa H, Ishii T, Tamura T, Fukami-Kobayashi K, Nakajima Y, Tabata SJP, Physiology C (2014) Plant genome DataBase Japan (PGDBj): a portal website for the integration of plant genome-related databases. Plant Cell Physiol 55(1):e8

Asmatulu E, Subeshan B, Twomey J, Overcash M (2020) Increasing the lifetime of products by nanomaterial inclusions—life cycle energy implications. Int J Life Cycle Assess 25(9):1783–1789

Atanassova A, Keiper F (2018) Plant breeding innovation: a global regulatory perspective. Cereal Chem 95. https://doi.org/10.1002/cche.10021

Benítez Candia N, Fernández Ríos D, Vicién C (2020) Paraguay's path toward the simplification of procedures in the approval of GE crops. Front Bioeng Biotechnol 8(1023). https://doi.org/10.3389/fbioe.2020.01023

Bhatia U, Robinson K, Gilbert W, Klenk H-P, White O, Venter J (1997) Dealing with database explosion: a cautionary note. Science 276(5319):1724–1725

Blakeney M (2020) Legislative support for agricultural innovation in India. In: Blakeney M, Siddique KM (eds) Local knowledge, intellectual property and agricultural innovation. Springer, pp 173–198

Bolger M, Schwacke R, Gundlach H, Schmutzer T, Chen J, Arend D, Oppermann M, Weise S, Lange M, Fiorani FJ (2017) From plant genomes to phenotypes. J Biotechnol 261:46–52

Bolser DM, Staines DM, Perry E, Kersey PJ (2017) Ensembl plants: integrating tools for visualizing, mining, and analyzing plant genomic data. In: van Dijk ADJ (ed) Plant genomics databases. Springer, pp 1–31

Borevitz JO, Ecker JR (2004) Plant genomics: the third wave. Annu Rev Genomics Hum Genet 5:443–477

Branford S (2013) Peru: a 10-year ban on GMOs. Lat Am Bur https://lab.org.uk/peru-a-10-year-ban-on-gmos/

Bratlie S, Halvorsen K, Myskja BK, Mellegård H, Bjorvatn C, Frost P, Heiene G, Hofmann B, Holst-Jensen A, Holst-Larsen T (2019) A novel governance framework for GMO: A tiered, more flexible regulation for GMO s would help to stimulate innovation and public debate. EMBO Rep 20(5):e47812

Brookes M (1996) Hybridisation rules OK! Reed Business Information Ltd, Quadrant House, The Quadrant, Sutton

Brown JL, Ping Y (2003) Consumer perception of risk associated with eating genetically engineered soybeans is less in the presence of a perceived consumer benefit. J Am Diet Assoc 103(2):208–214

Callebaut S (2015) New developments in modern biotechnology: a survey and analysis of the regulatory status of plants produced through New breeding techniques. Master Thesis. Faculty of Law Ghent University, Ghent

Campos-de Quiroz H (2002) Plant genomics: an overview. Biol Res 35(3–4):385–399

Canada Go (2020) Novel Feeds. https://www.inspection.gc.ca/animal-health/livestock-feeds/novel-feeds/eng/1370227088259/1370227136675

Canada H (2013) Novel food information—Cibus Canola Event 5715 (Imidazolinone and Sulfonylurea Herbicide Tolerant). https://www.canada.ca/en/health-canada/services/food-nutrition/genetically-modified-foods-other-novel-foods/approved-products/novelfood-information-cibus-canola-event-5715-imidazolinone-sulfonylureaherbicide-tolerant.html

Cao C (2018) GMO China: how global debates transformed China's agricultural biotechnology policies. Columbia University Press

Carr S, Levidow L (1997) How biotechnology regulation separates ethics from risk. SAGE J 26(3): 145–150

CFIA (2013) DD 2013–100: Determination of the Safety of Cibus Canada Inc.'s Canola (Brassica napus L.) Event 5715

CFIA (2020) Plants with novel traits. https://inspection.canada.ca/plant-varieties/plants-with-novel-traits/eng/1300137887237/1300137939635

Chandrasekhar A (2016) Government approves GMO ban extension. https://www.swissinfo.ch/eng/society/genetically-modified-organisms_government-approves-gmo-ban-extension/42260828

Close TJ, Wanamaker S, Roose ML, Lyon M (2005) HarvEST. In: Edwards D (ed) Plant bioinformatics. Springer, Cham, pp 161–177

Cohen J (2019) To feed its 1.4 billion, China bets big on genome editing of crops. https://www.sciencemag.org/news/2019/07/feedits-14-billion-china-bets-big-genome-editing-crops

Commission E (2019) EC study on new genomic techniques

Cremer J (2020) China moves toward commercialization of GMO corn and soy. Alliance Science, Ithaca, NY

Dhanapal AP, Govindaraj M (2015) Unlimited thirst for genome sequencing, data interpretation, and database usage in genomic era: the road towards fast-track crop plant improvement. Genet Genom Era 2015:684321

Dicks J, Anderson M, Cardle L, Cartinhour S, Couchman M, Davenport G, Dickson J, Gale M, Marshall D, May S (2000) UK CropNet: a collection of databases and bioinformatics resources for crop plant genomics. Nucleic Acids Res 28(1):104–107

Dondanville TW, Dougherty ML (2020) Porousness and Peru's moratorium on genetically modified organisms: stakeholder epistemologies and neoliberal science. Environ Sociol 6(1): 107–119. https://doi.org/10.1080/23251042.2019.1690726

Dowd-Uribe B, Schnurr MA (2016) Briefing: Burkina Faso's reversal on genetically modified cotton and the implications for Africa. Afr Aff 115(458):161–172. https://doi.org/10.1093/afraf/adv063%JAfricanAffairs

Duvick J, Fu A, Muppirala U, Sabharwal M, Wilkerson MD, Lawrence CJ, Lushbough C, Brendel V (2007) PlantGDB: a resource for comparative plant genomics. Nucleic Acids Res 36 (suppl_1):D959–D965

ECJ (2018) Judgment of 25 July 2018, Confédération Paysanne a.o., C-528/16, ECLI:EU:C:2018:583. https://eur-lex.europa.eu/legal-content/DE/TXT/?uri=CELEX:62016CJ0528

Eckerstorfer MF, Engelhard M, Heissenberger A, Simon S, Teichmann H (2019) Plants developed by new genetic modification techniques—comparison of existing regulatory frameworks in the EU and non-EU countries. Front Bioeng Biotechnol 7:26

ENGL (2019) Detection of food and feed plant products obtained by new mutagenesis techniques. https://gmo-crl.jrc.ec.europa.eu/doc/JRC116289-GE-report-ENGL.pdf

EPA (1986) Indian Environment (Protection) Act, 1986

Eriksson D, Kershen D, Nepomuceno A, Pogson BJ, Prieto H, Purnhagen K, Smyth S, Wesseler J, Whelan A (2019) A comparison of the EU regulatory approach to directed mutagenesis with that

of other jurisdictions, consequences for international trade and potential steps forward. New Phytol 222(4):1673–1684
FAS (2016) Russia bans cultivation and breeding of GE crops and animals. https://www.fas.usda.gov/data/russia-russia-bans-cultivation-and-breeding-ge-crops-and-animals
FAS (2018) GAIN report: agricultural biotechnology annual—Venezuela. https://apps.fas.usda.gov/newgainapi/api/Report/DownloadReportByFileName?fileName=AgriculturalBiotechnologyAnnual_Caracas_Venezuela_12-20-2018
FAS (2020a) GAIN report: agricultural biotechnology annual—Chile. https://apps.fas.usda.gov/newgainapi/api/Report/DownloadReportByFileName?fileName=AgriculturalBiotechnologyAnnual_Santiago_Chile_10-20-2019
FAS (2020b) GAIN report: Agricultural biotechnology annual—Japan. https://apps.fas.usda.gov/newgainapi/api/Report/DownloadReportByFileName?fileName=AgriculturalBiotechnologyAnnual_Tokyo_Japan_10-20-2019
FAS (2020c) GAIN report: New food security doctrine adopted—Russian Federation. https://apps.fas.usda.gov/newgainapi/api/Report/DownloadReportByFileName?fileName=NewFoodSecurityDoctrineAdopted_Moscow_RussianFederation_02-03-2020
Fernandes V (2020) India's genome-editing draft guidelines are needlessly restrictive. India: Smart Indian Agriculture
FOEF (2018) Biotechnology: in brief. https://www.bafu.admin.ch/bafu/en/home/topics/biotechnology/in-brief.html
Frewer L (2004) The public and effective risk communication. Toxicol Lett 149(1–3):391–397
Fritsche S, Poovaiah C, MacRae E, Thorlby G (2018) A New Zealand perspective on the application and regulation of gene editing. Front Plant Sci 9(1323). https://doi.org/10.3389/fpls.2018.01323
Gakpo JO (2019) Egypt poised to again lead Africa in ag biotech innovation. Alliance Science, Ithaca, NY
Gakpo JO (2020) Burkina Faso renews commitment to GM crops with Bt cowpea. Alliance Science, Ithaca, NY
Galata Bickell E (2019) The framing effect of the media in the regulation of GMOs: a case study of Russia. Russian J Commun 11(3):240–252
Garcia S, Gálvez F, Gras A, Kovařík A, Garnatje TJD (2014) Plant rDNA database: update and new features. Database (Oxford) 2014:bau063
Garnatje T, Canela MA, Garcia S, Hidalgo O, Pellicer J, Sánchez-Jiménez I, Siljak-Yakovlev S, Vitales D, Vallès J (2011) GSAD: a genome size in the Asteraceae database
Gatica-Arias A (2020) The regulatory current status of plant breeding technologies in some Latin American and the Caribbean countries. Plant Cell Tissue Organ Culture 141:229–242. https://doi.org/10.1007/s11240-020-01799-1
Glover D, Paarlberg R (2009) Starved for science: how biotechnology is being kept out of Africa. J Modern Afr Stud 47(3):485
Gómez V, Borges, P., Montaño Astros, I., and Casañas, O. (2010) National strategy for the conservation of biological diversity of the Bolivarian Republic of Venezuela.
Goodstein DM, Shu S, Howson R, Neupane R, Hayes RD, Fazo J, Mitros T, Dirks W, Hellsten U, Putnam N, Rokhsar DS (2012) Phytozome: a comparative platform for green plant genomics. Nucleic Acids Res 40(D1):D1178–D1186
Grant D, Imsande MARCE, Shoemaker R (1996) SoyBase, a soybean genome database (No. RESEARCH)
Greely HT (2015) Genomics, ethical issues in. In: Wright JD (ed) International encyclopedia of the social & behavioral sciences, 2nd edn. Elsevier, Oxford, pp 32–41. https://doi.org/10.1016/B978-0-08-097086-8.82011-5
Hansson SO (2019) A science-informed ethics for agricultural biotechnology. Crop Breed Genet Genom 1(1):e190006
Hawtin GA (1997) Question of ethics: definition and issues relevant to equitable conservation and use of genetic resources. In Ethics and equity in conservation and use of genetic resources for

sustainable food security: proceedings of a workshop to develop guidelines for the CGIAR, 1997. IPGRI Foz do Iguacu
Herrera FF, Domené-Painenao O, Cruces JM (2017) The history of agroecology in Venezuela: a complex and multifocal process. Agroecol Sustain Food Syst 41(3–4):401–415. https://doi.org/10.1080/21683565.2017.1285842
Hudson M, Mead ATP, Chagné D, Roskruge N, Morrison S, Wilcox PL, Allan AC (2019) Indigenous perspectives and gene editing in Aotearoa New Zealand. Front Bioeng Biotechnol 7(70). https://doi.org/10.3389/fbioe.2019.00070
Hundleby PA, Harwood WA (2019) Impacts of the EU GMO regulatory framework for plant genome editing. Food energy Secur 8(2):e00161
ISAAA (2017) Biotech country facts and trends: China. http://isaaa.org/resources/publications/biotech_country_facts_and_trends/download/FactsandTrends-China.pdf
ISAAA (2018) Global status of commercialized biotech/GM crops in 2018: executive brief. ISAAA, Ithaca, NY. https://www.isaaa.org/resources/publications/briefs/54/
ISAAA (2020a) GM approval database. https://www.isaaa.org/gmapprovaldatabase/default.asp
ISAAA. (2020b) ISAAA brief 55-2019: executive summary. https://www.isaaa.org/resources/publications/briefs/55/executivesummary/
Isaac N (2019) African union mulls harmonized biosafety system framework. Alliance Science, Ithaca, NY
ISF (2020) Seed statistics. https://www.worldseed.org/resources/seed-statistics/
Ishii T, Araki M (2017) A future scenario of the global regulatory landscape regarding genome-edited crops. GM Crops Food 8(1):44–56. https://doi.org/10.1080/21645698.2016.1261787
Jawo E, Abubakar AS, Entonu ME, Yahaya J, Camara J (2020) Development, adoption and commercialization of GM crops: a needed commodity in Nigeria. Afr J Biol Sci 2:1–8
Jin Y, Drabik D, Heerink N, Wesseler J (2019) Getting an imported GM crop approved in China. Trends Biotechnol 37(6):566–569
Jones S, Balasubramanian D (1995) The language of the genes. 74(3):145–146
Jorasch P (2020) Will the EU stay out of step with science and the rest of the world on plant breeding innovation? Plant Cell Rep 39(1):163–167
Kanz C, Aldebert P, Althorpe N, Baker W, Baldwin A, Bates K, Browne P, van den Broek A, Castro M, Cochrane G, Duggan K, Eberhardt R, Faruque N, Gamble J, Diez FG, Harte N, Kulikova T, Lin Q, Lombard V, Lopez R, Mancuso R, McHale M, Nardone F, Silventoinen V, Sobhany S, Stoehr P, Tuli MA, Tzouvara K, Vaughan R, Wu D, Zhu W, Apweiler R (2005) The EMBL nucleotide sequence database. Nucleic Acids Res 33(suppl_1):D29–D33
Kato-Nitta N, Maeda T, Inagaki Y, Tachikawa M (2019) Expert and public perceptions of gene-edited crops: attitude changes in relation to scientific knowledge. Palgrave Commun 5(1):1–14
Kealey T, Nelson RR (1996) The economic laws of scientific research. Macmillan, London
Kistler L, Bieker VC, Martin MD, Pedersen MW, Ramos Madrigal J, Wales N (2020) Ancient plant genomics in archaeology, herbaria, and the environment. Annu Rev Plant Biol 71:605–629
Komen J, Tripathi L, Mkoko B, Ofosu DO, Oloka H, Wangari D (2020) Biosafety regulatory reviews and leeway to operate: case studies from sub-Sahara, Africa. Front Plant Sci 11:130. https://doi.org/10.3389/fpls.2020.00130
Kumar S, Misra A, Verma AK, Roy R, Tripathi A, Ansari KM, Das M, Dwivedi PD (2011) Bt Brinjal in India: A long way to go. GM Crops 2(2):92–98
Künne C, Lange M, Funke T, Miehe H, Thiel T, Grosse I, Scholz U (2005) CR-EST: a resource for crop ESTs. Nucleic Acids Res 33(suppl_1):D619–D621
Kurata N, Yamazaki Y (2006) Oryzabase. An integrated biological and genome information database for rice. Plant Physiol 140(1):12–17
Kuzma J (2016) Policy: reboot the debate on genetic engineering. Nature 531(7593):165
Lassoued R, Phillips PWB, Smyth SJ, Hesseln H (2019) Estimating the cost of regulating genome edited crops: expert judgment and overconfidence. GM Crops Food 10(1):44–62. https://doi.org/10.1080/21645698.2019.1612689

Lombardo L, Grando MS (2020) Genetically modified plants for nutritionally improved food: a promise kept? Food Rev Int 36(1):58–76
Luna JK (2020) Peasant essentialism in GMO debates: Bt cotton in Burkina Faso. J Agrar Change 20(4):579–597
Mabaya E, Fulton J, Simiyu-Wafukho S, Nang'ayo F (2015) Factors influencing adoption of genetically modified crops in Africa. Dev South Afr 32(5):577–591. https://doi.org/10.1080/0376835X.2015.1044078
Marchant GE, Stevens YA (2015) A new window of opportunity to reject process-based biotechnology regulation. GM Crops Food 6(4):233–242
Martin-Laffon J, Kuntz M, Ricroch AE (2019) Worldwide CRISPR patent landscape shows strong geographical biases. Nat Biotechnol 37(6):613–620. https://doi.org/10.1038/s41587-019-0138-7
Mathaiyan J, Chandrasekaran A, Davis S (2013) Ethics of genomic research. Perspect Clin Res 4(1):100
Matsushita A, Goto H, Takahashi Y, Tsuda M, Ohsawa R (2020) Consideration of familiarity accumulated in the confined field trials for environmental risk assessment of genetically modified soybean (Glycine max) in Japan. Transgenic Res 29(2):229–242. https://doi.org/10.1007/s11248-020-00193-z
McHughen A (2016) A critical assessment of regulatory triggers for products of biotechnology: Product vs. process. GM Crops Food 7(3–4):125–158
Medvedieva M, Blume YB (2018) Legal regulation of plant genome editing with the CRISPR/Cas9 technology as an example. Cytol Genet 52(3):204–212
MoEFCC (1989) Ministry of Environment, Forest and Climate Change (MoEFCC), G.S.R. 1037 (E). Rules for manufacture, use/import/export & storage of hazardous microorganisms/genetically engineered organisms or cells. 1989
MoEFCC (2015) MoEFCC and Biotech Consortium India Limited. Regulatory framework for GE plants in India In: Phase II capacity building project on biosafety
Montaguth S (2020) Battle over 15-year GMO ban extension rages in Peru as farmers breed and cultivate illegal biotech seed. https://geneticliteracyproject.org/2020/12/01/battle-over-15-year-gmo-banrages-in-peru-as-farmers-breed-and-cultivate-illegal-biotech-seed/
New PBP (2007) BASF launches AgCelence
Norero D (2017) Ecuador passes law allowing GMO crop research. Genetic Literacy Project, Atlanta, GA
Norero D (2018) Top 15 advances on GM crops and gene editing in Latin America during 2017. Alliance Science, Ithaca, NY
O'Sullivan H (2007) GrainGenes. Methods Mol Biol 406:301–314. Plant Bioinformatics. Springer
OGTR (2020) 2019 Amendments to the regulations. http://www.ogtr.gov.au/internet/ogtr/publishing.nsf/Content/amendments C to C the C regs C 2019
Omnibus GE (n.d.) The National Center for biotechnology information (NCBI), NIH
Paarlberg R (2010) GMO foods and crops: Africa's choice. New Biotechnol 27(5):609–613
Paarlberg R (2014) A dubious success: the NGO campaign against GMOs. GM Crops Food 5(3):223–228
Pellicer J, Leitch IJ (2020) The Plant DNA C-values database (release 7.1): an updated online repository of plant genome size data for comparative studies. New Phytologist 226(2):301–305
Poole RL (2005) The TAIR database. Methods Mol Biol 406:179–212. Plant Bioinformatics. Springer
Qaim M (2020) Role of new plant breeding technologies for food security and sustainable agricultural development. Appl Econ Perspect Policy 42(2):129–150
Raman R (2017) The impact of genetically modified (GM) crops in modern agriculture: a review. GM Crops Food 8(4):195–208. https://doi.org/10.1080/21645698.2017.1413522
Ramaswami B, Pray CE, Lalitha N (2012) The spread of illegal transgenic cotton varieties in India: biosafety Regulation, monopoly, and enforcement. World Dev 40(1):177–188. https://doi.org/10.1016/j.worlddev.2011.04.007

Rao CK (2013) Genetically engineered crops would ensure food security in India. Cambridge University Press, Cambridge, pp 167–183
Reiss MJ, Straughan R (2001) Improving nature? The science and ethics of genetic engineering. Cambridge University Press, New York
Reportlinker (2020) Global agricultural biotechnology industry: global agricultural biotechnology market to reach US$66.2 billion by the year 2027
Robinson J (1999) Ethics and transgenic crops: a review. Electron J Biotechnol 2(2):5–6
Rock J, Schurman R (2020) The complex choreography of agricultural biotechnology in Africa. Afr Aff 119(477):499–525. https://doi.org/10.1093/afraf/adaa021%JAfricanAffairs
Rosendal GK (2008) Interpreting sustainable development and societal utility in Norwegian GMO assessments. Eur Environ 18(4):243–256
Rosendal GK, Myhr AI (2009) GMO assessment in Norway: societal utility and sustainable development. EMBO Rep 10(9):939–940
Salazar MP, Valenzuela D, Tironi M, Gutiérrez RA (2019) The ambivalent regulator: the construction of a regulatory style for genetically modified crops in Chile. Tapuya: Latin American. Sci Technol Soc 2(1):199–219. https://doi.org/10.1080/25729861.2019.1611992
Sánchez MA (2020) Chile as a key enabler country for global plant breeding, agricultural innovation, and biotechnology. GM Crops Food 11(3):130–139. https://doi.org/10.1080/21645698.2020.1761757
Sánchez MA, León G (2016) Status of market, regulation and research of genetically modified crops in Chile. New Biotechnol 33(6):815–823. https://doi.org/10.1016/j.nbt.2016.07.017
Savadori L, Savio S, Nicotra E, Rumiati R, Finucane M, Slovic P (2004) Expert and public perception of risk from biotechnology. Risk Anal 24(5):1289–1299
Scheben A, Edwards D (2018) Bottlenecks for genome-edited crops on the road from lab to farm. Genome Biol 19(1):1–7
Schnurr MA (2019) Africa's gene revolution: genetically modified crops and the future of African agriculture. McGill-Queen's Press-MQUP, Montreal
Scoones I, Glover D (2009) Africa's biotechnology battle. Nature 460(7257):797–798
Shahbandeh M (2020) World cotton production by country 2019/2020. Statista Inc, New York, NY
Smyth SJ (2017) Canadian regulatory perspectives on genome engineered crops. GM Crops Food 8(1):35–43. https://doi.org/10.1080/21645698.2016.1257468
Smyth SJ (2020) Regulatory barriers to improving global food security. Glob Food Sec 26:100440. https://doi.org/10.1016/j.gfs.2020.100440
Smyth SJ, Lassoued R (2019) Agriculture R&D implications of the CJEU's gene-specific mutagenesis ruling. Trends Biotecnhol 37(4):337–340
Sprink T, Eriksson D, Schiemann J, Hartung F (2016) Regulatory hurdles for genome editing: process-vs. product-based approaches in different regulatory contexts. Plant Cell Rep 35(7): 1493–1506
Straughan R (1995a) Ethics, morality and crop biotechnology. 1. Intrinsic concerns. SAGE 24(3): 187–192
Straughan R (1995b) Ethics, morality and crop biotechnology 2 Extrinsic concerns about consequences. SAGE 24(4):233–240
The Moscow Times (2016) Russian state duma bans import and production of GMOs. . The Moscow Times,
Todhunter C (2019) Illegal Bt Brinjal growing in India: a call to initiate criminal proceedings against regulators and corporations. South Asia: Asia-Pacific Research
Tsuda M, Watanabe KN, Ohsawa R (2019) Regulatory status of genome-edited organisms under the. Japanese Cartagena Act 7:387. https://doi.org/10.3389/fbioe.2019.00387
Vandepoele K (2017) A guide to the PLAZA 3.0 plant comparative genomic database. In: Plant genomics databases. Springer. pp 183–200
Waltz E (2016) Gene-edited CRISPR mushroom escapes US regulation. Nature 532(7599): 293–293. https://doi.org/10.1038/nature.2016.19754

Wang X, Wu J, Liang J, Cheng F, Wang XJD (2015) Brassica database (BRAD) version 2.0: integrating and mining Brassicaceae species genomic resources. Database (Oxford) 2015: bav093

Ware DH, Jaiswal P, Ni J, Yap IV, Pan X, Clark KY, Teytelman L, Schmidt SC, Zhao W, Chang K (2002) Gramene, a tool for grass genomics. Plant Physiol 130(4):1606–1613

Wasmer M (2019) Roads forward for European GMO Policy—Uncertainties in wake of ECJ judgment have to be mitigated by regulatory reform. Front Bioeng Biotechnol 7:132

Weil V (1996) Biotechnology and ethics: a blueprint for the future. Biotechnology: social impact and quandaries. OECD

Wendel JF, Jackson SA, Meyers BC, Wing RA (2016) Evolution of plant genome architecture. Genome Biol 17(1):1–14

Whelan AI, Lema MA (2015) Regulatory framework for gene editing and other new breeding techniques (NBTs) in Argentina. GM Crops Food 6(4):253–265. https://doi.org/10.1080/21645698.2015.1114698

Whelan AI, Gutti P, Lema MA (2020) Gene editing regulation and innovation economics. Front Bioeng Biotecnhol 8(303). https://doi.org/10.3389/fbioe.2020.00303

Wolt JD, Wolf C (2018) Policy and governance perspectives for regulation of genome edited crops in the United States. Front Plant Sci 9:1606. https://doi.org/10.3389/fpls.2018.01606

World Health Organization (2020a) Summary tables and maps on global prevalence of vitamin a deficiency 1995–2005. https://www.who.int/vmnis/database/vitamina/status/en/

World Health Organization (2020b) Micronutrient deficiencies

Xiaodong W (2020) GMO crops set to pass biosafety tests. . https://www.chinadaily.com.cn/a/202001/02/WS5e0d28cfa310cf3e35581e9f.html

Yamazaki Y, Sugawara H (2009) National bioResource project information center. Exp Anim 58(2):75–84

Ye X, Al-Babili S, Klöti A, Zhang J, Lucca P, Beyer P, Potrykus I (2000) Engineering the provitamin A (beta-carotene) biosynthetic pathway into (carotenoid-free) rice endosperm. Science (New York, NY) 287(5451):303–305. https://doi.org/10.1126/science.287.5451.303

Yu J, Ke T, Tehrim S, Sun F, Liao B, Hua WJD (2015) PTGBase: an integrated database to study tandem duplicated genes in plants. Database (Oxford) 2015:bav017

Zhang Y, Pribil M, Palmgren M, Gao C (2020) A CRISPR way for accelerating improvement of food crops. Nature Food 1(4):200–205. https://doi.org/10.1038/s43016-020-0051-8

Zhao W, Canaran P, Jurkuta R, Fulton T, Glaubitz J, Buckler E, Doebley J, Gaut B, Goodman M, Holland J (2006) Panzea: a database and resource for molecular and functional diversity in the maize genome. Nucleic Acids Res 34(suppl_1):D752–D757

Zhu T, Liang C, Meng Z, Sun G, Meng Z, Guo S, Zhang R (2017) CottonFGD: an integrated functional genomics database for cotton. BMC Plant Biol 17(1):1–9

The manufacturer's authorised representative in the EU is Springer Nature Customer Service Centre GmbH, Europaplatz 3, 69115 Heidelberg, Germany. If you have any concerns regarding our products, please contact ProductSafety@springernature.com

Printed and bound by CPI Group (UK) Ltd, Croydon, CR0 4YY
30/09/2026
02225250-0014